Ecology and Management of Terrestrial Vertebrate Invasive Species in the United States

Ecology and Management of Terrestrial Vertebrate Invasive Species in the United States

Edited by
William C. Pitt, James C. Beasley,
and Gary W. Witmer

CRC Press
Taylor & Francis Group
Boca Raton London New York

CRC Press is an imprint of the
Taylor & Francis Group, an **informa** business

CRC Press
Taylor & Francis Group
6000 Broken Sound Parkway NW, Suite 300
Boca Raton, FL 33487-2742

First issued in paperback 2021

ISBN 13: 978-1-03-209616-2 (pbk)
ISBN 13: 978-1-4987-0482-3 (hbk)

Library of Congress Cataloging-in-Publication Data

Names: Pitt, William C., editor.
Title: Ecology and management of terrestrial vertebrate invasive species in the United States / William C. Pitt, James Beasley, and Gary W. Witmer.
Description: Boca Raton : Taylor & Francis, 2018. | "A CRC title, part of the Taylor & Francis imprint, a member of the Taylor & Francis Group, the academic division of T&F Informa plc." | Includes bibliographical references.
Identifiers: LCCN 2017023606 | ISBN 9781498704823 (hardback : alk. paper)
Subjects: LCSH: Biological invasions--United States. | Introduced organisms--United States. | Biological invasions--United States--Management.
Classification: LCC QH353 .E33 2018 | DDC 577/.18--dc23
LC record available at https://lccn.loc.gov/2017023606

Visit the Taylor & Francis Web site at
http://www.taylorandfrancis.com

and the CRC Press Web site at
http://www.crcpress.com

In memory of Daniel S. Vice (1971–2014), who dedicated his career to mitigating the impact of invasive species on native plants and animals. In doing so, he improved our natural world and created opportunities for his children to discover wildlife. Dan was an inspiring conservationist, fisherman, and a cherished friend to many.

Contents

Editors

William C. Pitt is the deputy director at the Smithsonian Conservation Biology Institute and the associate director of Conservation and Science at the Smithsonian's National Zoological Park. He manages the Smithsonian Conservation Biology Institute's science centers and research programs. He is responsible for the financial and facility programs, the operations of the Smithsonian–Mason School of Conservation, and SCBI's 3200-acre conservation and research facility and the Smithsonian Mason School of Conservation, in Front Royal, Virginia. He oversees more than 220 scientists, postdoctoral fellows, and students from universities around the world. Pitt works passionately to improve the conservation of endangered species through collaboration, education, and research. As a researcher, he spent more than a decade researching methods to reduce the effects of invasive vertebrates on native species, mediating human–wildlife interactions, and evaluating the effects of species management in Pacific islands ecosystems. He has worked extensively as a researcher for the USDA's National Wildlife Research Center in Hawaii. He has published more than 80 articles in peer-reviewed journals and science-related publications. He earned both a PhD in Ecology and an MS in Wildlife Ecology from Utah State University. He holds a BS in Fish and Wildlife Biology from the University of Minnesota. He spent 24 years in the military, working as an environmental science officer for the United States Army Reserve, where he identified and assessed potential environmental and entomological hazards to humans.

James C. Beasley is an assistant professor at the Savannah River Ecology Lab and the Warnell School of Forestry and Natural Resources at the University of Georgia. Beasley earned a BS in Wildlife Science from SUNY–Environmental Science and an MS and PhD in Wildlife Ecology from Purdue University. His research is focused on understanding the effects of anthropogenic activities on wildlife populations, ecology and management of wild pigs and other invasive species, carnivore ecology and management, and scavenging ecology. Beasley is actively involved in research on these topics, both nationally and internationally, including studies on large mammal populations in Chernobyl and Fukushima. He is a Certified Wildlife Biologist with the Wildlife Society and currently serves as the chair of the research subcommittee for the National Wild Pig Task Force. He also serves as the International Atomic Energy Agency's wildlife advisor to the Fukushima Prefecture in Japan, in response to the nuclear accident that occurred in 2011.

Gary W. Witmer is a supervisory research wildlife biologist and rodent research project leader with the USDA APHIS Wildlife Services' National Wildlife Research Center in Fort Collins, Colorado. He earned a PhD in Wildlife Science from Oregon State University with minors in statistics and forest management, an MS in Wildlife Ecology from Purdue University, and an MS and BS in Biology from the University of Michigan. His research focuses on resolving human–wildlife conflicts and has included ungulates, carnivores, and rodents. Most recently, he has been working on invasive species and has designed successful eradication strategies for invasive rodent species on several islands. He has also worked with a large number of native rodent species in a wide array of settings.

Contributors

Aaron Anderson
USDA APHIS Wildlife Services
National Wildlife Research Center
Fort Collins, Colorado

Julian D. Avery
Ecosystem Science and
    Management
Penn State University
University Park, Pennsylvania

Michael L. Avery
USDA APHIS Wildlife Services
National Wildlife Research Center
Gainesville, Florida

Karen H. Beard
Department of Wildland Resources
Utah State University
Logan, Utah

James C. Beasley
University of Georgia
Savannah River Ecology Laboratory
Warnell School of Forestry and Natural
    Resources
Aiken, South Carolina

Are R. Berentsen
USDA APHIS Wildlife Services
National Wildlife Research Center
Fort Collins, Colorado

Craig Clark
U.S. Department of Agriculture
Animal and Plant Health Inspection
    Service
Honolulu, Hawaii

Larry Clark
U.S. Department of Agriculture
National Wildlife Research Center
Fort Collins, Colorado

Michael R. Conover
Wildland Resources Department
Utah State University
Logan, Utah

Michael E. Dorcas
Professional Ecological Services LLC
Huntersville, North Carolina

David C. Duffy
Pacific Cooperative Studies Unit
and
Department of Botany
University of Hawai'i at Mānoa
Honolulu, Hawaii

William Haden-Chomphosy
USDA APHIS Wildlife Services
National Wildlife Research Center
Fort Collins, Colorado

Steven C. Hess
U.S. Geological Survey
Pacific Island Ecosystems Research
    Center
Hawaii National Park, Hawaii

Jason Holderieath
Colorado State University
Fort Collins, Colorado

Ron Johnson
Department of Forestry and Natural
    Resources
Clemson University
Clemson, South Carolina

Steve A. Johnson
Department of Wildlife Ecology and
    Conservation
University of Florida
Gainesville, Florida

Fred Kraus
Department of Ecology and
    Evolutionary Biology
University of Michigan
Ann Arbor, Michigan

Christopher A. Lepczyk
School of Forestry and Wildlife
    Sciences
Auburn University
Auburn, Alabama

Robert J. Lewis
Quinney College of Natural Resources
Utah State University
Logan, Utah

George Linz
USDA APHIS Wildlife Services
National Wildlife Research Center
Fargo, North Dakota

Julie L. Lockwood
Ecology, Evolution, and Natural
    Resources
Rutgers University
New Brunswick, New Jersey

John J. Mayer
Savannah River National Laboratory
Aiken, South Carolina

William C. Pitt
Smithsonian Conservation Biology
    Institute
Smithsonian Institute
Front Royal, Virginia

Shannon E. Pittman
Department of Biology
Davidson College
Davidson, North Carolina

Daniel Rubinoff
Department of Plant and Environmental
    Protection Sciences
University of Hawai'i at Mānoa
Honolulu, Hawaii

Aaron B. Shiels
U.S. Department of Agriculture
National Wildlife Research Center
Fort Collins, Colorado

Stephanie Shwiff
USDA APHIS Wildlife Services
National Wildlife Research Center
Fort Collins, Colorado

Steven Shwiff
Texas A&M
Commerce, Texas

Shane Siers
U.S. Department of Agriculture
National Wildlife Research Center
Hilo, Hawaii

Robert T. Sugihara
USDA APHIS Wildlife Services
National Wildlife Research Center
Hilo, Hawaii

James Thiele
Nebraska Wildlife Services
USDA APHIS Wildlife Services
Lincoln, Nebraska

Dirk H. Van Vuren
Department of Wildlife, Fish, and
    Conservation Biology
University of California–Davis
Davis, California

Lori Williams (Retired)
Beaverton, Orgeon

John D. Willson
Department of Biological Sciences
University of Arkansas
Fayetteville, Arkansas

Gary W. Witmer
USDA APHIS Wildlife Services
National Wildlife Research Center
Fort Collins, Colorado

1 Introduction

Gary W. Witmer, William C. Pitt,
and James C. Beasley

Vertebrate species have been introduced to almost all parts of the world for thousands of years. Within the United States and its territories alone, over 1000 vertebrate species have been introduced since the early sailing ships explored the world. This includes at least 86 species of mammals, 127 species of birds, 126 species of reptiles, 53 species of amphibians, and over 673 species of fish (Witmer and Fuller 2011). Many of these species were native to the United States, but were moved to novel regions, often unintentionally or intentionally by humans. While invasive vertebrates have been introduced to all parts of the world, in this book, we focus on introduced terrestrial vertebrates in the United States and its territories, and the intention is to provide an overview of the complexity and challenges associated with managing invasive species within the United States. Often, the management of invasive species and the prevention of new species becoming established is largely a function of the regulatory framework established within a specific country. In this book, although historical management successes and failures are discussed, the focus is on current effective management options and potential future developments to minimize the effects of invasive species and prevent their spread into new areas.

Although plants and animals have been introduced into new areas for centuries, the increased volume of worldwide trade and transportation has accelerated the rate of species introductions over the last 150 years. Animals are introduced for many reasons, both purposeful and accidental. Intentional introductions include both legal and illegal activities such as the production of food and fur, work animals, sport hunting opportunities, companion animals, aesthetics, pets, pet trade propagation, religious purposes, and pest control. Accidental introductions occur because of stowaways in transport vehicles, hitchhikers or stowaways in or on other commodities, escapees, and, in some cases, because of range expansion of a species, often facilitated by human activities and land use. For example, a tropical storm is thought to have brought the cattle egret *Bubulcus ibis* to North America (Florida initially) from the Caribbean islands after they had crossed the Atlantic Ocean from Europe and Africa. However, it may have been agricultural land use that allowed its subsequent rapid range expansion westward across North America. Likewise, habitat fragmentation stemming from anthropogenic land use has facilitated the expansion of coyotes (*Canis latrans*) across the Eastern United States and far south into Central America over the last several decades, reaching areas where the species formerly did not occur.

Many introduced vertebrate species have provided important resources and economic gains for humans and do not cause undue adverse effects, especially with appropriate management. Others have caused serious adverse effects, including disease and safety hazards, predation and competition with native species, crop consumption and contamination both in the field and during storage, livestock predation, and, in some cases, significant environmental degradation. On many islands, such as Hawaii and Guam, invasive species have caused the extinction of many native species and produced cascading effects across ecosystems. Unfortunately, for many species of introduced vertebrates, it is not yet known if they are causing, or in the future will cause, significant harm to the environment or human resources.

While there are a number of biotic, abiotic, and environmental attributes that contribute to the establishment of an invasive species, there are several underlying themes shared by many successful invaders. For example, species that have high reproductive rates or few effective predators in their introduced range can often quickly establish new populations. Similarly, species with broad dietary or other niche requirements often are better equipped to take advantage of available resources in a new area. Species that are small or secretive generally do well because they can avoid interdiction efforts and go undetected while becoming established. Species that have a wide tolerance to climate have more opportunities to invade and are able to survive in transport. Because many transportation opportunities are in human population centers, species that live in close proximity to people are more frequently transported. Some places are also more vulnerable to invasive species than others. For example, transportation hubs allow for greater opportunities for invasion, and obviously, more species are able to adapt and thrive in areas with less extreme climates. Areas that are geographically isolated, such as islands, tend to have fewer native species that may compete or prey on a recently introduced species. Additionally, habitats that have been altered by people or other invasive species provide opportunities for invasive species to establish.

The President's Executive Order 13112 (1999) set the stage for formally addressing invasive species issues in the United States. This brought attention to the large number of invasive species in the United States, the increased rate of new introductions, and large amounts of economic and environmental damage caused by invasive species. In that document, invasive species are defined as "an alien species whose introduction does or is likely to cause economic or environmental harm or harm to human health." The National Invasive Species Council (NISC) was formed and directed to coordinate across federal agencies in the resolution of invasive species problems. As part of these efforts, NISC put together a National Invasive Species Management Plan which is periodically revised. The plan gives background information on invasive species and the damage they cause, but also makes specific recommendations on how those problems can be removed or lessened. The plan also details five strategic goals: prevention, early detection and rapid response, control and management, restoration, and organizational collaboration.

A wide array of methods is used to manage invasive vertebrates and the damage they cause in the United States. The methods vary somewhat by taxonomic group, and the long-term management of invasive vertebrates is conducted similarly to managing any wildlife damage situation. Methods include traps and snares, netting, shooting, frightening devices, decoys, toxicants, dogs, Judas animals, purposely introduced predators, habitat manipulation, barriers, and sterilants. In some cases, cultural methods may also be used (e.g., sanitation, the type of crop selected and the timing of planting and harvest, compensation and insurance programs, etc.). Details on most of these methods, how they are used, and their advantages and disadvantages are presented in the various chapters of this book and elsewhere.

Eradication strategies are more complex and are discussed using species from various taxonomic groups in the chapters that follow. Eradication generally involves careful planning, new methods development, appropriate timing, and an adequate allocation of personnel and resources. Having an informed and supportive public is also a critical component to achieving success in the eradication of an invasive species. Additionally, follow-up monitoring is essential to assure that a successful eradication has been achieved and to address and respond to a reinvasion, should that occur. Fortunately, there have been numerous successful eradications of invasive vertebrates in parts of the United States. This includes invasive rats and mice, feral goats, and feral cats.

Managing invasive species is a challenge because they often can be difficult to detect, and damage may go unnoticed for a period of time. Another challenge is the need to develop new methods to manage or detect species. We often turn to traditional tools to manage new species, but our ability to control a new invasive species often requires the development of novel tools. This was certainly the case with the brown tree snakes (*Boiga irregularis*) in Guam. There may be resistance to control efforts before effects are widespread because most people do not see the need. Further, some people do not like management of animals (e.g., feral cats, *Felis catus*) that are commonly kept as pets, regardless of the extent of damages they inflict upon native species or ecosystems. Additionally, funding for interdiction efforts and invasive species management is often limited—yet costs to mount large-scale efforts against well-established species are extremely high.

Most of the management and research for invasive species can be put into two broad categories of activities, prevention/detection and control/eradication. Prior to the establishment of a species, efforts are often focused on prevention and early detection, as well as awareness of potential impacts. During this time, communication among groups is critical to avoid duplication of effort and to maintain a clear channel for the reporting of new introductions or incipient populations. Unfortunately, often no action is taken until the invasive species is well established, after which efforts typically shift towards documenting impacts and the development of control or, potentially, eradication strategies. Public education is critical throughout both stages. However, the same line used to demarcate the establishment of a species could also be referred to as the money line because funding typically available to manage species is often only available after a species is well established and proven to be harmful. Before a species is established, there is limited funding for research and management due to a lack of public interest and knowledge of potential consequences. Why spend money now if we don't have to? However, the irony is that reactive management of invasive species after they have become well established is far more costly, and the probability of success is greatly reduced compared to proactive strategies. Nonetheless, once a species is established and is having ecosystem, economic, and/or human health effects, public interest increases and funding becomes available. Thus, if we can shift our efforts and resources to developing tools for interdiction and preventing new species from becoming established, we can manage invasive species much more effectively.

Given the global ecological and economic impacts of invasive species, as well as the complexities and costs associated with control and eradication efforts, invasive species represent one of the greatest challenges facing agencies and managers today. Through the integration of numerous and specific case studies, this book highlights the impacts and control efforts for many prominent invasive terrestrial vertebrates in the United States. The book is organized into three sections. The first section reviews the context in which management occurs and addresses the current U.S. policy and regulation, biosecurity threats and risk assessment, and the economics of vertebrate invasive species management. The second section provides an overview of environmental, agricultural, societal, and ecological impacts from invasive species. The last section provides case studies related to noteworthy and well-established terrestrial invasive amphibians, reptiles, mammals, and birds. Specifically, these case studies illustrate the impacts associated with these species, the unique challenges associated with management, and potential management options that could be developed. The management of established invasive species and reducing their impacts is extremely challenging, but significant progress has been made for many species as will be detailed in the chapters that follow.

REFERENCE

Witmer, G.W., and P.L. Fuller. 2011. Vertebrate species introductions in the United States and its territories. *Current Zoology* 57: 559–567.

2 Biosecurity and Risk Management

Fred Kraus

CONTENTS

INTRODUCTION

Vertebrates have been among the most spectacularly destructive alien species transported by humans, with many species creating huge losses to native species, ecosystems, human health, and economies in the regions to which they've been introduced (Morgan and Woods 1986; Ebenhard 1988; Case and Bolger 1991; Henderson 1992; Pimentel 2002; Pimentel et al. 2000, 2005; Blackburn et al. 2004; Jenkins et al. 2007; Kroeger 2007). Progress has been made in recent years toward reversing some of these impacts by removing certain of these species from some invaded regions (e.g., Nogales et al. 2004; Campbell and Donlan 2005; Clout and Russell 2006; Howald et al. 2007). In most instances, however, perpetual control of widespread or abundant alien vertebrates is required to minimize damage to especially important areas or resources. However, such eradication and control operations are achieved at high cost and can only be successful in limited circumstances, making them an unreliable foundation upon which to base a nation's ecological future. A much more efficient and cost-effective means of managing alien species is to prevent their introduction in the first place. This remains an important goal in alien-vertebrate management

5

because the majority of the world's vertebrate fauna has not yet been introduced everywhere, and most invaded regions are still afflicted by only a relatively few species. Furthermore, prevention is far more cost-effective than post hoc response (Naylor 2000; Touza et al. 2007). Consequently, prevention of future invasions should form the centerpiece of any region's alien species management program.

Successful prevention of alien species invasions can occur along any of the sequential steps of the invasion process (transport to the new location, introduction into the wild, population establishment, range expansion), each of which poses a different set of challenges and requires a different set of attributes allow the alien species to successfully transition through it. The logic behind managing invasions early in the invasion process is obvious, and its recognition is the reason that efforts to protect the economically valuable agriculture sector of the U.S. economy have always relied on border inspection, quarantine, and pre-shipping clearance of imported goods. Moreover, it is also clear that the more barriers to invasion that can be erected across the serial stages of the invasion process, the greater the chances of intercepting and thwarting potential invaders. This simply follows the additive nature of using multiple preventive actions—if one method doesn't serve to intercept a species at one stage, another may at a different stage. Ideally, a serious program to prevent biological invasions will involve separate actions taken at a number of different stages of the invasion process. So, *biosecurity* can be defined as *a comprehensive system of safeguards to protect a jurisdiction from the risk posed by further invasion of alien species*, and such a comprehensive program is referred to as a *biosecurity program*.

Two conceptual features of biosecurity are important to recognize. First, biosecurity should be applied to as broad an array of taxa as possible, with no taxon intentionally excluded a priori. This is important to recognize because, in most countries, biosecurity has most often been applied against disease organisms and pests threatening agriculture and human health, whereas taxa threatening natural areas have been ignored. As a result, biosecurity has rarely been applied to vertebrate species in any comprehensive fashion. Second, although it may seem inherent in the term -*security*, and obvious upon mention, biosecurity is based on recognizing that the threat posed by invasive species is a valid societal concern that merits meaningful political response. I mention this because recognition of invasive species as a valid societal concern has never had broad political acceptance in the United States, whose governments—state and federal—have largely exhibited collective avoidance of the subject for decades. The causes for this political evasion will be briefly discussed below.

The intent of a biosecurity program is to provide the most efficient, cost-effective, and reliable means of limiting additional species invasions and their consequent damages, economic or ecological. Consequently, effective coordination and explicit accountabilities among component programs is a must so that responsibility for new invasions cannot be avoided. Biosecurity is generally the purview of governments because protecting jurisdictions from unwanted ingress across borders is a governmental function. But this need not always be the case. Logically, biosecurity procedures could be applied to defend protected natural areas or even industrial sites—such as mines or forestry plantations—from unwanted pests. For example, the

Northwest Hawaiian Islands National Wildlife Refuge has a biosecurity program to limit new incursions to those islands from visitors and management personnel arriving from the main Hawaiian Islands. However, because biosecurity largely depends on defending incursions across defensible borders, most such measures will be taken by national governments.

In considering biosecurity and its relevance to invasive terrestrial vertebrates in the United States, I will first discuss the components of biosecurity programs in general terms and then explore how biosecurity has been applied to vertebrate invasions in the United States.

BIOSECURITY COMPONENTS

Biosecurity tools can be broadly categorized as (1) preventing introduction of new species, (2) preventing successful establishment and spread of those species that become introduced, or (3) providing long-term mitigation of damages from those species that become established. By *introduction*, I mean release into the wild, which is not to be confused with mere importation, for which human control is typically maintained. Each biosecurity goal involves correspondingly different tools: (1) border risk management, (2) rapid response to new incursions, and (3) population management of well-established species. A biosecurity program is ideally proactive so as to prevent invasions from occurring, so emphasis is logically focused on actions taken earlier in the process of invasion. When long-term damage mitigation must occur, it is implicit evidence that biosecurity has failed.

Central to the application of a biosecurity system is the role of risk assessment. It is not physically possible to inspect all imported goods for invasive species, nor is it possible to be absolutely certain that all permitted deliberate introductions are safe, or to detect and eradicate all new incursions. Hence, rational choices need to be made as to which goods, species, and areas to target for biosecurity activities, and these choices are made on the basis of assessed risk. So an intelligent triage system that directs resources toward those goods, pathways, species, and areas of greatest risk for successful invasion is an essential conceptual foundation for any biosecurity program, although politics can trump these considerations.

Risk tracks the same sequential series of stages as the invasion process itself, so overall risk can be apportioned into the component risks of introduction, establishment, and impact. A difficulty in deriving management decisions from risk assessment is that of determining aggregate risk across these three components. One proposal calculates an overall probability of establishment as the product of the risk of the organism being in a particular pathway × the risk of surviving transport in that pathway × the risk of that organism escaping from the pathway and establishing × the risk of the established population spreading. It also takes the hazard of establishment to be the sum of economic, ecological, and social/political costs (Orr 2003). This system is silent, however, as to how the risk of establishment and the hazard are to be combined to judge overall risk, an attribute taken by Orr (2003) to be a strength of the approach inasmuch as it allows for flexibility of judgment. It is also important to recognize that humans are typically poor at judging composite risk when one component probability is low (in this case, probability of establishment) but another (in this case, hazard) is high (Perrings et al. 2005). Consequently, care must be taken not to over- or underestimate risk magnitude when combining establishment and hazard risks, but a consensus on how best to do that for species invasions is not available and is an important flaw in current risk assessment for invasive organisms.

One final point about risk merits mention. The imprecise nature of risk prediction in invasion biology leads many biologists to urge adoption of a precautionary approach, which urges that in the face of scientific uncertainty, risk judgments err on the side of avoiding hazard. More precisely, it has been proposed that one "should act if a hypothesized effect is serious and there is good *prima facie* evidence of a causal nexus, even if the matter is not completely settled scientifically" (Simberloff 2005). However, it should be borne in mind that ecologists understand precaution differently than do economists and many managers, with ecologists interpreting precaution as preventing damage and economists/managers interpreting it as minimizing opportunity costs and maximizing marginal effectiveness (Finnoff et al. 2007). This can lead to both groups interpreting diametrically opposed actions as precautionary, with managers often stinting prevention actions in favor of control, which can lead to reduced social welfare (Finnoff et al. 2007).

BORDER RISK MANAGEMENT

Invasive species can be introduced via either intentional or unintentional pathways. The means for managing these two pathway modes differ, with some tools applicable only to intentional pathways, others to unintentional pathways. Successful border risk management must encompass both and logically entails the two elements of preventing species from being transported to the jurisdiction in the first place, and locating and removing those individuals that do reach the border. Each strategic element must cover both intentional and unintentional pathways, so multiple tools are relevant for each. Typically, providing biosecurity against accidental introductions involves a risk assessment of the particular pathways of introduction; providing biosecurity against intentional introductions involves a risk assessment of the particular species or higher taxa.

Accidental Introductions

Preventing accidental importation is the primary goal in protecting against agricultural pests and disease organisms, and this often entails pre-export inspection programs in cooperating countries and inspection and quarantine of imported products at high risk of harboring pests. If goods are inspected in the exporting country and found clean, they may then be allowed import without further inspection or quarantine at the port of entry. Such programs are rarely applied to terrestrial vertebrates, but the U.S. Department of Agriculture (USDA) has maintained an inspection program on Guam since 1993 to ensure that brown tree snakes (*Boiga irregularis*) are not accidentally exported in outbound cargo and spread to other locations.

Inspection and quarantine programs can be applied to any accidental hitchhiker are well understood conceptually, and do not need great elaboration here. Risk assessment for border protection activities relies on data collected from random searches of incoming vessels and commodities to prioritize inspecting those goods, vessels, or exporting localities that provide the greatest risk of unwanted species. Routine quality-assurance assessments for quarantine programs are designed to keep running track of how risks are changing through time, address highest-risk goods and export locations with increased inspection or agreements designed to reduce risk from those goods or locations, and identify new risks as they emerge. These programs have a long history of application to protect agricultural interests, but they are now sometimes applied more broadly. Within the United States, rather regular use of this tool against alien vertebrates has only been applied in Hawaii, as far as I know, but certain vertebrates, like snakes, are probably stopped by quarantine services wherever found. The USDA's brown tree snake program conducts quality assurance by recording snake captures in cargo and around ports on Guam so as to detect and respond to shifting risk patterns in meeting its programmatic goals (Engeman et al. 1998).

Intentional Introductions

Providing biosecurity against intentional imports can take either of two approaches, both of which involve banning a set of species. A "black list" approach is the more common and typically involves an importation ban on a particular species already known to be invasive or assessed to likely become invasive. The burden of proof under this system lies on the public (as represented by the government) to demonstrate that a species is likely to be invasive; otherwise, its importation is allowed. Because relatively few species undergo risk assessment, this approach can only limit a small pool of all potential invasive species. Releases and invasions are common under this paradigm because this approach cannot keep pace with the thousands of species in commercial trade at any given time. A contrary, precautionary approach presumes that scientific uncertainty in predicting invasiveness should not inhibit actions to protect the environment. Recognizing that many alien species will have some form of ecological impact but that data limitations and time constraints will not allow all to be evaluated, this "white list" or "clean list" approach allows importation of only those species that have undergone a risk assessment and been determined to be of low invasion risk; it prohibits importation of all other species, pending individual risk assessments of any proposed for importation. This risk-averse paradigm is less often used but is widely advocated by invasion biologists because it better avoids environmental damages and places the burden of proof—to show a species to be of low risk—on the proponent of importation, who benefits from the proposed importation.

Ideally, both approaches would rely on quantitative models to assess risk of invasiveness. Across diverse vertebrate taxa, it has been found that establishment risk is best predicted by (1) climate match between a species' range and the destination for proposed importation, (2) propagule pressure (the more animals released, or the more separate releases, the greater the chances of establishment), (3) whether the species has established elsewhere before, and (4) taxonomic group (some taxa establish with greater facility relative to the numbers of their introductions than do others). The first three of these factors have been found to be of explanatory importance for mammals, birds, reptiles, and amphibians (Bomford 2008; Hayes and Barry 2008; Capellini et al. 2015), although propagule pressure cannot readily be incorporated into predictive models since its future values cannot be known. Some taxa are especially prone to successful establishment, and these vary among groups. Mammals are more prone to establish than birds (Wolf et al. 1996; Bomford 2003, 2006), and within reptiles and amphibians, some taxa show greater establishment success relative to introduction effort (Bomford et al. 2005, 2009; Bomford and Kraus 2008; Kraus 2009; Fujisaki et al. 2010). Furthermore, for birds and mammals, overseas range size is also predictive of establishment risk (Bomford 2008); establishment success of anurans appears dependent on introduction mode (Rago et al. 2012), and that of amphibians and reptiles is greater in the presence of congeners (Tingley et al. 2011; Ferreira et al. 2012), whereas that of reptiles is greater with increased phylogenetic distance to native species in the introduced range (van Wilgen and Richardson 2012). It has also been shown that certain biological attributes are predictive of establishment success for certain taxa, for example, nonmigratory behavior and sexual monochromatism in birds (Veltman et al. 1996; Cassey 2002; Bomford 2003; Blackburn et al. 2009), dietary generalism in birds and mammals (Cassey 2002; Bomford 2003, 2008; Blackburn et al. 2009), ability to survive in human-disturbed habitats in birds and mammals (Bomford 2003, 2008), habitat breadth and large body mass in birds (Blackburn et al. 2009), reproductive output in mammals (Capellini et al. 2015), brain size (Amiel et al. 2011), and age to sexual maturity (van Wilgen and Richardson 2012) in reptiles and amphibians. Subsequent to successful establishment, spread rate in alien reptiles and amphibians has been shown to be positively related to congeneric species diversity and negatively correlated with topographic heterogeneity (Liu et al. 2014).

Models can also include information on risk of harm to humans or risk of economic or ecological impact. Birds generally pose little direct danger to humans, but a variety of mammals does. Avian and mammalian pest risk to agriculture and ecological values has been modeled using taxonomy (certain families are prone to creating damage), dietary generality, competition with native fauna for tree hollows, overseas pest status, and climatic match to areas having susceptible natural areas or agricultural-production values (Bomford 2008). Similarly, habitat generalism is correlated with impact in Europe and Australia (Evans et al. 2014). Information on ecological hazard posed by reptiles and amphibians has been less well studied; however, dangerously venomous snakes, large constrictors, and large crocodilians pose obvious risks to human safety, and some reptiles and amphibians can threaten certain agricultural values (Kraus 2009), suggesting that development of pest-risk metrics for these taxa may be feasible. A wide range of biotic and abiotic information may thus be useful in vertebrate risk assessments, but relatively few variables have been tested in more than a single study and shown to be robustly predictive within major vertebrate groups. More work is especially needed to identify relevant ecological variables that may predict establishment success or impact, but synthesis of the findings from existing studies is still desperately needed, with Bomford (2008) providing the best—though somewhat dated—attempt yet made.

Because climate matching is important for predicting establishment success across all terrestrial vertebrate groups (Bomford 2008; Hayes and Barry 2008), there has been a recent effort to predict potential global geographic range of some invasives using climate-derived ecological-niche models (e.g., Peterson and Vieglais 2001; Jiménez-Valverde et al. 2011). These species-distribution models typically use a few climate measures (but may also include other variables, such as a measure of human disturbance, e.g., Ficetola et al. 2007) from the known distribution of a species, construct a multivariate climatic niche "envelope" for the species, and map that envelope across the globe to predict the potentially suitable range for the alien. This is then often interpreted as invasion risk. But, the method typically does not include biotic variables, and the ecological context of a native range is necessarily different from that of an invaded range, so this method can be prone to both underpredicting (cf., Reed and Rodda 2009) and overpredicting risk (e.g., Tingley et al. 2015). Furthermore, the analytical methods used can prove highly sensitive to initial assumptions (Rodda et al. 2011), resulting in contested (Pyron et al. 2008; Rodda et al. 2009) predictions. Although the method can point to regions of the globe that may be climatically suitable for a particular species, it is a one-dimensional measure of risk and is of limited use for management. As best as I can determine, in the United States, this approach has only been used in a regulatory context in support of listing a handful of giant constricting snakes as injurious under the Lacey Act (Reed and Rodda 2009), although it was only one factor of many in that assessment.

It is well understood that predictive failures will occur with any modeling approach for risk assessment—some approved species will become invasive despite the fact they were predicted to be safe, and some species excluded due to perceived risk may in fact be safe. Nonetheless, a priori risk assessments provide a proactive means of providing biosecurity that is potentially much more comprehensive and more scientifically transparent than efforts that merely list a few widely known invasive species. Further, even granting that errors will be made, it has been shown that the benefits of excluding the invasives exceed the opportunity costs of the few beneficial organisms erroneously excluded (Keller et al. 2007; Springborn et al. 2011).

Species banned from importation that are nonetheless smuggled into a jurisdiction may be stopped from subsequent release (intentional or otherwise) only by competent wildlife law-enforcement programs. Even when well run, such programs probably only discover and prosecute a small fraction of scofflaws because successful prosecution typically requires directly witnessing the prohibited activities (e.g., ownership, release into the wild), which rarely occurs. This limitation applies regardless of the listing approach taken to intentional importations.

EARLY DETECTION AND RAPID RESPONSE

Should species elude border protection programs, the next option is to prevent their establishment and spread. This is most efficiently done if the incipient populations can be detected sufficiently early so that they can be eradicated. These efforts are referred to as early detection/rapid-response (EDRR) programs. As the name suggests, EDRR programs require means of identifying new incipient populations, followed by assessment to determine if eradication remains feasible, followed by eradication of susceptible populations. Obtaining reports of new incursions will usually rely on a two-pronged strategy: (1) systematized surveys using trained staff that target sites at high risk of receiving new infestations (such as the immediate environs of ports) or that house many alien species (such as nurseries, botanical gardens, or the environs of pet stores), and (2) establishing a "hotline" for the public to report new sightings from other locations. Because release sites for most vertebrates are relatively unpredictable, most new vertebrate incursions are probably identified via reports received from the general public.

Once discovered, incipient populations must be evaluated for eradication feasibility. This is done to determine the likelihood of eradication given available resources and social constraints and to identify what methods are required. Feasibility assessments are commonly undertaken for eradication operations in New Zealand, Galapagos, and elsewhere, but I know of no instances targeting vertebrates in the United States, although they may exist in agency files. Lack of such a feasibility study may be one reason for failure to eradicate purple swamphens (*Porphyrio porphyrio*) in Florida. Funding can be a major problem in implementing eradication programs. Ideally, such operations would rely on predictable base funding, but often agencies scramble to assemble sufficient funds for an emergency project, taking them from other programs. Having contingency funds in place for such emergencies would avoid this problem, with the rationale for such accounts being similar to that for fire-fighting forces: the forces are not always needed, but when they are, there is no substitute for having the staff and funding immediately available. It is widely recognized that eradication operations should not begin unless sufficient funding is guaranteed through to completion (Wittenberg and Cock 2001, 2005).

LONG-TERM MITIGATION

Species that evade border protection and rapid-response filters may often form deeply entrenched and geographically widespread populations that can then only be managed to mitigate damage to valued resources, such as natural areas or agricultural lands. In the United States, historical unconcern with vertebrate introductions has ensured that this is the most common response to invasive vertebrates. In all instances of which I am aware, default reliance on this final management action has meant that invasive vertebrates spread widely, cause tremendous cumulative damage, but may only be controlled in a few high-value areas. Fundamentally, though, sole reliance on this final management tool is no biosecurity program at all, which is, by definition, a comprehensive program of protection via prevention.

RATIONALE FOR BIOSECURITY

It will be obvious from the outline above that providing meaningful biosecurity protection to a jurisdiction is an involved affair, and it costs money. This expense is often used to justify slighting biosecurity, arguing that the cost is too high and the benefits inessential. To the contrary, the benefits of biosecurity are high because the costs of failing to take action are high and often much higher than the costs of action (Keller et al. 2007; Springborn et al. 2011; Keller and Springborn 2014). Within the

United States alone, estimates more than a decade old showed the costs of invasive species to the U.S. economy were conservatively $120 billion/year (Pimentel et al. 2000, 2005; Pimentel 2002). These estimates include damage costs, mitigation costs, and costs of lost economic opportunities; and the estimates were conservative because not all invasive species could be provided with cost estimates. Of course, it is also true that there are economic benefits to certain economic sectors from importation of alien species, including those that prove invasive upon escape or release. However, whereas the costs of invasive species are a public harm—and the benefits of biosecurity a public good—dispersed across the citizenry as a whole, the benefits of importation of invasives accrue to a small number of economic actors, not to the general public. The rationale for biosecurity is that its provision provides a service to the public as a whole and the economy as a whole, without undue preference given to any particular economic sector. And, as stated above, it has been found to be a cost-effective means of protecting the public's interests (Keller et al. 2007; Springborn et al. 2011; Keller and Springborn 2014). Nonetheless, particular economic sectors (e.g., the pet trade, recreational fishing interests) may have some activities limited by a biosecurity program, so those sectors may resist such action. This status quo is identical to industry resistance in the 1960s to controlling chemical pollution—benefits from dumping toxins in the environment accrued to a small sector of the industrial economy; the (high) costs of compromised health and environment were dispersed across society as a whole. Currently in the United States, the benefits and costs of invasive species are apportioned with the same dishonest economic reckoning that was accorded to environmental pollution before it was regulated. Because of this similarity, it is often useful to view invasive species as biological pollution (Bederman 1991)—a form of pollution that is more insidious than chemical pollution because it is self-replicating.

CURRENT STATUS OF INVASIVE TERRESTRIAL VERTEBRATES IN THE UNITED STATES

Hundreds of terrestrial alien vertebrate species have been introduced into the United States or its territories, comprising at least 86 mammals, 207 birds, 281 reptiles, and 77 amphibians (Long 1981; Kraus 2009, updated; Witmer et al. 2007). Of these, at least 33 mammals, 97 birds, 100 reptiles, and 36 amphibians currently have established populations in the United States or its territories (Temple 1992; Long 2003; Kraus 2009, updated). That number is probably larger, however, given the age of the bird and mammal estimates, occasional uncertainty about the status of some populations, and lag times in discovering new invasive species. Many of these species are hugely destructive to ecological and economic values, leading to extinctions of natives, biotic homogenization, disruptions to food webs, changes to primary productivity of ecosystems, changes in soil formation, alterations of community structure, wholesale conversion or replacement of ecosystems, changes in nutrient-cycling dynamics, degradation of watersheds, increases in erosion and flooding rates, losses to agriculture, damage to human structures, power outages, disease epidemics, and degradation of human quality of life (Ebenhard 1988; Office of Technology Assessment 1993; Wilcove et al. 1998; Pimentel 2002; Pimentel et al. 2000, 2005;

Mooney and Cleland 2001; Mooney 2005; Kraus 2009, 2015). Details of many of the worst impacts in the United States are available in these citations and the following chapters, so they need not be repeated here. Suffice it to say that the problems created by invasive vertebrates in the United States are large and diverse, and there is no diminution in sight inasmuch as mitigation is difficult and introduction pressure is not decreasing.

Global introductions for all major groups of terrestrial vertebrates have been increasing exponentially or approximately exponentially for many years, with mammal introductions booming after the mid-1700s, bird introductions after the mid-1800s, and reptile and amphibian introductions after the mid-1900s (Kraus 2003, 2009). The same exponential pattern applies within the United States for reptiles and amphibians (Kraus 2003, 2009), but it remains uncertain whether U.S. bird and mammal introductions are also exponential in recent decades, although introductions in both groups continue to increase (e.g., Perry et al. 2006; Pranty and Garrett 2011). Exponential or not, species introductions remain a growing problem in the United States and its territories, with more species added every year.

Pathways of species introduction are diverse, but each vertebrate class has a few dominant historical pathways that account for the majority of introductions, and these pathways differ among those groups (Kraus 2003). Most intentional introductions of mammals have been carried out to provide food or game; however, in the Twentieth century, introductions to provide furs and for aesthetic reasons were also important. Small rodents form an exception to these generalities: they rarely have been introduced intentionally but are ubiquitous commensals with humans; most populations probably predate the Twentieth century. Birds have primarily been introduced intentionally, either as game animals or as pet releases. Reptiles and amphibians are mostly introduced intentionally via the pet trade or unintentionally as cargo stowaways. What is clear across all of these groups is that the majority of introduction pathways is intentional (Table 2.1) and the majority of introductions has occurred via intentional pathways (Kraus 2003, 2009). Of course, not all introductions result in establishment; nevertheless, it is important to note that, with the

---

TABLE 2.1

Pathways of Introduction for Terrestrial Vertebrates in the United States: Pathways in Italics Are Intentional Introductions

Amphibians	Reptiles	Birds	Mammals
Aesthetic	*Aesthetic*	*Aesthetic*	*Aesthetic*
Aquaculture	*Biocontrol*	*Biocontrol*	*Biocontrol*
Bait use	Cargo stowaway	*Domestication*	Cargo stowaway
Biocontrol	*Food*	*Game*	*Domestic*
Cargo stowaway	Nursery trade	*Pet trade*	*Food*
Food	*Pet trade*		*Fur*
Nursery trade	*Religious use*		*Game*
Pet trade	*Research*		*Tourist attraction*
Research	Zoo/exhibit		*Work*

TABLE 2.2

Major Invasive Vertebrate Pests in the United States and Their Primary Pathways of Introduction: Pathways in Italics Are Intentional Introductions

Taxon	Species	Introduction Pathway(s)
Mammals	Horses and donkeys (*Equus* spp.)	*Work*
	Pig (*Sus scrofa*)	*Food*
	Goat (*Capra hircus*)	*Food*
	Sheep and mouflon (*Ovis* spp.)	*Food, game*
	Cattle (*Bos taurus*)	*Food*
	Axis deer (*Axis axis*)	*Game*
	Mongoose (*Herpestes auropunctatus*)	*Biocontrol*
	Cat (*Felis catus*)	*Domestic, biocontrol*
	Dog (*Canis familiaris*)	*Domestic*
	Rabbit (*Oryctolagus cuniculus*)	*Food*
	Rats (*Rattus* spp.)	Cargo stowaway, *game/food*
	House mouse (*Mus musculus*)	Cargo stowaway
	Nutria (*Myocastor coypus*)	*Fur*
Birds	Pigeon (*Columba livia*)	*Domestic*
	Collared dove (*Streptopelia decaocto*)	Range expansion
	European starling (*Sturnus vulgaris*)	*Aesthetic*
	House sparrow (*Passer domesticus*)	*Aesthetic*
	Parrots and parakeets (Psittacidae)	*Pet trade*
	Bulbuls (*Pycnonotus* spp.)	*Pet trade*
	Myna (*Acridotheres tristis*)	*Biocontrol*
Reptiles	Brown anole (*Anolis sagrei*)	Cargo stowaway, *aesthetic, pet trade*
	Iguana (*Iguana iguana*)	*Pet trade*
	Burmese python (*Python molurus*)	*Pet trade*
	Brown tree snake (*Boiga irregularis*)	Cargo stowaway
Amphibians	Bullfrog (*Lithobates catesbeianus*)	*Food*
	Coqui (*Eleutherodactylus coqui*)	Nursery trade
	Cuban tree frog (*Osteopilus septentrionalis*)	Cargo stowaway, nursery trade, *pet trade*
	Tiger salamander (*Ambystoma tigrinum*)	*Bait use*

exception of rats, mice, and a handful of reptiles and amphibians, the most damaging invasive terrestrial vertebrates with established populations in the United States have been introduced deliberately for some perceived amenity or use value (Table 2.2). The centrality of intentional pathways in vertebrate introductions has important implications for designing biosecurity strategy to prevent additional introductions.

FUTURE INVASION RISK

Because the greatest risk of future vertebrate introductions to the United States is via intentional human importation and subsequent intentional or inadvertent release, limiting those actions can mitigate the risk. However, limiting release of legally imported animals cannot be ensured because of unforeseen accidents (e.g., hurricanes) or

TABLE 2.3
Some Terrestrial Vertebrates Repeatedly Successful as Invasives Elsewhere That Have Yet to Establish Populations in the United States

Amphibians	Reptiles	Birds	Mammals
Amietophrynus gutturalis	Anolis extremus	Acridotheres fuscus	Bubalus bubalis [b]
Duttaphrynus melanostictus	Carlia ailanpalai[b]	Cacatua galerita	Cercopithecus aethiops
Eleutherodactylus antillensis	Chamaeleo chamaeleon	Corvus splendens	Cercopithecus mona
Eleutherodactylus johnstonei	Emys orbicularis	Estrilda astrild	Cervus timorensis
Glandirana rugosa[a]	Gehyra mutilata [a,b]	Foudia madagascariensis	Erythrocebus patas
Hoplobatrachus chinensis	Gymnophthalmus underwoodi	Padda oryzivora[a]	Herpestes edwardsi
Hoplobatrachus tigerinus	Hemidactylus brookii	Pycnonotus cafer[a]	Mustela putorius (M. furo)
Hyla meridionalis	Lampropholis delicata[a]		Suncus murinus [b]
Ichthyosaura alpestris	Lycodon aulicus		Trichosurus vulpecula
Kaloula pulchra	Mauremys reevesi		
Lissotriton vulgaris	Pelodiscus sinensis [a,b]		
Litoria aurea	Phelsuma cepediana		
Litoria ewingi	Phelsuma dubia		
Polypedates leucomystax	Phelsuma laticauda[a]		
Ptychadena mascareniensis	Podarcis pityusensis		
Pelophylax nigromaculatus	Testudo graeca		
Silvirana guentheri[b]	Testudo hermanni		
Scinax ruber	Trachemys stejnegeri		
Triturus carnifex	Trachemys terrapen		
	Tupinambis teguixin		
	Varanus indicus[b]		
	Xenochrophis vittatus		

[a] Established in Hawaii but not on the mainland.
[b] Established on Guam but not on the mainland.

deliberate release by unconcerned individuals (common with pet-trade animals), so prohibiting importation of risky species is a more efficacious means of providing biosecurity protection than trying to control human behavior or natural disasters after importation. Of 1704 species of alien vertebrates imported into the United States from 2000 to 2004, 13% were found to be potentially risky as invasive or disease carriers

(Jenkins 2007), highlighting the huge task involved in comprehensively screening traded species. This will be a conservative estimate of invasion risk, however, because riskiness was largely based on knowledge of already invasive taxa.

Given that most vertebrate introductions have involved deliberate importation and release, and given the elastic and rapidly changing nature of the pet trade (Romagosa 2015), it is obvious that virtually any attractive vertebrate may be liable to introduction and that future risk remains very high overall. I provide examples of a few known invasive terrestrial vertebrates not yet established in the mainland United States so as to illustrate some of the obvious current exposure (Table 2.3). However, this excludes species already established in limited areas of the U.S. mainland but not geographically widespread, it excludes relatives of taxa that are invasive whose biological characteristics make them likely to be invasive themselves, and it excludes all species not yet introduced into the wild. Hence, it must be viewed as only a minimal illustration of some of the taxa that could pose a future threat to the country's ecological values or economic resources.

This list highlights another important point. As mentioned above, most invasive vertebrates have been deliberately imported and introduced, and most invasive reptiles and amphibians have been introduced via the pet trade. However, most of the high-risk reptiles and amphibians listed in Table 2.3 have been introduced elsewhere either as cargo stowaways or in nursery plants (an important subcategory of the former). This makes clear that screening proposed intentional importations, although absolutely necessary for preventing further vertebrate invasions into the United States, is insufficient to entirely meet that goal. Preventing unintentional introductions via cargo and nursery-trade pathways is also necessary, at least for reptiles and amphibians. And risk assessments are lacking for these pathways as well.

BIOSECURITY APPLICATIONS IN THE UNITED STATES

In the United States, biosecurity protection against terrestrial vertebrates has been highly porous. Prevention of intentional vertebrate introductions is federally regulated under the "injurious wildlife" provision of the Lacey Act, which adopts the black list approach of prohibiting importation and interstate transport or sale[*] of a relative handful of vertebrate, crustacean, and mollusk species deemed high-risk, except for permitted scientific, medical, or educational purposes; it allows unhindered movement and trade of the remainder of the world's wildlife (Fowler et al. 2007; Alexander 2013). As of January 11, 2016, 105 terrestrial vertebrate species (92 mammals, 4 birds, 9 reptiles) were banned under this act (Alexander 2013; U.S. Fish and Wildlife Service 2015). On January 12, 2016, the Fish and Wildlife Services (FWS) took the broader proactive measure of banning importation and interstate movement of 201 salamander species in an effort to prevent arrival into the United States of the potentially devastating fungal parasite *Batrachochytrium salamandrivorans* (http://www.fws.gov/injuriouswildlife/salamanders.html), which has led to

---

[*] Note added in press: The United States Court of Appeals for the District of Columbia ruled on April 7, 2017 (Decision 15–5199) that the FWS lacks the authority to regulate movement of listed injurious wildlife among the 49 continental states, cf., *U.S. Association of Reptile Keepers v. Zinke.*

a decline of salamander populations in Europe (Martel et al. 2013). Most states follow a similar black list approach that only bans a handful of vertebrates deemed to pose an unacceptable risk of harm (Kurdila 1988), although at least two states use a white list approach for terrestrial vertebrates. Alaska maintains a "clean" list of animals—mostly domestic and livestock species—that may be imported without a permit, but not released into the wild; all other species require a permit (Otts 2015). And Hawaii allows importation of almost 500 species approved for commercial sale, allows almost 1000 additional species for use by scientific and educational institutions, and bans all remaining species (Hawaii Revised Statutes, Chapter 150A). Both lists are highly idiosyncratic and clearly not based on formal risk assessment. Otherwise, white list approaches seem to have rarely been applied in the United States, at least for vertebrates, although the Lacey Act was originally a white list law that was changed to its current black list status by Congress in 1949 (Alexander 2013). The FWS attempted to get the Lacey Act modified to readopt the white list approach in 1973, and again in 1975, but Congress refused to pass the necessary legislation (Kurdila 1988; Fowler et al. 2007).

Because of concern with disease spread by foreign vertebrates, species can also be prohibited entry into the United States by the USDA under the Animal Health Protection Act and Plant Protection Act, which allow for regulation of any animal that is a pest or disease carrier that threatens livestock or plants; and the Public Health Service Act allows the Department of Health and Human Services (DHHS) to regulate any animal that presents a human disease risk. USDA bans imports of a few species from particular regions known to harbor diseases dangerous to wildlife, but it also bans importation of three species of African tortoises because they vector ticks that carry ruminant heartwater disease. DHHS makes it illegal to import primates into the country for the pet trade, but this is unenforced, and licensed importing zoos and research centers frequently redistribute animals to private dealers (Green 1999). Quarantine of arriving animals is required for commercially imported livestock, birds, and a few other animals (Jenkins 2007) but this doesn't cover diseases of other animals or prohibit import of quarantined species; it merely tries to ensure that arriving animals are free of disease.

Federal procedures to prevent unintentional introductions of vertebrates appear largely lacking in the United States except for the USDA program on Guam to stop the spread of brown tree snakes from that island (Engeman and Vice 2001). No doubt, occasional vertebrate hitchhikers found in arriving cargo are lethally dispatched, but the USDA lacks a stated mission to prevent such introductions, so actions are ad hoc. Interceptions are also made by state and territorial agricultural services to prevent new accidental introductions, with Hawaii probably adopting the most systematic effort to intercept accidental vertebrate introductions.

Preventive use of risk-assessment models for management purposes remains unevenly developed. Reed (2005) assessed invasion risk for 23 large pythons and boids using an index based on six ecological and trade attributes of the snakes. He quantitatively ordered species from lowest to highest risk but did not categorize them into risk ranks (e.g., "high," "medium," or "low" risk). This assessment was expanded by Reed and Rodda (2009) for nine of the largest constrictors, using a diverse combination of ecological and reproductive attributes to assess establishment and impact potential following the approach advocated by the Aquatic Nuisance Species Task Force (ANSTF, cf., Orr 2003). They found the overall risk of five of these species to be high and that of the remaining four to be medium. The FWS has subsequently acted on recommendations to list most of these species as "injurious," thereby limiting importation and interstate trade of eight of these nine species (U.S. Fish and Wildlife Service 2012, 2015), but curiously not listing that species (*Boa constrictor*) most successful in establishing invasive populations. Fujisaki et al. (2010) assessed establishment risk in Florida for 33 reptile species not yet reported from the wild, using models developed to discriminate between established versus failed reptile species reported in that state. They found 12 of these to pose a risk of establishment, and at least four were a danger to either humans or the ecosystem. Rolan (2003) assessed the risk of 24 commonly imported amphibians, also using the ANSTF approach (Orr 2003). Only two of these were deemed to pose high risk of establishment, 10 were rated "medium," and the remainder "low." In the Rolan (2003) and Fujisaki et al. (2010) studies, however, many candidate species with high invasion success elsewhere were not considered. Springborn et al. (2015) assessed 165 species in the live-bird trade using a model combining establishment risk and disease risk, although they offered no management recommendations for particular species. Such integrated risk assessments are a sensible approach for future work on vertebrates given that alien vertebrates are liable to transport diseases of economic and ecological significance. Nonetheless, *a priori* risk assessments for terrestrial vertebrates in the United States have been minimal.

The FWS has recently applied a model to evaluate risk of invasiveness for aquatic fish, crayfish, and mollusks using data on climate matching and history of invasiveness (https://www.fws.gov/injuriouswildlife/pdf_files/ERSS-SOP-Final-Version.pdf). Risk of impact is assigned as "high," "low," or "uncertain" based on documented impacts and disease risk, and degree of climate match in the United States is measured as the areal extent of the mainland United States having a high climate-matching score, categorized as "high," "medium," or "low." These two categories are then combined for an overall risk assessment of "high" if the impact assessment is "high" and the climate-match assessment is "high"; and risk is assessed as "low" if the impact assessment is "low" and the climate-match assessment is "low." Species for which information is limited are assessed as "uncertain." However, it is unclear how species with other combinations of impact and climate-match assessment are to be rated overall. This simple model could also be applied to terrestrial vertebrates, but its reliance on only two risk-assessment parameters limits the generality of its application. Species with no information on impact or likely risk cannot be assigned, and those with uncertain assignment are left *incertae sedis*. Nonetheless, its application to terrestrial vertebrates could provide a means of at least screening from importation of those species of greatest risk (see also Jenkins 2013).

Rapid response against new incursions of novel alien vertebrates seems to have been rarely tried in the United States and, when tried, has often failed due to insufficient funding or initial underestimates of pest numbers. Responsibility for vertebrate EDRR lies with individual states, which have authority over wildlife management within their borders, including authority to release alien vertebrates (Dentler 1993). Hawaii has adopted EDRR for a diversity of invasive species, including some vertebrates (Kraus and Duffy 2010; Penniman et al. 2011). Incipient populations of pest vertebrates have been eradicated there (e.g., guinea pigs, rabbits, coqui frogs; cf., Kraus and Duffy 2010; Beachy et al. 2011), and removal of mitred conures (*Psittacara mitrata*) from Maui may be nearing success (T. Penniman, pers. comm.). Interest in such work is growing in Florida as well, as evidenced by recent programs to eradicate new incursions of sacred ibis (*Threskiornis aethiopicus*) and purple swamphens (*Porphyrio porphyrio*) in the Everglades region (Hardin et al. 2011; South Florida Ecosystem Restoration Task Force 2015). Isolated populations of invasive frogs have been eradicated in California, North Carolina, and Virginia (Tinsley and McCoid 1996; Kraus 2009; Measey et al. 2012), but those seem not to have been part of any general EDRR programs. I am unaware of other systematic applications of EDRR against vertebrates in the United States.

A handful of well-established, widespread invasives are subject to control operations at high-value locations in many states. Rats and mice, of course, are routinely subject to control in a diversity of residential, industrial, and agricultural settings. And the Animal Damage and Control Act provides the USDA authority to conduct research and control operations on invasive vertebrates that damage agriculture interests, buildings, natural resources, or human health. Relatively few species are subject to control under this authority, some examples being rats, nutria, and starlings. Examples of responses to some of the other more problematic invaders are given in chapters throughout this book. But because most established invaders must be controlled in perpetuity, those activities are typically effected in only a small fraction of their invaded ranges, such as agricultural fields or high-value natural areas. More cost-effective than long-term control would be to eradicate the alien where possible. That can generally be done only in very geographically circumscribed areas, such as islands, that have boundaries inherently liable to defense against reinvasion. In the United States, destructive rodents, rabbits, and ungulates have been eradicated from a diversity of islands and fenced conservation areas, mostly in California and Hawaii (Taylor and Katahira 1988; Loope et al. 1991; Anderson and Stone 1993; Katahira et al. 1993; Tunison et al. 1995; Loh and Tunison 1999; Campbell and Donlan 2005; Howald et al. 2005, 2007; Morrison 2007; McCann and Garcelon 2008; Hess and Jacobi 2011; Witmer et al. 2011; Cole and Litton 2014); Arctic foxes have been removed from over 40 Aleutian islands (Ebbert 2000); and wild turkeys were removed from Santa Cruz Island, California (Morrison et al. 2014). However, it is evident that successful eradication is rarely achieved, relative to the number and extent of invasions, and when achieved, applies only to small areas relative to the entire infested landscape. More commonly, control operations are billed as "eradications" until it becomes apparent that the size of the target populations is larger than expected and that available resources are not commensurate with those needed for eradication (e.g., Rosen and Schwalbe 1996; Campbell 2005).

BIOSECURITY GAPS IN THE UNITED STATES

Given the large pool of invasive vertebrates in the United States, continued increase in numbers of new introductions, and paucity of successful (and attempted) eradications, it is clear that meaningful biosecurity in the United States remains as great a need as ever. The continuing onslaught of invasions in the United States suggests that the tools needed to provide meaningful biosecurity are either lacking or are available but ineffectively implemented. Hence, a closer look at biosecurity gaps in the United States is warranted.

Biosecurity as it relates to terrestrial vertebrates in the United States is overwhelmingly an issue of regulating the live-animal trade, and legislative tools to regulate that trade provide minimal effectiveness. At the federal level, the Lacey Act is a limited and inefficient tool for preventing vertebrate invasions. This is because it requires a tedious administrative rule-making procedure, is difficult to enforce, is often used reactively only after a species is irrevocably established, and doesn't include authority to prevent escape of animals already in the United States or to manage populations of listed species so as to stop or reverse their spread (Kurdila 1988; Dentler 1993; Fowler et al. 2007; Jenkins 2007). It is also doubtful that the act can be helpful for addressing unintentional imports (Alexander 2013). Species currently listed as injurious under the act comprise 0.9% of the world's 33,546 species of terrestrial vertebrates and a small percentage of known invasives. Indeed, 60% of these are listed not because they are invasive in their own right but because they are potential carriers of a disease that could be devastating to native salamanders. Eight of the 105 species listed for invasiveness in their own right are already established in the United States or its territories (true as well for several listed fish and invertebrates), although most have restricted ranges. Perhaps most problematic is that the Lacey Act black list approach guarantees that the large majority of vertebrates in trade will never be screened for invasiveness prior to importation because the slow pace of listing (an average of four years to complete a new listing, Fowler et al. 2007) cannot meaningfully respond to the thousands of animals in trade (Jenkins 2013). Individual states can prohibit species that the federal government ignores, but, as noted above, they largely employ similarly ineffective black list approaches, and their lists are not yet based on scientific risk assessment, so far as I can determine. Even should a state ban a particular species, the contiguous 48 states are at the mercy of permissive decisions made by neighboring states, inasmuch as species released in adjacent states will not respect political boundaries as they spread (e.g., Kurdila 1988).

Even if FWS and the states were suddenly authorized to pursue a white list approach, so little work has been done on identifying scientifically robust, repeatable models to screen vertebrates for the United States that any assessment would continue to be either ad hoc or poorly parameterized. Lack of such models is one reason for the long listing process under current FWS regulations. As of the end of 2015, FWS is employing a two-parameter screening process for aquatic animals (http://www.fws.gov/injuriouswildlife/Injurious_prevention.html), but application to terrestrial vertebrates has not yet been made. Even were it applied to terrestrial vertebrates, the model is based only on extent of climate match and known impacts elsewhere. The latter precludes its application to species not yet introduced anywhere or those introduced but not yet investigated for impacts. Hence, it is a useful beginning but a rather crude tool in need of further refinement. Without credible and general risk-assessment models, systematic screening of vertebrates proposed for introduction is unlikely to be done with sufficient efficiency or sensitivity to stem the rate of continued invasion, no matter which listing paradigm is adopted.

USDA and DHHS regulations prohibit importation of a few additional species of vertebrates, but because those laws are designed to stop importation of animal diseases into the country, they have minimal relevance to preventing vertebrate invasions per se, inasmuch as most animals are imported following appropriate quarantine and permitting. Further, they do not quarantine highly infectious diseases—such as *Batrachochytrium dendrobatidis* and *B. salamandrivorens*—that could decimate large sectors of native wildlife in the United States. This highlights another limitation of biosecurity in the United States: it has largely been concerned with protecting agriculture or human health, not the health of native biota. Consequently, species (or their diseases) that potentially threaten natural areas or ecological values have largely been ignored. Because USDA has primary authority for inspection of arriving goods, hitchhiking vertebrates discovered during inspections may be ignored because they are not part of that agency's mandate. Focus on species of direct harm to humans or agriculture has not served to protect native biota in the United States. Even when damages are acknowledged for the few regulated species, existing tort law in the United States is inadequate to compensate for those damages, in part because they do not allow private plaintiffs to take action to protect common property (Dentler 1993). This removes one potential incentive to achieve improved regulation.

A further, little-discussed limitation is the widespread failure to discriminate political from biogeographic boundaries in invasive species management. Much of the protective management that exists against alien invasives focuses only on excluding species native to areas outside the United States. Ignored is the fact that the United States is a large and biogeographically diverse country in which species trans-shipped from one region to another are just as alien (and potentially invasive) as species coming from foreign countries. This is especially obvious when contrasting, say, Hawaii to the mainland: virtually every mainland species is alien and potentially invasive in Hawaii. But the same holds true for mainland species moved from east to west or vice versa—most taxa native to, say, Florida will not be shared with, say, California. These aliens should also be prevented from introduction outside their native ranges pending risk assessments. Many of the most damaging invasive vertebrates in the United States are native to some portion of the country, but federal laws generally do not prohibit their interstate shipment. Regulation is left to state management instead, which is inherently difficult because most states lack any form of border protection against invasive species.

A related problem is federal quarantine policy that preempts stricter state laws if the latter are interpreted as interfering with interstate commerce. Weak federal quarantine laws typically list as actionable pests only species that threaten major interests or portions of the country, and preemption of stricter state laws results in species risky to only a small region of the United States being allowed unhindered importation and interstate shipment (Lodge et al. 2006). This federal "preemption" policy has been a problem for the state of Hawaii's attempts to prohibit species or products risky there but of no concern to the mainland (Warren 2006; Ewalt 2013), it is a limiting factor in protecting the Great Lakes from ballast discharges (Stewart 2007), and other regions are potentially vulnerable to this same problem. This policy has mainly affected importation of invertebrate pests and pathways (plant media), and it is unclear whether it would present a problem for preventing introduction of any alien vertebrate. But it serves again to highlight that biogeographic concerns run counter to the tenor of U.S. commercial policy. In contrast, invasions need to be conceived and managed using meaningful biogeographic criteria, not merely artificial national boundaries, and these considerations need to be included in risk assessments.

Most alien terrestrial vertebrates not in the live-animal trade arrive in cargo and nursery shipments (Kraus 2003, 2009). Means to treat shipments to remove these hitchhikers are poorly developed. The USDA may make inspections, but that agency is only authorized to treat or seize shipments containing pests to plants or livestock, which does not include most vertebrates. Hence, there are presently no meaningful biosecurity barriers whatsoever to those introduction pathways.

Few states have EDRR programs to control incipient pests, and those that do usually target plants (e.g., Hawaii, cf., Kraus and Duffy 2010; Penniman et al. 2011), although a few vertebrate eradications have been successful, as noted above. A major limitation of EDRR programs in the United States is the frequent inability to access all lands on which an infestation may occur because most governments demur from mandating landowner cooperation. Landowner recalcitrance can lead to failed eradication attempts because the requirement that all individuals be put at risk (Bomford and O'Brien 1995) cannot be met (e.g., Penniman et al. 2011; Witmer and Hall 2011). Furthermore, some invasive vertebrates are protected by federal law: feral burros and horses are protected from control programs by the federal Wild Horse and Burro Act, and birds native to North America but introduced into Hawaii are protected under the Migratory Bird Treaty Act. EDRR programs can also be undermined by absence of funding dedicated to that purpose. Many governments are reluctant to fund programs that may be intermittently needed and that may leave dedicated funds lying fallow between operations.

It is clear from all this that the answer to the question of why biosecurity is so poor in the United States primarily lies in the absence of effective preventive tools. Biosecurity in the United States is not risk averse; gaps in its structure have been identified for decades (Kurdila 1988; Dentler 1993; Office of Technology Assessment 1993), and the tools needed to correct those gaps have long been known (e.g., National Invasive Species Council 2001; Wittenberg and Cock 2001) and are obvious in any event, but the needed changes have not been adopted. Dozens of federal and innumerable state agencies have some responsibility over this issue (Corn and Johnson 2013), and large gaps exist in interagency authorities and coordination among the agencies (Office of Technology Assessment 1993; Kraus 2009: 116–121); hence, it has long been embarrassingly clear that a single, comprehensive biosecurity law is needed to protect the United States from further invasions. Yet the U.S. Congress has long eschewed pursuing that option, and even the piecemeal legislation it has adopted to improve the country's tatterdemalion biosecurity approach has been infrequent. Of the eight laws enacted by Congress since 1990 that treat some aspect of species invasions, all deal with aquatic organisms, plants, or animal diseases, and most provide limited authorities or treat only a portion of the biosecurity problem even for those taxa.

Given the comprehensive and long-standing nature of these gaps, the tremendous costs inflicted on the United States by failure to respond, and the ability of other countries to effectively develop invasive species programs, it would seem worthwhile to consider some of the possible reasons these gaps remain unaddressed. This requires delving into issues of political structure and cultural mores, but a brief consideration of those topics will make clear the challenges that remain in providing the United States with meaningful biosecurity.

POLITICAL BARRIERS TO BIOSECURITY

A variety of institutional and cultural habits shed light on the political failure to develop meaningful biosecurity in the United States despite decades of attention to the topic. Together they sum to create a daunting set of roadblocks to achieving meaningful reform, and these contrast with the political and cultural contexts of countries that have been more responsive on this issue. Both the legislative and executive branches of government abet this inaction, but I will focus my remarks on the former because legislative relief is of primary importance in providing a comprehensive biosecurity program. Executive action can only improve administration of the existing incomplete patchwork of laws and policies; hence, it is of secondary importance at present.

1. *Federal government prone to gridlock.* The framers of the U.S. Constitution designed the government to be slow moving (the famed "checks and balances") because of their concern to contain a potential surfeit of democracy: "We ... underestimate the extent to which our system was designed for deadlock and inaction" (Burns 1973: 6). When the U.S. government does act, it is most often reactionary, responding to problems only when they can no longer be ignored, and not always then. Even if legislative gridlock was not the intentional design of the framers of the Constitution, it is certainly its frequent result, and partisan polarization makes gridlock more likely (Binder 2003). Given the system's design for factionalization (Burns 1973), it most often responds usefully when there is broad public and political consensus across interest groups on an issue that can lead to creation of multifaction coalitions. However, there is no public consensus in the United States on species invasions and biosecurity, and partisan polarization in the country is extreme, so progress in developing comprehensive biosecurity legislation has been sparse. Although piecemeal reforms have appeared sporadically over the past few decades, many and major gaps remain, the pace of response is incommensurate with the size of the problem, and reform lags well behind the pace taken by several other countries.

2. *No important constituency for biosecurity.* For obvious political reasons, Congress is most likely to act when there is a strong voting or funding constituency pressing an issue. Obviously, this criterion is not met for biosecurity. Biologists and conservationists concerned with the issue total few votes and provide no important funding or other political support; hence, they are ignored at no political peril. And the large conservation organizations in the country have mostly neglected this issue. Furthermore, there are rarely electoral consequences to individual legislators for Congressional gridlock (Binder 2003). The result of both is that there is little incentive for

Congress to address biosecurity. Agricultural interests could, theoretically, serve as such a constituency—as they did in the passage of New Zealand's Biosecurity Act—but that has not happened in the United States, probably, in part, for the cultural reasons given below.

3. *Important constituencies opposed to biosecurity.* More problematic, there are organized, vocal constituencies opposed to enacting meaningful biosecurity programs. Attempts in the 1970s to revert the Lacey Act's injurious-wildlife provision to its original white list approach were defeated in part by strong opposition from the pet industry (Fowler et al. 2007). Similarly, attempts by FWS to implement President Carter's Executive Order 11987 on exotic species were stymied by strong opposition from pet-trade, agricultural, and other vested interests (Kurdila 1988; Office of Technology Assessment 1993). These interests have only grown larger in the meantime, although FWS was recently successful in listing several giant constrictor snakes as injurious (U.S. Fish and Wildlife Service 2012, 2015) despite strong opposition from pet-trade interests and pet fanciers. Congress would face similar pressures should it ever choose to act on related topics.

4. *Cultural preference for unhindered commerce.* Early in the history of the United States, the notion of free, unhindered trade became deeply intertwined with that of individual liberty and equality (Appleby 1978). Freedom to pursue economic opportunity was conflated with the public good, and it has remained so ever since. Consequently, there is a strong cultural disposition among the general public, regulators, and legislators toward minimal interference with commerce and only when given strong justification of the broader public good (e.g., the general public is not allowed to freely trade purified uranium). This freedom is codified in the Commerce Clause of the U.S. Constitution, which is designed to protect unfettered flow of interstate commerce. Nonetheless, state regulations banning imports to protect native wildlife have been upheld by the Supreme Court, although states have a heavy burden to demonstrate that such bans cannot be met by less restrictive means (Dentler 1993). Despite this potential for constitutional conformance, many people do not believe that invasive species constitute a compelling justification for regulatory hindrance of trade (e.g., restrictions on animal trade), which makes passing such measures difficult.

5. *Cultural preference for presumption of innocence.* The United States has a judicial system based on presuming an accused to be innocent until proven otherwise, and this attitude of fair-mindedness pervades many nonjudicial judgments in the culture. A black list approach to removing species from trade accords nicely with this paradigm; the more precautionary white list approach instead presumes liability until safety is demonstrated, which is, superficially, reminiscent of institutionalized suspicion by a number of Twentieth century political tyrannies toward their citizens. Although these parallels are merely superficial and do not reflect underlying political or regulatory similarities, to a public ignorant of invasive species issues, they can often lead to an emotional bias against white list regulations when first learning of that approach.

6. *Marginalization of long-term thinking.* The widely adopted paradigm of economism in the United States has led many people to decide values and understand cause and effect largely in terms of short-term consequences, consistent with a culture focused on yearly budgets and quarterly profit statements. Phenomena that develop over longer time scales tend to be discounted or down-weighted under this viewpoint (at the very least, they are subject to errors in estimating discount rates used to model long-term cost/benefit analyses). This is dangerous for accurately evaluating hazard because damages that develop over longer periods of time may easily be elided relative to the short-term costs of action. Most invasions develop slowly, making them prone to this sort of perceptual bias.

BIOSECURITY NEEDS IN THE UNITED STATES

Although the cultural biases just discussed make biosecurity reform in the United States difficult, it remains important to note several items needed to provide the United States with a functioning biosecurity program capable of protecting the country from future vertebrate (and other) invasions. The importance of and justification for several of these actions is discussed in greater detail elsewhere (Kraus 2009: 116–121). It remains to be seen whether creative means of meeting these needs can be devised that will overcome the structural and cultural roadblocks just presented.

1. *Adoption of a precautionary listing approach.* Preventing arrival of additional invasive vertebrates is most efficiently done by adopting the precautionary approach that prohibits entry to all species not screened and found to be of low risk for invasiveness. This white list approach would provide immediate protection against all species not already in the United States, and future imports would be less likely to be invasive, having been explicitly assessed for risk. Adoption of this approach requires honest accounting of who gains and who pays the economic and environmental costs of the current system of unfettered importation. Implicit in a revised, precautionary paradigm is that all pests—not just those of risk to agriculture or human health—are included under its purview. This is the single most-important improvement needed for U.S. biosecurity.

2. *Development of compelling risk-assessment models.* Any listing program—whether white list or black list in nature—is only as good as the means of assessing species risk. Models for this purpose remain poorly developed, poorly parameterized, and minimally tested for all terrestrial vertebrates. Although studies have shown a number of factors to correlate with invasion risk in different taxa, models that combine all relevant factors are lacking for most. Further, there are likely additional parameters yet to be evaluated for risk prediction that could make models more precise, and some parameters identified in scientific studies are too complicated for ready use by managers. Without scientifically justifiable, readily determined parameters for assessing invasion risk, regulations prohibiting certain species are opaque and potentially liable to challenge. The FWS has recently made

some progress in applying risk modeling to invasive species prediction, but the models used remain either ad hoc or poorly parameterized. Studies to develop more comprehensive models predictive across broad taxonomic groups (including species not yet known to be invasive elsewhere) are critically needed and should be a funding priority in developing better biosecurity in the United States.

3. *Expanded coordination, authorities, and resources.* The current system of having responsibility for invasive species partitioned among innumerable federal agencies, councils, and task forces (Corn and Johnson 2013) has created numerous authority gaps and, ultimately, biosecurity failure. The same dispersed authorities likely characterize most state governments. Better coordination and support between agencies is needed to overcome these obstacles and try to compensate for some missing authorities. For example, USDA port inspectors should be allowed to prevent the arrival of any hitch-hiking vertebrates or, generally, taxa of concern to other agencies. Further, biosecurity to protect native wildlife and ecosystems needs to be made a higher governmental priority, with authorities and resources provided that are commensurate with those afforded protection of agricultural interests. Theoretically, Executive Order 13112 could be used to help meet some of these needs, but direct congressional action will be needed for acquiring new authorities and funding new mandates.

4. *Development of EDRR programs.* Successful EDRR programs will all need a reliable source of funds, a means of receiving and responding to reports from the general public, and personnel to respond to reports and incursions. They also need to operate successfully across multiple land-ownerships, which requires a level of interagency cooperation that is often unfamiliar to agency personnel. Coordinating effective actions across multiple agencies and securing funding to respond to new, unbudgeted threats can be difficult, although successful models for achieving both results are available (Kraus and Duffy 2010). Given the large pool of invasions that merit response in many states, establishing permanent programs and hiring full-time staff to meet these needs will often be needed. Governmental requirements will be needed as well for landowner cooperation with EDRR programs, as seen in other countries. Provision of sanctuary to a species by even one landowner opposed to control operations can lead to operational failure (e.g., Witmer and Hall 2011).

5. *Development of new control techniques.* Means to remove populations of a number of the worst mammalian invaders are well developed (cf., Veitch and Clout 2002; Veitch et al. 2011), but that is not true for most other terrestrial vertebrates, and this has led to failure of some eradication attempts (Kraus 2009: 107–108). For some species, control is not even attempted for lack of reliable tools. Funding is needed to create useful control methods for eradicating a diversity of bird, reptile, and amphibian invasions, perhaps in the form of a short-term (5–10 years) intensive program to develop and test improved tools. In a similar vein, development of control methods to sterilize high-risk pathways—such as shipments of nursery plants—should

be pursued. This would allow for efficiently excluding a diversity of species using one or a few treatment methods, much as has been developed for treating ballast water or raw logs.

6. *Education.* The need for public education in the United States about biosecurity is critical. The topic is poorly understood, and it is especially important to convey to policy makers and the general public the importance of the precautionary principle, the costs to society of unfettered trade in animals, and the dishonest accounting system that externalizes those costs upon the public. It is particularly important to discuss these topics in ways that distance them from the default cultural biases favoring unhindered trade and presumptions of innocence. Both of those concepts have their rightful social places, but not in the realm of biosecurity.

ACKNOWLEDGMENTS

I thank P. Jenkins, C. Martin, K. Minami, T. Penniman, R. Reed, and C. Romagosa for pointing me to helpful supporting information or references.

REFERENCES

Alexander, K. 2013. *Injurious Species Listings Under the Lacey Act: A Legal Briefing.* CRS Report No. R43170, Congressional Research Service, Washington, DC.

Amiel, J.J., R. Tingley, and R. Shine. 2011. Smart moves: Effects of relative brain size on establishment success of invasive amphibians and reptiles. *PLoS ONE* 6: e18277.

Anderson, S.J., and C.P. Stone. 1993. Snaring to control feral pigs *Sus scrofa* in a remote Hawaiian rain forest. *Biological Conservation* 63: 195–201.

Appleby, J. 1978. The social origins of American revolutionary idealogy. *Journal of American History* 64: 935–958.

Beachy, J.R., R. Neville, and C. Arnott. 2011. Successful control of an incipient invasive amphibian: *Eleutherodactylus coqui* on O'ahu, Hawai'i. Pp. 140–147, in Veitch, C.R., Clout, M.N., and Towns, D.R. (eds.). *Island Invasives: Eradication and Management.* IUCN, Gland, Switzerland.

Bederman, D.J. 1991. International control of marine "pollution" by exotic species. *Ecology Law Quarterly* 18: 677–717.

Binder, S.A. 2003. *Stalemate: Causes and Consequences of Legislative Gridlock.* The Brookings Institute, Washington, DC.

Blackburn, T.M., P. Cassey, and R.P. Duncan. 2004. Avian extinction and mammalian introduction on oceanic islands. *Science* 305: 1955–1958.

Blackburn, T.M., P. Cassey, and J.L. Lockwood. 2009. The role of species traits in the establishment success of exotic birds. *Global Change Biology* 15: 2852–2860.

Bomford, M. 2003. *Risk Assessment for the Import and Keeping of Exotic Vertebrates in Australia.* Bureau of Rural Sciences, Canberra, Australia.

Bomford, M. 2006. *Risk Assessment for the Establishment of Exotic Vertebrates in Australia: Recalibration and Refinement of Models.* Bureau of Rural Sciences, Canberra, Australia.

Bomford, M. 2008. *Risk Assessment Models for Establishment of Exotic Vertebrates in Australia and New Zealand.* Invasive Animals Cooperative Research Centre, Canberra.

Bomford, M., and F. Kraus. 2008. Exotic reptiles and amphibians. Pp. 34–91, in Bomford, M. *Risk Assessment Models for Establishment of Exotic Vertebrates in Australia and New Zealand.* Invasive Animals Cooperative Research Centre, Canberra.

Bomford, M., and P. O'Brien. 1995. Eradication or control for vertebrate pests? *Wildlife Society Bulletin* 23: 249–255.

Bomford, M., F. Kraus, M. Braysher, L. Walter, and L. Brown. 2005. *Risk Assessment Model for the Import and Keeping of Exotic Reptiles and Amphibians*. Bureau of Rural Sciences, Canberra, Australia.

Bomford, M., F. Kraus, S. Barry, and E. Lawrence. 2009. Determinants of establishment success in introduced reptiles and amphibians: A role for climate matching. *Biological Invasions* 11: 713–724.

Burns, J.M. 1973. *The Deadlock of Democracy: Four-Party Politics in America*. Prentice-Hall, Englewood Cliffs, New Jersey.

Campbell, K., and C.J. Donlan. 2005. Feral goat eradications on islands. *Conservation Biology* 19: 1362–1374.

Campbell, T.S. 2005. Eradication of introduced carnivorous lizards from Southwest Florida. Final report to the National Fish and Wildlife Foundation, 1120 Connecticut Avenue NW, Suite 900, Washington, DC 20036. 30 pp.

Capellini, I., J. Baker, W.L. Allen, S.E. Street, and C. Venditti. 2015. The role of life history traits in mammalian invasion success. *Ecology Letters* 18: 1099–1107.

Case, T.J., and D.T. Bolger. 1991. The role of introduced species in shaping the distribution and abundance of island reptiles. *Evolutionary Ecology* 5: 272–290.

Cassey, P. 2002. Life history and ecology influences establishment success of introduced land birds. *Biological Journal of the Linnean Society* 76: 465–480.

Clout, M.N., and J.C. Russell. 2006. The eradication of mammals from New Zealand islands. Pp. 127–141, in Koike, F., M.N. Clout, M. Kawamichi, M. De Poorter, and K. Iwatsuki (eds.). *Assessment and Control of Biological Invasion Risks*. Shoukadoh Book Sellers, Kyoto, Japan and the World Conservation Union (IUCN), Gland, Switzerland.

Cole, R.J., and C.M. Litton. 2014. Vegetation response to removal of nonnative feral pigs from Hawaiian tropical montane wet forest. *Biological Invasions* 16: 125–140.

Corn, M.L., and R. Johnson. 2013. *Invasive Species: Major Laws and the Role of Selected Federal Agencies*. Congressional Research Service. R43258.

Dentler, J.L. 1993. Noah's farce: The regulation and control of exotic fish and wildlife. *University of Puget Sound Law Review* 17: 191–242.

Ebbert, S. 2000. Successful eradication of introduced Arctic foxes from large Aleutian Islands. *Proceedings of the Vertebrate Pest Conference* 19: 127–132.

Ebenhard, T. 1988. Introduced birds and mammals and their ecological effects. *Viltrevy* 13: 1–107.

Engeman, R.M., M.A. Linell, D.S. Vice, and M.E. Pitzler. 1998. Efficacy of the methods used in an integrated program to deter the dispersal of brown tree snakes from Guam. *Proceedings of the Australian Vertebrate Pest Conference* 11: 435–440.

Engeman, R.M., and D.S. Vice. 2001. Objectives and integrated approaches for the control of brown tree snakes. *Integrated Pest Management Reviews* 6: 59–76.

Evans, T., S. Kumschick, E. Dyer, and T. Blackburn. 2014. Comparing determinants of alien bird impacts across two continents: Implications for risk assessment and management. *Ecology and Evolution* 4: 2957–2967.

Ewalt, S.A. 2013. Hawaii's freight inspection fee and avoiding preemption. *The Air & Space Lawyer* 25(4) 8–11, 17.

Ferreira, R.B., K.H. Beard, S.L. Peterson, S.A. Poessel, and C.M. Callahan. 2012. Establishment of introduced reptiles increases with the presence and richness of native congeners. *Amphibia-Reptilia* 33: 387–392.

Ficetola, G.F., W. Thuiller, and C. Miaud. 2007. Prediction and validation of the potential global distribution of a problematic alien invasive species: The American bullfrog. *Diversity and Distributions* 13: 476–485.

Finnoff, D., J.F. Shogren, B. Leung, and D. Lodge. 2007. Take a risk: Preferring prevention over control of biological invaders. *Ecological Economics* 62: 216–222.

Fowler, A.J., D.M. Lodge, and J.F. Hsia. 2007. Failure of the Lacey Act to protect US ecosystems against animal invasions. *Frontiers in Ecology and the Environment* 5: 353–359.

Fujisaki, I., K.M. Hart, F.J. Mazzotti, K.G. Rice, S. Snow, and M. Rochford. 2010. Risk assessment of potential invasiveness of exotic reptiles imported to south Florida. *Biological Invasions* 12: 2585–2596.

Green, A. 1999. High-risk monkey business. *Mother Jones Magazine* 24(6): 50–55.

Hardin, S., E. Donlan, M. Ward, and D. Eggeman. 2011. Attempted eradication of *Porphyrio porphyrio* Linnaeus in the Florida Everglades. *Management of Biological Invasions* 2: 47–55.

Hayes, K.R., and S.C. Barry. 2008. Are there any consistent predictors of invasion success? *Biological Invasions* 10: 483–506.

Henderson, R.W. 1992. Consequences of predator introductions and habitat destruction on amphibians and reptiles in the post-Columbus West Indies. *Caribbean Journal of Science* 28: 1–10.

Hess, S.C., and J.D. Jacobi. 2011. The history of mammal eradications in Hawai'i and the United States associated islands of the Central Pacific. Pp. 67–73, in Veitch, C.R., M.N. Clout, and D.R. Towns (eds.). *Island Invasives: Eradication and Management.* IUCN, Gland, Switzerland.

Howald, G.R., C.J. Donlan, J.P. Galván, J.C. Russell, J. Parkes, A. Samaniego, Y. Wang, D. Veitch, P. Genovesi, M. Pascal, A. Saunders, and B. Tershy. 2007. Invasive rodent eradications on islands. *Conservation Biology* 21: 1258–1268.

Howald, G.R., K.R. Faulkner, B. Tershy, B. Keitt, H. Gellerman, E.M. Creel, M. Grinnel, S. Ortega, and D. A. Croll. 2005. Eradication of black rats from Anacapa Island: Biological and social considerations. Pp. 299–312, in Garcelon, D. K., and C. A. Schwemm (eds.). *Proceedings of the Sixth California Islands Symposium.* Institute for Wildlife Studies, Arcata, California.

Jenkins, P.T. 2007. The failed regulatory system for animal imports into the United States—And how to fix it. Pp. 85–89, in Witmer, G.W., W.C. Pitt, and K.A. Fagerstone (eds.). *Managing Vertebrate Invasive Species: Proceedings of an International Symposium.* USDA/APHIS/WS National Wildlife Research Center, Fort Collins, Colorado.

Jenkins, P.T. 2013. Invasive animals and wildlife pathogens in the United States: The economic case for more risk assessments and regulation. *Biological Invasions* 15: 243–248.

Jenkins, P.T., K. Genovese, and H. Ruffler. 2007. *Broken Screens: the Regulation of Live Animal Imports in the United States.* Defenders of Wildlife, Washington, DC.

Jiménez-Valverde, A., A.T. Peterson, J. Soberón, J.M. Overton, P. Aragón, and J.M. Lobo. 2011. Use of niche models in invasive species risk assessments. *Biological Invasions* 13: 2785–2797.

Katahira, L.K., P. Finnegan, and C.P. Stone. 1993. Eradicating feral pigs in montane mesic habitat at Hawaii Volcanoes National Park. *Wildlife Society Bulletin* 21: 269–274.

Keller, R.P., D.M. Lodge, and D.C. Finnoff. 2007. Risk assessment for invasive species produces net bioeconomic benefits. *Proceedings of the National Academy of Sciences, USA* 104: 203–207.

Keller, R.P., and M.R. Springborn. 2014. Closing the screen door to new invasions. *Conservation Letters* 7: 285–292.

Kraus, F. 2003. Invasion pathways for terrestrial vertebrates. Pp. 68–92, in Carlton, J., G. Ruiz, and R. Mack (eds.). *Invasive Species: Vectors and Management Strategies.* Island Press, Washington, DC.

Kraus, F. 2009. *Alien Reptiles and Amphibians: A Scientific Compendium and Analysis.* Springer Science and Business Media B.V., Dordrecht, Netherlands. 563 pp.

Kraus, F. 2015. Impacts from invasive reptiles and amphibians. *Annual Review of Ecology, Evolution, and Systematics* 46: 75–97.

Kraus, F., and D.C. Duffy. 2010. A successful model from Hawaii for rapid response to invasive species. *Journal for Nature Conservation* 18: 135–141.

Kroeger, T. 2007. *Economic Impacts of Live Wild Animal Imports in the United States.* Defenders of Wildlife, Washington, DC.

Kurdila, J. 1988. The introduction of exotic species into the United States: There goes the neighborhood! *Boston College Environmental Affairs Law Review* 16: 95–118.

Liu, X., X. Li, Z. Liu, R. Tingley, F. Kraus, Z. Guo, and Y. Li. 2014. Congener diversity, topographic heterogeneity and human-assisted dispersal predict spread rates of alien herpetofauna at a global scale. *Ecology Letters* 17: 821–829.

Lodge, D.M., S. Williams, H.J. MacIsaac, K.R. Hayes, B. Leung, S. Reichard, R.N. Mack, P.B. Moyle, M. Smith, D.A. Andow, J.T. Carlton, and A. McMichael. 2006. Biological invasions: Recommendations for U.S. policy and management. *Ecological Applications* 16: 2035–2054.

Loh, R.K., and J.T. Tunison. 1999. Vegetation recovery following pig removal in 'Ola'a-Koa Rainforest Unit, Hawaii Volcanoes National Park. Cooperative National Park Resources Studies Unit, University of Hawaii at Manoa, Department of Botany. *PCSU Technical Report* 123: 1–31.

Long, J.L. 1981. *Introduced Birds of the World.* Universe Books, New York.

Long, J.L. 2003. *Introduced Mammals of the World: Their History, Distribution, and Influence.* CSIRO, Collingwood, Australia.

Loope, L.L., A.C. Medeiros, and B.H. Gagné. 1991. Recovery of vegetation of a montane bog following protection from feral pig rooting. Cooperative National Park Resources Studies Unit. University of Hawaii at Manoa, Department of Botany. *PCSU Technical Report* 77: 1–23.

Martel A., A. Spitzen-van der Sluijs, M. Blooi, W. Bert, R. Ducatelle, M.C. Fisher, A. Woeltjes et al. 2013. *Batrachochytrium salamandrivorans* sp. nov. causes lethal chytridiomycosis in amphibians. *Proceedings of the National Academy of Sciences USA* 110: 15325–15329.

McCann, B.E., and D.K. Garcelon. 2008. Eradication of feral pigs from Pinnacles National Monument. *Journal of Wildlife Management* 72: 1287–1295.

Measey, G.J., D. Rödder, S.L. Green, R. Kobayashi, F. Lillo, G. Lobos, R. Rebelo, and J.-M. Thirion. 2012. Ongoing invasions of the African clawed frog, *Xenopus laevis*: A global review. *Biological Invasions* 14: 2255–2270.

Mooney, H.A. 2005. Invasive alien species: The nature of the problem. Pp. 1–15, in Mooney, H.A., R.N. Mack, J.A. McNeely, L.E. Neville, P.J. Schei, and J.K. Waage (eds.). *Invasive Alien Species: A New Synthesis.* Island Press, Washington, DC.

Mooney, H.A., and E.E. Cleland. 2001. The evolutionary impact of invasive species. *Proceedings of the National Academy of Sciences, USA* 98: 5446–5451.

Morgan, G.S., and C.A. Woods. 1986. Extinction and the zoogeography of West Indian land mammals. *Biological Journal of the Linnean Society* 28: 167–203.

Morrison, S.A. 2007. Reducing risk and enhancing efficiency in non-native vertebrate removal efforts on islands: A 25 year multi-taxa retrospective from Santa Cruz Island, California. Pp. 398–409, in Witmer, G.W., W.C. Pitt, and K.A. Fagerstone (eds.). *Managing Vertebrate Invasive Species: Proceedings of an International Symposium.* USDA/APHIS/WS National Wildlife Research Center, Fort Collins, Colorado.

Morrison, S.A., A.J. DeNicola, K. Walker, D. Dewey, L. Laughrin, R. Wolstenholme, and N. Macdonald. 2014. *An Irruption Interrupted: Eradication of Wild Turkeys Meleagris gallopavo from Santa Cruz Island.* California. Oryx doi: 10.1017/S0030605314000428

National Invasive Species Council. 2001. *National Management Plan: Meeting the Invasive-Species Challenge.* Washington, DC. http://www.invasivespeciesinfo.gov/council/nmp toc.shtml

Naylor, R.L. 2000. The economics of alien species invasions. Pp. 241–259, in Mooney, H.A., and R.J. Hobbs (eds.). *Invasive Species in A Changing World.* Island Press, Washington, DC.

Nogales, M., A. Martín, B.R. Tershy, C.J. Donlan, D. Veitch, N. Puerta, B. Wood, and J. Alonso. 2004. A review of feral cat eradication on islands. *Conservation Biology* 18: 310–319.

Office of Technology Assessment. 1993. *Harmful Non-Indigenous Species in the United States, OTA-F-565.* U.S. Government Printing Office, Washington, DC. 391 pp.

Orr, R. 2003. Generic nonindigenous aquatic organisms risk analysis review process. Pp. 415–431, in Carlton, J., G. Ruiz, and R. Mack (eds.). *Invasive Species: Vectors and Management Strategies.* Island Press, Washington, DC.

Otts, S.S. 2015. Confronting the marine invasive species threat: Practical and legal challenges. Pp. 215–234, in Abate, R.S. (ed.). *Climate Change Impacts on Ocean and Coastal Law: U.S. and International Perspectives.* Oxford University Press, New York.

Penniman, T.M., L. Buchanan, and L.L. Loope. 2011. Recent plant eradications on the islands of Maui County, Hawai'i. Pp. 325–331, in Veitch, C.R., M.N. Clout, and D.R. Towns (eds.). *Island Invasives: Eradication and Management.* IUCN, Gland, Switzerland.

Perrings, C., S. Dalmazzone, and M. Williamson. 2005. The economics of biological invasions. Pp. 16–35, in Mooney, H.A., R.N. Mack, J.A. McNeely, L.E. Neville, P.J. Schei, and J.K. Waage (eds.). *Invasive Alien Species: A New Synthesis.* Island Press, Washington, DC.

Perry, N.D., B. Hanson, W. Hobgood, R.L. Lopez, C.R. Okraska, K. Karem, I.K. Damon, and D.S. Carroll. 2006. New invasive species in southern Florida: Gambian rat (Cricetomys gambianus). *Journal of Mammalogy* 87: 262–264.

Peterson, A.T., and D.A. Vieglais. 2001. Predicting species invasions using ecological niche modeling: New approaches from bioinformatics attack a pressing problem. *BioScience* 51: 363–371.

Pimentel, D. (ed.). 2002. *Biological Invasions: Economic and Environmental Costs of Alien Plant, Animal, and Microbe Species.* CRC Press, Boca Raton, Florida.

Pimentel, D., L. Lach, R. Zuniga, and D. Morrison. 2000. Environmental and economic costs of nonindigenous species in the United States. *BioScience* 50: 53–65.

Pimentel, D., R. Zuniga, and D. Morrison. 2005. Update on the environmental and economic costs associated with alien-invasive species in the United States. *Ecological Economics* 52: 273–288.

Pranty, B., and K.L. Garrett. 2011. Under the radar: "non-countable" exotic birds in the ABA area. *Birding* September: 46–58.

Pyron, R.A., F.T. Burbrink, and T.J. Guiher. 2008. Claims of potential expansion throughout the U.S. by invasive python species are contradicted by ecological niche models. *PLoS ONE* 3: e2931.

Rago, A., G.M. White, and T. Uller. 2012. Introduction pathway and climate trump ecology and life history as predictors of establishment success in alien frogs and toads. *Ecology & Evolution* 2: 1437–1445.

Reed, R.N. 2005. An ecological risk assessment of nonnative boas and pythons as potentially invasive species in the United States. *Risk Analysis* 25: 753–766.

Reed, R.N., and G.H. Rodda. 2009. Giant constrictors: biological and management profiles and an establishment risk assessment for nine large species of pythons, anacondas, and the boa constrictor. *U.S. Geological Survey Open-File Report 2009–1202.* 302 pp.

Rodda, G.H., C.S. Jarnevich, and R.N. Reed. 2009. What parts of the US mainland are climatically suitable for invasive alien pythons spreading from Everglades National Park? *Biological Invasions* 11: 241–252.

Rodda, G.H., C.S. Jarnevich, and R.N. Reed. 2011. Challenges in identifying sites climatically matched to the native ranges of animal invaders. *PLoS ONE* 6: e14670.

Rolan, R.G. 2003. Invasive species risk assessment for amphibians. *Report to the United States Fish and Wildlife Service*. Division of Scientific Authority, Washington, DC.

Romagosa, C. 2015. Patterns of live vertebrate importation into the United States: Analysis of an invasion pathway. In: Keller, R.P., M.W. Cadotte, and G. Sandiford (eds.). *Invasive Species in A Globalized World: Ecological, Social, and Legal Perspectives on Policy.* University of Chicago Press, Chicago, Pp. 115–146.

Rosen, P.C., and C.R. Schwalbe. 1996. A critical interim evaluation of the effectiveness of bullfrog removal methods at San Bernardino National Wildlife Refuge. *Final report to Arizona Game and Fish Department*, Phoenix, Arizona, and U.S. Fish and Wildlife Service, Douglas, Arizona.

Simberloff, D. 2005. The politics of assessing risk for biological invasions: The USA as a case study. *Trends in Ecology and Evolution* 20: 216–222.

South Florida Ecosystem Restoration Task Force. 2015. EDRR case study: The sacred ibis project. http://www.evergladesrestoration.gov/content/ies/docs/Goal2_Case_Study_Sacred_Ibis.pdf

Springborn, M.R., R.P. Keller, S. Elwood, C.M. Romagosa, C. Zambrana-Torrelio, and P. Daszak. 2015. Integrating invasion and disease in the risk assessment of live bird trade. *Diversity and Distributions* 21: 101–110.

Springborn, M.R., C.M. Romagosa, and R.P. Keller. 2011. The value of nonindigenous species risk assessment in international trade. *Ecological Economics* 70: 2145–2153.

Stewart, R.D. 2007. Do institutional mechanisms exist to set and enforce standards for the Great Lakes that exceed the ballast water performance and exchange standards established by the International Maritime Organization? *Transportation Research Board Special Report 29*. http://onlinepubs.trb.org/onlinepubs/sr/sr291_stewart.pdf

Taylor, D., and L. Katahira. 1988. Radio telemetry as an aid in eradicating remnant feral goats. *Wildlife Society Bulletin* 16: 297–299.

Temple, S.A. 1992. Exotic birds: A growing problem with no easy solution. *Auk* 109: 395–397.

Tingley, R., B.L. Phillips, and R. Shine. 2011. Establishment success of introduced amphibians increases in the presence of congeneric species. *American Naturalist* 177: 382–388.

Tingley, R., M.B. Thompson, S. Hartley, and D.G. Chapple. 2015. Patterns of niche filling and expansion across the invaded ranges of an Australian lizard. *Ecography* 38: 1–11.

Tinsley, R.C., and M.J. McCoid. 1996. Feral populations of *Xenopus* outside Africa. Pp. 81–94, in Tinsley, R.C., and H.R. Kobel (eds.). *The Biology of Xenopus*. Oxford University Press, New York.

Touza, J., K. Dehnen-Schmutz, and G. Jones. 2007. Economic analysis of invasive species policies. Pp. 353–366, in Nentwig, W. (ed.). *Biological Invasions*. Springer, Berlin.

Tunison, J.T., A.A. McKinney, and W.L. Markiewicz. 1995. The expansion of koa forest after cattle and goat removal, Hawai'i Volcanoes National Park. Honolulu (HI): Cooperative National Park Resources Studies Unit, University of Hawaii at Manoa, Department of Botany. *PCSU Technical Report* 99: 1–32.

U.S. Fish and Wildlife Service. 2012. Injurious wildlife species; listing three python species and one anaconda species as injurious reptiles; final rule. *Federal Register* 77: 3329–3366.

U.S. Fish and Wildlife Service. 2015. Injurious wildlife species; listing three anaconda species and one python species as injurious reptiles; final rule. *Federal Register* 80: 12702–12745.

van Wilgen, N.J., and D.M. Richardson. 2012. The roles of climate, phylogenetic relatedness, introduction effort, and reproductive traits in the establishment of non-native reptiles and amphibians. *Conservation Biology* 26: 267–277.

Veitch, C.R., and M.N. Clout (eds.). 2002. *Turning the Tide: The Eradication of Invasive Species*. IUCN SSC Invasive Species Specialist Group, Gland, Switzerland.

Veitch, C.R., M.N. Clout, and D.R. Towns (eds.). 2011. *Island Invasives: Eradication and Management*. IUCN, Gland, Switzerland.

Veltman, C.J., S. Nee, and M.J. Crawley. 1996. Correlates of introduction success in exotic New Zealand birds. *American Naturalist* 147: 542–557.

Warren, P. 2006. *Biosecurity Systems of Hawái: An Evaluation and Recommendations for Reform*. Honolulu: Hawai Conservation Alliance. http://www.hawaiiconservation.org/images/uploads/pages/warren_report_2006.pdf

Wilcove, D.S., D. Rothstein, J. Dubow, A. Phillips, and E. Losos. 1998. Quantifying threats to imperiled species in the United States. *BioScience* 48: 607–615.

Witmer, G.W., and P. Hall. 2011. Attempting to eradicate invasive Gambian giant pouched rats (Cricetomys gambianus) in the United States: Lessons learned. Pp. 131–134, in Veitch, C.R., M.N. Clout, and D.R. Towns (eds.). *Island Invasives: Eradication and Management*. IUCN, Gland, Switzerland.

Witmer, G.W., J. Pierce, and W.C. Pitt. 2011. Eradication of invasive rodents on islands of the United States. Pp. 135–138, in Veitch, C.R., M.N. Clout, and D.R. Towns (eds.). *Island Invasives: Eradication and Management*. IUCN, Gland, Switzerland.

Witmer, G.W., W.C. Pitt, and M.L. Avery. 2007. Management of invasive vertebrates in the United States: An overview. Pp. 127–137, in Witmer, G.W., W.C. Pitt, and K.A. Fagerstone (eds.). *Managing Vertebrate Invasive Species: Proceedings of an International Symposium*. USDA/APHIS/WS National Wildlife Research Center, Fort Collins, Colorado.

Wittenberg, R., and M.J.W. Cock. 2001. *Invasive Alien Species: A Toolkit for Best Prevention and Management Practices*. CAB International, Wallingford, U.K.

Wittenberg, R., and M.J.W. Cock. 2005. Best practices for the prevention and management of invasive alien species. Pp. 209–232, in Mooney, H.A., R.N. Mack, J.A. McNeely, L.E. Neville, P.J. Schei, and J.K. Waage (eds.). *Invasive Alien Species: A new Synthesis*. Island Press, Washington, DC.

Wolf, C.M., B. Griffith, C. Reed, and S.A. Temple. 1996. Avian and mammalian translocations: Update and reanalysis of 1987 survey data. *Conservation Biology* 10: 1142–1154.

3 Economics of Invasive Species Damage and Damage Management

Stephanie Shwiff, Steven Shwiff, Jason Holderieath,
William Haden-Chomphosy, and Aaron Anderson

CONTENTS

INTRODUCTION

Annually, the estimated damage caused by invasive species in the United States has exceeded $100 billion, becoming one of the leading causes of environmental change and global biodiversity loss (Wilcove et al. 1998; Mack et al. 2000; Sala et al. 2000; Pimentel et al. 2005). Invasions by nonnative species highlight the undeniable link and feedback loops between ecological and economic systems (Perrings et al. 2002; Julia et al. 2007).

35

Ecological systems determine if the conditions are suitable for invasion by nonnative species; however, economic systems help fuel the introduction of nonnative species and are themselves affected by invasive species when the ecosystem's ability to provide services is diminished or when livestock or crops are made unmarketable (Julia et al. 2007).

Invasive species have played an important role in U.S. agriculture. While some of the goods cultivated by the U.S. agricultural sector are indigenous plant and animal species, many are introduced; a minimum of 4542 species currently existing in the United States originated from outside its borders (Office of Technology Assessment 1993). Introduced species, such as corn, wheat, rice, as well as cattle, poultry, and other livestock, are all important commodities produced by the U.S. agricultural sector. Some introduced species have potential conservation values as well, providing food and shelter for native species, acting as catalysts for restoration, serving as substitutes for extinct species, and augmenting ecosystem services (Schlaepfer et al. 2011). A distinction can be drawn, then, between introduced species and invasive species. Like introduced species, invasive species are nonnative to that ecosystem; however, invasive species have the potential to cause harm, whether measured economically, environmentally, or as a human health hazard (The White House 1999).

Vertebrate invasive species (VIS) are a subset of nonnative invasive species that can include bony fish, sharks, rays, amphibians, reptiles, mammals, and birds. They are exemplified by such species as the Burmese python (*Python molurus bivittatus*), brown tree snake (*Boiga irregularis*), European starling (*Sturnus vulgaris*), and wild boar (*Sus scrofa*).

Wild boar (*Sus scrofa*) is actually just one species among several categorized more broadly as "feral swine"; other species that fall within this category include feral domestic pigs (*Sus scrofa domestica*), Eurasian wild boar (*Sus scrofa linnaeus*), and hybrids between the two. Feral swine are the most abundant free-ranging, exotic ungulate in North America, so a significant amount of literature has been published regarding their impacts. Given the substantial amount of attention paid to feral swine, as well as their unique ability to create damage, we will examine separately the impacts of feral swine from other VIS in this chapter.

Earlier chapters have provided evidence that suggests that the frequency of VIS invasions may be increasing and creating significant environmental, ecological, and agricultural damages. Estimating the total economic impact and potential future economic impact of VIS is crucial to targeted prevention, management, and control efforts (McNeely 2001; National Invasive Species Council 2001). Commonly, to generate funding to fight an established VIS or to prevent the expansion of a VIS, it is necessary first to understand the full range of potential economic impacts.

Studies examining the full scope of economic impacts of VIS are relatively recent. Most of the early studies simply examined the direct economic impact, typically to agricultural production, associated with a specific VIS already established in a limited geographical region (Engeman et al. 2010). Very few studies have used these direct economic impacts to examine or forecast the broader macroeconomic (indirect and induced) impacts. Even fewer studies have combined biology and economics into a bioeconomic model to predict impacts before a VIS actually becomes established and estimate the value of preemptive versus reactive management strategies (Kolar and Lodge 2002). The challenge facing policy makers, of course, is to determine biologically effective and economically feasible methods of prevention, control, and damage mitigation of invasive species (Burnett et al. 2008).

In this chapter, we provide a general overview of the economic impact of both the presence and management of VIS in the United States. We begin by framing the general role of economics in determining the overall impact of VIS. We then examine current published estimates of damage and management costs. Finally, we discuss ways to improve economic estimation of VIS impacts.

FRAMING THE ECONOMIC IMPACTS OF VERTEBRATE INVASIVE SPECIES (VIS) WITHIN AN ECOLOGICAL CONTEXT

Although published estimates of impacts exist, the contextual roadmap that links economic impact and the ecology of VIS is not described extensively in the literature. One example of a methodology for determining the ecological impact of invasive species is provided by Parker et al. (1999) who derived the simple equation $I = R \times A \times E$, where I = impact, R = range size, A = abundance, and E = effect per individual. While R and A are a function of a suite of biological factors, E is a function of the ability of a VIS to create economic damage.

The direct economic damage or harm created by a VIS typically falls into three broad categories: destruction, depredation, and disease transmission. Total economic damage (D) of a VIS is the sum across these three categories and across time.

Destruction refers to destroyed property (e.g., statues, golf courses, buildings, bridges, power lines), equipment (e.g., vehicles, farm equipment, cables, irrigation equipment), crops (e.g., nonconsumptive impacts associated with rooting behavior), habitat, and associated recreational opportunities (e.g., lost tourism or hunting) (Daszak et al. 2000; Kaller and Kelso 2006; Hartin et al. 2007; Engeman et al. 2008; Jones et al. 2008; Campbell and Long 2009; Shwiff et al. 2010; Depenbusch et al. 2011; Loss et al. 2013; Bevins et al. 2014; Doody et al. 2014; Yang et al. 2014). Depredation refers to the consumption of crops, livestock, wildlife species, or companion animals by a VIS. Disease refers to mortality or morbidity in humans, companion animals, livestock, or wildlife caused by a VIS-associated pathogen (Witmer and Sanders 2003; Campbell et al. 2008; Hall et al. 2008).

In general, most invasive species impose damages that fall within two of the three categories. Reptiles (e.g., Burmese pythons and Brown tree snakes) and other aquatic nonnatives typically cause economic impact through depredation and environmental destruction but rarely through disease transmission (Greene et al. 2007; Snow et al. 2007). Some avian species, such as starlings, can create impacts in all three categories through depredation of crops, destruction of property (e.g., statues, bridges, buildings), and disease transmission (e.g., fecal contamination of livestock feed) (Shwiff et al. 2012). A substantial portion of the overall impact of avian VIS tends to be through depredation of crops, while the other two categories of damage tend to contribute significantly less to the overall impact. Many rodent VIS are similar to avian VIS in that the majority of the impact comes from depredation on crops and significantly less from destruction and disease transmission. These latter impacts are still important, but often dwarfed by the impact of depredation to crops. Feral swine, in contrast, can create significant impact in all three categories. Research has focused largely on the impact of feral swine to crop depredation (Seward et al. 2004; Pimentel et al. 2005; Ober et al. 2011; Mengak 2012); however, it has provided substantial estimates of other damage categories as well (Frederick 1998; Engeman et al. 2003; Mayer and Johns 2011; Higginbotham 2013). Valuing the damages caused by a VIS in each of these three categories requires an understanding of the implications of a biological impact for different sectors of the economy. To do this, both the primary and secondary impacts must be quantified.

METHODS OF VALUATION

Primary Impacts

Valuation of the primary damage caused by VIS—through destruction, depredation, and disease transmission—is usually accomplished by estimating the market, loss, repair, or restoration values associated with the affected resource. Market values are commonly used when monetizing impacts to livestock or crops (Cumming et al. 2005; Engeman et al. 2010; Gebhardt et al. 2011). Loss values are often used in the case of death related to disease transmission, or predation of things not actively bought and sold in markets, including humans, companion animals, and sometimes wildlife. Repair costs and restoration costs are typically used as the valuation method for damages categorized as destruction (Engeman et al. 2008). Finally, restoration costs, rehabilitation costs, lost recreational opportunities, or nonmarket values are often used to quantify economic damages to ecosystems and wildlife (Engeman et al. 2004a,b, 2005).

Nonmarket valuation of wildlife can occur through survey methods such as the contingent valuation method (CVM) and travel cost method (TCM), as well as nonsurvey methods, such as benefit transfer. CVM is a survey-based, stated preference approach that solicits responses from individuals regarding their willingness to pay (WTP) for various use and nonuse values associated with wildlife (Loomis 1990; Kotchen and Reiling 1998). Several factors can affect WTP for wildlife, including the species' usefulness and likeability, information level of respondents, level of economic damage created by the species, and questionnaire design (Brown 1994; Brown et al. 1996; Nunes and van den Bergh 2001; Bateman et al. 2002; Tisdell and Wilson 2006; Martín-López et al. 2007, 2008). Criticisms of CVM include the hypothetical nature of the questionnaire and the inability to validate responses, causing some to question its usefulness for determining value (Eberle and Hayden 1991; Boyle 2003). Additionally, this type of valuation typically understates the true nonmarket value (Pearce and Moran 1994; Balmford et al. 2002).

TCM is another survey approach which uses costs incurred for travel to quantify demand for recreational activities that are sometimes linked to a species of interest (Kotchen and Reiling 1998). TCM is based on the idea that as some environmental amenity changes (e.g., the size of a wildlife population), the amount people are willing to pay to use it will change, which is revealed by a change in travel costs (see Loomis and Walsh 1997 for an extensive discussion and examples of this method). Criticisms of this method include concerns about the assumption that visitors' values equal or exceed their travel costs. Critics argue that travel costs are simply costs, not an accurate representation of value. Another concern is that this method requires values to be assigned to the time individuals spend traveling to a site. It is difficult to assign accurate values to the opportunity cost of travelers' time because people value their time differently, depending on their occupation or the activity they gave up in order to travel to the site.

The benefit-transfer method relies on benefit values derived from CVM and TCM studies in one geographical location and species, which are then transferred to another location and similar species. Adjustments to these values can be made by factoring in differences in incomes or prices from one area to an other. Typical criticisms of this method focus on the reliability of the original value estimates derived from CVM or TCM (Brouwer 2000; Smith et al. 2002).

Primary damages can generate secondary impacts due to economic factors that create linkages to established economic sectors. For example, primary damages arising from the destruction of an ecosystem may be measured by multiplying the number of acres damaged by the restoration price per acre. However, if the ecosystem destruction also reduces economic activity that would have been generated from tourist expenditures in a nearby town, this would represent the secondary impact (Shwiff et al. 2010).

Secondary Impacts

Regional economic analysis (REA) is an accepted methodology for estimating the secondary impacts in an economy based on the most current economic and demographic data available (BEA 2008). Regional economic models attempt to quantify the impacts on output as a result of input changes in a regional economy. These models are developed by constructing a mathematical replica of a regional economy (city, county, state, etc.) that contains all the linkages between existing economic sectors (e.g., agricultural, manufacturing, and industrial). The model then uses existing estimates of primary impacts to quantify secondary impacts, thereby calculating the total effect on jobs and revenue in a specified regional economy.

REA allows for the estimation of secondary (indirect and induced) impacts associated with primary VIS damages in units of measure that are important to the general public (e.g., revenue, income, and jobs). These secondary impacts are also known as upstream and downstream impacts. For example, when a VIS depredates crops, the reduction in yield per acre translates into less yield delivered to the processor and eventually to retail; these are downstream impacts. Additionally, the producer may buy fewer inputs (e.g., fuel and parts for equipment) because there are fewer acres to harvest; these are upstream impacts. These upstream and downstream impacts can be measured through the use of regional economic models, also known as input-output (IO) models, such as Impact Analysis for Planning (IMPLAN, Minnesota IMPLAN Group) and Regional Economic Modeling (REMI Inc.).

IO models are the most widely used tool for modeling the linkages and leakages of a regional economy. These models use transaction tables to illustrate how outputs from one industry may be sold to other industries as intermediate inputs or as final goods to consumers, and how households can use wages from their labor to purchase final goods (Richards 1972). This allows for the tracking of annual monetary transactions between industry sectors (processing), payments to factors of production (value added), and consumers of final goods (final demand). This complex network of transactions is summarized in the form of "multipliers" which measure how changes in economic activity relate to changes in final demand for a particular good. Many regional economic models are static; that is, they estimate economic impacts only within a single time period.

Arguably, economic impacts generated by VIS are dynamic, and therefore require a regional economic model that can account for complex interactions among economic sectors over multiple time periods. A dynamic regional economic model has been developed to generate annual forecasts and simulate behavioral responses to compensation, price, and other economic factors (REMI: Model Documentation – Version 9.5). The REMI model incorporates interindustry transactions, endogenous final-demand feedbacks, substitution among factors of production in response to changes in the relative factor prices, wage responses to changes in labor-market conditions, and changes in the share of local and export markets in response to changes in regional profitability and production costs (Treyz et al. 1991). The dynamic nature of REMI enables it to create a control (baseline) forecast that projects economic conditions within a region on the basis of trends in historical data. Economic impacts are then examined by comparing the control forecast to simulations that account for changes in variables such as industry-specific income, value added, and employment. Modeling impacts in this way can translate the primary impacts of a VIS into regional impacts on revenue and jobs, expanding the general public's perception of the potential benefits of preventing or combatting a VIS. These secondary impacts not only help estimate the total impact of a VIS, but also help engage a broader audience by highlighting the implications of a VIS for local communities and economies.

CURRENT PUBLISHED ESTIMATES OF PRIMARY DAMAGE

Below, we summarize current published estimates of damage. Many damage estimates are aggregated across the three damage categories; a related tendency is to report destruction and depredation impacts as a single number. Studies that do this are often not replicable and difficult to extend or extrapolate to other areas. In the case of studies that simply list damage as an aggregated estimate, we listed those impacts under the destruction category.

Examining published estimates of economic damage created by invasive species excluding feral swine, it is clear there is a paucity of research in this area. This explains why the most widely cited estimate of total damage from bird, mammal, reptile, and amphibian invasive species is $39.4 billion annually (Pimentel et al. 2000, 2005). Additionally, Pimentel et al. (2005) estimate the annual control costs are $11.5 million, although this only includes feral pig and brown tree snake control costs. Below, we dig deeper into the published literature, beyond Pimentel et al. (2000, 2005), to determine what other damage estimates exist.

VIS Except Feral Swine

Destruction

Marbuah et al. (2014) estimated damage costs from 79 harmful species to be $185 billion in the United States in 1993, including a cost of $46 billion per year for invasive mammals and birds. During the federal fiscal years from 1990 to 1997, damages reported to U.S. Department of Agriculture (USDA) Animal and Plant Health

TABLE 3.1
Estimated Annual VIS Damage

VIS	Annual Damage Estimate (in Millions USD)
Wild horses	$5
Mongooses	$50
Rats	$19,000
Cats	$17,000
Dogs	$250
Pigeons	$1100
Starlings	$800
Brown tree snakes	$1

Source: Office of Technology Assessment. 1993. *Harmful Non-Indigenous Species in the United States, OTA-F-565.* U.S. Government Printing Office, Washington, DC, USA.

Inspection Service (APHIS) Wildlife Services (WS) included $1,226,717 from invasive reptiles, $14 million from invasive mammals, and $28 million from invasive birds (Bergman et al. 2002). The annual cost of introduced rats alone was estimated at $21.2 million (Cusack et al. 2009).

In 1993, the Office of Technology Assessment estimated that terrestrial VIS caused $39.4 billion in damages annually in the United States; however, feral swine estimates are included in that amount. Specific estimates of damage from that report are detailed in Table 3.1.

The brown tree snake (*Boiga irregularis*) is capable of causing significant damages to property and productivity in the north Pacific through its tendency to create power outages (Fritts 2002) and impact tourism. Shwiff et al. (2012) used data from the snake's invasion on Guam, along with survey information from Hawaii, to estimate the cost of a potential invasion into Hawaii. Results suggested that total annual damage from such an invasion would be between $593 million and $2.14 billion.

Depredation

One of the most common forms of damage by VIS is agricultural loss due to depredation. We refer to depredation as both crop and livestock losses attributable to VIS activities, excluding losses associated with diseases transmitted by VIS. Invasive bird species are common culprits of agricultural depredation because they frequently forage in crop-intensive areas. Pimentel et al. (2005) estimate that the European starling (*Sturnus vulgaris*) is responsible for $800 million in crop losses annually. This figure reflects both the starling's large population in the United States as well as their ability as individuals to inflict crop losses. This estimate is based on losses not only from grain fields, but also in fruit production, such as cherries.

Bergman et al. (2002) calculate that, from fiscal years 1990 to 1997, the most frequent requests for assistance with invasive mammals in the United States were related to livestock predation by invasive canines. Invasive dogs (*Canis* spp.) were responsible for 20% of the total damage reported to the USDA WS during that time frame. Invasive dogs are introduced species of canines that cause economic, environmental, or human harm. This definition excludes native species like wolves or coyotes. In the case of invasive dogs, the important part of the definition is that the dogs are causing harm. The most frequent occurrence is livestock depredation by dogs that have become feral.

Disease

Published estimates that detail the economic impacts of VIS-associated disease transmission are scant. While it is generally known that VIS can play a substantial role in the transmission of transboundary diseases between humans, wildlife, and domestic animals, it is difficult to translate that impact into dollar terms. It is estimated that wildlife—some but not all of which are VIS—play a role in 79% of the reportable domestic animal diseases and, of those diseases, 40% are zoonotic. For example, the common pigeon and European starling are known carriers of dozens of diseases that pose a threat to human and livestock health and safety (Weber 1979). In another example, across the Gulf Coast in the United States, invasive nutria (*Myocastor coypus*) may carry tuberculosis, septicemia, and a variety of parasites that represent a health hazard to water supplies and recreation (USDA APHIS WS 2010). While there are many examples of diseases that VIS may host, the need for economic estimates of the impact caused by VIS-introduced disease represents an important area of future research.

Control

In 2011 alone, the U.S. Department of the Interior spent $100 million on invasive species prevention, early detection, rapid response, control, management, research, outreach, international cooperation, and habitat restoration. In 2005, the U.S. Fish and Wildlife Services (FWS) and its partners spent $2 million working with 15 trappers to eradicate over 8000 nutria from Maryland's Blackwater National Wildlife Refuge. FWS, in partnership with many organizations, has spent more than $6 million since 2005 on finding and applying solutions to the growing problem of Burmese pythons and other large invasive constrictor snakes in Florida. FWS spent $604,656 over a three-year period (2007–2009) to design python traps, deploy and maintain them, and educate the public in the Florida Keys to prevent the potential extinction of the endangered Key Largo woodrat and other vulnerable endangered species. From 1999 to 2009, federal and state agencies spent $1.4 million on Key Largo woodrat recovery and $101.2 million on wood stork recovery to combat python impacts (U.S. Fish & Wildlife Service 2012a,b,c). The National Park Service has spent $317,000 annually on various programs related to constrictor snake issues, such as researching snake biology for removal purposes in Everglades National Park (U.S. Fish & Wildlife Service 2012a,b). Research and control of brown tree snakes requires nearly $4 million per year; this is in addition to normal operating costs for management of Guam's National Wildlife Refuge and military environmental programs (USGS).

FERAL SWINE: A NOTORIOUS VIS

As noted earlier, feral swine are the most abundant free-ranging, exotic ungulate in North America. Enough literature has been published about their impacts to justify a more in-depth review. Feral swine have existed in pockets of the southeastern United States, California, and Hawaii for nearly five hundred years, and recent trends indicate a general northward expansion of populations. Feral swine have experienced significant range expansion over the past 30 years, in part because a subset of the human population wants to hunt them closer to home (Spencer et al. 2005; Acevedo et al. 2006; Saito et al. 2012; Bevins et al. 2014). This expansion has increased conflicts with agriculture and humans, triggering several assessments of the costs and benefits of feral swine in different locations (e.g., Higginbotham et al. 2008; Campbell and Long 2009; Siemann et al. 2009; Ober et al. 2011; Engeman et al. 2012; Mengak 2012; Campbell et al. 2013; Higginbotham 2013; Bevins et al. 2014). In addition, there has been considerable research on the increasing management conflicts stemming from feral swine expansion (e.g., Weeks and Packard 2009; Honda and Kawauchi 2011; Koichi et al. 2013; Warner and Kinslow 2013).

Destruction

The most commonly cited publication about feral swine damage is Pimentel et al. (2005), which reports an estimated annual impact of $800 million ($941 million 2012 USD) resulting from crop and environmental damage. Environmental damages associated with feral swine include erosion due to rooting, grubbing, and wallowing (Engeman et al. 2004a,b; Seward et al. 2004). Their impact is significant enough that, for example, they were found to "dominate the disturbance regime" of the Northern California Coast Range Preserve (Kotanen 1995). Feral swine have also contributed to the decline of 22 species of plants and four species of amphibians, in addition to the predation of marine turtles and their nests (Seward et al. 2004). Damage to marshes and parks by feral swine has also been noted (Engeman et al. 2003, 2004a; Pimentel et al. 2005), including damage to priceless archaeological sites (Engeman et al. 2012). Another commonly reported form of property damage is vehicle collisions involving feral swine. One study examined 179 vehicle collisions in South Carolina involving feral swine and found an average damage estimate of $1173 per collision (Mayer and Johns 2011).

Table 3.2 summarizes a wide variety of damage estimates found in the feral swine literature, adjusted for inflation to 2012 using BLS (2014). The base year of 2012 was chosen to put these figures on par with the most recent USDA Census of Agriculture. When possible, the data were converted into annualized costs. Given the diverse circumstances underlying each research project, the comparison of costs across different locations and time scales is problematic. For example, Higginbotham et al. (2008) found feral swine cause $58 million/year in damage to the whole of Texas agriculture, an area of 59 million acres. Meanwhile, Mengak (2012) reported a similar $58 million/year for crop damages to 9.7 million acres in Georgia, only part of which was agricultural land. This area is responsible for approximately 1% of U.S. total crop sales (USDA NASS 2014). Jerrolds et al. (2014) conducted a survey of agricultural groups and resource managers in Tennessee and found that 94% of counties had swine populations, and most complaints related to crop and pasture damage. There is also some anecdotal evidence of considerable losses realized in New York. Hall (2012) discusses a farm in Clinton County suffering $25,000 in losses from corn, apple, and strawberry depredation. Westenbroek (2011) discusses a farm in

TABLE 3.2
Estimates of Feral Swine Destruction

Crops—Single Incidents

Geographical Area	Description	Estimates
Texas (7)	Peanuts	$64,803
New York (4)	Corn	$15,157
New York (5)	Corn, apples, and strawberries	$25,000

Crops—Annual Aggregates

Geographical Area	Description	Estimates
Texas (1)	Peanuts	$225,518/yr.
Texas (1)	N/A	$15,492–$464,765/yr.
Texas (17)	Corn, soybeans, wheat, rice, sorghum, peanuts	$89,817,000/yr.
Alabama (17)	Corn, soybeans, wheat, rice, peanuts	$21,322,000/yr.
Arkansas (17)	Corn, soybeans, wheat, rice, peanuts	$19,575,000/yr.
Florida (17)	Corn, soybeans, wheat, rice, peanuts	$5,985,000/yr.
North Florida (3)	Corn, cotton, peanuts, and soybeans	$1,921,224/yr..
Georgia (6)	Reported crops—Mengak (2012, p. 13) SW Extension District	$58,180,000/yr.
Georgia (17)	Corn, soybeans, wheat, rice, peanuts	$5,150,000/yr.
Louisiana (17)	Corn, soybeans, wheat, rice, peanuts	$15,670,000/yr.
Mississippi (17)	Corn, soybeans, wheat, rice, peanuts	$18,518,000/yr.
Missouri (17)	Corn, soybeans, wheat, rice, peanuts	$485,000/yr.
North Carolina (17)	Corn, soybeans, wheat, rice, peanuts	$4,684,000/yr.
South Carolina (17)	Corn, soybeans, wheat, rice, peanuts	$8,747,000/yr.

Property

Geographical Area	Description	Estimates
New York (12)	Two Lawns	$421 each
Georgia (6)	Property damage in SW Extension District	$24,500,000/yr.
California (8)	31 residential properties and 1 golf course	$93,652/yr.
Nationwide (13)	Avg. property damage from feral swine—vehicle collisions	$1,197/per car

Total Uncategorized

Geographical Area	Description	Estimates
Texas (9)	"Economic loss since feral swine appeared on the respondent's property" (Adams, et al. 2005, p. 1316)	$32,25,796
Texas (10)	Cost to Texas agriculture	$57,580,650/yr.
Texas (10)	Repairing damage and control	$7,751,242/yr.
California (8)	Total reported damage to hay, forage, ponds, lawns, drainage, orchards, vineyards, Irrigation, livestock, crops, trees, fruits, and nuts	$2,634,343/yr.

(*Continued*)

TABLE 3.2 (*Continued*)
Estimates of Feral Swine Destruction

Geographical Area	Description	Environmental Estimates
Florida (14)	Value of damaged area of Savannas Preserve State Park	$1,545,717–$5,036,456
Florida (15)	Damage to 3 Florida state parks at the end of the study period	$6,652–28,384/ha
California (16)	Damage and control	$400,169/yr.

Sources: 1. Tolleson, D.R., et al., 1995, *Great Plains Wildlife Damage Control workshop Proceedings*, June 2–3, Fort Worth, Texas, p. 454.; 3. Ober, H.K., et al., 2014. *Farmer Perceptions of Wildlife Damage to Row Crops in North Florida*. Department of Wildlife Ecology and Conservation, Florida Cooperative Extension Service, Institute of Food and Agricultural Sciences, University of Florida.; 4. Westenbroek, T. Letter to P. Anderson. September 25, 2011. Estimate of Damage due to feral swine. Cornell University Cooperative Extension, Sullivan County; 5. Hall, W. 2012. *Wayne's World: Many Folks Despise them but Feral Hogs are Smart*. The Times Herald Record, Middletown, NY.; 6. Mengak, M.T., 2012, Georgia Wild Pig Survey, Final Report, University of Georgia, Athens, GA; 7. Beach, R., 2013, *Texas Natural Wildlife*, San Angelo, TX; 8. Frederick, J.M., 1998, *18th Vertebrate Pest Conference*, University of California, Davis; 9. Adams, C.E., et al., 2005, *Wildlife Society Bulletin* 33,1312–1320; 10. Higginbotham, B., G. Clary, L. Hysmith, and M. Bodenchuk. 2008. Statewide Feral Hog abatement pilot project, 2006–2007. Texas AgnLife Extension Service. Available online at http://feralhogs.tamu.edu/files/2010/05/06-07-Feral-Hog-Abatement-Pilot-Project.pdf. Accessed December 12, 2016; 12. USDA APHIS VS, 2010, *National Brucellosis Surveillance Strategy*, Riverdale, Maryland, 20737; 13. Mayer, J.J. and P.E. Johns. 2011. *Characterization of Wild Pig-Vehicle Collisions*. Washington Savannah River Company, Aiken, SC and Carolina Wildlife Consultants, New Ellenton, SC. May 23, 2011; 14. Engeman, R.M., et al., 2004b, *Journal for Nature Conservation* 12, 143–147; 15. Engeman, R.M., et al., 2003, *Environmental Conservation* 30, 319–324; 16. Sweitzer, R.A. and B.E. McCann. 2007. Natural areas ecological damage and economic costs survey report. Unpublished report submitted to all interested survey respondents. Prepared by R.A. Sweitzer and B.E. McCann. Department of Biology, University of North Dakota, Grand Forks, North Dakota, USA, 37pp.

Note: All figures have been adjusted to 2012 USD using BLS (2014).

Delaware County that lost $14,850 to feral swine consumption of corn fields. It is difficult to compare in a meaningful way two areas so different in size and context, even though they experienced similar levels of reported damage.

Depredation

There is very little quantitative data published about the predatory behavior of feral swine. However, what is lacking in quantitative data is offset by what is known in qualitative terms. Surveys, reports describing feral swine attacks, and anecdotal evidence are available from several sources. Survey respondents have experienced or are concerned about danger to humans from attack, livestock depredation, and damage or injury to pets (Barrett and Pine 1981; Rollins 1993; Sweitzer and McCann 2007; Mengak 2012). Several popular press articles describe actual attacks on humans (Moore Jr. 2008; Roberts 2011; Sanchez 2011). Love (2013) details the case of an inmate on a work crew who was attacked by a feral swine. Mayer (2013) found that up to 15% of reported attacks on humans by feral swine are fatal.

Feral swine are known to prey on livestock, primarily sheep (*Ovis aries*) and goats (*Capra hircus*), but also cows (*Bos taurus*) and exotic game species (Frederick 1998; Seward et al. 2004; Christie et al. 2014). Barrios-Garcia and Ballari (2012) reported that around 30% of feral swine diets consist of animal matter, depending on the ecosystem and season.

Seward et al. (2004) report that feral swine cause greater than $1.2 million in goat losses annually. Some feral swine kills may be mistakenly reported as coyote kills, leading to under reporting of feral swine depredation (Seward et al. 2004). Anecdotally, though, a rancher in Texas experienced a 15%–20% reduction in goat kid production on property where feral swine resided (Beck 1999). In 1990, Texas authorities documented 1243 head of sheep and goats lost to feral swine, at a value of $110, 669 in 2012 (Rollins 1993). In 1991, Texas and California reported 1473 sheep, goats, and exotic game animals killed by feral swine (Barrett and Birmingham 1994). Feral swine density has actually been found to be a good predictor of ewes losing lambs (Choquenot et al. 1997).

Without a larger body of quantitative work, it is difficult to know the extent of the economic threat that feral swine pose to livestock. However, the available qualitative research reveals that feral swine depredation is a real problem for agricultural producers. Further research and more robust data collection will be necessary to effectively quantify feral swine depredation costs.

Disease

Feral swine are a potential reservoir of both zoonotic and nonzoonotic diseases that could impact the U.S. economy through a number of channels (Roger 1988; Paarlberg 2002). Of the 42 serious pathogens with a wildlife component reported by Miller et al. (2013), feral swine are explicitly involved in seven. Survey respondents indicated concern or experience with feral swine spreading disease to livestock or acting as a potential disease reservoir (Barrett and Pine 1981; Rollins 1993). They have also been known to carry diseases dangerous to humans (Bengsen et al. 2013). For example, feral swine are a potential vector for new forms of influenza because they have the required receptors for both avian and human strains of the virus, which provides an opportunity for the viruses to combine (Hall et al. 2008).

While the disease threat posed by feral swine is clearly recognized within the literature, it has thus far been difficult to accurately model their role as vectors during a disease outbreak. Current disease transmission models are largely focused on the spread of a single disease between a limited number of species (e.g., Ward et al. 2007, 2009). However, the complexity of the feral swine problem requires a model flexible enough to accommodate the potential for transmission of multiple pathogens across multiple species.

The cost of one outbreak of foot-and-mouth disease (FMD) in the United States involving feral swine is estimated to range from $7.5 million to $5.8 billion USD for a single state (Cozzens 2010; Cozzens et al. 2010). Feral swine have also been identified as an important reservoir for transboundary animal diseases such as classical swine fever virus, African swine fever virus, and porcine reproductive and respiratory syndrome (Jori and Bastos 2009; Reiner et al. 2009; Müller et al. 2011). Additionally, there is concern over potential losses in cattle associated with transmission of pseudorabies (Aujeszky's disease) from feral swine (Bitsch 1975; Hagemoser et al. 1978; Crandell et al. 1982).

Research on pathogen transmission between feral swine and livestock has been making progress. Pineda-Krch et al. (2010) developed a disease transmission model to simulate the spread and control of FMD among feral swine and beef and dairy herds in California. The model incorporates elements of space and randomness. Results show that introduction of FMD from feral swine to livestock could result in a large and rapidly moving outbreak. However, tested containment strategies showed potential to reduce the size and duration of the outbreaks.

Ward et al. (2007, 2009) built a disease spread model that explicitly models the potential for FMD spread between domestic cattle, feral swine, and white-tailed deer in Texas. The model considered geographic relationships between the species and found that densities, distributions, and the resulting potential for contact between affected species were important in determining the extent of the outbreak (Ward et al. 2007, 2009).

Beyond the modeling of an outbreak, the next challenge is valuing potential damage to the agricultural sector and economy as a whole. The potential damage to commercial livestock production is related to the number of exposed animals. U.S. livestock sales totaled $90 billion in 2012 (USDA NASS 2014), with $5 billion in beef exports (USDA ERS 2013), and $6.3 billion in pork exports (MEF 2014). Almost 13% of total beef production, and 27% of pork production is exported (MEF 2013). Even limited outbreaks can be exceptionally costly, due to the potential for international banning of U.S. imports of the affected species, which triggers price effects for the entire U.S. herd. Coffey et al. (2005) estimate that the single reported case of bovine spongiform encephalopathy (BSE) in December 2003 (which did not involve feral swine) cost the U.S. beef industry between $3.9 billion and $5.7 billion in lost exports alone in 2004.

Some of the only studies on the economic impacts of disease transmission involving feral swine are Cozzens (2010) and Cozzens et al. (2010). Cozzens (2010) found that potential producer losses in Kansas due to feral swine transmission of FMD to domestic livestock could be as much as $6.1 billion. Total economic impact for a hypothetical transmission of FMD to livestock from infected feral swine in Missouri was estimated at $12.6 million (Cozzens et al. 2010).

There are also concerns about contamination of the human food supply by feral swine. The deadly September 2006 outbreak of *E. coli* O157:H7 was traced back to feral swine–contaminated spinach (Kreith 2007). In response to the outbreak, consumer expenditures on leafy greens declined by $69 million; spinach producers in particular lost $234 million because lettuce and similar produce were substituted for spinach (Arnade et al. 2009). This example illustrates the economy-wide impacts that disease outbreaks can generate, affecting both consumers and producers.

In addition to food safety issues tied legitimately to feral swine, consumers are also sensitive to perceived but unproven disease threats associated with feral swine. In 2009, an outbreak of H1N1 influenza was initially called "swine flu" by authorities. This mislabeling led to substantial negative consumer response, even though Attavanich et al. (2011) determined that pork remained safe to consume throughout the entire event. Agricultural sector losses of $159 million were attributed to media coverage of "swine flu" (Attavanich et al. 2011).

The ability to measure both the epidemiologic and economic impacts of a multi-species, multipathogen outbreak induced by feral swine is still beyond the scope of currently available models. However, evaluation of the costs associated with single-disease outbreaks of FMD or BSE between feral swine and other species shows the damaging potential of even small-scale disease transmission events, including costs from both real and perceived food safety threats. While the full magnitude of feral swine disease impacts are not currently known, it is clear from available evidence that they pose a legitimate threat to the U.S. agricultural sector.

Control

Given our discussion of the damages feral swine cause, it is no surprise that considerable effort and resources have been devoted to the control and management of feral swine populations. There is substantial interest in obtaining an accurate measure of feral swine management costs, to compare with the damages incurred, which are the implicit cost of failing to control existing feral swine populations. The feral swine herd in Texas has been estimated at two million (Higginbotham et al. 2008). Current nationwide population estimates range between four and five million feral swine (Pimentel 2007; Higginbotham et al. 2008; USDA APHIS 2013). However, census is extremely difficult and few studies have generated a reliable national population estimate.

Feral swine are incredibly prolific, capable of speeding up their reproductive cycles under pressure (Hanson et al. 2009) and increasing their reproduction rates when population is below the local carrying capacity (Bengsen et al. 2013). All of these factors create unique and costly challenges in the management and control of feral swine. Saunders and Bryant (1988) found an inverse relationship between control efforts and control success. Specifically, the more feral swine were shot from a helicopter (within a fixed study area), the more difficult it became to detect and shoot the remaining individuals. This confirms the potential infeasibility of eradicating established populations. In fact, studies have shown that lethal control efforts must result in mortality rates ranging between 60% and 80% in order to impair the ability of feral swine to maintain their population (Hone and Pedersen 1980; Barrett and Pine 1981; Kreith 2007; Bengsen et al. 2013). Cost estimates for feral swine control are presented in Table 3.3.

The difference in average removal costs between the two studies based in Australia (Hone and Pedersen 1980; Saunders and Bryant 1988) may be due to a couple of different reasons. First, the two studies used different control methods as the primary method of control. Hone and Pedersen (1980) placed poison baits at water sources known to be frequented by feral swine and then observed the baits to record any nontarget species take. Saunders and Bryant (1988) used helicopter

TABLE 3.3
Control Costs

Geographical Area (Source)	Description	Estimates (USD)
California (16)	Feral swine related costs incurred from management within natural areas in California	$4.29M/yr.
California (16)	Feral swine eradication efforts during 3-yr. study period	$3.89M/yr.
California (16)	Per km construction and maintenance cost of exclusion fence at pinnacles national monument (~20-yr. life span)	$58,403/km
California (18)	Total construction cost of exclusion fence at Pinnacles National Monument (~20-yr. life span)	$1,871,690
California (18)	Eradication efforts at Pinnacles National Monument	$1,053,138 (over ~3 yrs)
California (18)	Annual maintenance cost of exclusion fence at Pinnacles National Monument	$68,629/yr.
Florida (14)	Average removal cost	$41.18/head
Texas (11)	Average removal cost	$69.61/head
Australia (24)	Average removal cost	$91.60/head
Australia (25)	Average removal cost	$16.52/head

Sources: 11. Higginbotham, B., G. Clary, L. Hysmith, and M. Bodenchuk. 2008. Statewide Feral Hog abatement pilot project, 2006–2007. Texas AgnLife Extension Service. Available online at http://feral-hogs.tamu.edu/files/2010/05/06-07-Feral-Hog-Abatement-Pilot-Project.pdf. Accessed December 12, 2016; 14. Engeman, R.M., et al., 2004b, *Journal for Nature Conservation,* 12, 143–147; 16. Sweitzer, R.A. and B.E. McCann. 2007. Natural areas ecological damage and economic costs survey report. Unpublished report submitted to all interested survey respondents. Prepared by R.A. Sweitzer and B.E. McCann. Department of Biology, University of North Dakota, Grand Forks, North Dakota, USA, 37pp.; 18. Kreith, M., 2007, Wild pigs in California: The issues, University of California Agricultural Issues Center, Davis, California; 24. Hone J. and Pedersen H. 1980. *Proceedings of the Ninth Vertebrate Pest Conference*, pp. 176–182. University of California, Davis; 25. Saunders, G., et al., 1988, *Wildlife Research,* 15, 73–81.

Note: All figures have been adjusted to 2012 USD using BLS (2014).

shooting in an effort to eradicate feral swine from a specific eradication zone. The labor requirements of the additional bait observation in Hone and Pedersen (1980) are a likely contributor to the additional cost; indeed, labor made up over half of the total estimated project cost. Additionally, the Saunders and Bryant (1988) study killed a much larger number of pigs (946 compared to 120 in the other study), suggesting there may be some economies of scale driving down the average pig removal cost.

Methods of feral swine control deemed acceptable differ by stakeholder groups. Acceptability of management practices is influenced by stakeholder group identification (e.g., residents vs. tourists), awareness of a feral swine problem, and social factors (Koichi et al. 2013). For example, feral swine are so well established in the local culture around a national park in Texas that residents do not consider them nonnative (Weeks and Packard 2009). Control efforts are met with considerable resistance, especially when professional hunters are hired. Similarly, feral swine control efforts conducted by "outsiders" in Hawaii (e.g., U.S. federal agencies), without public consent, have been met with considerable public opposition (Weeks and Packard 2009). Stakeholders' conflicting views of control strategies are one of the primary hurdles to effective feral swine management.

DISCUSSION: IMPROVING ESTIMATES OF DAMAGE FROM VIS

CURRENT KNOWLEDGE GAPS

Our review of the literature has revealed an incomplete understanding of the economic damages and control costs arising from VIS. Improvements to this understanding can occur through several pathways. First, there is a need for improved data collection using methods that allow for replication and extrapolation. Second, these improved data and associated research insights need to be integrated into future management decisions to identify economically efficient (or at least the most cost-effective) management strategies for VIS. Last, regional economic models should be used to rigorously link primary damage impacts to the appropriate economic sector in order to estimate secondary impacts.

To improve data collection, a nationwide surveillance effort is needed to estimate feral swine damage to agricultural products and livestock. Ober et al. (2011) and Mengak (2012) both used survey methods that are scientifically replicable. And several recent surveys have been conducted that could be used as foundations for developing a nationwide questionnaire related to feral swine damages and society's attitudes about them (Hamrick 2013; Adams et al. 2005; Higginbotham et al. 2008; Ober et al. 2011; Mengak 2012).

Each category of damage—destruction, depredation, disease—suffers different data challenges that are difficult to overcome. Destruction has been the most thoroughly addressed damage category, but its estimates vary in scope and approach. This makes comparison between studies difficult and calculation of an accurate national aggregate nearly impossible. Geographic scales range from as small as a single farm to as large as the entire state of Texas. Furthermore, destruction estimates sometimes include only crop damage, and other times include only environmental damage or control costs. It is therefore very difficult to generalize findings from one study on environmental damages to total damages across the larger United States. It is possible that GIS (Geographical Information Systems) could be used to combine and extrapolate disparate data, but estimates will be ad hoc at best, relying on rules-of-thumb and heuristics. Ideally, there would be common survey questions, agreed-upon units of measure, and standard reporting protocols (including mean and variance of estimates to enable inference), such that companion studies could be undertaken to inform a nationwide estimate.

Regarding the next damage category, depredation of livestock by VIS, verifiable data are currently lacking. There is considerable qualitative information, however, that may help researchers identify incidents of VIS predation. For example, insurance companies may have data on depredation losses, but presumably only for producers who carry coverage and file a claim. A state-level policy requiring the universal reporting of livestock killed by feral swine would provide more complete data, which would then enable better-informed response. Some states, such as Texas, could serve as a model for reporting livestock losses in the field (Higginbotham et al. 2008).

Regarding the final damage category, disease losses attributable to VIS, we also lack a complete picture of current and potential disease risk. This is by far the most difficult category of damage to measure, but potentially the most important that needs to be addressed, due in part to potential implications for international trade. The severity of an outbreak depends on probabilities of infection and transmission between individuals and between species, as well as the medical severity of the disease itself.

Unfortunately, the probabilities of an outbreak and how feral swine density and distribution affect those probabilities are largely unknown. As disease spread models incorporate new data, or develop more flexibility to account for uncertainty in existing data, this daunting goal will become more achievable. Groups like the National Institute for Mathematical and Biological Synthesis (NIMBioS) are currently working to develop simulation models that fully capture the livestock–wildlife interface. Scientists across disciplines are also making strides toward the One Health Initiative approach of combined human and veterinary medicine. These advancements will help uncover the true scope of risks stemming from feral swine diseases.

In an attempt to mitigate damage caused by feral swine, substantial resources have been committed to management and control efforts. These efforts impose both direct costs (in terms of outlays of actual dollars on lethal and nonlethal control efforts) as well as indirect costs (in terms of lost time and resources devoted to controlling feral swine), both of which represent resources that could have been allocated elsewhere. However, management and control costs are categorically different than damages inflicted by feral swine. Management and control costs are a choice made in response to damages or potential damages. These two forms of expense should therefore be recorded separately. By erroneously combining damage estimates with management and control efforts, granularity in information that is needed for effective decision-making is lost.

A Way Forward: Bioeconomic Modeling

Bioeconomic modeling is another analytical tool that can be used to address some of the knowledge gaps about VIS impacts and efficient management strategies. Bioeconomic models describe biological processes and predict the effects of management decisions on those processes. Therefore, they can be used to determine the most cost-effective management policies given biological constraints and bioeconomic feedback loops. Development and use of these models is constrained, however, by limited budgets and time, as well as gaps in our biological understanding.

The combined use of bioeconomic models with regional economic models can provide the most comprehensive estimate of total economic impact of a VIS, as well as net benefits of alternative management strategies. Such modeling would benefit significantly, however, from improved estimates of damage, depredation, and disease, along with associated animal density and population control data to determine the mathematical relationships that exist between them.

SUMMARY

In reviewing the existing research, a number of gaps in our knowledge about VIS abundance, damage, and efficient control strategies have become clear. These gaps represent opportunities to expand upon the knowledge needed for meaningful VIS management. Currently, the literature does not contain adequate estimates of VIS populations, ranges, expansion, current levels of damage within any of the three categories (destruction, depredation, and disease), or measures of control costs. For example, feral swine are known to damage timber production (Jackson 1990; Whitehouse 1999; Mengak 2012), yet no actual measure of the economic impact of this destruction exists. Many of the estimates of VIS damage come from a single publication, Pimentel et al. (2005). Such limited results highlight the need for more impact studies, although research about feral swine damage seems to be increasing at a much greater rate than research about other VIS.

The lack of economic impact estimates is problematic because they are necessary to determine the efficient level of control and management effort. One obvious factor missing in most VIS analyses is a discussion of potential trade implications of disease transmission from VIS to livestock. Disease transmission can not only restrict animal movements within the United States, but also restrict our ability to export livestock commodities. Such restrictions can inflict significant damage to the U.S. economy, as evidenced during the 2003–2004 outbreak of BSE (albeit unrelated to VIS).

Impact estimates summarized in this chapter also highlight the need for more comprehensive national estimates of damage from VIS. The precise size of nationwide populations of various VIS and their rate of expansion are not known with certainty. Alternative methods are needed to identify areas in which VIS occur and to estimate their prevalence. If nationwide data were available, modeling exercises could then be used to determine a national estimate of potential damage from one or more VIS. Arguably the most ambitious goal would be to develop a national disease spread model that includes wildlife populations alongside domestic animals and humans. A quality estimate of the nationwide potential for VIS-transmitted diseases would provide a significant step forward toward understanding and ranking potential VIS impacts, as well as developing prevention or control strategies.

REFERENCES

Acevedo, P., M. Escudero, R. Muńoz, and C. Gortázar. 2006. Factors affecting wild boar abundance across an environmental gradient in Spain. *Acta Theriologica* 51: 327–336.

Adams, C.E., B.J. Higginbotham, D. Rollins, R.B. Taylor, R. Skiles, M. Mapston, and S. Turman. 2005. Regional perspectives and opportunities for feral hog management in Texas. *Wildlife Society Bulletin* 33: 1312–1320.

Anderson, A., C. Slootmaker, E. Harper, J. Holderieath, and S.A. Shwiff. 2016. Economic estimates of feral swine damage and control in 11 US states. *Crop Protection* 89: 89–94.

Arnade, C., L. Calvin, and F. Kuchler. 2009. Consumer response to a food safety shock: The 2006 food-borne illness outbreak of *E. coli* O157: H7 linked to spinach. *Applied Economic Perspectives and Policy* 31: 734–750.

Attavanich, W., B.A. McCarl, and D. Bessler. 2011. The effect of H1N1 (Swine Flu) media coverage on agricultural commodity markets. *Applied Economic Perspectives and Policy* 33: 241–259.

Balmford, A., A. Bruner, P. Cooper, R. Costanza, S. Farber, R.E. Green, M. Jenkins et al. 2002. Economic reasons for conserving wild nature. *Science* 2975583: 950–953.

Barrett, R.H., and G.H. Birmingham. 1994. Wild pigs. In S.E. Hygnstrom, R.M. Timm, and G.E. Larson eds. *The Handbook: Prevention and Control of Wildlife Damage*, University of Nebraska, Lincoln. P. 51.

Barrett, R.H., and D. Pine. 1981. History and status of wild pigs (*Sus scrofa*) in San Benito County, California. *California Fish and Game* 67: 105–117.

Barrios-Garcia, M.N., and S. Ballari. 2012. Impact of wild boar (*Sus scrofa*) in its introduced and native range: A review. *Biological Invasions* 14: 2283–2300.

Bateman, I., R. Carson, B. Day, W. Hanemann, N. Hanley, T. Hett, M. Lee et al. 2002. *Economic Valuation with Stated Preference Techniques: A Manual*. Department for Transport, Edward Elgar, UK.

BEA. 2008. Regional Economic Accounts. http://www.bea.gov/regional/. Accessed 5/25/2016.

Beach, R. 2013. Depredation problems involving feral hogs. In C. Wayne Hanselka, J.F. Cadenhead eds. *Texas Natural Wildlife*. District 7 AgriLife Research and Extension Center, San Angelo, TX. http://agrilife.org/texnatwildlife/feral-hogs/depredation-problems-involving-feral-hogs/. Accessed 5/24/2016.

Beck, R. 1999. Eden—Texas. In *First National Feral Swine Conference*. June 2–3, 1999. Ft. Worth, Texas, p. 23.

Bengsen, A.J., M.N. Gentle, J.L. Mitchell, H.E. Pearson, and G.R. Saunders. 2013. Impacts and management of wild pigs *Sus scrofa* in Australia. *Mammal Review* 44(2): 135–147.

Bergman, D., M. Chandler, and A. Locklear. 2002. The economic impact of invasive species to wildlife services" cooperators. In L. Clark, J. Hone, J., Shivik, K. VerCauteren, R. Watkins, and J. Yoder eds. *Human Wildlife Conflicts: Economic Considerations*. National Wildlife Research Center, Fort Collins, Colorado, pp. 169–178.

Bevins, S.N., K. Pedersen, M. Lutman, T. Gidlewski, and T.J. Deliberto. 2014. Consequences associated with the recent range expansion of nonnative feral swine. *BioScience* 64: 291.

Bitsch, V. 1975. Study of outbreaks of Aujeszky's disease in cattle. I.-Virological and epidemiological findings. *Acta Veterinaria Scandinavica* 16: 420–433.

BLS. 2014. CPI inflation calculator. In *Tables Calculators by Subject*. Databases ed. U.S. Bureau of Labor Statistics, Washington, DC.

Boyle, K.J. 2003. Introduction to revealed preference methods. In P. A. Champ, K. J. Boyle, and T. C. Brown eds. *A Primer on Nonmarket Valuation*. Kluwer Academic Publisher, Dordrecht. The Netherlands.

Brouwer, R. 2000. Environmental value transfer: State of the art and future prospects. *Ecological Economics* 21(1): 137–152.

Brown, T.C. 1994. Experiments on the difference between willingness to pay and willingness to accept. *Land Economics* 70(4): 520–522.

Brown, T.C., P.A. Champ, R.C. Bishop, and D.W. McCollum. 1996. Which response format reveals the truth about donations to a public good? *Land Economics* 152–166.

Burnett, K. M., S. D'evelyn, and B. A. Kaiser. 2008. Beyond the lamppost: Optimal prevention and control of the brown tree snake in Hawaii. *Ecological Economics* 67(1): 66–74.

Campbell, T.A., R.W. DeYoung, E.M. Wehland, L.I. Grassman, D.B. Long, and J. Delgado-Acevedo. 2008. Feral swine exposure to selected viral and bacterial pathogens in southern Texas.

Campbell, T.A., J.A. Foster, M.J. Bodenchuk, J.D. Eisemann, L. Staples, and S.J. Lapidge. 2013. Effectiveness and target-specificity of a novel design of food dispenser to deliver a toxin to feral swine in the United States. *International Journal of Pest Management* 59: 197–204.

Campbell, T.A., and D.B. Long. 2009. Feral swine damage and damage management in forested ecosystems. *Forest Ecology and Management* 257: 2319–2326.

Choquenot, D., B. Lukins, and G. Curran. 1997. Assessing lamb predation by feral pigs in Australia's semi-arid rangelands. *Journal of Applied Ecology* 34: 1445–1454.

Christie, J., E. DeMarco, E. Hiroyasu, A. Kreger, and M. Ludington. 2014. *Wild Pig Management at Tejon Ranch*. University of California, Santa Barbara.

Coffey, B., J. Mintert, S. Fox, T. Schroeder, and L. Valentine. 2005. *The Economic Impact of BSE on the U.S. Beef Industry: Product Value Losses, Regulatory Costs, and Consumer Reactions*. Kansas State University Agricultural Experimental Station and Cooperative Extension. Serial no. MF-2678, April.

Cozzens, T. 2010. *Economic Impact of Feral Swine Transmitting Foot-and-Mouth Disease to Livestock in Kansas*. Unpublished master's thesis, Colorado State University, Fort Collins, Colorado.

Cozzens, T., K. Gebhardt, S. Shwiff, M. Lutman, K. Pedersen, and S. Swafford. 2010. Modeling the economic impact of feral swine-transmitted foot-and-mouth disease: A case study from Missouri. In R.M. Timm, and K.A. Fagerstone eds. *Proc. 24th Vertebr. Pest Conf.* Univ. of Calif., Davis, pp. 308–311.

Crandell, R., G. Mesfin, and R. Mock. 1982. Horizontal transmission of pseudorabies virus in cattle. *American Journal of Veterinary Research* 43(2): 326–328.

Cummings, J.L., S.A. Shwiff, and S.K. Tupper. 2005. Economic impacts of blackbird damage to the rice industry. In D.L. Nolte and K.A. Fagerstone eds. *Proceedings of the 11th Wildlife Damage Management Conference*, Wildlife Damage Management Working Group of The Wildlife Society, Fort Collins, Colorado, pp. 317–322.

Cusack, C., M. Harte, and S. Chan. 2009. The Economics of Invasive Species. Prepared for the Oregon Invasive Species Council. http://www.oregon.gov/oisc/docs/pdf/economics_invasive.pdf. Accessed 7/5/2013.

Daszak, P., A.A. Cunningham, and A.D. Hyatt. 2000. Emerging infectious diseases of wildlife: Threats to biodiversity and human health. *Science* 287: 443–449.

Depenbusch, B.E., J.S. Drouillard, and C.D. Lee. 2011. Feed depredation by European starlings in a Kansas feedlot. *Human–Wildlife Interactions* 5: 58–65.

Doody, J.S., P. Mayes, S. Clulow, D. Rhind, B. Green, C.M. Castellano, D. D'Amore, et al. 2014. Impacts of the invasive cane toad on aquatic reptiles in a highly modified ecosystem: The importance of replicating impact studies. *Biological Invasions* 16: 2303–2309.

Eberle, W.D., and F.G. Hayden. 1991. Critique of contingent valuation and travel cost methods for valuing natural resources and ecosystems. *Journal of Economic Issues* 25(3): 649–687.

Engeman, R.M., K.J. Couturier, R.K. Felix Jr, and M.L. Avery. 2012. Feral swine disturbance at important archaeological sites. *Environmental Science and Pollution Research* 20: 4093–4093.

Engeman, R.M., J.A. Duquesnel, E.M. Cowan, H.T. Smith, S.A. Shwiff, and M. Karlin. 2008. Assessing boat damage to seagrass bed habitat in a Florida park from a bioeconomics perspective. *Journal of Coastal Research*, 24(2): 527–532.

Engeman, R.M., J.E. Laborde, B.U. Constantin, S.A. Shwiff, P. Hall, A. Duffiney and F. Luciano. 2010. The economic impacts to commercial farms from invasive monkeys in Puerto Rico. *Crop Protection* 29(4): 401–405.

Engeman, R.M., R.E. Martin, H.T. Smith, J. Wolloard, C.K. Crady, S.A. Shwiff, B. Constantin, et al. 2005. Dramatic reduction in predation on marine turtle nests through improved predator monitoring and management. *Oryx* 39(3): 318–326.

Engeman, R.M., H.T. Smith, R. Severson, M.A. Severson, J. Woolard, S.A. Shwiff, B. Constantin, et al. 2004a. Damage reduction estimates and benefit–cost ratios for feral swine control from the last remnant of a basin marsh system in Florida. *Environmental Conservation* 31: 207–211.

Engeman, R.M., H.T. Smith, R. Severson, M. Ann Severson, S.A. Shwiff, B. Constantin, and D. Griffin. 2004b. The amount and economic cost of feral swine damage to the last remnant of a basin marsh system in Florida. *Journal for Nature Conservation* 12: 143–147.

Engeman, R.M., H.T. Smith, S.A. Shwiff, B. Constantin, J. Woolard, M. Nelson, and D. Griffin. 2003. Prevalence and economic value of feral swine damage to native habitat in three Florida state parks. *Environmental Conservation* 30: 319–324.

Frederick, J.M. 1998. Overview of wild pig damage in California. In R.O. Baker, and A.C. Crabb eds. *18th Vertebrate Pest Conference*. University of California, Davis, pp. 82–86

Fritts, T.H. 2002. Economic costs of electrical system instability and power outages caused by snakes on the island of Guam. *International Biodeterioration & Biodegradation* 49(2): 93–100.

Gebhardt, K., A.M. Anderson, K. Kirkpatrick, and S.A. Shwiff. 2011. A review and synthesis of bird and rodent damage to select California crops. *Crop Protection* 30(9): 1109–1116.

Greene, D.U., J.M. Potts, J.G. Duquesnel, and R.W. Snow. 2007. Geographic distribution: *Python molurus bivittatus* (Burmese python). *Herpetological Review* 38: 355.

Hagemoser, W., H. Hill, and E. Moss. 1978. Nonfatal pseudorabies in cattle. *Journal of the American Veterinary Medical Association* 173(2): 205.

Hall, J.S., R.B. Minnis, T.A. Campbell, S. Barras, R.W. DeYoung, K. Pabilonia, M.L. Avery et al. 2008. Influenza exposure in United States feral swine populations. *Journal of Wildlife Diseases* 44: 362–368.

Hall, W. 2012. *Wayne's World: Many Folks Despise them but Feral Hogs are Smart*. The Times Herald Record, Middletown, NY.

Hanson, L.B., M.S. Mitchell, J.B. Grand, B. Jolley, B.D. Sparklin, and S.S. Ditchkoff. 2009. Effect of experimental manipulation on survival and recruitment of feral pigs. *Wildlife Research* 36: 185–191.

Hartin, R.E., M.R. Ryan, and T.A. Campbell. 2007. Distribution and disease prevalence of feral hogs in Missouri. *Human–Wildlife Conflicts* 1: 186–191.

Higginbotham, B. 2013. Wild Pig Damage Abatement. Education and Applied Research Activities. Texas A&M AgriLife Research and Extension Center-Overton.

Higginbotham, B., G. Clary, L. Hysmith, and M. Bodenchuk. 2008. Statewide Feral Hog abatement pilot project, 2006–2007. Texas AgnLife Extension Service. Available online at http://feralhogs.tamu.edu/files/2010/05/06-07-Feral-Hog-Abatement-Pilot-Project.pdf. Accessed December 12, 2016.

Honda, T., and N. Kawauchi. 2011. Methods for constructing a wild boar relative-density map to resolve human–wild boar conflicts. *Mammal Study* 36: 79–85.

Hone, J., and H. Pedersen. 1980. Changes in a feral pig population after poisoning. In J.P. Clark ed. *Proceedings of the Ninth Vertebrate Pest Conference*, pp. 176–182. University of California, Davis.

Jackson, J.J. 1990. Controlling vertebrate animal damage in southern pines. In L.R. Davis, and R.E. Marsh eds. *14th Vertebrate Pest Conference*. University of California, Davis, pp. 199–202.

Jerrolds W.R., E.C. Pelren, B.A. Darroch, and R.G. Anderson. 2014. A survey to estimate population distribution of and damage caused by Feral Swine in Tennessee. *Journal of the Southeastern Association of Fish and Wildlife Agencies* 1: 167–174.

Jones, K.E., N.G. Patel, M.A. Levy, A. Storeygard, D. Balk, J.L. Gittleman, and P. Daszak. 2008. Global trends in emerging infectious diseases. *Nature* 451: 990–993.

Jori, F., and A.D. 2009. Bastos. Role of wild suids in the epidemiology of African swine fever. *EcoHealth* 6(2): 296–310.

Julia, R., D.W. Holland, and J. Guenthner. 2007. Assessing the economic impact of invasive species: The case of yellow starthistle (Centaurea solsitialis L.) in the rangelands of Idaho, USA. *Journal of Environmental Management* 85(4): 876–882.

Kaller, M.D., and W.E. Kelso. 2006. Swine activity alters invertebrate and microbial communities in a coastal plain watershed. *The American Midland Naturalist* 156(1): 163–177.

Koichi, K., A. Cottrell, K.K. Sangha, and I.J. Gordon. 2013. What determines the acceptability of wildlife control methods? A case of feral pig management in the Wet Tropics World Heritage Area, Australia. *Human Dimensions of Wildlife* 18: 97–108.

Kolar, C.S., and D.M. Lodge. 2002. Ecological predictions and risk assessment for alien fishes in North America. *Science* 298(5596): 1233–1236.

Kotanen, P.M. 1995. Responses of vegetation to a changing regime of disturbance: Effects of feral pigs in a Californian coastal prairie. *Ecography* 18: 190–199.

Kotchen, M.J., and S.D. Reiling. 1998. Estimating and questioning economic values for endangered species: An application and discussion. *Endangered Species Update* 15(5): 77–83.

Kreith, M. 2007. Wild pigs in California: The issues. University of California Agricultural Issues Center, Davis, California. AIC Issues Brief.

Loomis, J.B. 1990. Comparative reliability of the dichotomous choice and open ended contingent valuation techniques. *Journal of Environmental Economics and Management* 18(1): 78–85.

Loomis, J.B., and R.G. Walsh. 1997. *Recreation Economic Decisions; Comparing Benefits and Costs*. Ed.2 Venture Publishing Inc.

Loss, S.R., T. Will, and P.P. Marra. 2013. The impact of free-ranging domestic cats on wildlife of the United States. *Nature Communications* 4: 1396.

Love, J.D. 2013. Robert Anthony Burrough, #1443625 vs. Brad Livingston et al. In T.D. United States District Court for the Eastern District of Texas ed. Civil Action No. 6:12cv339. 2013 U.S. Dist. LEXIS 122748.

Mack, R.N., D. Simberloff, W.M. Lonsdale, H. Evans, M. Clout, and F.A. Bazzazz. 2000. Biotic invasions: Causes, epidemiology, global consequences, and control. *Ecological Applications* 10(3): 689–710.

Marbuah, G., I.M. Gren, and B. McKie. 2014. Economics of harmful invasive species: A review. *Diversity* 6(3): 500–523.

Martin-Lopez, B., C. Montes, and J. Benayas. 2007. The non-economic motives behind the willingness to pay for biodiversity conservation. *Biological Conservation* 139(1): 67–82.

Martin-Lopez, B., C. Montes, and J. Benayas. 2008. Economic valuation of biodiversity conservation: The meaning of numbers. *Conservation biology* 22(3): 624–635.

Mayer, J.J. and P.E. Johns. 2011. *Characterization of Wild Pig-Vehicle Collisions*. Washington Savannah River Company, Aiken, SC and Carolina Wildlife Consultants, New Ellenton, SC. May 23, 2011.

Mayer, J.J. 2013. Wild pig attacks on humans. In J.B. Armstrong, and G.R. Gallagher eds. *15th Wildlife Damage Management Conference*. 230 Madren Center Dr., Clemson, SC 29634, USA, Lincoln, NE, pp. 17–35.

McNeely, J.A. 2001. In H.A. Mooney, L.E. Neville, P. Schei, J.K. Waage eds. *Global Strategy on Invasive Alien Species*. IUCN on behalf of the Global Invasive Species Programme, Gland, Switzerland, pp. 86–120.

MEF, U. 2013. Beef, Pork Exports Set New Records in 2012. In. *U.S. Meat Export Federation*. Denver, CO.

MEF, U. 2014. Total U.S. Pork EXPORTS. In. *U.S. Meat Export Federation*. Denver, CO. https://www.agweb.com/article/beef_pork_exports_set_new_records_in_2012. Accessed December 2016; https://www.usmef.org/news-statistics/press-releases/value-records-for-beef-pork-exports-in-2014-volumes-also-higher-year-over-year/.

Mengak, M.T. 2012. *Georgia Wild Pig Survey. Final Report. Warnell School of Forestry and Natural Resources*, University of Georgia, Athens, GA 30602, USA.

Miller, R.S., M.L. Farnsworth, and J.L. Malmberg. 2013. Diseases at the livestock–wildlife interface: Status, challenges, and opportunities in the United States. *Preventive Veterinary Medicine* 110: 119–132.

Moore, C. Jr. 2008. Wild hog attacks rare but scary. In PAnews.com. Port Author, TX, PAnews.com.

Müller, T., E.C. Hahn, F. Tottewitz, M. Kramer, B.G. Klupp, T.C. Mettenleiter, and C. Freuling 2011. Pseudorabies virus in wild swine: A global perspective. *Archives of Virology.* 156(10): 1691–1705.

National Invasive Species Council. 2001. *Meeting the Invasive Species Challenge: National Invasive Species Management Plan.* National Invasive Species Council, Washington, DC.

Nunes, P.A., and J.C. van den Bergh. 2001. Economic valuation of biodiversity sense or nonsense? *Ecological Economics* 39(2): 203–222.

Ober, H.K., G.R. Edmondson, W.M. Giuliano, D.L. Wright, J. Atkins, A. Anderson, S. Eubanks, et al. 2014. *Farmer Perceptions of Wildlife Damage to Row Crops in North Florida.* Department of Wildlife Ecology and Conservation, Florida Cooperative Extension Service, Institute of Food and Agricultural Sciences, University of Florida.

Office of Technology Assessment. 1993. *Harmful Non-Indigenous Species in the United States, OTA-F-565.* U.S. Government Printing Office, Washington, DC, USA.

Paarlberg, P.L., J.G. Lee, and A.H. Seitzinger. 2002. Potential revenue impact of an outbreak of foot-and-mouth disease in the United States. *Journal of the American Veterinary Medical Association* 220(7): 988–992.

Parker, I.M., D. Simberloff, W.M. Lonsdale, K. Goodell, M. Wonham, P.M. Kareiva, M.H. Williamson et al. 1999. Impact: Toward a framework for understanding the ecological effects of invaders. *Biological Invasions* 1(1): 3–19.

Pearce, D.W., and D. Moran. 1994. *The Economic Value of Biodiversity.* Earthscan, London.

Perrings, C., M. Williamson, E.B. Barbier, D. Delfino, S. Dalmazzone, J. Shogren, P. Simmons, et al. 2002. Biological invasion risks and the public good: An economic perspective. *Conservation Ecology* 6(1): 1.

Pimentel, D., L. Lach, R. Zuniga, and D. Morrison. 2000. Environmental and economic costs of nonindigenous species in the United States. *BioScience* 50(1): 53–65.

Pimentel, D., R. Zuniga, and D. Morrison. 2005. Update on the environmental and economic costs associated with alien-invasive species in the United States. *Ecological Economics* 52: 273–288.

Pimentel, D. 2007. Environmental and economic costs of vertebrate species invasions into the United States. In G.W. Witmer, W.C. Pitt, and K.A. Fagerstone eds. *Managing Vertebrate Invasive Species.* USDA National Wildlife Research Center, Fort Collins, CO, August 7–9.

Pineda-Krch, M., J.M. O'Brien, C. Thunes, and T.E. Carpenter. 2010. Potential impact of introduction of foot-and-mouth disease from wild pigs into commercial livestock premises in California. *American Journal of Veterinary Research* 71: 82–88.

Reiner, G. et al. 2009. Porcine Reproductive and Respiratory Syndrome Virus (PRRSV) infection in wild boars. *Veterinary Microbiology* 136(3): 250–258.

Richards, H.W. 1972. *Input-Output and Regional Economics.* Weidnefield and Nicolson, London.

Roberts, C. 2011. *Wild Pig Suspected in Sonoma Attack.* NBC Bay Area, San Francisco, CA. http://www.nbcbayarea.com/news/local/Wild-Pig-Suspected-in-Sonoma-Campground-Attack-126731243.html. Accessed October 11, 2016.

Roger, P., and J. Hone. 1988. A model of the dynamics and control of an outbreak of foot and mouth disease in feral pigs in Australia. *Journal of Applied Ecology* 63–77.

Rollins, D. 1993. Statewide attitude survey on feral hogs in Texas. In C.W. Hanselka, and J.F. Cadenhead eds. *Feral Swine: A Compendium for Resource Managers.* Texas Agricultural Extension Service, Kerrville, TX, USA, pp. 1–8.

Saito, M., F. Koike, H. Momose, T. Mihira, S. Uematsu, T. Ohtani, and K. Sekiyama. 2012. Forecasting the range expansion of a recolonising wild boar *Sus scrofa* population. *Wildlife Biology* 18: 383–392.

Sala, O.E., F.S. Chapin, J.J. Armesto, E. Berlow, J. Bloomfield, R. Dirzo, E. Huber-Sanwald et al. 2000. Global biodiversity scenarios for the year 2100. *Science* 287(5459): 1770–1774.

Sanchez, R. 2011 Wild boars invade farms, attack pets. In *Reuters.* Reuters. https://www.reuters.com/article/us-wildboar-newyork-odd-idUSTRE76S4DP20110729. Accessed October 11, 2016.

Saunders, G., and H. Bryant. 1988. The Evaluation of a feral pig eradication program during a simulated exotic disease outbreak. *Wildlife Research* 15: 73–81.

Schlaepfer, M.A., D.F. Sax, and J.D. Olden. 2011. The potential conservation value of non-native species. *Conservation Biology* 25: 428–437.

Seward, N.W., K.C. VerCauteren, G.W. Witmer, and R.M. Engeman. 2004. Feral swine impacts on agriculture and the environment. *Sheep & Goat Research Journal* 12: 12.

Shwiff, S.A., J.C. Carlson, J.H. Glass, J. Suckow, M.S. Lowney, K.M. Moxcey, and G.M. Linz. 2012. Producer survey of bird–livestock interactions in commercial dairies. *Journal of Dairy Science*, 95(11): 6820–6829.

Shwiff, S.A., K. Gebhardt, K.N. Kirkpatrick, and S.S. Shwiff. 2010. Potential economic damage from introduction of brown tree snakes, *Boiga irregularis* (Reptilia: Colubridae), to the Islands of Hawai'i. *Pacific Science* 64: 1–10.

Siemann, E., J.A. Carrillo, C.A. Gabler, R. Zipp, and W.E. Rogers. 2009. Experimental test of the impacts of feral hogs on forest dynamics and processes in the southeastern US. *Forest Ecology and Management* 258: 546–553.

Smith, V.K., G. Van Houtven, and S.K. Pattanayak. 2002. Benefit transfer via preference calibration: Prudential algebra for policy. *Land Economics* 78(1): 132–152.

Snow, R.W, L.L. Krysko, K.M. Enge, L. Oberhofer, A. Walker-Bradley, and L. Wilkins. 2007. Introduced populations of *Boa constrictor* (Boidae) and *Python molurus bivittatus* (Pythonidae) in southern Flora. In R.W. Henderson, and R. Powell eds. *The Biology of Boas and Pythons.* Eagle Mountain Publishing, Eagle Mountain, Utah, pp. 416–438.

Spencer, P.B.S., J.O. Hampton, and Hall. 2005. Illegal translocation and genetic structure of feral pigs in Western Australia. *Journal of Wildlife Management* 69: 377–384.

Sweitzer, R.A., and B.E. McCann. 2007. *Natural Areas Ecological Damage and Economic Costs Survey Report.* Department of Biology, University of North Dakota, Grand Forks, North Dakota, USA, pp. 1–36.

Tisdell, C., and C. Wilson. Information, wildlife valuation, conservation: Experiments and policy. *Contemporary Economic Policy* 24(1): 144–159.

Tolleson, D.R., W.E. Pinchak, D. Rollins, and L.J. Hunt. 1995. Feral hogs in the rolling plains of Texas: Perspectives, problems, and potential. In *Great Plains Wildlife Damage Control workshop Proceedings*, June 2–3, Fort Worth, Texas p. 454.

Treyz, G.I., D.S. Rickman, and G. Shao. 1991. The REMI economic-demographic forecasting and simulation model. *International Regional Science Review* 14(3): 221–253.

USDA. 2014. 2012 *Census of Agriculture. Summary and State Data.* vol. 1. Geographic Area Series. Part 51. AC-12-A-51, USDA, United States Department of Agriculture, Washington DC, p. 695.

USDA APHIS. 2013. *Feral Swine: Damage and Disease Threats.* In USDA APHIS ed. Publications Distribution, Riverdale, MD.

USDA APHIS WS. 2010. Nutria an Invasive Rodent. October 2010 Factsheet. https://www.aphis.usda.gov/publications/wildlife_damage/content/printable_version/fs_nutria10.pdf. Accessed 5/24/2016.

USDA ERS. 2013. *Home/Topics/Animal Products/Cattle & Beef/Statistics & Information.* USDA ERS, Washington, DC.

USDA NASS. 2014. *USDA/NASS QuickStats Ad-hoc Query Tool.* United States Department of Agriculture. http://quickstats.nass.usda.gov.

U.S. Fish & Wildlife Service. 2012a. Chesapeake Bay Nutria Eradication Project. http://www.fws.gov/chesapeakenutriaproject/FAQs.html. Accessed 5/25/2016.

U.S. Fish & Wildlife Service. 2012b. The cost of invasive species. January 2012 Factsheet. http://www.fws.gov/home/feature/2012/pdfs/CostofInvasivesFactSheet.pdf. Accessed 7/8/2013.

U.S. Fish & Wildlife Service. 2012c. The economic cost of large constrictor snakes. January 2012 Factsheet. http://www.fws.gov/home/feature/2012/pdfs/EconImpact.pdf

USGS. Economic damages from the brown tree snake on Guam. http://www.fort.usgs.gov/resources/education/bts/impacts/economic.asp. Accessed 7/8/2013.

Ward, M.P., S.W. Laffan, and L.D. Highfield. 2007. The potential role of wild and feral animals as reservoirs of foot-and-mouth disease. *Preventive Veterinary Medicine* 80: 9–23.

Ward, M.P., S.W. Laffan, and L.D. Highfield. 2009. Modelling spread of foot-and-mouth disease in wild white-tailed deer and feral pig populations using a geographic-automata model and animal distributions. *Preventive Veterinary Medicine* 91: 55–63.

Warner, K.D., and F. Kinslow. 2013. Manipulating risk communication: Value predispositions shape public understandings of invasive species science in Hawaii. *Public Understanding of Science* 22: 203–218.

Weber, W.J. 1979. *Health Hazards from Pigeons, Starlings, and English Sparrows: Diseases and Parasites Associated With Pigeons, Starlings, and English Sparrows Which Affect Domestic Animals.* Thomson Publications, Fresno, CA.

Weeks, P., and J. Packard. 2009. Feral hogs: Invasive species or nature's bounty? *Human Organization* 68: 280–292.

Westenbroek, T. Letter to P. Anderson. September 25, 2011. Estimate of damage due to feral swine. Cornell University Cooperative Extension, Sullivan County.

The White House. 1999. *Executive Order 13112 of February 3, 1999. Invasive Species.* Government Printing Office, Washington, DC, United States.

Whitehouse, D.B. 1999. Impacts of feral hogs on corporate timberlands in the south eastern U.S. In *First National Feral Swine Conference.* Fort Worth, TX.

Wilcove, D.S., D. Rothstein, J. Dubow, A. Phillips, and E. Losos. 1998. Quantifying threats to imperiled species in the United States. *BioScience* 48(8): 607–615.

Witmer, G.W., R.B. Sanders, A.C. 2003. *Taft. Feral Swine: Are They a Disease Threat to Livestock in the United States?* USDA National Wildlife Research Center—Staff Publications, p. 292.

Yang, X., W. Huang, B. Tian, and J. Ding. 2014. Differences in growth and herbivory damage of native and invasive kudzu (*Peuraria montana* var. *lobata*) populations grown in the native range. *Plant Ecology* 215: 339–346.

4 Policy and Regulation

Lori Williams

CONTENTS

INTRODUCTION

In May of 2003, I received an urgent phone call from an official with the Centers for Disease Control and Prevention (CDC). The CDC was hard at work with other federal and state agencies to trace a recent outbreak of monkeypox virus in the United States. Monkeypox is a rare, viral disease that occurs primarily in central and western African rainforests that can cause rashes similar to those seen in smallpox, as well as fever, chills, cough, and other symptoms. The illness was first noted in monkeys but also infects other primates, rodents, and other mammals. Investigators identified the primary source of the 2003 outbreak as prairie dogs that had contracted monkeypox from diseased African rodents imported from Ghana, in April of 2003, by a Texan animal distributor [1].

The CDC official contacted me at the National Invasive Species Council (NISC) to find out which agency with authority over invasive species currently regulated nonnative rodents, or had emergency authority to regulate the importation of rodents that could spread a disease to both native animals and humans. It was critical to stop the importation and movement of African rodents and other known hosts of monkey-pox in order to stop the spread of the disease, and hopefully eradicate it in the United States. The CDC official found it difficult to believe my answer, that no agency had authority to regulate the importation or movement of African rodents as an invasive species. This was the case regardless if they posed an immediate threat of spreading a zoonotic disease to other animals and possibly humans.

Fortunately, in this instance, the CDC and the Federal Drug Administration (FDA) were able to issue a joint regulation under Section 361 of the Public Health Service Act (42 U.S.C. 264) to ban the importation and regulate the movement of African rodents and other host species, first under an emergency rule and later as a final rule [2]. As a result of this action and a well-coordinated effort among numer-ous federal and state agencies, the monkeypox outbreak was successfully contained because monkeypox was a known threat to human health. This case illustrates the challenge of navigating the complex array of federal laws and regulations governing invasive species, as well as the importance of timely coordination among federal and state agencies to address a potential crisis.

CHAPTER OVERVIEW

This chapter provides a general guide to the national laws and regulations governing Terrestrial Vertebrate Invasive Species (TVIS). The focus is on U.S. federal laws including highlights of the most important relevant U.S. laws and policies. The first section provides the context and major reasons for the development of the complex and fragmented framework of law and policy governing TVIS. Section two pro-vides a brief analysis of the effectiveness of domestic law and policy in preventing the introduction and spread of TVIS. The last section highlights some of the most promising steps that could be taken to address weaknesses of the current system, by focusing on the key benefits and caveats of these approaches.

Given that such a broad topic could easily comprise an entire book, this chap-ter provides an overview of federal invasive species law and policy, rather than a complete catalog of all invasive species laws and regulations. This chapter includes minimal information about state invasive species laws or international laws and trea-ties, which are critical components of the legal regime governing invasive species. In addressing a specific invasive species issue, it is always important to consult the exact text of the relevant laws, statutes, and regulations. In most situations, it is criti-cal to examine the state (and sometimes local or county) law in addition to federal and (if appropriate) international legal requirements.

CONTEXT FOR THE DEVELOPMENT OF U.S. INVASIVE SPECIES LAW AND POLICY

Commentators have criticized U.S. federal invasive species law as incomplete, frag-mented, ad hoc, scattered, and ineffective [3]. Even a partial list of federal invasive species statutes includes over 30 laws, numerous regulations and policies, and sev-eral executive orders. At least 13 federal departments and agencies have authority over invasive species or invasive species pathways [4]. Yet, there is no single, comprehensive law that deals with all types of invasive species, and their pathways, at all stages of invasion. This is in contrast to other areas of environmental law, including the regulation of air pollution, water pollution, pesticide regulation, and reviews of environmental impacts for which comprehensive, overarching national laws have been enacted [5]. This chapter explores how and why such a complex array of invasive species laws have developed in order to improve the capacity to utilize these laws to prevent and minimize the impacts of TVIS.

First, consider the great variety of invasive species and the myriad means by which they can be introduced (as described in Chapter 2). Invasive species may be animals, plants, or pathogens and may be terrestrial or aquatic. These diverse types of species are regulated by numerous different agencies and governed by specific and separate authorities. In fact, a species may be invasive in one part of the United States and not in another, due to the large size of the United States which includes over 25 different ecoregions [6]. In addition, different entities and agencies regulate the many and diverse pathways for introduction of invasive species, such as ballast water, agricultural imports, cargo ships, air shipments, horticulture, and the pet trade. Given these factors, it is hardly surprising that a complex array of invasive species laws and regulations have evolved over the past decades to deal with species as varied as zebra mussels, cheat grass, and Burmese pythons.

Another key component is the particular purpose of the various laws and regulations that address invasive species. These laws were often designed to protect a valued economic activity or resource (such as agriculture, timber harvest, migratory birds, or native species) rather than to target or eliminate specific invasive species, many of which were considered valuable or benign when first introduced. For example, Vermont enacted what was likely the first noxious weed legislation within the United States in 1795 in order to protect crop and pasture lands from Canada thistle; New York enacted similar legislation in 1831, and Nebraska in 1873 [7].

The first major federal invasive species legislation, the Plant Quarantine Act of 1912, was enacted to protect nursery stock from invasive plants (Pub. L. 62-275, 37 Stat. 315 [Aug. 20, 1912]). Given that 90% of Americans were engaged in agriculture as their primary economic activity during this period, it is not surprising that early invasive species legislation focused on the protection of agricultural resources [8].

In 1900, Congressman Lacey and others in Congress became concerned about threats to native wildlife (especially birds) due to unregulated commerce and introduced, nonnative species [9]. Section 2 of the Lacey Act authorized the Secretary to ban the importation and interstate transport of wildlife species that are "injurious" or harmful to humans, agriculture, horticulture, forestry, wildlife, or wildlife resources (Lacey Act, Ch. 553, 31 Stat. 187, 188 [May 25, 1900] [now codified at 18 U.S.C. 42]).

Other laws were enacted to protect public and private lands and waters by focusing on particular vectors for species introduction (such as ballast water), specific species (such as nutria), or by more generally authorizing the management of invasive species on public lands (i.e., land management statutes and organic acts discussed below).

Finally, the role of the states in regulating wildlife within their borders and the importance of state invasive species laws must be considered. Under the tenth amendment to the U.S. Constitution, powers not delegated to the federal government are reserved to the states [10]. In addition, there is a long tradition of state (and local) control of wildlife as best positioned to oversee wildlife resources within their borders. As a result of these legal principles and traditions, state governments, rather than the federal government, exercise jurisdiction over wildlife, including invasive terrestrial species, absent a compelling federal interest or relevant authority [11].

OVERVIEW OF U.S. FEDERAL LAW GOVERNING INVASIVE SPECIES

The current, national invasive species laws and regulations are primarily intended to do one of four things: protect human health and U.S. agriculture (including crops, livestock, and poultry), protect valued native wildlife and natural resources, address key pathways for the introduction of invasive species, or provide resources or direction for research and control of specific invasive species; or manage invasive species on federal lands. This overview looks at invasive species laws through the lens of how and why they were developed to help explain which laws serve as the best tools for different invasive species and invasive species problems.

Laws Designed to Protect Human Health and Agriculture

Public Health Service Act of 1944

Section 361 of the Public Health Service Act of 1944 (PHSA, 42 U.S.C. 264) authorizes the Surgeon General, with the approval of the Secretary of Health and Human Services, to develop and enforce regulations "to prevent the introduction, transmission, and spread of communicable diseases from foreign countries" into the United States or from one state to another state. Section 361 further provides, "For purposes of carrying out and enforcing such regulations, the Surgeon General may provide for such inspection, fumigation, disinfection, sanitation, pest extermination, destruction of animals or articles found to be so infected or contaminated as to be sources of dangerous infection to human beings ..." Thus, the FDA or CDC (or both) may take a broad range of actions to prevent the importation and/or movement of TVIS if the invasive species is a known host to a zoonotic disease that threatens human health, such as the 2003 response to the outbreak of the monkeypox virus (described above).

Plant Protection Act of 2000

As mentioned above, the Plant Quarantine Act of 1912 was the first major federal law to protect agricultural crops from plant pests. It narrowly focused on protecting nursery stock from certain plant pests (Pub. L. 62-275, 37 Stat. 315 [now codified at 7 U.S.C. 151 et seq.]). Subsequently, a series of other laws addressed other types of plant pests [12]. Federal law did not regulate invasive plants themselves until the passage of the Federal Noxious Weed Act in 1974, and then only those listed as noxious weeds (Pub. L. 93-629, 88 Stat. 2148 [Jan. 3, 1975][originally codified at 7 U.S.C. §§ 2801–2813], repealed by the Plant Protection Act § 438[a][4], Pub. L. 106-244, 114 Stat. 358 [June 20, 2000][now codified at 7 U.S.C. 7701–7786]).

In 2000, Congress consolidated most of these laws into a more comprehensive plant protection measure, the Plant Protection Act of 2000 (PPA, 7 U.S.C. 7701–7786). The PPA authorizes the U.S. Department of Agriculture (USDA) to prohibit or restrict the importation or interstate movement of any plant, plant product, biological control organism, noxious weed, or means of conveyance if the Secretary determines that the prohibition or restriction is necessary to prevent the introduction or movement of a plant pest or noxious weed.

The Act also includes a number of important attributes that strengthen the federal government's authority to address invasive species. First, the findings recognize that plant pests harm not only agricultural and economic resources, but also the environment (Section 7701). Second, plant pests are defined broadly as any living stage of a protozoan, nonhuman animal, parasitic plant, bacterium, fungus, virus, or pathogen that directly or indirectly can cause damage or disease in any plant or plant product (Section 7702[14]). A "noxious weed" is defined to include plants that harm the environment, natural resources of the United States, as well as agriculture (Section 7702[10]). Under the PPA, USDA is authorized to hold, seize, quarantine, treat, or (if necessary) destroy any plant pest or noxious weed in order to stop the introduction or movement of a species that is new to or not known to be widely distributed in the United States (Section 7714). The PPA also includes: emergency authority (Section 7715), strong penalty and enforcement provisions (Sections 7734 and 7735), as well as authority to charge fees to pay for inspections (Section 7759). It also authorizes assistance to state programs to detect species before they spread broadly and become more difficult and costly to control (Section 7721). Finally, the PPA authorizes USDA to transfer funds from the Commodity Credit Corporation (or other USDA programs) to control a specific plant pest of concern. In practice, transfers must first be approved by the Office of Management and Budget [13].

The Animal Damage Control Act of 1931

The Animal Damage Control Act (ADCA) (codified at 7 U.S.C. 426) provides broad authority for research and control of injurious mammals, rodents, and birds as well as mammals and birds that serve as reservoirs for zoonotic disease [14]. These actions are authorized in order to protect agriculture, buildings, natural resources, people, or endangered species. Thus, activities under this law may be undertaken for the protection of agriculture and natural resources, including wildlife.

In 1985, this authority was transferred from the Secretary of the Interior (U.S. Fish and Wildlife Service [FWS]) to the Secretary of Agriculture (Pub. L. 99-190, 99 Stat. 1185, Dec. 19, 1985). In 1991, Congress amended the Act directing the Secretaries of Defense and Agriculture to take actions to prevent the spread of brown tree snakes from Guam to other areas of the United States (Pub. L. 102-190, Div. A, Title III, Sec. 348, 105 Stat. 1348 [Dec. 5, 1991], and Pub. L. 102-237, Title X, Sec. 1013[d], 105 Stat. 1901, [Dec. 13, 1991]).

The Wildlife Services Program within the Animal and Plant Health Inspection Service (APHIS) administers the ADCA, including assistance for efforts to protect aircraft against collisions with birds and public health against wildlife-borne diseases such as rabies and West Nile virus. Wildlife Services provides advice and technical assistance and operates specific lethal and nonlethal control programs in cooperation with local and state governments as well as other federal agencies, and publishes an annual Program Data Report on its wildlife damage management activities [15].

Wildlife Services also operates the National Wildlife Research Center, which has an active invasive species program that has produced research on island invasive terrestrial species, including the brown tree snake, nutria, and other TVIS [16]. For example, Wildlife Services received $20 million in 2014 to implement a national feral swine management program in the 39 states with a significant feral swine population to protect human and animal health as well as agriculture and natural resources [17].

Animal Health Protection Act

The Animal Health Protection Act authorizes USDA to prohibit or restrict the importation and interstate movement of any animal "if the Secretary determines that the prohibition or restriction is necessary to prevent the introduction into or dissemination within the United States of any pest or disease of livestock" (Sec. 8303 of 7 U.S.C. 8301–8322, Pub. L. 107-171, 116 Stat. 494 [May 13, 2002]). The statute also authorizes the Secretary to seize, quarantine, remove, or destroy the animal as long as the purpose of regulation is to protect livestock and poultry.

Wildlife Protection Laws That Regulate Invasive Species

The Lacey Act

The Lacey Act of 1900, as noted above, was the first major federal wildlife protection statute (31 Stat. 187, 188 [May 25, 1900] [now codified at 18 U.S.C. 42]). Congressman Lacey was concerned about the nation's unregulated wildlife trade and resulting decline in wildlife populations, primarily the massive take of migratory birds to decorate ladies' hats. He was also concerned about introduced, nonnative mammals and birds, including crop-eating birds such as the introduced starling and English sparrow he described in his supporting statement as "that rat of the air" [18]. The stated purposes of the law (which was initially implemented by the Secretary of Agriculture) were: (1) to authorize the introduction and preservation of game, song, and insectivorous wild birds; (2) to prevent the "unwise" introduction of foreign birds and animals; and (3) to supplement state laws for the protection of game and birds [19]. The original Act's provisions regarding invasive species were quite broad, prohibiting the importation "of any foreign wild animal or bird except under special wildlife permit" issued by the Department of Agriculture, as well as four specific species (the mongoose, fruit bat, English sparrow, and starling) (31 Stat. 187, 188).

However, Congress amended this part of the Lacey Act in 1949 and significantly narrowed it by banning only those species specifically listed as "injurious" by Congress or by administrative action. This changed the law's approach from a so-called "white list" law, banning the introduction of all species not specifically allowed, to a "black list" law, banning only those species listed as prohibited (Pub. L. 81-72, 63 Stat. 89). Further amendments transferred authority from Secretary of Agriculture to Secretary of the Interior and added authority to list amphibians, reptiles, mollusks, and crustaceans in 1969 (Pub. L. 91-135, 83 Stat. 275, 279–281), and fish in 1981 (Pub. L. 97-79, 95 Stat. 1073).

Under current law, the FWS administers the injurious wildlife provisions of the Lacey Act (18 U.S.C. 42) on behalf of the Secretary of the Interior. The law prohibits the importation into the United States or shipment between states or territories of certain categories of animal species determined to be "injurious" to people, agriculture, horticulture, forestry, wildlife, or wildlife resources of the United States. Only wild mammals, wild birds, fish, mollusks, crustaceans, amphibians, and reptiles, and their offspring or eggs can be added to the injurious species list. FWS is authorized to export or destroy any injurious species at the expense of the importer. The Service may issue permits for injurious species imported for zoological, educational, medical, and scientific purposes (18 U.S.C. 42; 50 CFR Pt. 16).

The Act does not regulate possession, use, or transport of injurious species within a state. The relevant state is responsible for any regulation of species in these cases. A violation is classified as a Class B misdemeanor, and is punishable by no more than six months in jail and/or up to a $5000 fine for an individual, or $10,000 for an organization (18 U.S.C. 42[b]; 18 U.S.C. 3571[b][6] and [c][6]).

A species may be listed as injurious by Congress (by legislation) or by the FWS, which may either initiate a listing determination or respond to a citizen petition. Species are listed by regulation and the Service must comply with all of the usual regulatory requirements. These include public notice and comment as called for by the Administrative Procedure Act, compliance with the National Environmental Policy Act (NEPA), and a number of regulatory analyses required by statute and executive order. FWS prepares a detailed biological assessment of the species, the basis for their determination, and an analysis of the economic impact of the listing, if required. Listing of species as injurious by regulation utilizing this process has taken an average of four years [20]. The Lacey Act does not authorize listing a species on an emergency basis until the regulatory process can be completed; and many injurious species are already present and established in the United States before they are listed [21].

The "Other" Lacey Act

Another provision of the Lacey Act (16 [U.S.C. 3371—3378]) may also be used to regulate invasive species in certain circumstances. Briefly, this law (which is administered by the Secretaries of the Interior, Commerce, and Agriculture) makes it unlawful for any person to import, export, transport, sell, receive, acquire, or purchase in interstate or foreign commerce any fish, wildlife, or plant taken or sold in violation of any federal, state, tribal, or foreign law (16 U.S.C. 3372). Thus, if a covered action (such as transporting a species listed by a state as invasive into another state) is taken with an invasive species in violation of a state, federal, or tribal law or a treaty, that action would also be in violation of federal law and be subject to civil forfeiture and criminal penalties under this part of the Lacey Act (16 U.S.C. 3373, 3374). This increases the importance of an invasive species being listed as injurious or invasive by one or more states, whether or not it is federally listed.

Endangered Species Act

Experts consider invasive species second only to habitat loss as threats to biodiversity, and they are cited as a cause of endangerment for 48% of species listed as threatened or endangered under the Endangered Species Act (ESA) [22]. Island ecosystems are considered even more vulnerable to invasive species than continental ecosystems, with invasive species constituting the greatest threat to rare species [23]. The purpose of the ESA is "... the conservation of ecosystems upon which threatened and endangered species of fish, wildlife, and plants depend" (16 U.S.C. 1531). The ESA is jointly administered by the Secretaries of the Interior (terrestrial and freshwater species) and Commerce (marine species). Section 7 of the ESA requires federal agencies to ensure that their actions are not likely to jeopardize the continued existence of listed species or adversely modify their critical habitat (16 U.S.C. 1536[a][2]). If an agency action may affect a listed species, the agency must consult with either the FWS or the National Marine Fisheries Service (NMFS) (for marine species) to determine how to avoid significant harm to the species (50 C.F.R. § 402.01).

In those cases that a federal action or authorization would cause the introduction or spread of an invasive species that may harm a listed species, consultation under Section 7 of the ESA may be required. For Section 7 to be relevant, the invasive species must be shown to have a significant impact (may cause jeopardy to the continued existence of a listed species) as a result of a federal action or authorization (including grants). For actions of private parties that would result in the taking (harming or killing) of a listed species, the party carrying out the action may be required to get an incidental take permit from FWS or NMFS under Section 9 of the Act (16 U.S.C. 1538).

In addition, recovery plans and Habitat Conservation Plans may call for the control or eradication of invasive species in order to conserve listed species. In most cases, the implementation of these plans is voluntary and funding for the implementation of recovery plans is often inadequate or unavailable [24]. Nevertheless, for those TVIS that can be shown to harm endangered or threatened species, the ESA is potentially an important legal tool, especially if control or eradication of the invasive species would significantly assist the recovery or conservation of the affected species. For example, in 2010, APHIS/Wildlife Services, in cooperation with the Hawaiian Department of Land and Natural Resources and the Hawaii Army National Guard and in coordination with FWS, established an invasive species control program targeting rats, house mice, mongoose, and other introduced species to protect and assist in the recovery of endemic and endangered Hawaiian birds [25].

Laws Relating to Specific Pathways of Invasive Species

Although many TVIS are intentionally introduced for food, fur, as pets, or for other purposes (as discussed in Chapter 2), numerous invasive species are introduced accidentally or unintentionally. The most effective way to prevent or reduce these introductions is by regulating the pathways (vectors) for introduction; and a number of invasive species laws focus primarily on key pathways.

Nonindigenous Aquatic Nuisance Prevention and Control Act
of 1990 and National Invasive Species Act of 1996

The Nonindigenous Aquatic Nuisance Prevention and Control Act of 1990 (NANPCA, Pub. L. 101-646, Title I, sec. 1002, 104 Stat. 4761 [Nov. 29, 1990], codified at 16 U.S.C. 4701 et seq.) focused primarily on preventing and slowing the spread of nonnative, aquatic species through ballast water releases into the Great Lakes. Ships had introduced the tiny, but extremely harmful, zebra mussel into U.S. waters through the discharge of ballast water, which is the seawater taken on at foreign ports to steady the ships when cargo is offloaded.

In 1996, Congress passed the National Invasive Species Act, which reauthorized and expanded the scope of NANPCA to address aquatic invasive species and their pathways more generally (NISA, Pub. L. 104-332, 110 Stat. 4073, 4091 [Oct. 26, 2006]).

The legislation defines aquatic nuisance species (ANS) broadly as "a nonindigenous species that threatens the diversity or abundance of native species or the ecological stability of infested waters, or commercial, agricultural, aquacultural, or recreational activities dependent of such waters…" (16 U.S.C. 4702[1]). Given this broad definition, species including the nutria, American bullfrog, and cane toad are all considered ANS under NISA. In addition, Congress amended the law to include the terrestrial brown tree snake, due to concern about the introduction of this species into Guam and the harm it had caused to native species and infrastructure (Pub. L. 101-646, Title I, §1209, 104 Stat. 4772 [Nov. 29, 1990] codified at 16 U.S.C. 4728).

The legislation, as amended, also:

- Established the Aquatic Nuisance Species Task Force (ANSTF) cochaired by FWS and the National Ocean and Atmospheric Administration (NOAA) and including high-ranking officials of defense, transportation (Coast Guard), Environmental Protection Agency (EPA), and others to take action to prevent and control ANS;
- Created regional panels to coordinate activities throughout the United States;
- Directed the Coast Guard to create a ballast water discharge management program to prevent the establishment and spread of ANS in U.S. waters;
- Encouraged exploration of international measures and possible agreements to regulate ballast water;
- Provided for ANS research;
- Established a brown tree snake management program; and
- Authorized matching funds for approved state ANS management plans.

NANPCA, as amended by NISA, applies broadly to all aquatic invasive species and provides for several significant coordination mechanisms (ANSTF and the five regional panels). The Director of FWS and the Undersecretary of Commerce for Oceans and Atmosphere serve as cochairs of the Task Force, which includes representatives from 13 federal agencies and includes 12 organizations as ex officio members.

Wild Bird Conservation Act of 1992

The Wild Bird Conservation Act (16 U.S.C. 4901–4916), administered by FWS, was enacted to conserve wild birds by regulating the trade of wild birds caught in foreign countries and imported into the United States primarily for the pet trade, and by encouraging captive breeding. Although it does not directly apply to invasive species, it incidentally regulates one possible pathway for the introduction of invasive species. The Act restricts the importation of certain wild birds that could carry invasive pathogens or that could escape (or be released) into the wild and become invasive [26].

Alien Species Prevention and Enforcement Act

The Alien Species Prevention and Enforcement Act (ASPEA, Pub. L. 102-393, 106 Stat. 1729, 1774 [Oct. 6, 1992], codified at 39 U.S.C. 3015) is narrowly focused on the U.S. mail as a potential pathway for introduction of invasive species. It clarifies that the shipment through the mail of otherwise prohibited species is illegal. Prohibited species include those identified as injurious under the Lacey Act, as well as plants regulated under various plant pest and plant quarantine laws. The U.S. Postal Service administers the ASPEA.

Laws Providing Authority to Manage Invasive Species on Federal Lands

As discussed above, unless a federal law or specific federal interest is implicated, the states generally manage fish and wildlife (including invasive species) within their borders. However, a number of federal statutes provide federal agencies broad authority to manage all species, including invasive species, on their (federal) lands. For example, the Organic Administration Act (16 U.S.C. 551 et seq.) provides broad authority to USDA's Forest Service to protect national forests from destruction, including harm caused by invasive species. Other land management statues that provide authority for agencies to manage invasive species on their lands and waters include the Multiple Use Sustained Yield At of 1960 (16 U.S.C. 528-531); the Federal Land Policy and Management Act of 1976, which applies to lands managed by the Bureau of Land Management (43 U.S.C. 1701-1787); the National Park Service Organic Act (now codified at 54 U.S.C. 100101 et seq.); and the National Wildlife Refuge System Administration Act of 1996 (16 U.S.C. 668dd).

It is not necessary to list all of the applicable federal land management authorities. The central point is that federal agencies have broad authority to manage invasive species present on their lands and waters under their general land management authorities in order to protect federal natural resources, or to manage their lands and waters for specific purposes such as biodiversity, recreation, grazing, or other purposes as set out in the applicable statute [27]. This is true whether or not invasive species are specifically mentioned in the applicable law.

Laws Governing Specific TVIS

Often, the press, the public, and Congress focus on individual species that have become established, spread, and cause easily visible harm, including species such as the brown tree snake in Guam, the nutria in the Mid-Atlantic and Southeast, or the Burmese pythons invading the Everglades. For this reason, a number of invasive species laws focus on one or only a few invasive species. Even for those species that are the subject of specific legislation, it may be too late to eradicate or effectively control the species if it has become established and spread, or if inadequate funds are provided. Those federal laws that focus on TVIS include:

The Brown Tree Snake Control and Eradication Act of 2004 provides for a coordinated multidepartmental effort to control and eradicate the brown tree snake on the island of Guam, as well as to prevent the spread of the snake to other areas (including territories) of the United States (Pub. L. 108-384, Sec. 4, 118 Stat. 2222 [Oct. 30, 2004], now codified at 7 U.S.C. 8503).

The Nutria Eradication and Control Act of 2003 (Pub. L. 108-16, 117 Stat. 621 [April 23, 2003]) authorized the Secretary of the Interior to provide financial assistance to the states of Maryland and Louisiana for programs to control and (if possible) eradicate the nutria and restore impacted wetlands.

Applicability of the National Environmental Policy Act of 1969

NEPA requires all federal agencies to assess the environmental and related social and economic effects of their proposed actions prior to making decisions, including approving permits, adopting federal land management actions, and approving construction of federally owned facilities. NEPA also provides an opportunity for public review and comment (42 U.S.C. 4321 et seq.). Invasive species issues may be raised and considered during the NEPA process; and possible ways to avoid or mitigate invasive species harm may be included in the discussion of alternatives. Federal agencies are required to comply with NEPA for actions to prevent, eradicate, or control invasive species. In emergency situations that call for an immediate response, an agency can work out an alternative NEPA process with the Council on Environmental Quality, the agency responsible for developing NEPA guidance (40 C.F.R. 1506.11).

Invasive Species Executive Orders

In addition to legislation, the executive branch has addressed invasive species through executive orders. On February 3, 1999, President Clinton issued Executive Order 13112 on Invasive Species (EO 13112) in response to the growing concern about the economic impacts of invasive species such as the zebra mussel, and a letter sent by 500 scientists and stakeholders expressing concern about the ecological harm caused by invasive species as well as the need for a comprehensive and coordinated national strategy.

EO 13112 replaced a shorter, less specific executive order on invasive species signed by President Carter in 1977 (EO 11987). EO 13112 has three important components. First, it directs all federal agencies (within existing authorizations and budgets) to prevent and control invasive species within their lands and as part of their programs. Second, EO 13112 also directs federal agencies to avoid any action or authorization that is likely to cause the introduction or spread of invasive species, absent a compelling need for the action. Third, it set up the NISC made up of the leadership of (currently 13) federal departments and agencies and cochaired by the Secretaries of the Interior, Agriculture, and Commerce to provide coordination, planning, and leadership on invasive species. It also established the Invasive Species Advisory Committee to provide recommendations and advice from nonfederal experts and stakeholders.

ANALYSIS OF EFFECTIVENESS OF FEDERAL INVASIVE SPECIES LAW

This section briefly examines the adequacy of the extensive list of federal laws that cover one or more aspects of invasive species, focusing on national regulation of TVIS. Most recent studies, reports, and analyses since the 1990s have concluded that the complex and confusing "patchwork" of U.S. invasive species laws are inadequate to address the harm caused by invasive species to the economy, environment, and human health [28]. Despite the long list of relevant federal laws and regulations, this conclusion seems difficult to challenge given the large number of invasive species that have been introduced and become established in the last two decades. A number of experts and organizations have called for the adoption of a comprehensive framework law to address all or most invasive species [29]. However, I would argue that by looking more closely at the strengths, as well as the weaknesses, of current law, we may be able to chart a more realistic and timely path forward to enhance our national response to invasive species, through improved implementation, funding, and modest amendments to existing authorities.

The major critiques of the federal legal regime are that the laws and regulations (1) are too narrow and overly focused on protecting agriculture or other narrow interests, (2) cover only one or more taxa or type of invasive species rather than all taxa, (3) are implemented by more than one department or agency calling for extensive coordination, and (4) are reactive and based on a forbidden list of species that are often only prohibited once firmly established and already causing harm [30]. Many of these reviewers advocate the passage of a single, overarching statute that would govern all taxa and types of invasive species and all aspects of invasive species actions including prevention, eradication, and control. As the passage of an overarching, comprehensive bill seems highly unlikely in the near (or even medium-term) future, my analysis will focus on key issues with the major federal invasive species laws and regulations that could be addressed without the passage of a comprehensive framework law [31].

Analysis of Laws Governing Plants and Plant Pests (Primarily Insects, Diseases, and Noxious Weeds)

The passage of the PPA of 2000 consolidated and strengthened measures to protect plants from invasive species. Congress provided USDA broad authority to protect plants from plant pests, primarily insects, diseases, and noxious weeds. This includes authority to protect the environment and thus to address invasive species that harm natural resources in addition to agriculture. Many believe that implementation of the law, which has focused on harm to agriculture, has undermined the PPA as a tool to deal with invasive species that harm natural resources [32]. However, APHIS has taken significant steps to address one of the most (if not the most) important pathways for the introduction of environmentally harmful invasive plants and plant pests, the importation of horticultural plants (the nursery trade). Under the PPA, APHIS has established the NAPPRA regulatory program that stands for "Not Authorized Pending Pest Risk Analysis." If a plant is classified as NAPPRA, it cannot be imported until a pest risk analysis is requested and completed for that plant [33]. Providing enhanced staffing and resources for the existing noxious weed program and the NAPPRA risk assessment regulatory program would seem to be a more efficient and effective way to address more invasive plants and plant pests than the difficult and time-consuming process of enacting additional legal authorities or regulations.

Addressing Aquatic Species (Including Some TVIS)

Federal aquatic invasive species legislation (NANPCA as reauthorized by NISA) has been criticized as narrow, dealing only with aquatic invasive species and regulating only one pathway, ballast water [34]. However, the legislation has a number of strengths. It gives the ANSTF a broad mandate to implement a program to prevent the establishment and spread of all aquatic nuisance species. ANSTF includes 13 federal agencies, including virtually all the major agencies involved with aquatic nuisance species. ANSTF has established regional panels made up of state, tribal, and local officials, stakeholders, and experts that can coordinate prevention and control actions. Under the Act, there are 38 approved state ANS management plans and three interstate approved plans [35].

Despite progress made at the federal, regional, and state level under NANPCA/ NISA, the regional panels, implementation of the state plans, and ANSTF all suffer from anemic staffing and funding, especially in relation to the high economic impact caused by some ANS (e.g., zebra and quagga mussels, nutria, and the brown tree snake) [36]. In addition, this legislation has not been reauthorized since 1996 and is long overdue for updating. The ANS workshop on NISA in 2002 (sponsored by the EPA's Great Lakes Program office) recommended that legislation to reauthorize this program could and should address: developing programs for additional major pathways (including hull fouling), authorizing a program for early detection and rapid response, and strengthening the regional panels and state planning provisions [36].

Considering the significant number of coordinating and planning bodies at the federal, regional, and state levels that now exist to address aquatic invasive species, it is more efficient to enhance and improve the existing laws and regulations, rather than invent a new legal construct to deal with aquatic invasive species. Furthermore, NISA reauthorization legislation could also address the critical need to improve our ability to prevent the introduction of aquatic nuisance species before importation. This could be done either by including a screening process for planned introductions in the NISA reauthorization, or by updating or revising the injurious wildlife provisions of the Lacey Act (discussed below).

Addressing (Nonaquatic) Animal Species That Harm Native Species and Natural Resources

Given the long list of authorities discussed above, invasive species shown to harm agriculture, livestock, or human health, as well as aquatic invasive species, could be adequately addressed under existing laws and regulations—especially if these laws were better funded, implemented, and (in some cases) updated. However, nonnative animal species (including many TVIS) that primarily harm natural resources and native species are rarely prohibited under these statutes. Unless listed as an injurious species under Title II of the Lacey Act, any live wild mammal, bird, amphibian, or reptile may be imported subject only to administrative requirements, including a customs declaration [37], while the annual volume of live animal imports have roughly doubled since 1991 [38].

The weaknesses and problems associated with the injurious wildlife provisions of the Lacey Act have been well documented [39]. The most important include: (1) the reactive nature of how species are listed, (2) the time-consuming and cumbersome listing process, (3) absence of an emergency or rapid listing option, and (4) lack of resources to administer an effective program.

Once an invasive species has been introduced and established, it is very difficult and costly to eradicate or control. Yet, under current U.S. law, most nonnative animal species may be imported unless specifically listed as injurious (the "black list" approach). A few nations, including Australia and New Zealand, have adopted a different approach, one that prohibits the importation of a species unless it has been listed as allowed, or until the risk that it may become invasive has been evaluated ("white list" approach) [40].

In order to list a species as injurious under the Lacey Act, FWS must go through the rulemaking process in accordance with the Administrative Procedures Act. This involves numerous steps, including: gathering extensive information about the species and why it should be listed invasive, providing for public notice and opportunity to comment, preparing an environmental assessment or impact statement if required under NEPA, and complying with multiple executive orders on regulatory rulemaking (including providing an analysis of the economic impact of the listing). In recent years, it has taken an average of four years for FWS to list a species as injurious given this extensive process [41]. At the end of calendar year 2015, about 240 species had been listed as injurious, including 92 mammals, six birds, and nine snakes [42].

Nor does the law include a mechanism to list a species on a temporary or emergency basis while the listing determination is being made. Although Congress can list a species by legislation, this has occurred only a few times and the legislative process can be just as slow, if not slower, than federal rulemaking [43]. The 2007 Ecological Society of America research study on the effectiveness of the Lacey Act found that for the taxa listed, more than half were already present in the United States when listed, and most spread after listing [44].

Another factor, especially given the complexity of the listing process, is the low level of funding for the Injurious Wildlife Program within FWS. Just one or (at the most) two full-time staff positions have generally been devoted to the injurious listing program in recent years. This not only impacts the current listing effort, but also causes concern among affected stakeholders about whether FWS could efficiently manage an acceleration or streamlining of the injurious listing process in order to prevent the introduction and spread of invasive species.

In addition to enhancing the FWS invasive species program office, a number of actions could aid the current program until a new or revised law can be enacted, such as support for ongoing streamlining efforts by FWS for the injurious listing process. These include adoption of a categorical exclusion under NEPA for many (or most) injurious species listings [45]. Further, FWS has utilized a rapid screening process to provide a prediction of invasive potential for certain nonnatives, where adequate information is available, to allow proactive listing of species before they are introduced and become established [46]. FWS has listed an additional 201 species of salamanders, 10 fish, and one crayfish species in the last few years utilizing these and other tools to streamline the listing process. FWS is also working with stakeholders (including the pet trade) to craft guidelines to prevent the importation of potential invasive species on a voluntary basis [47].

A number of bills have been introduced to update and improve the injurious wildlife provisions of the Lacey Act, but none of these bills have passed either the House or the Senate [48]. Only a broad and well-organized coalition that includes the mainstream, responsible elements of the pet trade, environmentalists and aquaculture, can help pass significant legislation regulating animal imports. In addition, the issue must be considered urgent, due to one or more high-profile invasions (such as the Burmese python) or an animal that serves as host to an invasive pathogen (such as the Gambian rat).

Importance of State Laws and Working Coordination with State and Local Governments

A detailed discussion of state invasive species laws is beyond the scope of this chapter, and state laws vary greatly from state to state. In many cases, states have a complex array of laws addressing different invasive species issues. Those states with the most significant number of invasive species tend to have the most developed set of laws, which in some cases have provisions absent in federal law. For example, the states of Hawaii and New York have functioning early detection and rapid response programs, for the most part lacking at the federal level [49]. For most invasive species issues, it is critical to first consult and consider the relevant state laws by exploring state invasive species websites (where available) and several broader resources for state invasive species laws including: the National Caucus of State Legislators (NCEL), and the National Agricultural Library Invasive Species Information website [50].

RECOMMENDATIONS TO STRENGTHEN INVASIVE SPECIES LAWS AND REGULATIONS

Given the current political and budgetary climate, it is an extremely difficult task to pass any significant national environmental legislation. In fact, most of the major comprehensive or organic environmental statutes were passed in the 1960s and 1970s [51]. Rather than attempt to pass one comprehensive bill or launch competing efforts to pass many specialized invasive species bills, I recommend that diverse stakeholders whose interests are harmed by invasive species (states, counties, affected industries, recreationists, environmentalists, hunters, fishing groups, etc.) should work together to advocate the passage of select (five or six) legislative and administrative actions. Only by uniting and advocating for a realistic set of goals is significant legislative and regulatory progress on invasive species possible. I suggest the following set of actions, based on my research and experience directing the NISC from 2000 to 2014.

1. Support additional matching funds for states (total of $100 million or more over several years) to help prevent and manage invasive species on their lands and waters, primarily by enhancing existing programs. For example, fully fund ANSTF State ANS Plan grants; enhance invasive species grant opportunities in the conservation title of the Farm bill [52]; boost ESA recovery plan funding to reduce threats to endangered species through control and eradication of invasive species; enhance wildlife habitat enhancement programs to address invasive species such as the FWS Partners for Wildlife funding [53]; increase high-priority funding to APHIS/Wildlife Services to assist with key efforts to prevent and control priority TVIS species. This can be done through the budget and appropriations processes first, or in conjunction with other actions. These actions recognize the importance of states as partners and could address some of most problematic species by enhancing existing grant programs.

2. Accelerate and support the USDA/APHIS/NAPPRA program to examine which plants for planting (horticultural plants) should be regulated to prevent the introduction of new invasive plants and plant pests [54].

3. Enhance existing forest and rangeland management programs to address priority invasive animals, insects, and diseases.

4. Utilize both regulatory and voluntary means developed by FWS and industry to reduce imports of problematic invasive species, including TVIS. Provide the Injurious Wildlife program with a reasonable level of staffing and resources to operate effectively. Pass legislation to reform and update the Injurious Wildlife Provisions of the Lacey Act, working with all stakeholders including the mainstream pet trade.

5. Enhance and empower existing coordinating groups administratively and if possible: (1) reauthorize, strengthen, and address weaknesses of NANPCA/NISA and thus ANSTF; and (2) legislatively authorize a streamlined NISC. Charge these bodies with increasing and enhancing public education and targeted public outreach programs.

6. Authorize (and fund) both ANSTF and NISC to implement pilot and then permanent early detection and rapid response (EDRR) programs, based on providing matching grants to states, private organizations, and cooperative efforts including federal agencies. EDRR programs are more cost-effective than long-term control.

7. Enhance international efforts (beyond the scope of this chapter) to address key pathways including international trade (all taxa) and hull fouling (aquatic taxa) and consider expanding the mandate of the Office International des Epizooties (OIE) to address invasive animals that do not threaten livestock or poultry.

8. Align land and water conservation, restoration, and management goals with invasive species prevention and control objectives, and explore use of ecosystem and adaptive management to mitigate invasive species in order to enhance resiliency against climate change.

CONCLUSION

It is critical to prevent and control invasive species, not because they are inherently bad, but because they are species out of place that cause harm whether to crops, livestock, recreation, biodiversity, or habitats for endangered species. While it would be preferable if the United States had one comprehensive framework invasive species law, the passage of such a law is unlikely in the near future.

The most serious gap in U.S. invasive species law is our limited ability to stop the importation and spread of nonnative wildlife species that primarily threaten natural resources. This gap should be addressed by streamlining the listing process, and subsequently, passing a new law or updating Title II of the Lacey Act as soon as possible. Most other invasive species are arguably covered by one of the patchwork of federal laws outlined in this chapter. Coordinating mechanisms such as the ANSTF and the NISC, as well as state and regional coordinating bodies, have and should assist practitioners with utilizing the legal tools we have to protect our economy and environment from the impacts of terrestrial vertebrate invasive species.

REFERENCES AND NOTES

1. "Restrictions on the import, capture, transport, sale, barter, exchange, distribution, and release of African rodents, prairie dogs, and certain other animals; Interim final rule," 68 Federal Register 213 (November 4, 2003), pp. 62353–62372.
2. "Restrictions on the import, capture, transport, sale, barter, exchange, distribution, and release of African rodents, prairie dogs, and certain other animals; Interim final rule," 68 Federal Register 213 (November 4, 2003), p. 62355.
3. General Accounting Office, Invasive Species: Federal and Selected State Funding to Address Harmful Nonnative Species, GAO/RCED-00-219 (August 24, 2000); Miller, M. 2015, There Ought to Be a Law! The Peculiar Absence of Broad Federal Harmful Nonindigenous Species Legislation. In: *Invasive Species in a Globalized World: Ecological, Social and Legal Perspectives on Policy*, edited by Keller R. P et al. University of Chicago Press, 2015:pp. 327–355.
4. National Invasive Species Council. Management Plan National Invasive Species Council, Appendix 3, 2001, http://www.invasivespecies.gov/main_nav/mn_NISCC_ManagementPlan.html
5. Miller 2015 at 327.
6. U.S. Environmental Protection Agency. Ecoregions of North America. https://www.epa.gov/eco-research/ecoregions-north-america (last updated March 22, 2016).
7. University of Nebraska, Lincoln Extension. Noxious Weeds of Nebraska, Canada thistle, by Robert Wilson, EC 171, p. 1 (revised May 2009).
8. McCubbins, J.S.N et al. Frayed seams in the "patchwork quilt" of American Federalism: an empirical analysis of invasive plant species regulation. *Environmental Law* 43(35), p. 43, 2013.
9. Fowler, A. J., Lodge D. M., and Hsia J. F. Failure of the Lacey Act to Protect US Ecosystems against Animal Invasions. *Frontiers in Ecology and the Environment, Ecological Society of America* 5(7), pp. 353–359, 2007.
10. U.S. Const. amend. X; McCubbins. pp. 41–42, 2013.
11. McCubbins. pp. 41–42, 2013.
12. McCubbins. 2013, p. 43.
13. Congressional Research Service. Invasive species: Major Laws and the Role of Selected Federal Agencies, by M.L Corn and R. Johnson, CRS 7-5700 (R43258) (October 4, 2013).

14. Original law: 71 Pub. L. 776, 46 Stat. 1468 (March 2, 1931)(now codified at 7 U.S.C. 426); Tobin, M.E. U.S. Department of Agriculture Wildlife Services: Providing federal leadership in managing conflicts with wildlife. *Proceedings of the Wildlife Damage Conference* 14, pp. 1–2, 2012.
15. USDA/APHIS. Wildlife Damage Management website, https://www.aphis.usda.gov/aphis/ourfocus/wildlifedamage (last visited July 21, 2017).
16. USDA/APHIS. National Wildlife Research Center website. https://www.aphis.usda.gov/aphis/ourfocus/wildlifedamage/programs/nwrc (last visited July 21, 2017).
17. USDA/APHIS. Wildlife Damage Management, feral swine website. https://www.aphis.usda.gov/aphis/resources/pests-diseases/feral-swine, (last visited July 21, 2017).
18. Congressional Record 4871, 1900.
19. Strifling, D. A. An ecosystems approach to slowing the synergistic effects of invasive species and climate change. *Duke Environmental Law and Policy Forum*, 22(145), p. 161, 2011.
20. U.S. House of Representatives, Committee on Natural Resources, Subcommittee on Fisheries, Wildlife, Oceans and Insular Affairs. Hearing on Fish and Wildlife Service proposal for a categorical exclusion for the listing of species as injurious wildlife. Testimony of David Hoskins, on behalf of FWS (September 20, 2013).
21. Fowler 2007, pp. 353–354.
22. Wilcove, D.S. et al. Quantifying threats to imperiled species in the U.S. *BioScience.* 48(8) (American Institute of Biological Sciences) pp. 607–615, 1998.
23. Wilcove, D. S., Endangered species management: The US experience. In: *Conservation Biology for All*, edited by N.S. Sodhi and P.R. Erlich, Oxford University Press, pp. 220–235, 2010.
24. Wilcove, D. S., Endangered species management: the US experience. In: *Conservation Biology for All*, edited by N.S. Sodhi and P.R. Erlich, Oxford University Press, p. 230, 2010.
25. United States Department of Agriculture/APHIS/Wildlife Services. Final Environmental Assessment, Predator Damage Management to Protect Avian Wildlife in Hawaii, p. 1 (February 2010).
26. CRS. p. 16, 2013.
27. CRS. pp. 28–29, 2013; Miller. p. 536, 2015.
28. Miller. p. 327, 2015; McCubbins, p. 43, 2013; Shannon, M. *From Zebra Mussels to Coqui Frogs: Public Nuisance Liability as a Method to Combat the Introduction of Invasive Species.* University of California, Davis, 32(1), pp. 42–43, 2008.
29. Miller. p. 343, 2015.
30. Miller, M., The Paradox of U.S. Alien Species Law. *Environmental Law Review*, 35(10179), p. 10204, 2005; Fowler. 2007; Strifling. 2011; McCubbins. 2013.
31. CRS. pp. 5–6, 2013; Miller. pp. 350–351, 2015.
32. Miller 2015; Shannon. p. 47, 2008.
33. USDA/Animal Plant Health Inspection Service. Not Authorized Pending Pest Risk Analysis (NAPPRA) program website. https://www.aphis.usda.gov/aphis/ourfocus/planthealth/import-information/permits/plants-and-plant-products-permits/plants-for-planting/ct_nappra/ (last visited July 21, 2017).
34. Miller. 2015, pp. 332–333; Shannon 2008, pp. 44–45.
35. Aquatic Nuisance Species Task Force (ANSTF) website. www.anstaskforce.gov/stateplans.php (last visited August 2016).
36. EPA Great Lakes National Program Office Symposium, An Evaluation of NISA to Support Its Reauthorization. In: *Proceedings Document of the Symposium Looking Back: Assessing ANS Prevention and Control*, pp. 33–38, 2002.
37. Fowler 2007, p. 353.

38. Defenders of Wildlife. *Broken Screens, The Regulation of Live Animal Imports In the U.S.* (principal author, Peter Jenkins), Defenders of Wildlife, Washington DC, p. 4, 2007.
39. Fowler. 2007; Miller 2015; Ott, Stephanie S. and Terra Bowling, Legislative and Regulatory Efforts to Minimize Expansion of Invasive Mussels through Watercraft Movements. *The Arizona Journal of Environmental Law and Policy*, 3:61, 2013.
40. Simberloff, D. *Invasive Species, What Everyone Needs to Know.* Oxford University Press, pp. 162–168, 2013.
41. Fowler 2007; U.S. House of Representatives, 2013.
42. FWS. Species Listed as Injurious Wildlife under the Lacey Act, 50 CFR 16. https://www.fws.gov/injuriouswildlife/pdf_files/Current_Listed_IW.pdf (last visited July 21, 2017).
43. Congressional Research Service. Injurious Species Listings Under the Lacey Act: A Legal Briefing, by Kristina Alexander, CRS 7-5700 (R43170) (August 1, 2013).
44. Fowler 2007, p. 357.
45. FWS. Invasive species website explaining categorical exclusion for injurious species listings under NEPA. http://www.fws.gov/injuriouswildlife/catex.html
46. "Injurious Wildlife Species; Listing 10 Freshwater Fish and 1 Crayfish." 81 Federal Register 190, September 30, 2016, pp. 67864–67899, 67862.
47. FWS. Invasive species website describing voluntary program to prevent invasive species. https://www.fws.gov/injuriouswildlife/Injurious_prevention.html (last visited July 21, 2017).
48. Miller. pp. 350–351, 2015; see S. 3278 (114th Congress) introduced by Senator Gillibrand (D-NY) on July 14, 2016, and H.R. 5864, (112th Congress) introduced on May 31, 2012, by Rep. Louise Slaughter (D-NY), The Invasive Fish and Wildlife Prevention Act, (neither bill had passed either the House or the Senate as of October 17, 2016).
49. Hawaii Department of Land and Natural Resources, Funding for county-based Invasive Species Committees to carry out early detection and rapid response. http://dlnr.hawaii.gov/hisc/projects/fy16. 2016; New York Department of Environmental Conservation, Rapid Response Framework for Invasive Species (DLF-16-1). http:/www.dec.ny.gov/docs/lands_forests_pdf/invasivespeciesrapidresponse/pdf; New York Environmental Protection fund available for invasive species control. http://www.dec.ny.gov/about/92815.html.
50. National Agricultural Library, invasive species information website. https://www.invasivespeciesinfo.gov/laws/statelaws.shtml
51. Miller. p. 327, 2015.
52. Zulauf, C., 2014 Farm Bill Conservation (Title II) Programs. *Farmdoc Daily* (4): 89, Department of Agricultural and Consumer Economics, University of Illinois at Urbana-Champaign, (May 14, 2014). http://farmdocdaily.illinois.edu/2014/05/2014-farm-bill-conservation-title-ii-programs.html
53. FWS. Partners for wildlife website. http://fws.gov/invasives/programs.html, (last visited July 21, 2017); National Agricultural Library, invasive species information website, https://www.invasivespeciesinfo.gov/toolkit/main.shtml (last visited July 21, 2017).
54. USDA/Animal Plant Health Inspection Service. Not Authorized Pending Pest Risk Analysis (NAPPRA) program website, https://www.aphis.usda.gov/aphis/ourfocus/planthealth/import-information/permits/plants-and-plant-products-permits/plants-for-planting/ct_nappra/ (last visited July 21, 2017).

5 Ecological Issues

Christopher A. Lepczyk and Daniel Rubinoff

CONTENTS

INTRODUCTION

Terrestrial vertebrates make up a relatively small percentage (~1.92%; 33,278 of 1,730,725) of the world's biodiversity in comparison to plants, invertebrates, and microbes (The World Conservation Union 2014). Though lower in diversity, terrestrial vertebrates have played a large role in human societies as a source of food, fiber, labor, and in culture. In particular, because humans have domesticated a number of animals and have great affinity for many others, we have inadvertently initiated a grand experiment by moving them around the world with us, either intentionally or accidentally. While many of these animal introductions did not lead to invasions, other times, introductions have led to animals becoming invasive. These invasions have caused a wide range of ecological problems ranging from alteration of basic ecological processes to the extinction of species (e.g., cats and rats exterminating birds, lizards, and insects on oceanic islands; Priddel et al. 2003; Stolzenburg 2011). Moreover, these invasive animals cause a wide range of problems to human society (Chapter 3), costing the U.S. economy billions of dollars each year (Pimentel et al. 2005; Chapter 3)

While managing any invasive species poses a number of challenges, managing invasive terrestrial vertebrates is easier in comparison to other groups, like invasive insects and pathogens, in part because of their large size and basic biology (e.g., mode of reproduction, nutritional needs, foraging behavior). Though perhaps easier, managing invasive vertebrates, particularly removing them from an ecosystem, can be challenging in large part because of people's fondness for many of them (e.g., wild horses [*Equus caballus*], feral cats [*Felis catus*]), and often resultant opposition to lethal control, as well as political will. For instance, removing wild horses in the

FIGURE 5.1   The ecological impact of invasive herbivores can be easily noted when comparing fenced to unfenced areas. (Photo credit Edith Adkins.)

western United States is primarily an issue embroiled in politics and culture, not biology (Linklater et al. 2002). As a result, understanding and managing invasive terrestrial vertebrates often requires understanding both the ecology and the human dimensions of the problem.

Invasive terrestrial vertebrates both directly and indirectly affect ecosystems. For instance, invasive predators directly reduce populations of prey species (Lepczyk et al. 2004; Dorcas et al. 2012; Loss et al. 2013; McCleery et al. 2015), which in turn can lead to trophic cascades (Crooks and Soulé 1999). Likewise, invasive vertebrates may affect a variety of species indirectly through processes such as alteration of the physical environment (e.g., Barrios-Garcia and Ballari 2012; Figure 5.1). As a result, invasive vertebrates can cause state changes in systems and act as ecosystem engineers (Cuddington and Hastings 2004; Pietrek and González-Roglich 2015). In other words, invasive vertebrates impact the basic ecology of invaded systems at multiple scales from the individual to the landscape.

ECOLOGICAL SCALE OF PROBLEM

A useful framework for considering how invasive vertebrates operate with regard to the ecology of a system is the traditional ecological hierarchy of individuals, populations, communities, and ecosystems. The value of this traditional framework is that it can elucidate how invasive species affect each level of the hierarchy and thereby provide a useful guide for how management can be implemented at different scales. Furthermore, it is critical to understand how invasive species affect different ecological scales before attempting to fully disentangle their cross-scale effects.

INDIVIDUALS

At the level of individuals, one notable way in which invasive vertebrates interact with native species is through interbreeding, resulting in the creation of hybrid populations. Hybridization requires enough genetic compatibility between two species such that they produce viable offspring. Infertile hybrids do not pose a threat to the genetic integrity of native species. But the more closely related species are, the chance of producing fertile hybrids increases. Hybridization can negatively impact native species through two mechanisms (Wolf et al. 2001). First, if hybrids have reduced fertility or viability, such that they have lower fitness, relative to the parent species (i.e., outbreeding depression), then the growth rate of the species with the smaller population can decline below replacement level. Second, and perhaps of greater concern, is that fertile hybrids may backcross with native species, leading to introgression. If hybrids have increased fitness over the native species (i.e., hybrid vigor) or even superior numbers due to an influx of the invasive species, then introgression can result in genetic dilution of the native species so extensive as to lead to "genetic extinction" of the native species (Rhymer and Simberloff 1996; Simberloff 2013). Introgression is not simply a problem posed by nonnative species, but also between native species when one is of conservation concern.

Though humans intentionally create hybrids for pets and agriculture, unintentional hybridization between native and invasive animals occurs in a range of species. For instance, barred tiger salamander (*Ambystoma tigrinum mavortium*) larvae were caught in the U.S. continental interior and shipped west, outside of their native range, to be sold as fishing bait in the mid-twentieth century (Collins and Crump 2009). Fishermen released these larvae in such places as California multiple times, allowing for their introduction into new aquatic systems where they bred with the native California tiger salamanders (*Ambystoma californiense*). As a result of hybrid vigor, genetically pure California tiger salamanders may become extinct in the future (Fitzpatrick and Shaffer 2007). Furthermore, these hybrids have also been demonstrated to compete with and affect other native amphibians (Ryan et al. 2009). Likewise, mallard ducks (*Anas platyrhynchos*) are a common game and food species in much of the Northern Hemisphere and, as a result, have been introduced into a number of localities around the world. Even in native parts of their native range, mallards are highly aggressive breeders and produce hybrids with closely related species, such as northern pintails (*A. acuta*) and American black ducks (*A. rubripes*). However, mallards have been introduced far outside of their natural range, such as the Hawaiian Islands, where they were first introduced in the late 1800s followed by additional large numbers in the 1950s and 1960s for hunting (Fowler et al. 2009). Mallards readily hybridize with the endemic Hawaiian duck or koloa (*A. wyvilliana*), and now hybridization poses the greatest threat to existence of the koloa (Engilis et al. 2002; U.S. Fish and Wildlife Service 2005).

POPULATIONS

Just as populations form a core aspect of ecological study in the broad sense, the population scale has traditionally been the focal point for invasion biology research (Parker et al. 1999). In essence, when considering population-level effects of invasive terrestrial vertebrates, we consider those that impact the population dynamics of a species, particularly population growth and survival. In practice what this means is that the invasive species is either directly or indirectly influencing birth, death, immigration, or emigration. The outcome of such effects can be either reduced populations of native species or their extirpation (either locally or globally). In general, these outcomes are the result of three main types of interactions with native species, namely, predation, competition, and disease transmission.

Predation is the most commonly discussed impact of invasive species (Figure 5.2), likely due to the severe impacts seen on island systems (Lockwood et al. 2007), where predators have been either unintentionally introduced, as is the case with many rat species, or intentionally introduced for biocontrol (e.g., small Indian mongoose [*Herpestes javanicus*], cane toads [*Rhinella marina*]) or recreational hunting (e.g., Kalij pheasants [*Lophura leucomelana*; Lewin and Lewin 1984]; Lepczyk et al. 2011). Perhaps the most dramatic and devastating example of invasive species predation is the brown tree snake (*Boiga irregularis*), which has caused the near-total extirpation of Guam's avian biodiversity (Fritts and Rodda 1998). Of course, predators are a problem not only on islands, but also across continental areas (e.g., McCleery et al.

FIGURE 5.2   Invasive barn owl (*Tyto alba*) depredation of an endangered Hawaiian gallinule (*Gallinula galeata sandvicensis*). (Photo credit K. Uyehara/USFWS.)

2015). Importantly, from a broader ecological perspective, predation is a process and not simply the focus of specific taxa, such that animals that eat seeds (i.e., seed predators) are just as important to consider when discussing predation as animal predators (Janzen 1969; Silman et al. 2003). Hence, invasive rat species consume not only bird eggs and juvenile birds, but also seeds. For instance, invasive *Rattus rattus* forage upon native Hawaiian plant species, such as *Pritchardia maideniana*, supporting the hypothesis that rat species "may have assisted in the demise of native insular palm forests" (Shiels and Drake 2015). In fact, recent work has suggested that the collapse of human society on Rapa Nui (Easter Island) was largely due to seed consumption by the Polynesian rat (*Rattus exulans*), which obliterated the native forest (Hunt 2006; Hunt and Lipo 2011). Considering this broader perspective, it is easy to understand how invasive vertebrates can impact many native species simply through foraging. But, unlike native species that coevolved in the ecosystem together, native prey species may have little to no defenses against novel predators. Though with time, some native species may partially adapt to novel predators (Bailey et al. 2007; Vanderwerf 2012), most native species are unable to either persist long enough or adapt.

Competition between invasive and native species is another situation that can affect the population dynamics of native species. In general, competition operates either via interference competition, whereby individuals of one species prevent other species from accessing resources through intimidation or fighting, or resource competition, whereby two species using the same resource lower the supply, making it hard for either to meet their needs. Notably, under resource competition the two species need not ever directly interact as they may access the resource at different times, such that there is no temporal overlap. For instance, feral asses (*Equus asinus*) co-occur with bighorn sheep (*Ovis canadensis*) in the western United States, and evidence suggests that the asses are outcompeting the sheep, taking vital resources (Marshal et al. 2008). Competition can occur through a number of different specific mechanisms. Perhaps the least intuitive example might be in Florida where the Cuban tree frog (*Osteopilus septentrionalis*) competes acoustically with native green tree frogs (*Hyla cinerea*), such that the native species is modifying its calling behavior, which may ultimately affect the species fitness (Tennessen et al. 2016). Ultimately, in areas where an invasive competitor is present, the effect of competition on native species is typically a decrease in population size or extirpation.

One mechanism by which invasive vertebrates can markedly affect a native population is via the introduction of a novel disease or pathogen. When nonnative vertebrates are introduced to new ecosystems, they often house a host of microfauna, including diseases, to which the native species have no prior exposure. As a result, the novel disease can spread to naïve host species, impacting their survival and long-term population dynamics. For example, toxoplasmosis, caused by the parasite *Toxoplasma gondii*, has led to deaths in a number of native animal species (e.g., sea otters [*Enhydralutris nereis*; Kreuder et al. 2003], Hawaiian crows [*Corvus hawaiiensis*; Work et al. 2000], Hawaiian geese [*Branta sandvicensis*; Work et al. 2002], Hawaiian monk seals [*Monachus schauinslandi*; Honnold et al. 2005], and red-footed boobies [*Sula sula*; Work et al. 2002]) and requires felids, including feral cats, as definitive hosts to complete its life cycle. Hence, the movement of feral cats to new locations has inadvertently led to the introduction of toxoplasmosis to many of the world's ecosystems. Thus, cats have direct impacts, through predation, and indirect impacts as vectors of disease. Similarly, the release of bait salamanders (*A. tigrinum*), mentioned earlier, led to the spread of iridoviruses throughout the western United States (Jancovich et al. 2005; Picco and Collins 2008). Likewise, invasive wild pigs (*Sus scrofa*) harbor a variety of diseases (e.g., swine brucellosis [*Brucella suis*] which can persist in populations for long periods [Gresham et al. 2002]) that can be transmitted to native wildlife species as well as livestock.

COMMUNITY

When invasive vertebrates move into an existing community, they not only bring a set of ecological interactions with them, but inherently change the species composition, and hence the biodiversity of the community. Most simply, the addition of these invaders results in a change in species richness of the community. Often times the addition of the invader inflates the overall biodiversity of the community through an increase in species richness, at least in the short term. Because many biodiversity metrics (Magurran 2001) are based upon unique species and their relative abundances, they make no distinction on whether species are native or not. One outcome of this artifact is that invaded communities can be assessed as more diverse (which they are) and potentially pose challenges for conservation and management. In terms of both invasion biology and basic ecology, it is preferred to either only consider native species (e.g., Lepczyk et al. 2008) or divide species into guilds and classes that evaluate invasive species separately. Without such consideration, superficial biodiversity measures can be misleading and potentially affect conservation and management decisions.

Because communities are often the level in the ecological hierarchy at which we evaluate biodiversity, it is critical to consider the spatial and temporal scales of assessments. Without acknowledging these two factors, we run the risk of having a shifting baseline problem (*sensu* Pauly 1995). Though Pauly considered shifting baselines from a fisheries stock and management perspective, the principle is the same for terrestrial systems and invasive species. In essence, a shifting baseline occurs because each generation adopts the present condition in which they live as "normal." For instance, consider that someone living in the eastern third of the United States in the 1890s would have considered American chestnut (*Castanea dentate*), white ash (*Fraxinus americana*), eastern hemlock (*Tsuga canadensis*), and American elm (*Ulmus americana*) trees to be important elements of their natural environment, and passenger pigeons (*Ectopistes migratorius*) and Carolina parakeets (*Conuropsis carolinensis*) to be at least occasional visitors. Their children would have grown up not knowing either bird, and with chestnut and then elm becoming very rare. But this would have seemed normal to this second generation. Two generations later, we are on the verge of losing ash trees, and in 20 years, forests lacking all six of these once prominent tree and bird species may seem like the natural state to the people who live near them. Fighting to restore long-absent species may seem unimportant to those who never knew them. Even worse, invasive species can take their place and become part of the new, desirable "normal." Such "new normals" may also promulgate further promotion of novel, invasive ecosystems and a lack of concern over the spread of nonnative species (but we will return to that later when we discuss socioecology).

Perhaps the most intractable example of a shifting baseline is found in the Hawaiian Islands, regarding invasive wild pigs. When Polynesian settlers crisscrossed the Pacific, they brought with them all the elements they needed to establish their agricultural and cultural life on uninhabited islands. These elements included a variety of useful plants and pigs. These Polynesian pigs were relatively small, docile, and largely remained near the settlements where they were fed and provided Polynesians with a convenient source of protein. When Europeans made contact in the late 1700s, they introduced a much larger, more aggressive breed of pig from Europe. This European breed has a long history of reverting to a feral stage with significant negative impacts on native ecosystems. Perhaps because Hawaii lacked any native ungulates, in the 1950s the hunting of feral animals like pigs, goats, and deer was actively promoted (Tomich 1986; Duffy 2010). In the process, the public came to accept these animals, particularly pigs, as normal components of Hawaiian forests. Now, in addition to a widespread misconception that pigs are native to Hawaii, there is a vocal minority garnering the support of state agencies, seeking to preserve and increase the populations of invasive wild pigs across the state for recreational hunting, despite overwhelming evidence of the serious damage they cause (Figure 5.3). The original Polynesian settlers would not recognize the invasive wild pigs present in the forests today, and the experiences of Hawaii residents just 60 years ago, when invasive wild pigs were still considered pests to be eliminated, contrast sharply with the promotion and stature invasive wild pigs now enjoy. Invasive wild pigs have benefitted from a shifting baseline in public perception such that much of the public regards them as viable components of Hawaiian forests to be maintained, rather than as pests to be eliminated.

Ecosystem

When invasive vertebrates alter populations and communities of native species, they can, in turn, affect a number of ecosystem processes, such as material and energy flux and disturbance regimes (e.g., Bratton 1975). For instance, invasive wild pigs introduced into Californian coastal prairies were found to be the dominant source of soil disturbance (Kotanen 1995). Similarly, coqui frogs (*Eleutherodactylus coqui*) introduced in Hawaii may alter nutrient cycling rates (Sin et al. 2008). Beyond just processes, invasive vertebrates can influence the ecosystem through their effects on other trophic levels. Specifically, these animals can impact trophic levels through either top-down (predation) or bottom-up (resource availability) effects. In terms of invasive terrestrial vertebrates, most effects are top-down because most invasive vertebrates affect trophic levels below them through predation, herbivory, granivory, and so on. This top-down effect can be magnified in cases where a native apex predator is reduced or absent, resulting in an increase in the number of individuals (either native or invasive) of species at lower trophic levels that are released from their predators (Wallach et al. 2015; Figure 5.4), potentially resulting in a trophic cascade. One example of trophic cascade is in urban areas of California where coyotes (*Canis latrans*) were eliminated. In the absence of coyotes, mesopredators, especially cats,

FIGURE 5.3 Damage to native wiliwili (*Erythrina sandwicensis*) by invasive goats; note the bark has been stripped and consumed from the base of the trees. (Photo credit Edith Adkins.)

occurred in greater numbers, which resulted in a decrease in songbird populations (Crooks and Soulé 1999). Notably, cats, an invasive species in these systems, had more negative impacts than the native mesopredators with which the birds evolved.

Bottom-up processes can occur when factors such as food or habitat availability increase the populations of the higher-level trophic levels. For instance, in a review of invasive herpetofauna, Kraus (2015) noted that one primary way invaders affect the system is via competition that removes resources from natives. Alternatively, animals like invasive wild pigs, which root in the soil and greatly alter the physical conditions of the ecosystem, affect both the plant communities (i.e., food resource) and the habitat quality (habitat resource), which in turn influences populations of native species. For instance, in Hawaiian forests, pig wallows provide an essential resource for invasive mosquitoes to breed and transmit novel diseases to endangered forest birds (van Riper and Scott 2001).

Beyond top-down and bottom-up processes, there is also the potential for positive feedback loops when both predators and prey are invasive species. For instance,

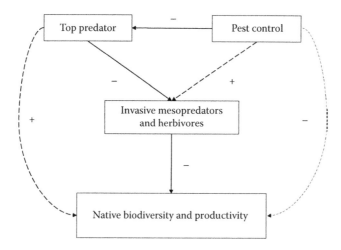

FIGURE 5.4 Theoretical model of a trophic cascade in which an invasive species negatively affects biodiversity due to control of a native top predator. (From Wallach, A. D. et al. 2010. Predator control promotes invasive dominated ecological states. *Ecology Letters* 13:1008–1018.)

recent experimental work in Hawaii has demonstrated that invasive vertebrates readily scavenge carrion of other invasive vertebrates (Abernethy et al. 2016). As a result, multitrophic levels of invasive vertebrates may result in markedly different rates of decomposition and affect other ecosystem processes.

One of the main impacts from invasive vertebrates altering an ecosystem are changes or elimination of ecosystem services. For instance, invasive ungulates can greatly alter soil erosion rates in steeper areas (i.e., areas with more relief), resulting in increased levels of particulate matter in water bodies upon which humans depend (e.g., Nogueira-Filho et al. 2009). Likewise, invasive vertebrates shed feces and diseases into these systems which become water-borne. As a result, a stream or other aquatic system may not be able to provide the same level of service in the form of clean water to people that it could without the presence of the invasive animals (e.g., Kaller et al. 2007). Another case in point is the disruption of plant–animal mutualisms, when either the plants or their animal associates are negatively impacted by invasive species, ultimately affecting services such as pollination, with downstream impacts as multiple species disappear in tandem from ecosystems (see Traveset and Richardson 2006 for a review).

The damage caused by an invasive vertebrate can be so severe as to lead to the extinction of ecosystem processes. For instance, the damage caused by the replacement of native pollinators with nonnative pollinators can lead to the loss of ecosystem processes. One dramatic example is the Hawaiian honeycreeper birds which, over millions of years, evolved a range of bill shapes, including some that were extraordinarily long and curved, to match the corollas of different flowers which increasingly relied on them for pollination. Many native birds went extinct rather suddenly around the turn of the nineteenth century due to the combined impacts of invasive mosquitoes and newly arrived invasive birds infected with, but resistant to, avian pox and avian malaria (Warner 1968; Woodworth et al. 2005). Pollination services for the native plants are now severely limited and may, in some cases, be insufficient for pollination of the remaining plants (Cox and Elmqvist 2000). More insidious is that, as some native plant species decline, the animals that have evolved to rely on them also decline. For example, insects that ate the plants are now disappearing along with their hosts (Zimmerman 1978), as well as the birds that relied on those insects for food, unraveling intricate evolutionary relationships that form the backbone of ecosystems. Over millennia, species evolve in the context of other species, building up defenses especially tailored to counter the attacks of familiar predators, or the defenses of familiar prey, or become dependent on specific resources, like particular plant species, that are available for their use. When new invaders arrive, the calculus changes, defenses that were effective against native predators or herbivores may not work on an invasive predator or grazer, and the newcomers might change the availability of resources, either through direct competition for that resource or indirectly, by outcompeting and replacing the original resource in the ecosystem. It is these downstream, indirect effects that begin a feedback loop called ecological meltdown (also referred to as an invasional meltdown), whereby one invasive species facilitates the impacts of one or more other invasives (Simberloff and Von Holle 1999). As more invasive species establish, they may operate in concert, causing the complete disintegration of the native ecosystem, even if these invaders had never been in contact before and come from different parts of the world. The combined impact exterminates suites of native species, leaving the invaded ecosystem open to ever more invaders. For example, invasive wild pigs knock down endemic tree ferns in Hawaiian rainforests to eat the starchy core, leaving a hollow for water to gather and introducing mosquitoes to breed (Lapointe et al. 2009). These mosquitoes spread avian pox and avian malaria which cause the decline of native Hawaiian honeycreepers (Atkinson and Lapointe 2009), opening opportunities for invasive birds. Furthermore, invasive wild pigs feast on the fruits of invasive strawberry guava (*Psidium cattleianum*), and then defecate the guava's seeds in areas that the pigs have cleared of native vegetation as they root through the soil. The guava replaces diverse native forest with monotypic stands that nourish the invasive wild pigs and little else. In this way, a single invader, the feral pig, provides a pathway for the invasion of multiple other invaders, which assist each other and together dismantle the native ecosystem, making it easier for additional invaders to establish. Once an invasional meltdown begins, it generates its own momentum and can be very difficult to reverse.

NOVEL ECOSYSTEMS

Today, nearly all of the world's ecosystems are impacted by humans (Vitousek et al. 1997), which has led to the establishment of a mixture of native and nonnative species in many ecosystems. When these new combinations of species are facilitated by people due to species introductions, anthropogenic land use change, and climate change, they create novel ecosystems (Hobbs et al. 2006). While the concept of a novel ecosystem has great attraction, it has had relatively little scrutiny in the literature (Murcia et al. 2014). Because novel ecosystems have promoted new discussions and directions for conservation and management (Hobbs et al. 2009), this lack of criticism can set a dangerous precedent for conserving relatively intact ecosystems and managing invasive species. Concomitant with inadequate evaluation has been a further challenge to invasion biology from some ecologists who argue that invasive species should be considered part of the ecosystems they invade, since many do not appear to have detrimental effects (Davis et al. 2011). Both novel ecosystems and the view that invasive species are not detrimental to ecosystems also get conflated with the ongoing debate over what conservation should be and what it should do (i.e., the so-called "new conservation science"; Kareiva and Marvier 2012). Ultimately, while these debates pose interesting academic discussions, at their heart is an acceptance of nonnative species, including invasive species, which can redirect resources away from addressing pressing problems regarding the control of invasives.

THREATENED AND ENDANGERED SPECIES

Invasive species, including many terrestrial vertebrates, contribute greatly to species endangerment in the United States (Wilcove et al. 1998). While predation and competition are often noted as contributory factors for endangerment, nearly every mechanism presented from ecosystem meltdown to mesopredator extinction has been cited as well. Locations such as California, Florida, and Hawaii, which house a large fraction of endangered species (Dobson et al. 1997; Rutledge et al. 2001), are also areas that house notable invasive vertebrates. For instance, cats in Florida are one of the main reasons why the Lower Keys rabbit (*Sylvilagus palustris hefneri*) was listed as endangered (Federal Register 1990).

CLIMATE CHANGE

Climate change is already having notable impacts on both native and nonnative species. Not only are some species moving poleward (La Sorte and Thompson 2007), many are shifting their ranges up in elevation, or in response to changing temperature and precipitation regimes. Because groups of species do not operate as single units, but rather move independently across the landscape, there is already a reshuffling of species as ranges shift in what is considered no-analog futures. That is, each species is moving in different directions and at different rates, such that new locations do not support all of the same species in the community as before, resulting in assemblages of species with which we have no corresponding example and are thus novel in the sense of having no analogs with which to compare (Stralberg et al. 2009). In fact, while climate change may exacerbate the damage from current invasive species, it could cause others to decline or have diminished impacts, and allow other, previously noninvasive, nonnative, species to become invasive (Hellmann et al. 2008). Thus, climate change is likely to have profound, but unpredictable effects on invasive species distributions and impacts.

Much of the research on climate change in relation to invasive species has focused on niche or species distribution models that seek to determine where a species is likely to be found now and in the future based on environmental factors (Peterson 2003). More broadly, however, models project that as climate changes, there are

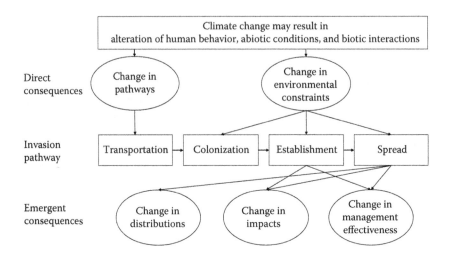

FIGURE 5.5 The potential consequences of climate change on invasive species. (Adapted from Hellmann, J. J. et al. 2008. Five potential consequences of climate change for invasive species. *Conservation Biology* 22:534–543.)

five possible consequences for invasive species (Figure 5.5; Hellmann et al. 2008). First, climate change may lead to alteration of how invasive species are transported and introduced to new locations on the planet. Second, climate change may alter the biotic and abiotic environmental conditions that constrain an invasive species. Third, as previously noted, invasive species distributions may be changed. Fourth, the impacts of invasive species may change. As a result of these four possible scenarios, the ultimate consequence is a potential for changes in the effectiveness of management for controlling invasive species.

While a great deal of uncertainty lies ahead in terms of what climate change will mean for invasion biology, the reality is that invasions will continue to occur and may challenge our current approaches to restoration and management. The unpredictability of climate change makes planning for future invasive species difficult and uncertain. The rapidity with which climate changes are already affecting ecosystems suggests there may not be much time to act to mitigate future invasive species invasions and damage. In the short term, we still can practice proactive management and quarantine, and these tools may become ever more important as climate change alters the dynamics of invasion biology.

SOCIOECOLOGY

The socioecology of invasive species is an increasingly important subfield of invasion biology. Because invasive species, by definition, are the result of a human-mediated problem, we are also intrinsic to the resolution of the problem through quarantine and control measures. Furthermore, as Aldo Leopold noted in relation to native white-tailed deer, "The real problem is one of human management. Wild life management is comparatively easy; human management is difficult" (Leopold 1943). The point here is that developing and implementing invasive species management and control requires understanding and including the stakeholders involved. Often times there may even be support for a particular management option once the general public is included in the process, as was the case with feral cats where there were clear views on the need to manage them and using lethal control was acceptable (Lohr and Lepczyk 2014; Lohr et al. 2014). Notably, however, this does not mean that addressing invasive species problems offers the luxury of taking more time.

It has become increasingly clear that no conservation plan or invasive species control effort anywhere, even in the national parks, can succeed without the support of the local community. Perhaps more than any other science, invasion biology depends on an informed and supportive public to fund invasive species research, control efforts, and legislation to improve quarantine and importation restrictions. Yet, the public's definition of "nature" and what is perceived as a threat to the environment may not be congruent with the best scientific information. This incongruence can lead not only to apathy regarding the removal of invasive species, but also to active opposition based on the perception that some charismatic invasive (e.g., feral cats, wild horses) species belong in the environment. Coupled with the problem of shifting baselines, it thus can be challenging to discuss what constitutes a natural or desirable condition for the environment. When such challenges occur, former allies in the fight against invasive species become adversaries intent on preserving degraded, but familiar, ecosystems or particular invasive species that now infest them.

The longer invasive species such as wild horses remain part of the landscape, the more entrenched their place as part of the natural world will become, and the harder it will be to garner support for their replacement with now-unfamiliar natives. One of the most striking examples of this change in perception is the state bird of South Dakota, the ring-necked pheasant (*Phasianus colchicus*). South Dakota, home to a number of native grouse species, instead has a species native to Asia and introduced across North America and Hawaii as the bird representative of their state. This belies a public psyche that identifies more closely with a widespread invasive game bird than with their endemic fauna. Convincing such a public to prioritize nature restoration that might not favor a prized game bird suggests just one of the challenges shifting baselines present to invasive species.

The fight against shifting baselines and the diluted perception of what native ecosystems should look like must be fought on two fronts. Not only is restoration of the original fauna essential to demonstrate to the public the difference in species richness and composition as native ecosystems recover, but also the public must accept a less familiar-looking image of nature, one which lacks the invasive trees, feral horses, or pigs to which they have become accustomed. Education is the only realistic way to bring about this change in values. However, getting a public to see beyond their shifted baseline of invasive species and support restoration is something that has very rarely been accomplished to date.

CONCLUSION

Invasive terrestrial vertebrates are neither a new phenomenon nor problem (e.g., Elton 1958), yet the field of invasion biology as a whole still lacks the maturity of other ecological sciences (Simberloff, pers. comm.). Thus, today many invasive terrestrial vertebrates are inadequately studied (e.g., Lohr and Lepczyk 2010). Given the known impacts that invasive species have on other species, ecosystems, and humans, this lack of knowledge is somewhat surprising. Hence, when Kraus (2015) noted that disease transmission "is likely the most widely damaging impact of any invasive reptile or amphibian, but pathogen spillover beyond these few examples remains unstudied," we should perhaps not be surprised. However, while there is ample evidence demonstrating the ecological impacts of invasives, the lack of controlled experiments, new theories and syntheses, and quite simply the evaluation of many species create challenges for effective conservation and management in a world of novel ecosystems.

The damage caused by invasive vertebrate species to both natural and agricultural ecosystems is irrefutable. These losses will only increase as additional species are introduced, and those already present exert ever-increasing pressure on native species, threatened or otherwise. Encouragingly, it is possible to control invasive vertebrates, given public will. In fact, there are many examples of rare native species and ecosystems recovering after invasive vertebrates have been eliminated (e.g., removing goats from Hawaii Volcanoes National Park; Hess and Jacobi 2011). A new, ominous movement poses a significant challenge to future success of eradication efforts as there are those who argue that nonnative species be considered native or pose no significant impact and thus be considered part of the natural ecosystem (e.g., Davis et al. 2011). A component of this argument stems from the idea that invasive species may benefit some native species. However, resources in virtually all ecosystems are limited. Thus, when invasive species are brought in, resource allocation changes and native species have to make do with less. Any position which argues that invasive species be considered "native" must, necessarily, be devaluing native ecosystems and the species that inhabit them. This view treats the millennia of evolution which brought about the complex interactions in native ecosystems as equal to the sudden amalgamation of invasives brought about by human activity in the past few hundred years. If we cannot see a difference between these two processes, we are doomed to live in a simplified, homogeneous world in which a few widespread species dominate, and ecosystems provide fewer and less-diverse resources to both us and the animals with whom we share the planet. Once we make the decision to let native species disappear and ecosystems fall apart, there is no simple way to reconstitute them; it is a one-way and final decision.

REFERENCES

Abernethy, E. F., K. L. Turner, J. C. Beasley, T. L. DeVault, W. C. Pitt, and O. E. Rhodes, Jr. 2016. Carcasses of invasive species are predominantly utilized by invasive scavengers in an island ecosystem. *Ecosphere*, 7(10): e01496.

Atkinson, C. T., and D. A. Lapointe. 2009. Ecology and pathogenicity of avian malaria and pox. In Pratt, T. K., Atkinson, C. T., Banko, P. C., Jacobi, J. D., and Woodworth, B. L. (eds.). *Conservation Biology of Hawaiian Forest Birds*. Yale University Press, New Haven, pp. 234–252.

Bailey, J. K., J. A. Schweitzer, B. J. Rehill, D. J. Irschick, T. G. Whitham, and R. L. Lindroth. 2007. Rapid shifts in the chemical composition of aspen forests: An introduced herbivore as an agent of natural selection. *Biological Invasions* 9: 715–722.

Barrios-Garcia, M. N., and S. A. Ballari. 2012. Impact of wild boar (*Sus scrofa*) in its intro-duced and native range: A review. *Biological Invasions* 14: 2283–2300.

Bratton, S. P. 1975. The effect of the European wild boar, *Sus scrofa*, on gray beech forest in the Great Smoky Mountains. *Ecology* 56: 1356–1366.

Collins, J. P., and M. L. Crump. 2009. *Extinction in Our Times: Global Amphibian Decline.* Oxford University Press, Oxford, UK.

Cox, P. A., and T. Elmqvist. 2000. Pollinator extinction in the Pacific Islands. *Conservation Biology* 14: 1237–1239.

Crooks, K. R., and M. E. Soulé. 1999. Mesopredator release and avifaunal extinctions in a fragmented system. *Nature* 400: 563–566.

Cuddington, K., and A. Hastings. 2004. Invasive engineers. *Ecological Modelling* 178: 335–347.

Davis, M. A., M. K. Chew, R. J. Hobbs, A. E. Lugo, J. J. Ewel, G. J. Vermeij, J. H. Brown et al. 2011. Don't judge species on their origins. *Nature* 474: 153–154.

Dobson, A. P., J. P. Rodriquez, W. M. Roberts, and D. S. Wilcove. 1997. Geographic distribu-tion of endangered species in the United States. *Science* 275: 550–553.

Dorcas, M. E., J. D. Willson, R. N. Reed, R. W. Snow, M. R. Rochford, M. A. Miller, W. E. Meshaka, Jr. et al. 2012. Severe mammal declines coincide with proliferation of invasive Burmese pythons in Everglades National Park. *Proceedings of the National Academy of Sciences* 109: 2418–2422.

Duffy, D. J. 2010. An historical analysis of hunting in Hawai'i. Thesis. University of Hawai'i at Mānoa.

Elton, C. 1958. *The Ecology of Invasions by Animals and Plants.* Methuen, London.

Engilis, A. Jr, K. J. Uyehara, and J. G. Giffin. 2002. Hawaiian duck (*Anas wyvilliana*). In Poole, A., and Gill, F. (eds.). *The Birds of North America, No. 694.* The Birds of North America, Inc., Philadelphia.

Federal Register. 1990. Endangered and threatened wildlife and plants; endangered status for the Lower Keys rabbit and threatened status for the Squirrel Chimney cave shrimp. *Federal Register* 55: 25588–25591.

Fitzpatrick, B. M., and H. B. Shaffer. 2007. Hybrid vigor between native and introduced sala-manders raises new challenges for conservation. *Proceedings of the National Academy of Sciences* 104: 15793–15798.

Fowler, A. C., J. M. Eadie, and A. Engilis Jr. 2009. Identification of endangered Hawaiian ducks (*Anas wyvilliana*), introduced North American mallards (*A. platyrhynchos*) and their hybrids using multilocus genotypes. *Conservation Genetics* 10: 1747–1758.

Fritts, T. H., and G. H. Rodda. 1998. The role of introduced species in the degradation of island ecosystems: A case history of Guam. *Annual Review of Ecology and Systematics* 29: 113–140.

Gresham, C. S., C. A. Gresham, M. J. Duffy, C. T. Faulkner, and S. Patton. 2002. Increased prevalence of *Brucella suis* and pseudorabies virus antibodies in adults of an isolated feral swine population in coastal South Carolina. *Journal of Wildlife Diseases* 38: 653–656.

Hellmann, J. J., J. E. Byers, B. G. Bierwagen, and J. S. Dukes. 2008. Five potential conse-quences of climate change for invasive species. *Conservation Biology* 22: 534–543.

Hess, S. C., and J. D. Jacobi. 2011. The history of mammal eradications in Hawai'i and the United States associated islands of the central Pacific. In Veitch, C. R., Clout, M. N., and Towns, D. R. (eds.), *Island Invasives: Eradication and Management.* IUCN, Gland, Switzerland, pp. 67–73.

Hobbs, R. J., S. Arico, J. Aronson, J. S. Baron, P. Bridgewater, V. A. Cramer, P. R. Epstein et al. 2006. Novel ecosystems: Theoretical and management aspects of the new ecologi-cal world order. *Global Ecology and Biogeography* 15: 1–7.

Hobbs, R. J., E. Higgs, and J. A. Harris. 2009. Novel ecosystems: Implications for conservation and restoration. *Trends in Ecology and Evolution* 24: 599–605.

Honnold, S. P., R. Braun, D. P. Scott, C. Sreekumar, and J. P. Dubey. 2005. Toxoplasmosis in a Hawaiian monk seal (*Monachus schauinslandi*). *Journal of Parasitology* 91: 695–697.

Hunt, T. L. 2006. Rethinking the fall of Easter Island. *American Scientist* 94: 412–419.

Hunt, T., and C. Lipo. 2011. *The Statues That Walked: Unraveling the Mystery of Easter Island.* New York, Free Press.

Jancovich, J. K., E. W. Davidson, N. Parameswaran, J. Mao, V. G. Chinchar, J. P. Collins, B. L. Jacobs, et al. 2005. Evidence for emergence of an amphibian iridoviral disease because of human-enhanced spread. *Molecular Ecology* 14: 213–224.

Janzen, D. H. 1969. Seed-eaters versus seed size, number, toxicity and dispersal. *Evolution* 23: 1–27.

Kaller, M. D., J. D. Hudson III, E. C. Achberger, and W. E. Kelso. 2007. Feral hog research in western Louisiana: Expanding populations and unforeseen consequences. *Human–Wildlife Conflicts* 1: 168–177.

Kareiva, P., and M. Marvier. 2012. What is conservation science? *BioScience* 62: 962–969.

Kotanen, P. M. 1995. Responses of vegetation to a changing regime of disturbance: Effects of feral pigs in a Californian coastal prairie. *Ecography* 18: 190–199.

Kraus, F. 2015. Impacts from invasive reptiles and amphibians. *Annual Review of Ecology, Evolution, and Systematics* 46: 75–97.

Kreuder, C., M. A. Miller, D. A. Jessup, L. J. Lowenstine, M. D. Harris, J. A. Ames, T. E. Carpenter et al. 2003. Patterns of mortality in southern sea otters (*Enhydra lutris nereis*) from 1998–2001. *Journal of Wildlife Diseases* 39: 495–509.

Lapointe, D. A., C. T. Atkinson, and S. I. Jarvi. 2009. Managing disease. In Pratt, T. K., Atkinson, C. T., Banko, P. C., Jacobi, J. D., and Woodworth, B. L. (eds.). *Conservation Biology of Hawaiian Forest Birds.* Yale University Press, New Haven, pp. 405–242.

La Sorte, F. A., and F. R. Thompson. 2007. Poleward shifts in winter ranges of North American birds. *Ecology* 88: 1803–1812.

Leopold, A. 1943. Prejudice or science: That is the issue. *Milwaukee Journal.* Accessed April 4.

Lepczyk, C. A., C. H. Flather, V. C. Radeloff, A. M. Pidgeon, R. B. Hammer, and J. Liu. 2008. Human impacts on regional avian diversity and abundance. *Conservation Biology* 22: 405–446.

Lepczyk, C. A., S. C. Hess, and E. D. Johnson. 2011. Hawaii and the North American model of wildlife conservation: One size fits all? *The Wildlife Professional Fall* 2011: 64–66.

Lepczyk, C. A., A. G. Mertig, and J. Liu. 2004. Landowners and cat predation across rural-to-urban landscapes. *Biological Conservation* 115: 191–201.

Lewin, V., and G. Lewin. 1984. The kalij pheasant, a newly established game bird on the Island of Hawaii. *The Wilson Bulletin* 96: 634–646.

Linklater, W. L., K. J. Stafford, E. O. Minot, and E. Z. Cameron. 2002. Researching feral horse ecology and behavior: Turning political debate into opportunity. *Wildlife Society Bulletin* 30: 644–650.

Lockwood, J. L., M. F. Hoopes, and M. P. Marchetti. 2007. *Invasion Ecology.* Blackwell Scientific Press, UK.

Lohr, C. A., and C. A. Lepczyk. 2010. The impacts of introduced mammals in the Hawaiian Islands: A literature review. *Proceedings of the 24th Vertebrate Pest Conference* Timm, R. M., and Fagerstone, K.A. (eds.), pp. 32–40. University of California, Davis.

Lohr, C. A., and C. A. Lepczyk. 2014. Desires and management preferences of stakeholders regarding feral cats in the Hawaiian Islands. *Conservation Biology* 28: 392–403.

Lohr, C. A., C. A. Lepczyk, and L. J. Cox. 2014. Identifying people's most preferred management technique for feral cats in Hawaii. *Human–Wildlife Interactions* 8: 56–66.

Loss, S. R., T. Will, and P. P. Marra. 2013. The impact of free-ranging domestic cats on wildlife of the United States. *Nature Communications* 4: 1396.

Magurran, A. E. 2001. *Measuring Biological Diversity.* Wiley-Blackwell, Malden, MA, USA.

Marshal, J. P., V. C. Bleich, and N. G. Andrew. 2008. Evidence for interspecific competition between feral ass *Equus asinus* and mountain sheep *Ovis canadensis* in a desert environment. *Wildlife Biology* 14: 228–236.

McCleery, R. A., A. Sovie, R. N. Reed, M. W. Cunningham, M. E. Hunter, and K. M. Hart. 2015. Marsh rabbit mortalities tie pythons to the precipitous decline of mammals in the Everglades. *Proceedings of the Royal Society B* 282: 20150120

Murcia, C., J. Aronson, G. H. Kattan, D. Moreno-Mateos, K. Dixon, and D. Simberloff. 2014. A critique of the 'novel ecosystem' concept. *Trends in Ecology and Evolution* 29: 548–553.

Nogueira-Filho, S. L. G., S. S. C. Nogueira, and J. M. V. Fragoso. 2009. Ecological impacts of feral pigs in the Hawaiian Islands. *Biodiversity and Conservation* 18: 3677–3683.

Parker, I. M., D. Simberloff, W. M. Lonsdale, K. Goodell, M. Wonham, P. M. Kareiva, M. H. Williamson et al. 1999. Impact: Toward a framework for understanding the ecological effects of invaders. *Biological Invasions* 1: 3–19.

Pauly, D. 1995. Anecdotes and the shifting baseline syndrome of fisheries. *Trends in Ecology and Evolution* 10: 430.

Peterson, A. T. 2003. Predicting the geography of species' invasions via ecological niche modeling. *Quarterly Review of Biology* 78: 419–433.

Picco, A. M., and J. P. Collins. 2008. Amphibian commerce as a likely source of pathogen pollution. *Conservation Biology* 22: 1582–1589.

Pietrek, A. G., and M. González-Roglich. 2015. Post-establishment changes in habitat selection by an invasive species: Beavers in the Patagonian steppe. *Biological Invasions* 17: 3225–3235.

Pimentel, D., R. Zuniga, and D. Morrison. 2005. Update on the environmental and economic costs associated with alien-invasive species in the United States. *Ecological Economics* 52: 273–288.

Priddel, D., N. Carlile, M. Humphrey, S. Fellenberg, and D. Hiscox. 2003. Rediscovery of the 'extinct' Lord Howe Island stickinsect (*Dryococelus australis* (Montrouzier)) (Phasmatodea) and recommendations for its conservation. *Biodiversity and Conservation* 12: 1391–1403.

Rhymer, J. M., and D. Simberloff. 1996. Extinction by hybridization and introgression. *Annual Review of Ecology and Systematics* 27: 83–109.

Rutledge, D. T., C. A. Lepczyk, J. Xie, and J. Liu. 2001. Spatiotemporal dynamics of endangered species hotspots in the United States. *Conservation Biology* 15: 475–487.

Ryan, M. E., J. R. Johnson, and B. M. Fitzpatrick. 2009. Invasive hybrid tiger salamander genotypes impact native amphibians. *Proceedings of the National Academy of Sciences* 106: 11166–11171.

Shiels, A. B., and D. R. Drake. 2015. Barriers to seed and seedling survival of once-common Hawaiian palms: The role of invasive rats and ungulates. *AoB Plants* 7: plv057.

Silman, M. R., J. W. Terborgh, and R. A. Kiltie. 2003. Population regulation of a dominant-rain forest tree by a major seed-predator. *Ecology* 84: 431–438.

Simberloff, D. 2013. *Invasive Species: What Everyone Needs to Know.* Oxford University Press, New York, NY.

Simberloff, D., and B. Von Holle. 1999. Positive interactions of nonindigenous species: Invasional meltdown? *Biological Invasions* 1: 21–32.

Sin, H., K. H. Beard, and W. C. Pitt. 2008. An invasive frog, *Eleutherodactylus coqui*, increases new leaf production and leaf litter decomposition rates through nutrient cycling in Hawaii. *Biological Invasions* 10: 335–345.

Stolzenburg, W. 2011. *Rat Island: Predators in Paradise and the World's Greatest Wildlife Rescue.* Bloomsbury Publishing, New York.

Stralberg, D., D. Jongsomjit, C. A. Howell, M. A. Snyder, J. D. Alexander, J. A. Wiens, and T. L. Root. 2009. Re-shuffling of species with climate disruption: A no-analog future for California birds? *PLoS ONE* 4(9): e6825.

Tennessen, J. B., S. E. Parks, T. P. Tennessen and T. Langkilde. 2016. Raising a racket: Invasive species compete acoustically with native treefrogs. *Animal Behaviour* 114: 53–61.

The World Conservation Union. 2014. IUCN Red List of Threatened Species 2014.3. Summary Statistics for Globally Threatened Species. Table 1: Numbers of threatened species by major groups of organisms (1996–2014).

Tomich, P. Q. 1986. *Mammals in Hawai'i*, Second Edition. Bishop Museum Press, Honolulu, HI.

Traveset, A., and D. M. Richardson. 2006. Biological invasions as disruptors of plant reproductive mutualisms. *Trends in Ecology and Evolution* 21: 208–216.

U. S. Fish and Wildlife Service. 2005. *Draft Revised Recovery Plan for Hawaiian Waterbirds, 2nd Draft of the 2nd Revision.* U.S. Fish and Wildlife Service, Portland, OR, pp. 155.

Vanderwerf, E. A. 2012. Evolution of nesting height in an endangered Hawaiian forest bird in response to a non-native predator. *Conservation Biology* 26: 905–911.

van Riper, C. I., and J. M. Scott. 2001. Limiting factors affecting Hawaiian native birds. *Studies in Avian Biology* 22: 221–233.

Vitousek, P. M., H. A. Mooney, J. Lubchenco, and J. M. Melillo. 1997. Human domination of Earth's Ecosystems. *Science* 277: 494–499.

Wallach, A. D., C. N. Johnson, E. G. Ritchie, and A. J. O'Neill. 2010. Predator control promotes invasive dominated ecological states. *Ecology Letters* 13: 1008–1018.

Wallach, A. D., W. J. Ripple, and S. P. Carroll. 2015. Novel trophic cascades: Apex predators enable coexistence. *Trends in Ecology & Evolution* 30: 146–153.

Warner, R. E. 1968. The role of introduced diseases in the extinction of the endemic Hawaiian avifauna. *The Condor* 70: 101–120.

Wilcove, D. S., D. Rothstein, J. Dubow, A. Phillips, and E. Losos. 1998. Quantifying threats to imperiled species in the United States. *Bioscience* 48: 607–615.

Wolf, D. E., N. Takebayashi, and L. H. Riesberg. 2001. Predicting the risk of extinction through hybridization. *Conservation Biology* 15: 1039–1053.

Woodworth, B. L., C. T. Atkinson, D. A. LaPointe, P. J. Hart, C. S. Spiegel, E. J. Tweed, C. Henneman et al. 2005. Host population persistence in the face of introduced vector-borne diseases: Hawaii amakihi and avian malaria. *Proceedings of the National Academy of Sciences of the United States of America* 102: 1531–1536.

Work, T. M., J. G. Massey, D. S. Lindsay, and J. P. Dubey. 2002. Toxoplasmosis in three species of native and introduced Hawaiian birds. *Journal of Parasitology* 88: 1040–1042.

Work, T. M., J. G. Massey, B. A. Rideout, C. H. Gardiner, D. B. Ledig, O. C. H. Kwok, and J. P. Dubey. 2000. Fatal toxoplasmosis in free-ranging endangered 'Alala from Hawaii. *Journal of Wildlife Disease* 36: 205–212.

Zimmerman, E. C. 1978. *Insects of Hawaii. A Manual of the Insects of the Hawaiian Islands, Including an Enumeration of the Species and Notes on Their Origin, Distribution, Hosts, Parasites, etc. Microlepidoptera. Part II. Gelechioidea.* Vol. 9. University Press of Hawaii, Honolulu, pp. 883–1903.

6 Feeding a Hungry World
Threats to Agriculture by Invasive Vertebrate Species in the United States

Robert J. Lewis and Michael R. Conover

CONTENTS

INTRODUCTION

As the globalization of the world's economy continues, increasing numbers of species will be introduced to North America. While most introduced species do not flourish and become injurious, those that do often pose direct threats to agriculture. There are 165 million hectares of cropland and 248 million hectares of land dedicated to livestock production in the United States (U.S. Department of Agriculture, Economic Research Service [USDA ERS] 2007). Combined, this farmland not only produces foodstuffs to feed the 321 million people in the United States, but also helps feed the other 7 billion people inhabiting the earth. Some invasive species were purposefully released in the United States, but most were accidentally introduced. This chapter discusses various threats to agriculture and highlights the major invasive species. All dollar values presented in this chapter have been adjusted for inflation to year 2015 U.S. dollars.

PREDATION

In 2012, livestock production in the United States was valued at $182 billion (United States Department of Agriculture, National Agricultural Statistics Service [USDA NASS] 2015). Predation by both native and invasive species of wildlife poses a direct threat to the production of livestock. Feral swine (*Sus scrofa*), feral dogs (*Canis famil-iaris*), and feral cats (*Felis domesticus*) are known to prey upon livestock including goats (*Capra hircus*), sheep (*Ovis aries*), calves (*Bos taurus*), and poultry, in addition to unconventional livestock including such species as white-tailed deer (*Odocoileus virginianus*), llamas (*Lama glama*), and ostriches (*Struthio camelus*; Bergman et al. 2009). In 1993, feral swine caused over $100,000 in damage through depredation of goats and sheep in Texas (Seward et al. 2004).

In a review of cattle losses to carnivores, USDA Veterinary Services found that 12%–17% of cattle losses were due to feral dogs (Bergman et al. 2009). In 2010, dogs killed 21,800 head of cattle for an estimated loss of $11 million, or 9.9% of all U.S. cattle predator losses incurred in that year (USDA NASS 2011). During 1999, dogs killed 41,300 head of sheep (15% of all sheep predator losses) for estimated damages of $3 million to sheep producers. Dogs also killed 10,700 head of goats in the states Arizona, New Mexico, and Texas in 1999, accounting for 17% of all goat losses for an estimated value of $850,000 (USDA NASS 2000).

Feral cats will kill poultry, especially in free-range production operations. In the year 2000, USDA Wildlife Services calculated that it had received reports of feral cats causing $75,000 in damages in the past eight years. This figure only reflects damage that was reported to Wildlife Services by people requesting assistance and thus is likely much lower than the true value of damage inflicted by feral cats (Bergman et al. 2000).

CROP DAMAGE

In 2012, crops represented the largest monetary component of agriculture in the United States, valued at $212.4 billion (USDA NASS 2015). Losses due to invasive species, though minor when compared to the amount of damage that native species such as white-tailed deer and blackbirds (Icteridae) cause, are not an insignificant cause of damage (USDA NASS 2002). Invasive species may damage crops through direct consumption or by rendering them unfit for sale through one of several means.

Feral swine damage agricultural crops by rooting, consuming, and trampling crops (Seward et al. 2004). Pimentel et al. (2004) estimated that feral swine cause $1.7 billion in the United States each year.

Like coyotes (*Canis latrans*), feral dogs have been known to damage crops. They will eat melons and corn and they have also been reported as having damaged irriga-tion systems by chewing on them (Bergman et al. 2009). There are no estimates of the amount of damage that feral dogs do to crops each year.

Starlings (*Sturnus vulgaris*) are known to damage virtually all types of crops by pecking (Linz et al. 2007). In one instance, starling damage resulted in the loss of 17% of a cherry crop (Linz et al. 2007). Damage to grapes was estimated to be $5.5 million in 1972 (Linz et al. 2007). Starlings may also damage blue-berry orchards when they form large juvenile flocks (Conover and Dolbeer 2007). Starlings will also damage grain crops. This damage usually occurs near large winter roosts where starlings number in the thousands (White et al. 1985; Linz et al. 2007).

FEEDLOT DAMAGE

Starlings impose a large economic burden on livestock producers when they consume feed intended for livestock. Starlings will consume grain, protein pellets, and higher energy components from livestock rations, thereby reducing the quality of the diet that the livestock receive (USDA Animal Plant Health Inspection Service [USDA APHIS] 2000). During 1999, three feedlot operators reported combined losses of $856,000 due to starlings. Feedlot operators in Idaho have reported starlings consuming 13,600–18,100 kg of feed per day (Linz et al. 2007). White et al. (1985) found that starlings spent 95% of their time in Tennessee cattle and hog feedlots where they consumed an estimated average of 36,000 kg of corn per winter, or 0.2% of all corn fed to livestock.

STORED FOODS

Once crops have been harvested, they must be stored for a variable period of time prior to sale, distribution, or consumption. During this time, they are susceptible to consumption or contamination by several different invasive species.

Wachtel and McNeely (1985) estimated that there is one rat (*Rattus* spp.) for every person in urban and suburban areas in the United States, or approximately 300 million rats. Each rat will consume 9–18 kg of foodstuffs per year and can potentially contaminate up to 10 times that amount with feces and urine, rendering it unfit for human consumption (Timm 1994). Rat damage to foodstuffs and property exceeds $30 billion; however, the percentage of damage that is food is unknown (Pimentel et al. 2004). House mice (*Mus musculus*) also consume and contaminate stored foodstuffs, but little is known regarding the extent of agricultural damage that house mice cause in the United States.

Feral pigeons (*Columba livia*) are known to roost and nest in areas where grain is stored. Pigeons contaminate much larger quantities of grain than they consume when they roost or loaf above grain stored in storage facilities (Smith 1994). Pimentel et al. (2004) used the costs of controlling pigeons ($1.4 billion per year) as a proxy for the amount of damage caused by pigeons in the United States. Bergman et al. (2000) found that USDA Wildlife Services received reports of pigeon damage that averaged $2.2 million yearly. These only represent damages reported to USDA Wildlife Services and are thus lower than the true damage experienced.

House sparrows (*Passer domesticus*) consume grains, both in the field and in storage. Because house sparrows form flocks outside of the breeding season, damage can be intense in certain areas (Fitzwater 1994). Like pigeons, house sparrows commonly roost and nest in superstructures above stored grain and contaminate it with their feces. House sparrows commonly consume and contaminate poultry feed (Fitzwater 1994).

RANGE DAMAGE

There are currently 67,000 feral horses (*Equus caballus*) and burros (*Equus asinus*) in the western United States (Bureau of Land Management [BLM] 2016). McKnight

(1958) documented that feral burros competed with livestock for limited water and grazing; Miller (1983) documented that feral horses compete with cattle for limited water in southwest Wyoming. Krysl et al. (1984) found a high degree of dietary and spatial overlap between feral horses and cattle, suggesting the potential for direct competition for limited resources in areas where they coexist. In many areas of the western United States, rangeland is easily overgrazed by feral horses when they are in large numbers (Davies et al. 2014). This overgrazing harms the land for all species.

AGRICULTURAL DISEASE

Many wildlife species serve as reservoirs and vectors for pathogens that impact agricultural production. Fortunately, most wildlife species avoid close association with humans and livestock, but this is not true of many invasive vertebrate species. The latter are frequently in close proximity to humans and livestock, and this enhances the risk of pathogens spreading between livestock, invasive species, and wildlife. Once these pathogens become established in wildlife populations, they are virtually impossible to eradicate (Conover and Vail 2015).

As of 2011, all domestic swine herds in the United States were free of pseudorabies, and all domestic swine herds, except some in Texas, were brucellosis free (USDA APHIS 2009, 2011). These diseases are still endemic in feral swine populations in the United States. Wyckoff et al. (2009) found 24% of feral swine in eastern Texas had been exposed to *Brucella* spp., the pathogens causing brucellosis, and 36% of feral swine in southern Texas had been exposed to the pathogens causing pseudorabies (*Suid herpesvirus 1*). With feral swine populations serving as a reservoir for these diseases, there is a risk that these diseases will be transmitted to domestic swine herds, as many small producers have limited or no biosecurity measures in place (Wyckoff et al. 2009). In 2013, 112 million domestic hogs were slaughtered in the United States for an estimated export value of approximately $6 billion (National Pork Producers Council 2013). If the pathogens for pseudorabies or brucellosis were to be reintroduced to domestic swine herds, the pork industries of those respective states would be seriously impacted through embargos on exportation of swine and swine products. Feral swine have also been implicated in *E. coli* (*Escherichia coli*) contamination of crops (Jay et al. 2007; Jay and Wiscomb 2008). One particular contamination event resulted in widespread recalls of baby spinach, the hospitalization of 102 people, and three deaths (Centers for Disease Control and Prevention 2006).

Feral dogs are one of the species that carry the protozoan that causes neosporosis (*Neospora caninum*); this pathogen can complete its sexual life cycle in canids (dogs, gray wolves, and coyotes in the US) and shed oocysts in fecal matter. Neosporosis can cause calf abortion (Anderson et al. 2000). Other species that may serve as definitive hosts for the pathogen causing neosporosis include coyotes and gray wolves (*Canis lupus*; Gondim et al. 2003; Dubey et al. 2011).

Starlings have the potential to transmit pathogens that cause many diseases—including salmonellosis (*Salmonella* spp.), paratuberculosis (*Mycobacterium avium paratuberculosis*), vibriosis (*Vibrio* spp.), and chlamydiosis (*Chlamydophila* spp.)—to livestock (Linz et al. 2007). Paratuberculosis, also known as Johne's disease, costs cattle producers an estimated \$285–\$357 million per year in the United States (Ott et al. 1999).

CONCLUSION

The incidence of invasive species being introduced to new areas will increase in the future as the world becomes more interconnected through commerce and travel. As we have seen in the United States, invasive vertebrate species can cause great economic losses to food production. These impacts posed by invasive species are worrisome and, in some instances, frightening. These negative impacts should be cause for concern over the spread of exotic and potentially invasive species, and certainly warrant consideration when importing exotic species for the pet trade, research, or biological control; these concerns should be kept in mind when drafting legislation regarding inspecting cargo and the devices that convey it, as well as when disposing of unwanted animals. We should not accept invasive species as the new normal, nor should we allow them to continue to be propagated. Diligence must be exercised whenever there is the potential for an exotic species to be introduced and released. It is unlikely that we will suffer the full consequences of increasing numbers of invasive species, but rather our children and their children will shoulder that burden.

REFERENCES

Anderson, M.L., A.G. Andrianarivo, and P.A. Conrad. 2000. Neosporosis in cattle. *Animal Reproduction Science* 60:417–431.

Bergman, D., S. Breck, and S. Bender. 2009. Dogs gone wild: Feral dog damage in the United States. *Proceedings of the Wildlife Damage Management Conference*, Saratoga Springs, New York, USA, vol. 13, pp. 177–183.

Bergman, D.L., M.D. Chandler, and A. Locklear. 2000. The economic impact of invasive species to Wildlife Services' cooperators. *Proceedings of the Third National Wildlife Research Center Special Symposium: Human Conflicts with Wildlife: Economic Considerations.* August 1–3, 2000; Fort Collins, Colorado, USA, pp. 169–178.

Bureau of Land Management (BLM). 2016. Wild horses and burros on public rangelands now 2.5 times greater than 1971 when protection law was passed. Press Release. U.S. Department of the Interior, Bureau of Land Management, Washington DC, USA.

Centers for Disease Control and Prevention. 2006. Multistate outbreak of *E. coli* O157:H7 infections linked to fresh spinach (final update). http://www.cdc.gov/ecoli/2006/spinach-10-2006.html. Accessed November 14, 2016.

Conover, M.R., and R.A Dolbeer. 2007. Use of decoy traps to protect blueberries from juvenile European starlings. *Human–Wildlife Conflicts* 1:265–270.

Conover, M.R., and R.M. Vail. 2015. *Human Diseases from Wildlife*. CRC Press, Boca Raton, Florida, USA.

Davies, K.W., G. Collins, and C.S. Boyd. 2014. Effects of feral free-roaming horses on semi-arid rangeland ecosystems: An example from the sagebrush steppe. *Ecosphere* 5(10):1–14.

Dubey, J.P., M.C. Jenkins, C. Rajendran, K. Miska, L.R. Ferreira, J. Martins, O.C.H. Kwok et al. 2011. Gray wolf (*Canis lupus*) is a natural definitive host for *Neospora caninum*. *Veterinary Parasitology* 18:382–387.

Fitzwater, W.D. 1994. House sparrows. In *The Handbook: Prevention and Control of Wildlife Damage*. S.E. Hyngstrom, R.M. Timm, and G.E. Larson, editors. University of Nebraska Cooperative Extension Service, Lincoln, Nebraska, USA, pp. E-101–E-109.

Gondim, L.F.P., M.M. McAllister, W.C. Pitt, and D.E. Zmlicka. 2003. Coyotes (*Canis latrans*) are definitive hosts of *Neospora caninum*. *International Journal for Parasitology* 34:159–161.

Jay, M.T., M. Cooley, D. Carychao, G.W. Wiscomb, R.A. Sweitzer, L. Crawford-Miksza, J.A. Farrar et al. 2007. *Escherichia coli* O157:H7 in feral swine near spinach fields and cattle, central California coast. *Emerging Infectious Diseases* 13:1908–1911.

Jay, M.T., and G.W. Wiscomb. 2008. Food safety risks and mitigation strategies for feral swine (*Sus scrofa*) near agriculture fields. *Proceedings of the Vertebrate Pest Conference*, San Diego, California, USA, vol. 23, pp. 21–25.

Krysl, L.J., M.E. Hubbert, B.F. Sowell, G.E. Plumb, T.K. Jewett, M.A. Smith, and J.W. Waggoner. 1984. Horses and cattle grazing in the Wyoming Red Desert, I. Food habits and dietary overlap. *Journal of Range Management* 37:72–76.

Linz, G.M., H.J. Homan, S.M. Gaukler, L.B. Penry, and W.J. Bleier. 2007. European starlings: A review of an invasive species with far-reaching impacts. In *Managing Vertebrate Invasive Species: Proceedings of an International Symposium*. G.W. Witmer, W.C. Pitt, and K.A. Fagerstone, editors. USDA APHIS Wildlife Services, National Wildlife Research Center, Fort Collins, Colorado, USA.

McKnight, T.L. 1958. The feral burro in the United States: Distribution and problems. *Journal of Wildlife Management* 22:163–179.

Miller, R. 1983. Habitat use of feral horses and cattle in Wyoming's Red Desert. *Journal of Range Management* 36:195–199.

National Pork Producers Council. 2013. Pork facts. http://www.nppc.org/pork-facts/. Accessed June 23, 2015.

Ott, S.L., S.J. Wells, and B.A. Wagner. 1999. Herd-level economic losses associated with Johne's disease on U.S. dairy operations. *Preventive Veterinary Medicine* 40:179–192.

Pimentel, D., R. Zuniga, and D. Morrison. 2004. Update on the environmental and economic costs associated with alien-invasive species in the United States. *Ecological Economics* 52:273–278.

Seward, N.W., K.C. VerCauteren, G.W. Witmer, and R.M. Engeman. 2004. Feral swine impacts on agriculture and the environment. *Sheep and Goat Research Journal* 19:34–40.

Smith, R.H. 1994. Rodents and birds as invaders of stored-grain ecosystems. In *Stored-Grain Ecosystems*. D.S. Jayas, N.D.G. White, and V.E. Muir, editors. CRC Press, Boca Raton, Florida, USA, pp. 289–323.

Timm, R.M. 1994. Norway rats. In *The Handbook: Prevention and Control of Wildlife Damage*. S.E. Hyngstrom, R.M. Timm, and G.E. Larson, editors. University of Nebraska Cooperative Extension Service, Lincoln, USA, pp. B-105–B-121.

United States Department of Agriculture, Animal and Plant Health Inspection Service (USDA APHIS). 2000. *Bird Damage Management at Livestock Feeding Facilities in the Kansas Wildlife Services Program*. Environmental Assessment, Manhattan, Kansas, USA.

United States Department of Agriculture, Animal and Plant Health Inspection Service (USDA APHIS). 2009. Pseudorabies state status map. http://www.aphis.usda.gov/wps/wcm/connect/52273b22-5b68-45ec-bff1-bd0ec975a1f0/pseu_st_status_map_255.jpg?MOD=AJPERES. Accessed June 23, 2015.

United States Department of Agriculture, Animal and Plant Health Inspection Service (USDA APHIS). 2011. Swine brucellosis state status map. http://www.aphis.usda.gov/wps/wcm/connect/294e846c-3d5f-4031-b5a6-500c283ad650/br_status_map_257.jpg?MOD=AJPERES. Accessed June 23, 2015.

United States Department of Agriculture, Economic Research Service (USDA ERS). 2007. Total land, 1945–2007, by state: The sum of cropland, pasture/range, forest-use land, special uses, urban area, and other land. http://www.ers.usda.gov/data-products/major-land-uses.aspx. Accessed June 23, 2015.

United States Department of Agriculture, National Agricultural Statistics Service (USDA NASS). 2000. Sheep and Goats Predator Loss. National Agricultural Statistics Board, Washington DC, USA.

United States Department of Agriculture, National Agricultural Statistics Service (USDA NASS). 2002. *U.S. Wildlife Damage.* National Agricultural Statistics Board, Washington, DC, USA.

United States Department of Agriculture, National Agricultural Statistics Service (USDA NASS). 2011. Cattle Death Loss. National Agricultural Statistics Board, Washington DC, USA.

United States Department of Agriculture, National Agricultural Statistics Service (USDA NASS). 2015. Census of agriculture 2012 census highlights. https://www.agcensus.usda.gov/Publications/2012/Online_Resources/Highlights/Farm_Economics/. Accessed October 25, 2016.

Wachtel, S.P., and J.A. McNeely. 1985. Oh rats. *International Wildlife* 15:20–24.

White, S.B., R.A. Dolbeer, and T.A. Bookhout. 1985. Ecology, bioenergetics, and agricultural impacts of a winter-roosting population of blackbirds and starlings. *Wildlife Monographs* 93:3–42.

Wyckoff, A.C, S.E. Henke, T.A. Campbell, D.G. Hewitt, and K.C. VerCauteren. 2009. Feral swine contact with domestic swine: A serologic survey and assessment of potential for disease transmission. *Journal of Wildlife Diseases* 45:422–429.

7 Brown Tree Snakes
Methods and Approaches for Control

Larry Clark, Craig Clark, and Shane Siers

CONTENTS

INTRODUCTION

This chapter summarizes the existing and emerging tools and strategies for the control of the invasive brown tree snake (*Boiga irregularis*, or BTS) on Guam and the prevention of its accidental transport to, and subsequent establishment on, other snake-free Pacific islands. The brown tree snake has long served as an example of the ecological and economic damages that can be wrought by a single generalist vertebrate predator upon introduction to ecosystems that evolved in isolation and without native predators (Fritts and Rodda 1998; Wiles et al. 2003; Rodda and Savidge 2007). Our attempt to summarize the state of the art for control technology development and use is not intended to be an exhaustive survey of all the brown tree snake literature. Rather, it is intended to introduce the reader to the main concepts, methods, and strategic management uses of the tools in an effort to control brown tree snakes on the island landscape and interdict their passage at ports. Significant practical advances for brown tree snake control have been made since the first comprehensive summaries were published by Rodda et al. (1999e), and there is great promise for future refinements and broader implementation for control efforts on an islandwide scale.

INVASION AND IMPACTS

The brown tree snake (Figure 7.1) is native to northern and eastern Australia and parts of Indonesia and Melanesia. It is presumed that brown tree snakes were transported to Guam along with shipments of military equipment from the Admiralty Archipelago in Papua New Guinea shortly after World War II. Inference on the origins of the population invasive to Guam was initially based on morphometric analysis (Whittier et al. 2000). Subsequent determination of the source population is supported by genetic evidence. Data suggest that the Guam population was the result of a single introduction of perhaps fewer than 10 individuals (Richmond et al. 2015).

Unchecked by predators or competitors, brown tree snakes on Guam achieved densities reaching as high as 50–100 snakes per hectare at the height of the irruption (Savidge 1991; Rodda et al. 1999d). These densities were sufficient to exert suppression on prey populations such as never before observed in a snake species (Rodda et al. 1997). Originating at Naval Base Guam in the late 1940s or early 1950s, the invasion front spread from there to the southern tip of Guam by the late 1960s and

FIGURE 7.1   Photograph of the brown tree snake, *Boiga irregularis*.

reached the more remote northern limestone forests by the early 1980s (Savidge 1987). This invasion front coincided with a wave of precipitous declines in bird diversity and abundance (Savidge 1987; Wiles et al. 2003), and resulted in the extirpation of 11 of Guam's native forest birds and the extinction of the Guam flycatcher, *Myiagra freycineti*; the Guam subspecies of the bridled white-eye, *Zosterops conspicillatus*; and the rufous fantail, *Rhipidura rufifrons* (Savidge 1987; Wiles et al. 2003). The Guam rail, *Gallirallus owstoni*, and the Guam Micronesian kingfisher, *Todiramphus cinnamominus*, are extinct in the wild, though captive populations have been maintained in the hope of reintroduction subsequent to effective brown tree snake suppression actions. The consequent loss of avian ecosystem function has resulted in cascading ecological consequences, including disturbance of plant reproduction (Mortensen et al. 2008; Rogers 2011), forest regeneration (Perry and Morton 1999), and arthropod release (Rogers et al. 2012). Predation by brown tree snakes has also negatively impacted nearly all native vertebrate populations on Guam (Fritts and Rodda 1998) as well as nonnative and domestic animals (Fritts and McCoid 1991; Wiewel et al. 2009).

In addition to these ecological impacts, economic detriments of the brown tree snake invasion of Guam (reviewed in Rodda and Savidge 2007) include damage to electrical power infrastructure, loss of pet and domestic animals, human envenomations, higher costs of shipping from Guam, and threats to the tourism industry. Such economic damages are likely to be experienced manifold if a similar brown tree snake invasion were to occur in Hawaii (Burnett et al. 2008; Shwiff et al. 2010).

BIOLOGY AND ECOLOGY

Brown tree snakes are rear-fanged colubrid snakes that are well adapted to nocturnal and arboreal foraging, having large eyes with elliptical pupils and long, slender body forms. Because brown tree snakes are primarily arboreal, forests are considered to be their preferred habitats; however, they successfully utilize all of Guam's terrestrial habitats, including savannas and urban areas.

Hatching at a size of approximately 350 mm snout-vent length (SVL) and 5 g of mass, brown tree snakes may undergo a six-fold increase in length and a 400-fold increase in weight throughout their life, reaching up to 2000 mm SVL and 2000 g in weight, with males achieving larger sizes than females. However, subsequent to the collapses of large-bodied prey populations precipitated by brown tree snake predation, very large snakes are now rarely encountered. In a recent sample of 1800 snakes collected by visual detection and hand capture from 18 sites on Guam, stratified by six major habitat types, 84% of all snakes collected were less than 1000 mm SVL (Siers 2015).

Sexual maturity of brown tree snakes is strongly linked to body size. Savidge et al. (2007) found that 95% of female snakes matured between lengths of 910 and 1025 mm SVL, while males matured from 940 to 1030 mm. By these growth benchmarks, the previously mentioned sample of 1800 snakes was composed of 38% immature snakes, 49% maturing snakes, and 13% in the size class at which all snakes were expected to be reproductively mature (Siers 2015).

Brown tree snakes exhibit a pronounced ontogenetic shift in prey preference: smaller snakes prey exclusively on small lizards, with a shift toward feeding on larger endothermic prey (birds and rodents) by larger snakes (Savidge 1988; Greene 1989;

Shine 1991; Siers 2015). This shift goes beyond a mere correlation of snake size with prey size. In free-choice feeding trials, small snakes simultaneously offered a dead gecko, skink, and neonatal mouse of similar mass overwhelmingly preferred the lizards, particularly the geckos, to the mice (Lardner et al. 2009a). This change in prey is accompanied by an ontogenetic shift in venom composition, with greater toxicity to lizards as juveniles and a shift toward greater toxicity to birds and mammals as brown tree snakes grow larger (Mackessey et al. 2006). Smaller snakes are almost exclusively arboreal, with an increased tendency to travel and hunt on the ground as snakes increase in size (Rodda and Reed 2007; Siers 2015). While small snakes subsist almost exclusively on abundant arboreal geckos (Savidge 1988; Siers 2015), the transition of larger snakes to terrestrial foraging appears to be in response to the loss of arboreal bird and rodent prey and an increased dependence upon larger terrestrial lizards, such as the introduced *Carilia ailanpalai*, in the diets of larger snakes, particularly in forest habitats (Siers 2015). The nonresponsiveness of small brown tree snakes to rodent-based lures constitutes one of the major challenges in effectively targeting all size classes of snakes with existing control technologies, to be discussed further in this chapter.

More comprehensive reviews of the origins, biology, and impacts of the brown tree snake on the island of Guam are available in Rodda et al. (1992, 1999c) and Rodda and Savidge (2007).

COORDINATION OF PARTNERSHIPS FOR BROWN TREE SNAKE MANAGEMENT

Subsequent to the discovery that the brown tree snake was the cause of the decline and disappearance of much of Guam's native wildlife, federal, state, and territorial agencies were prompted to management action. Several pieces of federal legislation and interagency agreements have provided a regulatory framework, including the passage by the U.S. Congress of the Brown Tree Snake Control and Eradication Act of 2004 (Public Law 108–384, 118 Statute 2221, 7 U.S.C. 8501), which established the Brown Treesnake Technical Working Group (BTS TWG) to ensure that "efforts concerning the brown tree snake are coordinated, effective, complementary, and cost-effective." The BTS TWG has three overarching long-term goals: (1) preventing the escape of the brown tree snake from Guam to other locations (interdiction), (2) suppression and control of brown tree snake numbers to reduce their impact on the island of Guam and to restore the island's ecosystem, and (3) eradication of the brown tree snake from Guam.

The BTS TWG is formally composed of federal, state, and territorial partner agencies, with periodic informal participation by nongovernmental organizations. TWG membership includes: the U.S. Department of the Interior's (DOI) Office of Insular Affairs (OIA), U.S. Fish and Wildlife Service—Pacific Islands Fish and Wildlife Office (USFWS PIFWO), U.S. Fish and Wildlife Service—Guam National Wildlife Refuge (GNWR), U.S. Geological Survey (USGS), National Invasive Species Council (NISC), and National Park Service (NPS); the U.S. Department of Agriculture—Animal and Plant Health Inspection Service (USDA APHIS) Wildlife Services (WS) Operations and WS National Wildlife Research Center (NWRC); numerous commands under the Department of Defense (DoD); the U.S. Department of Transportation (USDOT); and multiple agencies within the U.S. Territory of Guam (GovGuam), Commonwealth of the Northern Mariana Islands (CNMI), State of Hawaii, and nongovernmental partners.

To more effectively and efficiently build on the research successes of the past, the BTS TWG chartered the BTS Research Committee (BTS RC) in 2012, with the primary goal of developing strategic long-term plans and short-term priorities for advancing research on brown tree snakes, focusing on developing the biological knowledge and technical ability required to meet the management goals of the TWG. The RC has identified three primary "research themes": (1) interdiction, early detection, and rapid response; (2) landscape-scale suppression; and (3) restoration. Priority research areas within these themes are summarized in Table 7.1.

Research on brown tree snakes has traditionally been funded by DOI through grants from OIA, but DoD has provided increasing support for research and operational control through specific grants and contracts. Brown tree snake research is designed to be complementary; NWRC focuses its research on methods development, whereas USGS concentrates on understanding the effects of BTS control and interdiction on natural ecosystems and control tool validation, with this approach clarifying roles while allowing the opportunity to collaborate on projects when necessary.

NWRC conducts research focused on product development and improved methodologies with the objective of improving the overall efficacy of BTS management. Our first commitment is to developing tools that will enhance the ability to conduct large-scale control of brown tree snakes. Second, we develop methods to augment

TABLE 7.1
Draft BTS TWG Research Committee Themes and Priority Research Areas

Interdiction, Early Detection, and Rapid Response	Landscape-Scale Suppression	Restoration (Dependent on Suppression R&D)
• Quantify and increase BTS interception rates • Develop methods to detect snakes at low density, including rapid response • Develop methods to detect satiated snakes in new locations • Develop tools for interdiction of BTS not susceptible to mouse-based methods • Develop and test new irritant and repellent methods • Assess new barriers (physical, chemical, behavioral) and reduce barrier costs	• Automate toxicant delivery (automated aerial broadcast system) • Study effect of suppression on BTS and nontarget species • Develop alternative attractant, lures, and baits • Develop tools for control of BTS not susceptible to mouse-based methods • Integrate current data operational data sets into research programs • Control of BTS in urban environments	• Determine level of BTS suppression required for persistence of native species • Determine size of exclosure required for persistence of various native species • Predict native ecosystem response to toxicant application • Improve barrier cost effectiveness & durability

Order does not imply importance.

the current suite of interdiction tools. We also review operational methods and tools and conduct program analyses to ensure control and interdiction efforts are as effective as possible.

USGS is also a major contributor to brown tree snake research, with a multifaceted program to inform managers about the ecology and biology of brown tree snakes toward enhancement of methods development and operational control. USGS efforts focus on understanding the effect of removal techniques on all segments of the brown tree snake population, including variation in response to control techniques among age or size classes. Additionally, USGS pursues research projects aimed at increasing the ability to detect brown tree snakes at low density and maximize the efficiency of early detection and rapid response.

Both NWRC and USGS are cooperating in research designed to evaluate ecosystem changes that occur through brown tree snake control methods to understand the impact of large-scale control efforts on the environment.

USFWS, GovGuam, and a number of the other parties to the BTS TWG have primarily been concerned with bird recovery, ecological restoration, and prevention of further ecological damages that would ensue from further accidental spread of brown tree snake populations.

Apart from the brown tree snake research agenda, which is the primary focus of this chapter, it should not be overlooked that Wildlife Services Operations on Guam has implemented a very successful integrated program of interdiction. As of 2014, over 150,000 snakes were removed from areas surrounding airports, seaports, and critical infrastructure (e.g., electrical substations). Snakes continue to be removed from these areas at a rate of approximately 10,000 per year. Over 99% of outbound cargo from Guam is inspected by Wildlife Services detector dog teams. Since the canine program's inception in 1993, no live snakes have been found in cargo arriving in Hawaii from Guam. This constitutes a tremendous conservation victory.

MANAGEMENT STRATEGIES AND GOALS

The current management strategies for brown tree snake control on Guam are primarily targeted at achieving the goal of interdiction, preventing the accidental export of snakes to currently snake-free environments. The methods described herein were developed to address the needs for each of these "filters," or interdiction stages, originally conceived for interdiction and control of brown tree snakes by the U.S. federal agencies.

Interdiction tactics, described elsewhere in this chapter, are employed at each stage of movement from areas with high snake density across successive barriers and anthropogenic features toward areas of low snake density (Figure 7.2). This approach is based on the premise that snake density is serially reduced at each step such that risk of "break through" diminishes in a cumulative probabilistic fashion. Prioritization of interdiction activities focuses on likely ports of exit from Guam and entry into vulnerable locations such as the CNMI and Hawaiian Islands.

As snakes transcend each of these stages, the intent is to intercept every individual. While complete interception of all snakes at any one stage is unlikely, the risk of accidental export is decreased greatly by the repeated reduction of potential

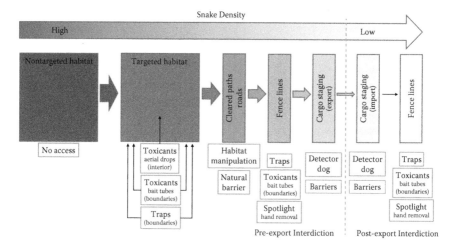

FIGURE 7.2 Stages of opportunity for interdiction of brown tree snakes. A schematic representation of the integration for brown tree snake control and interdiction. Shaded icons represent different habitats, boundaries, activities, or structures. Darker coloration indicates higher snake densities. Arrow size depicts decreasing probability of movement of brown tree snakes. Clear boxes depict the various control methods that might be used.

emigrants through the gauntlet of multiple interception points, to the extent where the odds of any snake penetrating all lines of defense becomes vanishingly small. As any snake penetrates each of these interdiction stages, the risk of potential invasion, and therefore the imperative for intercepting that snake at the next step, is greatly increased. If a hypothetical snake were to penetrate beyond all pre-export and post-export interdiction measures, it would move into the domain of early detection and rapid response. Some of these pre-export interdiction tactics are also implemented around important Guam infrastructure elements such as power plants.

CHEMICAL METHODS FOR BROWN TREE SNAKE CONTROL

ORAL AND DERMAL TOXICANTS

Toxicants are one in a suite of tools used by managers for the control of invasive and injurious animals (Witmer et al. 2007; El-Sayed et al. 2009), and their discovery and development involves the integration of numerous scientific, regulatory, and business activities (Figure 7.3; Fagerstone et al. 1990; Isman 2006; Ravensberg 2011). For brown tree snake control, toxicant discovery and development was identified as a high-priority need among U.S. federal agencies (Campbell et al. 1999). The goals were to identify a toxicant that acted rapidly, produced little apparent pain and suffering in the target animal, posed low nontarget and environmental risk, was commercially available, could be formulated into various delivery systems, and had a sufficient scientific, clinical, and environmental history such that different aspects of the regulatory requirements of the Federal Insecticide, Fungicide, Rodenticide Act (per Title 40 Code of Federal Regulations, Parts 150–189) could be addressed in the

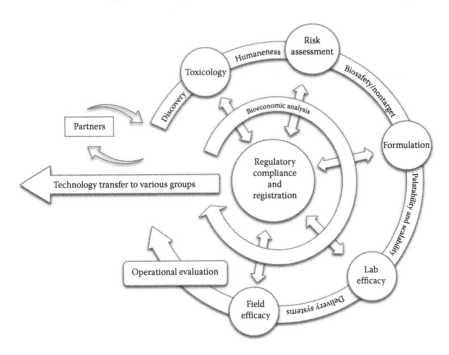

FIGURE 7.3 Schematic of the product registration process. A schematic of the general integration of disciplines and activities requirement to identify, develop, and register a brown tree snake toxicant or other chemical pesticide.

most cost-effective manner. The expense and effort to pass through each of these filters is enormous, and explains why so few chemically based wildlife damage management tools are available (Fagerstone et al. 1990; Fagerstone and Schafer 1998).

Numerous candidate oral and dermal toxicants were screened for efficacy during the discovery phase of product development (Brooks et al. 1998; Savarie and Bruggers 1999; Savarie et al. 2000; Johnston et al. 2001). For practical application reasons, delivery of a toxicant through an oral bait system rather than dermal applications was favored for development by USDA (addressed below: baits/lures). Early screening identified acetaminophen (CAS# 103-90-2) as the toxicant of choice.

Acetaminophen demonstrated efficacy in snakes (Savarie et al. 2000); its mode of action suggested that it would meet humaneness criteria of rapid time to death and low animal awareness (Sharp and Saunders 2011). In brown tree snakes, acetaminophen promotes the formation of methemoglobin, and thus deprives cells of oxygen (Mauldin et al., unpublished). The induced anoxia renders the animal inactive and eventually it passes into unconsciousness, analogous to carbon monoxide poisoning. The low dosage required to produce 100% mortality in brown tree snakes (80 mg/bait for the largest snakes tested; Savarie 2002) is well below U.S. Environmental Protection Agency (EPA) standards for safe human exposure (200 µg/L). Generally, environmental water sampling has shown acetaminophen and other analgesic pharmaceuticals to be two to four orders of magnitude lower than the no-observed-effects-levels (NOELs) for humans. Acetaminophen is rapidly bound and detoxified in soil, decreasing potential for off-site runoff or leaching into groundwater (Li et al. 2014). The environmental load for acetaminophen at anticipated field delivery rates was considered not to represent a human health or environmental risk (Schwab et al. 2005; Kim et al. 2007). For select nontarget, nonhuman standard test species, acetaminophen has an EPA worst-case scenario, acute hazard quotient >0.5, which is generally considered a significant risk (Johnston et al. 2001, 2002). Thus, precautions about toxicant placement and delivery method were considered for distribution on the landscape. Laboratory and field experimental and observation studies showed that nontarget wildlife encounters with the toxicant bait system could be reduced by mechanical exclusion, or that the nontarget species likely to be encountered on Guam either tolerated field delivery doses or ate the bait, but not the toxicant (Avery et al. 2004). The bioaccumulation potential (Primus et al. 2004) and risk quotient for exposure of nontarget organisms eating snake carcasses was considered low to nonsignificant (Johnston et al. 2002). Moreover, video and telemetry monitoring of baits treated with acetaminophen and placed in the field showed that snakes took the vast majority of baits and that nontarget bait take was rare (Savarie et al. 2000; Clark and Savarie 2012). Based on these and other field efficacy studies, U.S. EPA issued a pesticide label to the USDA for acetaminophen for control of brown tree snakes in 2003 (U.S. EPA Registration Number: 56228-34) with provisions for application in polyvinyl chloride (PVC) bait stations or traps and for hand or aerial broadcast.

CHEMICAL FUMIGANTS

Cargo shipments pose a risk for translocating invasive species (Work et al. 2005; Kraus 2007). Fumigation of cargo with a toxicant is used to mitigate against this risk and kill a variety of pest species, and in the case of commercial international trade, it is generally used to ensure phytosanitary standards (e.g., FAO 2006; GATFA 2012). Generally, gaseous pesticides are applied to a confined space (e.g., cargo container, tent enveloping cargo) and pests are suffocated or poisoned. The time period for exposure depends on the registered fumigant being used and the target pest species. Fumigation is a hazardous activity and can only be carried out by licensed pesticide applicators using registered pesticides. Fumigation also requires an operational infrastructure dedicated for its use. Despite these logistical constraints, the use of fumigants for brown tree snakes at cargo facilities was explored by USDA so as to provide managers with another set of tools in their effort to reduce the risk of snakes escaping Guam in outgoing cargo.

Several existing registered fumigant products were evaluated by USDA, with the rationale that the data support for supplementing existing pesticide labels and uses would prove to be a more cost-effective strategy in product development relative to registering a new active ingredient and finding a manufacturer to produce the product. The candidate products contained the following active ingredients: Meth-O-Gas® (methyl bromide); Metabrom Q® (methyl bromide); Vikane® (sulfuryl fluoride); Magtoxin® (magnesium phosphide); Fumi-Cell® (magnesium phosphide).

Fumigant treatments and dosages applied to the cargo containers (19.5 × 7.7 × 7.8 ft) were all within the preexisting registered label application rates for each of the active ingredients. All three active ingredients at label-specified rates killed brown

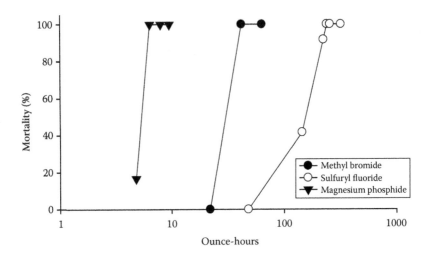

FIGURE 7.4   Brown tree snake mortality resulting from chemical fumigation. Mortality of brown tree snakes in cargo containers fumigated with three toxicants. Ounce-hours is the number of ounces of fumigant applied/1000 ft^3, multiplied by the exposure time in hours. (Data adapted from Savarie, P. J. et al. 1995. Unpublished report on file with the U.S. Department of Agriculture, Fort Collins, Colorado: Animal and Plant Health Inspection Service, National Wildlife Research Center and Savarie, P. J. et al. 2005. *International Biodeterioration & Biodegradation* 56:40–44.)

tree snakes (Figure 7.4). Based on these encouraging results, data were provided to the product owners, and two of the companies filed for amendment to their pesticide labels to include brown tree snakes: Meth-O-Gas (methyl bromide, Great Lakes Chemical Corporation, West Lafayette, IN; EPA Registration No. 5785-41) and Metabrom Q (methyl bromide, ICL-IP America, Inc., Gallipolis Ferry, WV; EPA Registration No. 8622-55). These products are now available for brown tree snake control. Despite the availability of products for the sanitation of cargo, we are not aware of any actual commercial operational use. Nonetheless, any cargo undergoing phytosanitation with any of these products or ingredients for other uses would also *de facto* be sanitizing for brown tree snakes.

THERMAL FUMIGANTS

There may be circumstances that favor nonchemical methods to mitigate against invasive species transport via cargo shipments. "Thermal fumigation" may provide a simple, chemical-free method to sanitize cargo (Heather and Hallman 2008; Hennessey et al. 2014). Because snakes lack the ability to physiologically thermoregulate, extremes in temperature exposure can prove lethal (Christy et al. 2007). For example, snakes as stowaways in aircraft wheel wells can freeze to death owing to the high altitude and low temperature of most commercial flights (Perry 2002). Reliance on passive heating or cooling of cargo, however, may not be a good strategy for reliable control, in that temperatures in surface cargo containers are likely to be too low to consistently kill snakes (Perry and Vice 2007). Recently, Kraus et al.

(2015a) demonstrated that snakes could be driven from simulated cargo refugia by applications of forced heated air streams. Application rates of temperatures between 48°C and 52°C at 3.4 m³/min were sufficient to induce escape from cargo within five minutes. While there remains work to be done to operationalize this method, it does appear to be a fruitful avenue for further exploration. While snakes demonstrated high variance in their response times to temperatures of forced air streams, "… only eight snakes of 160 tested failed to find the refugium exit, and those that failed died trying" (Kraus et al. 2015a). Ongoing research into thermal fumigation indicates that radiant energy may also be applied to heat cargo to lethal thermal limits, for applications where forced heated air may not be practical, such as in closed cargo containers (Kraus, unpublished data).

Repellents and Irritants

Chemicals have been used to protect food and crops from animal depredations and to repel animals away from areas (Mason and Clark 1995). This latter use pattern is the one most researchers have focused on in their search for brown tree snake control tools. Chemical barriers prevent snakes from entering refugia or cargo. When acting as a nonlethal fumigant, the chemical repellent is designed to elicit escape behavior of snakes hiding in confined spaces. The primary principle in operation in the applications of almost all repellents is stimulation of chemosensitive nociceptors (pain fibers) that then elicits avoidance behavior (i.e., reflective withdrawal/escape) by the animal (Clark and Avery 2013).

The embodiment of most chemical barrier systems is to expose a snake, through direct contact, to a chemical irritant by treating surfaces with compounds such as camphor, naphthalene, or sulfur (Ferraro 1995; summarized in Clark and Shivik 2002). Tactile products generally are sticky, such as polybutenes. The snake either avoids contact or becomes stuck (Takashi et al. 1992).

Another strategy would be to use the repellent as a "tear gas" (Stevens and Clark 1998). In this case, an aerosolized or vaporized repellent is introduced to an enclosed space. Snakes exposed to the irritant would then try to escape. This use pattern might be desirable once a detector dog identified cargo containing a snake, and would allow verification or capture without dismantling the cargo container or pallet.

A variety of single compounds and essential plant oils known to have high concentrations of chemicals irritating to birds and mammals were screened using brown tree snakes as a model (Clark and Shivik 2002, 2004). Some compounds proved to be toxic under high aerosol or vapor concentration, and some induced narcolepsy. Nonetheless, many compounds and mixtures demonstrated effectiveness in promoting escape behavior in simple simulated cargo configurations. Compounds such as cinnamon oil (containing cinnamaldehyde) were identified as promising because they did not require U.S. EPA registration for this use (USDA 2003). As encouraging as the direct exposures to these irritants were, several challenges remained. Many of the compounds had plasticizing properties (i.e., they would melt plastic), thus their use for certain types of cargo would be restricted. Dispersal of aerosols in the complex confined space of cargo is also problematic. Thus, using vapor would be the preferred embodiment of application because of its superior penetrating properties as opposed to aerosols, which would deposit on surfaces and have limited ability to penetrate the complex interstitial space of cargo. The greatest challenges for this method are vapor generation and penetration of complex cargo. To date, the reliability of the method, given the current technologies, is not sufficient to produce reliable escape behavior from simulated cargo (Kraus et al. 2015b).

Chemical Lures, Baits, and Delivery Systems

Chemical lures are odors/scents used to attract animals to a location or object (Linhart et al. 1997). For snakes, the sensory modality mediating these cues is primarily vomeronasal or olfactory (Chiszar 1990; Halpern 1992; Schwenk 1995). The first phase in the behavioral process exploits appetitive exploratory behavior by animals motivated by hunger or reproduction. The second phase in the behavioral process involves consumatory behavior either through ingestion (of a bait) or mating (in the case of pheromones).

As reflected by dietary studies, brown tree snakes are opportunistic feeders (Savidge 1988; Fritts et al. 1989; Fritts and Rodda 1998; Siers 2015). Moreover, brown tree snakes use visual, tactile (vibratory), and chemical cues to locate and capture their prey. Initial research focused on the attractiveness of odors for a variety of prey, for example, skinks, geckos, eggs, birds, bird feces, blood, and small mammals (Chiszar et al. 1988, 2001), to be incorporated as a lure for trapping efforts. Numerous studies focused on the sensory modalities used by brown tree snakes with the goal of designing an optimal lure/trap system. For example, in the laboratory, snakes were temporarily rendered blind by placing electrical tape over their eyes. Relative to controls, blinded snakes had similar accuracy for their strikes and capture, but took three times longer to initiate the strike, and the initial strike distance was one-third that of the sighted controls (Kardong and Smith 1991). Prey cues presented in concert are most effective in eliciting predatory appetitive behavior in the laboratory and capture success in traps in the field (Shivik 1998; Stark et al. 2002), with odor plus visual cues being twice as effective as either cue presented separately. When presented as a single stimulus, odor was slightly better than a visual cue alone.

Ultimately, for practical operational reasons (i.e., effectiveness as a lure and ease of maintenance), USDA settled on a live mouse lure system contained within a Wildlife Services Standard Trap design (Hall 1996). While effective at catching snakes, the expense of maintenance and operation is a major factor limiting the number of traps that can be deployed (Clark et al. 2012). For these reasons, efforts to replace the live lure system and exploit the natural prey preferences and sensory foraging modalities have been extensive (Savarie and Clark 2006).

While laboratory studies of alternative lures elicited interest from brown tree snakes, the rate of capture relative to the live mouse lure system was poor. The exception was the use of mouse carrion (Shivik and Clark 1997, 1999a,b; Shivik et al. 2000; DeVault and Krochmal 2002). As an aside, one interesting aspect about the attractiveness of dead neonatal mice (DNM), and as revealed by video analysis, traps with live mice attract more snakes, but are less efficient in their capture, whereas traps with DNM attract fewer snakes, but are more efficient in their capture (L. Clark, unpublished). This is a case of snakes exhibiting different foraging behaviors. For traps with live mice, snakes spend more time striking and are less likely to find the trap entrance (a one-way door), whereas snakes investigating DNM lures are more methodical in the investigatory behavior and find the one-way door without distraction. The appeal of DNM is that it could be used as both a lure (for trapping) and a bait (for toxicant delivery).

The attractiveness of DNM appears to be related to bacterial decomposition of the mouse skin (Jojola-Elverum et al. 2001). One- and two-day-old decomposed mouse skin yields optimal attractiveness to brown tree snakes. Other tissues (depelted mice; ground mice; other meats such as chicken, pork, and beef) all are less attractive than DNM (Savarie and Clark 2006; Savarie 2012).

Synthetic lures are desired because DNM have an effective usefulness of two days in the field and they are relatively expensive to obtain, ship, and store frozen. An improved lure/bait system would incorporate a synthetic lure that was inexpensively manufactured, shipped, stored, and deployed, and would be durable in hot, wet, humid tropical environs. Efforts to test the attractiveness of chemicals commonly associated with carrion yielded results that snakes were moderately attracted to a location, but could not hold the attention of foraging snakes sufficiently long enough to effect capture at traps or entice snakes to eat the bait at the levels observed for DNM (Savarie and Clark 2006; Kimball et al. 2016). It is presumed that a good synthetic lure can be incorporated into a matrix with controlled and timed release. Such a system would maintain effectiveness of traps while substantially reducing operating costs relative to traps with live mouse lures. Recently, Kimball et al. (2016) identified a promising system that uses a complex suite of compounds designed to more realistically mimic a two-day-old rotted mouse carcass. The delivery matrix identified (a commercial meat product) is also more amenable to retaining the synthetic lure for timed release and as a palatable bait that snakes accept at rates comparable to the DNM, thus it is suitable for toxicant delivery as well. From an operational perspective, the system is amenable for large-scale manufacture and production capacity that will be necessary if any large-area operational programs are to be successfully implemented in a cost-effective manner.

PHEROMONES

Use of pheromones is another lure-based strategy frequently used in the control and management of pest species (Howse et al. 2013; Pickett et al. 2014). For brown tree snake control, uses include luring snakes to traps or other control tools (Greene et al. 2001), reproductive inhibition through confusion or disruption of male snakes seeking female mates (Greene and Mason 2003), and early detection of nascent populations on islands at risk of snake invasion (Mathies et al. 2013).

Brown tree snakes do not demonstrate seasonal patterns of fertility or reproductive state on Guam (Savidge et al. 2007; Mathies et al. 2010). Yet, despite the continuous availability of reproductively receptive snakes at the population level, it also appears that the number of reproductively receptive brown tree snakes is low when compared to conspecifics in their native range, suggesting that overpopulation and competition for food resources are limiting reproductive opportunity (Moore et al. 2005). Male brown tree snakes readily follow the semiochemical trail of both male and female conspecifics (Greene et al. 2001). However, male courtship of females can be suppressed by cloacal secretions from females (Greene and Mason 2003). While this observation may have significance from an evolutionary signaling perspective, it is not year clear how this inhibitory mechanism could be adapted to a large-scale management scenario.

Using pheromones to attract animals to a trap is a more typical use pattern, that is to say, the lure and kill strategy. In laboratory studies, using a tongue flicking bioassay, the attractiveness of vitellogenic brown tree snake females to males is greater than nonvitellogenic females (Mathies et al. 2013). Exploiting this observation may also prove useful for the detection of reproductively active males on islands with low snake densities and where the attractiveness of food-based lures may not be sufficiently high (Mason et al. 2011). Regardless, more work is needed to identify the composition of the pheromone (Greene and Mason 1998) so that it could be synthetically produced at sufficient quantities, embedded in a sufficiently compatible matrix for distribution, and validated using trap capture success when used as the lure. Tongue flicking is a low threshold of success when compared to the sustained motivation and investigatory behaviors needed to capture a snake in a trap.

DEVICES AND OTHER STRATEGIES FOR BROWN TREE SNAKE CONTROL

TRAPS

Trapping is one of the most basic and universal forms of vertebrate pest control. Since the mid-1980s, numerous trap designs have been screened for their suitability for brown tree snake control (Rodda et al. 1999b). Incorporating preferred design components from these many years of testing by multiple agencies, and subsequent to rigorous testing (e.g., Linnell et al. 1998; Engeman and Vice 2000), USDA settled on the "Wildlife Services" Standard Trap (WSST) design, which employs components of a one-way door at the end of an inverted cone placed at both ends of a cylinder (Figure 7.5; Vice et al. 2005). The shape and orientation of the one-way door flap are such that gravity holds it closed until pushed open by an entering snake, with the flap then closing behind the snake and trapping it in the interior; this self-setting design allows for repeated captures of multiple snakes. Traps are typically suspended

FIGURE 7.5   Wildlife Services Standard Trap design. The modified minnow trap used by the USDA Wildlife Services program, known as the Wildlife Service Standard Trap (WSST).

at approximately waist to chest height for ease of checking and maintenance. Snakes are drawn to, and into, the traps by a lure, typically and most effectively a live mouse lure, which is contained within a separate wire mesh chamber containing provisions for food and moisture, which is enclosed within the body of the trap, and protects the mouse from snake strikes. The top half of the trap is typically covered by an opaque plastic shield provided to afford protection from rainfall and direct sunlight.

When using the live mouse lure, most snakes captured by the WSST design are within the size range of 800 to 1100 mm SVL (Vice et al. 2005). Such traps, using live mouse lures, are only partially effective at capturing snakes 700–900 mm SVL, and not effective for smaller juvenile snakes (Rodda et al. 2007b; Tyrrell et al. 2009) due to a strong ontogenetic prey preference for small lizards and nonresponsiveness to rodent-based lures (Savidge 1988; Lardner et al. 2009a; Siers 2015). Efficacy of different lure systems and field operational evaluations are covered below.

PERIMETER TRAPPING

Costs, logistics, and time always constrain management and control programs. From an operational perspective, these tradeoffs influence the calculus of which method to use under any set of circumstances. The fragmented nature of the islands' forested habitats represents an opportunity for control efforts. Fragments are interspersed with roads and trails, presenting the opportunity to easily place and maintain snake traps along forest edges. Snakes of size classes vulnerable to trapping can effectively be controlled in habitat blocks of up to 18 ha using the WSST with a live mouse lure and a perimeter trapping scheme (Engeman and Linnell 1998; Engeman et al. 1998a,c, 2000). In a similar habitat, spacing traps at 20-, 30-, and 40-m intervals did not affect snake capture rates (Engeman and Linnell 2004).

BAIT TUBES

Compared to live mice, DNM represent a lower cost bait matrix to deliver toxins to brown tree snakes (Clark et al. 2012). A convenient way to deliver the baits is to place the toxicant-treated DNM inside a PVC pipe bait station, or "bait tube," suspended along perimeters of forested habitat. This presentation minimizes exposure of the toxic bait to nontarget organisms. Savarie et al. (2001) demonstrated that delivery of acetaminophen-treated DNM in bait tubes was sufficient to reduce the population of snakes vulnerable to rodent-based tools by 83% (based on pre- and posttreatment mark recapture estimates using live mouse lures). One concern about using DNM is that this prey base may be selective to only a fraction of the population, owing to ontogenetic prey preferences attributable to size of the foraging snake, its foraging history, and satiety, among other factors. Small snakes (<843 mm) did not eat DNM from tubes (Lardner et al. 2013). Disappearance of DNM from bait tubes is often considered one index of snake abundance or snake foraging activity. The use of bait tubes alone as a method for monitoring snake populations should be viewed with some caution. Removal of DNM baits may indicate presence of certain size classes of snakes for a given satiety level, but does not provide a reliable index of population numbers. Bait disappearance may also result from takes by nontarget organisms such as rats or crabs, though photo and video evidence indicates that such events are rare (USDA, unpublished data). Snake control through applications of toxic DNM baits in bait tubes has been demonstrated to be 1.67 times more cost effective than trapping with live mouse lures (Clark et al. 2012).

AERIAL APPLICATION OF TOXIC BAITS

Trapping over moderately sized areas (~17 ha) can reduce snake populations significantly, but is logistically time consuming, costly, and frequently impractical when large areas or rugged terrain need to be managed (Clark et al. 2012). A lower-cost alternative to trapping has been proposed that uses dead mice baits treated with acetaminophen (Savarie et al. 2001; Clark et al. 2012). USDA developed prototype delivery systems (Savarie et al. 2007) and demonstrated their aerial delivery feasibility and target specificity (Shivik et al. 2002). Subsequently, DNM treated with toxic doses of acetaminophen were aerially broadcast over 6 ha at an application rate of 37.5 baits/ha. Take of unadulterated DNM baits from bait tubes, as an index of snake activity, was reduced approximately 85% relative to pretreatment levels and to isolated untreated reference plots (Figure 7.6; Clark and Savarie 2012; Dorr et al. 2016), thus demonstrating proof of concept and feasibility of large-scale area suppression of brown tree snake populations. Larger-scale aerial applications are

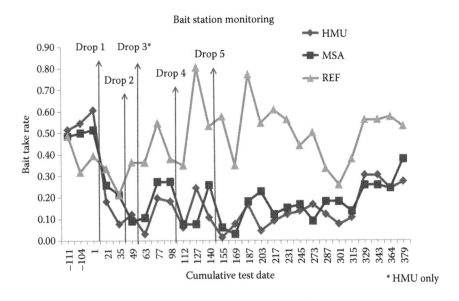

FIGURE 7.6 Reduction in snake activity following aerial application of baits. Reductions in brown tree snake activity, as indicated by takes of dead neonatal mouse baits from monitoring stations, following aerial application of acetaminophen-treated mice on Guam. (Data adapted from Dorr, B. S. et al. 2016. Aerial application of acetaminophen-treated baits for control of brown tree snakes [RC-200925; NWRC Study Number: QA-1828]. Final Report, U.S. Department of Defense ESTCP.)

planned by USDA that will incorporate automated bait packing and geospatially controlled delivery systems to further improve on the economics of large-area brown tree snake suppression.

VISUAL DETECTION, SPOT LIGHTING, AND FENCE LINE REMOVAL

Fences are often associated with property boundary roads and serve as a first line of security for properties surrounding airports, housing areas, and ports. Because snakes readily travel along the length of fences, using spotlights provides a simple method for visual inspection and removal which has been exploited by operational personnel (Engeman et al. 1999). In comparing trapping and visual spotlighting methods and their placement along forest edges and fences, Engeman and Vice (2001) found that trapping success rates were similar on fences and forest edges, and both were four to seven times more successful than visually directed hand removal efforts. However, studies on effectiveness of trapping show an exponential decline in trap success as local populations of snakes are depleted (Engeman and Linnell 1998), while the removal rate for spotlighting and hand capture remains constant over time (Engeman and Vice 2001).

Systematically searching for brown tree snakes directly on vegetative substrates along forested perimeters or along transects in forest interiors has proved more problematic given their overall drab coloration and very subtle movements. Christy et al. (2010) demonstrated that probabilities of visual detection of individual snakes are very low and influenced by multiple factors such as snake sex, size, and body condition; observer effects; and environmental conditions. Optimization of the way lighting is applied and interpreted can help to increase the probability that a searcher will detect a snake that is within his or her field of vision. Searching efficiency might be improved by selection of headlamps with desirable beam characteristics (Lardner et al. 2007, 2009b). While contrasts in background versus snake reflectance might theoretically be exploited, early simple efforts in this area have not proved successful (Siers et al. 2013).

Though more costly than trapping, an advantage of visual detection is that all size classes are exposed to detection, while size distributions of snakes captured by rodent lure-based methods are more biased toward larger snakes (Rodda et al. 2007b). This may be important when comparing snake size distributions among sampling locations (e.g., Siers 2015) and particularly important in early detection and rapid response efforts, when all snakes, including larger snakes not adequately attracted to baits (e.g., when satiated or when alternative prey are abundant), must be exposed to opportunities for detection.

BARRIERS

Constructed barriers can be used to enclose or exclude snakes on a temporary or permanent basis. Barrier designs include temporary, bulge, masonry, and vinyl (Rodda et al. 2002). Temporary barriers are used for excluding snakes at cargo staging areas and are 93%–99% effective at restricting snake movement across the barrier (Perry et al. 1998). Bulge barriers consist of 1/4-in. mesh hardware cloth attached to chain-link fence. Near the top, a hardware cloth bulge is formed which prevents further climbing by the snake; as it leans back, it does not have sufficient purchase to retain contact with the fence. Vinyl barriers are made from commercial interlocking sea-wall panels and can be cut to shape. The smooth surface of the material prevents adequate climbing purchase by snakes. A variation is use of fly-ash-covered wall coatings which can be glued to existing vertical structures (Rodda et al. 2007a). Walls of prestressed concrete panels are the sturdiest of the barrier systems, capable of withstanding typhoon-force winds. Under experimental tests, all four designs were 99%–100% effective at preventing snake movement across the barrier (Perry et al. 1998). Finally, electrical barriers have been outfitted onto individual trees and power structures to protect nesting birds (Aguon et al. 2002).

"Area 50" on Guam was enclosed using a bulge barrier, with snakes removed by trapping for the protection of Guam rails (*Gallirallus owstoni*) (Beauprez and Brock 1999). While snake densities remained low, there was a failure to eliminate snakes from this area. Many reasons have been postulated, from inadequate trapping effort to leakage in the barrier system or snakes being refractory to traps (attributable to a variety of causes) (Rodda et al. 2002). Given that severe suppression or elimination of snakes in small plots is feasible (Campbell et al. 1999; Rodda et al. 1999a), leakage and maintenance of the barrier may be the most likely contributory cause of failure in this case, demonstrating the commitment a barrier system needs to succeed. With high levels of maintenance, a similar barrier around the USGS Closed Population (Northwest Field, Andersen Air Force Base) has exceeded longevity expectations and is now 12 years old.

Roads and paths are barriers to wildlife movement in general and snake movement in particular (Trombulak and Frissell 2000). Small-scale studies on snakes fitted with radio transmitters demonstrated over short time periods that brown tree snakes only move short distances, and roads or runways were a partially effective barrier to movement across those substrates (Tobin et al. 1999; Clark and Savarie 2012). A more extensive analysis showed that traffic volume, gap width, and surface type all were negatively related to the probability of a snake crossing a road, and those snakes that did cross tended to be larger. Proximity to traps along a perimeter also decreased the probability of a snake crossing a road (i.e., possible demonstration of an intercepting effect) (Siers et al. 2014, 2016).

DETECTOR DOGS

Working dogs have been employed to detect a variety of targets (e.g., game, drugs, pests, explosives, pathogens) throughout history (Henry 1977; Crooks et al. 1983; Gordon and Haider 2004). As part of an interdiction and control effort, USDA Wildlife Services has used detector dogs to find brown tree snakes in cargo since 1993 (Hall 1996; Engeman et al. 1998b, 2002). Since then, other government agencies have begun to evaluate the utility of dogs in early detection and rapid response activities on islands not yet impacted by snakes. However, improvements are needed for effective detection and recovery of snakes from forested habitats (Savidge et al. 2011).

As used by the USDA, the detector dog effort is generally the last line of defense for inspecting high-risk outward-bound cargo (Linnell and Pitzler 1996; Perry and Vice 2009). Estimates of detection success under operational conditions at cargo ports using blind test challenges of handler-dog teams are in the 70% range (Engeman et al. 1998c). Adverse climatic events such as typhoons seem to promote the movement of snakes; subsequent to storms, detector dogs are especially useful as a line of defensive containment and interdiction because other methods such as traps are generally pulled from service during such events to prevent damage to them (Vice and Engeman 2000). With support from the USFWS and OIA, the Commonwealth of the Northern Mariana Islands—at high risk of accidental importation of snakes from Guam—employs detector-dog teams for the inspection of inbound cargo. These teams are also included in rapid response activities when snake sightings are reported. This dual strategy, covering points of entrance and egress, provides an added measure of effectiveness for interdiction and control.

OTHER CONTROL METHODS

BIOLOGICAL CONTROL

Introduction or augmentation of "natural enemies" (predators or parasites) to control invasive species has been effective against injurious or pest insect and plant systems, and has been considered as one avenue for brown tree snake control (Campbell et al. 1999). Engeman and Vice (2001) briefly review the limited potential and risks of biological control for management of brown tree snakes. Since that time, the potential of parasites for suppression of brown tree snakes has received further investigation— for example, Caudell et al. (2002) and surveys within the brown tree snake's native range for candidate pathogens or parasites (Richmond et al. 2012)—with no clear improvements in the prognosis for biological control. While theoretically holding the potential for a role in an Integrated Pest Management (IPM) program for brown tree snakes, this is currently not deemed likely by us to yield an effective control tool.

HARVEST INCENTIVES

Bounty systems have been used to promote conservation, damage management, and invasive species control (Parkes 1993; Pohja-Mykrä et al. 2005; Zabel and Roe 2009). A full discourse is beyond the scope of this review, however. Commercial markets for brown tree snake products are not feasible in that their skins do not have desirable characteristics for leather goods, and their lean bodies do not provide much meat (which is reported to be only marginally palatable). Bounties, or cash payments to individuals upon evidence of the collection of an organism, may be successful at recovering a large number of snakes, but to achieve effective control, a progressive bounty structure would be required to maintain participation when snakes become harder to obtain. At some point, the financial impetus to continue to produce snakes will create a "perverse incentive" to "cheat," for example, thru captive rearing or surreptitious importation to maintain this source of income; concern also exists that a ready revenue stream on Guam may incentivize the introduction of snakes to neighboring islands, for example, Saipan, Rota, or Tinian, in order to replicate this source of profit in these economically depressed markets. Ultimately, harvest incentives are incompatible with an eradication objective, as a program of harvest incentives is dependent upon a continued yield.

CONCLUSION

Great progress has been made in the understanding of brown tree snake biology and the development of control methods since the last formal synthesis by Rodda et al. (1999e). Control programs have been implemented and continue to undergo evaluation and adaptive management as new technologies develop. However, challenges still exist. Improvements in design and effectiveness are needed for existing technologies. Such improvements will be needed as operational considerations such as scalability (production and large-area implementation) become an issue under budgetary constraints. Efforts to monitor success and devise cost-effective strategies for monitoring and remedial treatments will also be needed. In short, formal systems analyses are needed for a variety of objectives: interdiction, containment, suppression, eradication, conservation, and ecosystem restoration. These analyses require better integration and communication among the various partners and will be critical for optimal resource allocation. The various working groups established by federal and state agencies are a good first step toward these goals.

REFERENCES

Aguon, C. F., E. W. Campbell III, and J. M. Morton. 2002. Efficacy of electrical barriers used to protect Mariana crow nests. *Wildlife Society Bulletin* 30:703–708.

Avery, M. L., E. A. Tillman, and P. Savarie. 2004. Responses of captive fish crows (*Corvus ossifragus*) to acetaminophen baits and bait stations for brown tree snake (*Boiga irregularis*) control on Guam. *Bird Behavior* 16:1–6.

Beauprez, G. M. and M. K. Brock. 1999. Establishment of populations of endangered species in snake-free areas. In *Annual Report, Fiscal*, eds. G. W. Davis, T. J. Pitlik, and G. J. Wiles, pp. 164–169. Mangilao, Guam: Guam Division of Aquatic and Wildlife Resources.

Brooks, J. E., P. J. Savarie, J. J. Johnston, and R. L. Bruggers. 1998. Toxicity of pyrethrin/pyrethroid logger products to brown tree snakes, *Boiga irregularis*, in cargo containers. *SNAKE-NITTAGUN* 28:33–36.

Burnett, K. M., S. D'evelyn, B. A. Kaiser, P. Nantamanasikarn, and J. A. Roumasset. 2008. Beyond the lamppost: Optimal prevention and control of the brown tree snake in Hawaii. *Ecological Economics* 67:66–74.

Campbell III, E. W., G. H. Rodda, and R. L. Bruggers. 1999. An integrated management plan for the brown tree snake (*Boiga irregularis*) on Pacific Islands. In *Problem Snake Management: The Habu and the Brown Treesnake*, eds. G. H. Rodda, Y. Sawai, D. Chiszar, and H. Tanaka, pp. 423–435. Ithaca: Cornell University Press.

Caudell, J. N., J. Whittier, and M. R. Conover. 2002. The effects of haemogregarine-like parasites on brown tree snakes (*Boiga irregularis*) and slatey-grey snakes (*Stegonotus cucullatus*) in Queensland, Australia. *International Biodeterioration & Biodegradation* 49:113–119.

Chiszar, D. A. 1990. The behavior of the brown tree snake: A study in applied comparative psychology. In *Contemporary Issues in Comparative Psychology*, ed. D. A. Dewsbury, pp. 101–123. Sunderland: Sinauer Associates.

Chiszar, D., T. M. Dunn, P. Stark, and H. M. Smith. 2001. Response of brown tree snakes (*Boiga irregularis*) to mammalian blood: Whole blood, serum, and cellular residue. *Journal of Chemical Ecology* 27:979–984.

Chiszar, D., K. Kandler, R. Lee, and H. M. Smith. 1988. Stimulus control of predatory attack in the brown tree snake (*Boiga irregularis*). *Amphibia-Reptilia* 9:77–88.

Christy, M. T., J. A. Savidge, R. Bischof, and G. H. Rodda. 2007. Can temperature be used as a tool for limiting brown tree snake invasion via transportation pathways? In *Managing Vertebrate Invasive Species: Proceedings of an International Symposium*, eds. G. W. Witmer, W. C. Pitt, and K. A. Fagerstone, pp. 246–253. Fort Collins: USDA/APHIS Wildlife Services, National Wildlife Research Center.

Christy, M. T., A. A. Yackel Adams, G. H. Rodda, J. A. Savidge, and C. L. Tyrell. 2010. Modelling detection probabilities to evaluate management and control tools for an invasive species. *Journal of Applied Ecology* 47:106–113.

Clark, L. and M. L. Avery. 2013. Effectiveness of chemical repellents in managing birds at airports. In *Wildlife in Airport Environments: Preventing Animal-Aircraft Collisions Through Science-Based Management*, eds. T. L. DeVault, B. F. Blackwell, and J. L. Belant, pp. 25–35. Baltimore: Johns Hopkins University Press in association with The Wildlife Society.

Clark, L. and P. J. Savarie. 2012. Efficacy of aerial broadcast baiting in reducing brown tree snake numbers. *Human–Wildlife Interactions* 6:212–221.

Clark, L., P. J. Savarie, J. A. Shivik, S. W. Breck, and B. S. Dorr. 2012. Efficacy, effort, and cost comparisons of trapping and acetaminophen-baiting for control of brown tree snakes on Guam. *Human–Wildlife Interactions* 6:222–236.

Clark, L. and J. A. Shivik. 2002. Aerosolized essential oils and individual natural product compounds as brown tree snake repellents. *Pest Management Science* 58:775–783.

Clark, L. and J. A. Shivik. 2004. Identification of snake repellents. U.S. Patent 6,689,397.

Crooks, E., K. Havel, M. Shannon, G. Snyder, and T. Wallenmaier. 1983. Stopping pest introductions. In *Exotic Plant Pests and North American Agriculture*, ed. C. Wilson, pp. 239–259. Amsterdam: Elsevier.

DeVault, T. L. and A. R. Krochmal. 2002. Scavenging by snakes: An examination of the literature. *Herpetologica* 58:429–436.

Dorr, B. S., C. S. Clark, and P. J. Savarie. 2016. Aerial application of acetaminophen-treated baits for control of brown tree snakes (RC-200925; NWRC Study Number: QA-1828). Final Report, US-Department of Defense ESTCP.

El-Sayed, A. M., D. M. Suckling, J. A. Byers, E. B. Jang, and C. H. Wearing. 2009. Potential of "lure and kill" in long-term pest management and eradication of invasive species. *Journal of Economic Entomology* 102:815–835.

Engeman, R. M. and M. A. Linnell. 1998. Trapping strategies for deterring the spread of brown tree snakes from Guam. *Pacific Conservation Biology* 4:348–353.

Engeman, R. M. and M. A. Linnell. 2004. The effect of trap spacing on the capture of brown tree snakes on Guam. *International Biodeterioration & Biodegradation* 54:265–267.

Engeman, R. M., M. A. Linnell, P. Aguon, A. Manibusan, S. Sayama, and A. Techaira. 1999. Implications of brown tree snake captures from fences. *Wildlife Research* 26:111–116.

Engeman, R. M., M. A. Linnell, P. A. Pochop, and J. Gamboa. 1998a. Substantial reductions of brown tree snake (*Boiga irregularis*) populations in blocks of land on Guam through operational trapping. *International Biodeterioration & Biodegradation* 42:167–171.

Engeman, R. M., D. V. Rodriquez, M. A. Linnell and M. E. Pitzler. 1998b. A review of the case histories of the brown tree snakes (*Boiga irregularis*) located by detector dogs on Guam. *International Biodeterioration & Biodegradation* 42:161–165.

Engeman, R. M. and D. S. Vice. 2000. Standardizing the evaluation of brown tree snake trap designs. *Integrated Pest Management Reviews* 5:205–212.

Engeman, R. M. and D. S. Vice. 2001. A direct comparison of trapping and spotlight searches for capturing brown tree snakes on Guam. *Pacific Conservation Biology* 7:4–8.

Engeman, R. M., D. S. Vice, G. Nelson, and E. Muña. 2000. Brown tree snakes effectively removed from a large plot of land on Guam by perimeter trapping. *International Biodeterioration & Bodegradation* 45:139–142.

Engeman, R. M., D. S. Vice, D. V. Rodriguez, K. S. Gruver, W. S. Santos, and M. E. Pitzler. 1998c. Effectiveness of the detector dogs used for deterring the dispersal of brown tree snakes. *Pacific Conservation Biology* 4:256–260.

Engeman, R. M., D. S. Vice, D. York, and K. S. Gruver. 2002. Sustained evaluation of the effectiveness of detector dogs for locating brown tree snakes in cargo outbound from Guam. *International Biodeterioration & Biodegradation* 49:101–106.

Fagerstone, K. A., R. W. Bullard, and C. A. Ramey. 1990. Politics and economics of maintaining pesticide registrations. *Proceedings of the Vertebrate Pest Conference* 14:8–11.

Fagerstone, K. A. and E. W. Schafer Jr. 1998. Status of APHIS vertebrate pesticides and drugs. *Proceedings of the Vertebrate Pest Conference* 18:319–324.

FAO. 2006. *International Standards for Phytosanitary Measures.* Food and Agriculture Organization of the United Nations, Rome, Italy.

Ferraro, D. M. 1995. The efficacy of naphthalene and sulfur repellents. *Great Plains Wildlife Damage and Control Workshop* 12:116–120.

Fritts, T. H. and M. J. McCoid. 1991. Predation by the brown tree snake, *Boiga irregularis*, on poultry and other domesticated animals in Guam. *Snake* 23:75–80.

Fritts, T. H. and G. H. Rodda. 1998. The role of introduced species in the degradation of island ecosystems: A case history of Guam. *Annual Review of Ecology and Systematics* 29:113–140.

Fritts, T. H., N. J. Scott, and B. E. Smith. 1989. Trapping *Boiga irregularis* on Guam using bird odors. *Journal of Herpetology* 2:189–192.

GATFA. 2012. *GTAS Fumigation and Pest Control Manual.* Vol. 5. London: Grain and Feed Trade Association.

Gordon, H. and D. Haider. 2004. The use of "drug dogs" in psychiatry. *Psychiatric Bulletin* 28:196–198.

Greene, H. W. 1989. Ecological, evolutionary, and conservation implications of feeding biology in Old World cat snakes, genus *Boiga* (Colubridae). *Proceedings of the California Academy of Sciences* 46:193–207.

Greene, M. J. and R. T. Mason. 1998. Chemically mediated sexual behavior of the brown tree snake, *Boiga irregularis*. *Ecoscience* 5:405–409.

Greene, M. J. and R. T. Mason. 2003. Pheromonal inhibition of male courtship behaviour in the brown tree snake, *Boiga irregularis*: A mechanism for the rejection of potential mates. *Animal Behaviour* 65:905–910.

Greene, M. J., S. L. Stark, and R. T. Mason. 2001. Pheromone trailing behavior of the brown tree snake, *Boiga irregularis*. *Journal of Chemical Ecology* 27:2193–2201.

Hall, T. C. 1996. Operational control of the brown tree snake on Guam. *Proceedings of the Vertebrate Pest Conference* 17:234–240.

Halpern, M. 1992. Nasal chemical senses in reptiles: Structure and function. *Biology of the Reptilia* 18:423–523.

Heather, N. W. and G. J. Hallman. 2008. *Pest Management and Phytosanitary Trade Barriers.* London: CABI.

Hennessey, M. K., L. Jeffers, D. Nendick et al. 2014. Phytosanitary treatments. In *The Handbook of Plant Biosecurity*, eds. G. Gords and S. McKirdy, pp. 269–308. New York: Springer.

Henry, J. H. 1977. A review of technology for detection of explosives, *Proceedings of SPIE 0108, Optics in Security and Law Enforcement I*:2–9. doi:10.1117/12.955473.

Howse, P., J. M. Stevens, and G. A. D. Jones. 2013. *Insect Pheromones and Their Use in Pest Management.* Berlin: Springer Science & Business Media.

Isman, M. B. 2006. Botanical insecticides, deterrents, and repellents in modern agriculture and an increasingly regulated world. *Annual Reviews of Entomology* 51:45–66.

Johnston, J. J., R. E. Mauldin, P. J. Savarie, J. E. Brooks, and T. M. Primus. 2001. Ecotoxicological risks of potential toxicants for brown tree snake control on Guam. In *Pesticides and Wildlife, ACS Symposium Series 771*, pp. 213–224. Washington DC: American Chemical Society.

Johnston, J. J., P. J. Savarie, T. M. Primus, J. D. Eisemann, J. C. Hurley, and D. J. Kohler. 2002. Risk assessment of an acetaminophen baiting program for chemical control of brown tree snakes on Guam: Evaluation of baits, snake residues, and potential primary and secondary hazards. *Environmental Science & Technology* 36:3827–3833.

Jojola-Elverum, S. M., J. A. Shivik, and L. Clark. 2001. Importance of bacterial decomposition, and carrion substrate to foraging, brown tree snakes. *Journal of Chemical Ecology* 27:1315–1331.

Kardong, K. V. and P. R. Smith. 1991. The role of sensory receptors in the predatory behavior of the brown tree snake, *Boiga irregularis* (Squamata: Colubridae). *Journal of Herpetology* 25:229–231.

Kim, Y., K. Choi, J. Jung, S. Park, P. G. Kim, and J. Park. 2007. Aquatic toxicity of acetaminophen, carbamazepine, cimetidine, diltiazem and six major sulfonamides, and their potential ecological risks in Korea. *Environment International* 33:370–375.

Kimball, B. A., S. A. Stelting, T. W. McAuliffe, R. S. Stahl, R. A. Garcia, and W. C. Pitt. 2016. Development of artificial bait for brown tree snake suppression. *Biological Invasions* 18:1–11.

Kraus, F. 2007. Using pathway analysis to inform prevention strategies for alien reptiles and amphibians. In *Managing Vertebrate Invasive Species: Proceedings of an International Symposium*, eds. G. W. Witmer, W. C. Pitt, and K. A. Fagerstone, pp. 94–103. Fort Collins: USDA/APHIS Wildlife Services, National Wildlife Research Center.

Kraus, F., R. Stahl, and W. Pitt. 2015a. Thermal fumigation provides a simple and effective solution for sanitizing cargo from invasive snakes. *Journal of Pest Science* 88:331–341.

Kraus, F., R. Stahl, and W. Pitt. 2015b. Chemical repellents appear non-useful for eliciting exit of brown tree snakes from cargo. *International Journal of Pest Management* 61:144–152.

Lardner, B., J. A. Savidge, and G. H. Rodda. 2007. Spotting cryptic animals in the dark: What light properties should a good headlamp have? In *Managing Vertebrate Invasive Species: Proceedings of an International Symposium*, eds. G. W. Witmer, W. C. Pitt, and K. A. Fagerstone, pp. 234–245. Fort Collins: USDA/APHIS Wildlife Services, National Wildlife Research Center.

Lardner, B., J. A. Savidge, G. H. Rodda, and R. N. Reed. 2009a. Prey preferences and prey acceptance in juvenile brown tree snakes (*Boiga irregularis*). *Herpetological Conservation and Biology* 4:313–323.

Lardner, B., J. A. Savidge, G. H. Rodda, R. N. Reed, and A. A. Yackel Adams. 2009b. The results of nocturnal visual surveys are influenced by lamp properties. *Applied Herpetology* 6:391–396.

Lardner, B., A. A. Yackel Adams, J. A. Savidge, G. H. Rodda, R. N. Reed, and C. S. Clark. 2013. Effectiveness of bait tubes for Brown tree snake control on Guam. *Wildlife Society Bulletin* 37:664–673.

Li, J., Q. Ye, and J. Gan. 2014. Degradation and transformation products of acetaminophen in soil. *Water Research* 49:44–52.

Linhart, S. B., A. Kappeler, and L. A. Windberg. 1997. A review of baits and bait delivery systems for free-ranging carnivores and ungulates. In *Contraception in Wildlife: Animal and Plant Health Inspection Service*, ed. T. J. Kreeger, pp. 69–132. United States Department of Agriculture, Washington, DC.

Linnell, M. A., R. M. Engeman, M. E. Pitzler, M. O. Warren, G. G. Whitehead, and R. C. Miller. 1998. An evaluation of two designs of stamped metal trap flaps for use in operational trapping of brown tree snakes (*Boiga irregularis*). *SNAKE-NITTAGUN* 28:14–18.

Linnell, M. A. and M. E. Pitzler. 1996. Summary of Guam's outbound cargo process with emphasis on preventing the spread of the brown tree snake off-island. Working Draft Report. Guam District: USDA/APHIS/ Animal Damage Control.

Mackessey, S. P., N. M. Sixberry, W. H. Heyborne, and T. Fritts. 2006. Venom of the brown treesnake, *Boiga irregularis*: Ontogenetic shifts and taxa-specific toxicity. *Toxicon* 47:537–548.

Mason, J. R. and L. Clark. 1995. Avian repellents: Options, modes of action, and economic considerations. In *Repellents in Wildlife, National Wildlife Research Center Conference Proceedings*, ed. J. R. Mason, pp. 371–390. Fort Collins: USDA/APHIS Wildlife Services, National Wildlife Research Center.

Mason, L. C., J. A. Savidge, G. H. Rodda, and A. A. Yackel Adams. 2011. Scented guide ropes as a method to enhance brown tree snake (*Boiga irregularis*) trap capture success on Guam. *Journal of Herpetology* 45:308–312.

Mathies, T., J. A. Cruz, V. A. Lance, and J. A. Savidge. 2010. Reproductive biology of male brown tree snakes (*Boiga irregularis*) on Guam. *Journal of Herpetology* 44:209–221.

Mathies, T., B. Levine, R. Engeman, and J. A. Savidge. 2013. Pheromonal control of the invasive brown tree snake: Potency of female sexual attractiveness pheromone varies with ovarian state. *International Journal of Pest Management* 59:141–149.

Moore, I. T., M. J. Greene, D. T. Lerner et al. 2005. Physiological evidence for reproductive suppression in the introduced population of brown tree snakes (*Boiga irregularis*) on Guam. *Biological Conservation* 121:91–98.

Mortensen, H. S., Y. L. Dupont, and J. M. Olesen. 2008. A snake in paradise: Disturbance of plant reproduction following extirpation of bird flower-visitors on Guam. *Biological Conservation* 141:2146–2154.

Parkes, J. P. 1993. The ecological dynamics of pest-resource-people systems. *New Zealand Journal of Zoology* 20:223–230.

Perry, G. 2002. Wheel-well and cargo compartment temperatures of large aircraft in flight: Implications for stowaways. *Aviation, Space, and Environmental Medicine* 73:673–676.

Perry, G., E. W. Campbell III, G. H. Rodda, and T. H. Fritts. 1998. Managing island biotas: Brown tree snake control using barrier technology. *Proceedings of the Vertebrate Pest Conference* 18:138–143.

Perry, G. and J. M. Morton. 1999. Regeneration rates of the woody vegetation of Guam's Northwest Field following major disturbance: Land use patterns, feral ungulates, and cascading effects of the brown tree snake. *MICRONESICA-AGANA* 32:125–142.

Perry, G. and D. S. Vice. 2007. An evaluation of passive thermal fumigation for brown tree snake control in surface transportation from Guam. In *Managing Vertebrate Invasive Species: Proceedings of an International Symposium*, eds. G. W. Witmer, W. C. Pitt, and K. A. Fagerstone, pp. 224–233. Fort Collins: USDA/APHIS Wildlife Services, National Wildlife Research Center.

Perry, G. and D. Vice. 2009. Forecasting the risk of brown tree snake dispersal from Guam: A mixed transport-establishment model. *Conservation Biology* 23:992–1000.

Pickett, J. A., S. Barasa, and M. A. Birkett. 2014. Vertebrate pheromones and other semiochemicals: The potential for accommodating complexity in signaling by volatile compounds for vertebrate management. *Biochemical Society Transactions* 42:846–850.

Pohja-Mykrä, M., T. Vuorisalo, and S. Mykrä. 2005. Hunting bounties as a key measure of historical wildlife management and game conservation: Finnish bounty schemes 1647–1975. *Oryx* 39:284–291.

Primus, T. M., D. J. Kohler, C. A. Furcolow, M. J. Goodall, J. J. Johnston, and P. J. Savarie. 2004. Determination of acetaminophen residues in whole body brown tree snakes. *Journal of Liquid Chromatography & Related Technologies* 27:897–909.

Ravensberg, W. J. 2011. *A Roadmap to the Successful Development and Commercialization of Microbial Pest Control Products for Control of Arthropods.* Vol. 10., Dordrecht, Netherlands: Springer Science & Business Media.

Richmond, J. Q., J. W. Stanford, D. A. Wood, and R. N. Fisher. 2012. An evolutionary approach to the biological management of invasive brown tree snakes (*Boiga irregularis*) *on Guam.* Final draft report to SERDP, Project No. 10 SISEED01-007 (SI-1733), approved for public release, accessed on October 11, 2015 at http://oai.dtic.mil/oai/oai?verb=getRecord&metadataPrefix=html&identifier=ADA600553.

Richmond, J. Q., D. A. Wood, J. W. Stanford, and R. N. Fisher. 2015. Testing for multiple invasion routes and source populations for the invasive brown tree snake (*Boiga irregularis*) on Guam: Implications for pest management. *Biological Invasions* 17:337–349.

Rodda, G. H., J. L. Farley, R. Bischof, and R. N. Reed. 2007a. New developments in snake barrier technology: Flyash covered wall offers a feasible alternative for permanent barriers to brown tree snakes (*Boiga irregularis*). *Herpetological Conservation and Biology* 2:157–163.

Rodda, G. H., T. H. Fritts, and E. W. Campbell III. 1999a. The feasibility of controlling the brown tree snake in small plots. In *Problem Snake Management: The Habu and the Brown Treesnake*, eds. G. H. Rodda, Y. Sawai, D. Chiszar, and H. Tanaka, pp. 469–477. Ithaca: Cornell University Press.

Rodda, G. H., T. H. Fritts, E. W. Campbell III, K. Dean-Bradley, G. Perry, and C. P. Qualls. 2002. Practical concerns in the eradication of island snakes. In *Turning the Tide: The Eradication of Invasive Species. Proceedings of the International Conference on Eradication of Island Invasives*, eds. C. R. Veitch and M. N. Clout, pp. 260–265. Gland, Switzerland and Cambridge, UK: IUCN.

Rodda, G. H., T. H. Fritts, and D. Chiszar. 1997. The disappearance of Guam's wildlife. *BioScience* 47:565–574.

Rodda, G. H., T. H. Fritts, C. S. Clark, S. W. Gotte, and D. Chiszar. 1999b. A state-of-the-art trap for the brown tree snake. In *Problem Snake Management: The Habu and the Brown Treesnake*, eds. G. H. Rodda, Y. Sawai, D. Chiszar, and H. Tanaka, pp. 268–305. Ithaca: Cornell University Press.

Rodda, G. H., T. H. Fritts, and P. J. Conry. 1992. Origin and population growth of the brown tree snake, *Boiga irregularis*, on Guam. *Pacific Science* 46:46–57.

Rodda, G. H., T. H. Fritts, M. J. McCoid, and E. W. Campbell III. 1999c. An overview of the biology of the brown tree snake (*Boiga irregularis*), a costly introduced predator on Pacific Islands. In *Problem Snake Management: The Habu and the Brown Treesnake*, eds. G. H. Rodda, Y. Sawai, D. Chiszar, and H. Tanaka, pp. 44–80. Ithaca: Cornell University Press.

Rodda, G. H., M. J. McCoid, T. H. Fritts, and E. W. Campbell III. 1999d. Population trends and limiting factors in *Boiga irregularis*. In *Problem Snake Management: The Habu and the Brown Treesnake*, eds. G. H. Rodda, Y. Sawai, D. Chiszar, and H. Tanaka, pp. 236–254. Ithaca: Cornell University Press.

Rodda, G. H. and R. N. Reed. 2007. Size-based trends and management implications of microhabitat utilization by brown tree snakes, with an emphasis on juvenile snakes. In *Managing Vertebrate Invasive Species: Proceedings of an International Symposium*, eds. G. W. Witmer, W. C. Pitt, and K. A. Fagerstone, pp. 257–267. Fort Collins: USDA/APHIS Wildlife Services, National Wildlife Research Center.

Rodda, G. H. and J. A. Savidge. 2007. Biology and impacts of Pacific Island invasive species. 2. *Boiga irregularis*, the brown tree snake (Reptilia: Colubridae). *Pacific Science* 61:307–324.

Rodda, G. H., J. A. Savidge, C. L. Tyrrell, M. T. Christy, and A. R. Ellingson. 2007b. Size bias in visual searches and trapping of brown tree snakes on Guam. *The Journal of Wildlife Management* 71:656–661.

Rodda, G. H., Y. Sawai, D. Chiszar, and H. Tanaka. 1999e. *Problem Snake Management: The Habu and the Brown Treesnake*. Ithaca: Cornell University Press.

Rogers, H. S. 2011. The fate of a silent forest: The effects of complete bird loss on the forest of Guam. PhD diss, Univ. of Washington.

Rogers, H., J. H. R. Lambers, R. Miller, and J. J. Tewksbury. 2012. 'Natural experiment' demonstrates top-down control of spiders by birds on a landscape level. *PloS One* 7:p. e43446.

Savarie, P. J. 2002. Acute oral toxicity of acetaminophen-treated mice to brown tree snakes. Unpublished report—QA-636. Fort Collins, Colorado: National Wildlife Research Center.

Savarie, P. J. 2012. Development of non-prey baits for delivery of acetaminophen to brown tree snakes (*Boiga irregularis*) on Guam. Final Report. Hawaii Invasive Species Council.

Savarie, P. J. and L. Bruggers. 1999. Candidate repellents, oral and dermal toxicants, and fumigants for brown tree snake control. In *Problem Snake Management: The Habu and the Brown Treesnake*, eds. G. H. Rodda, Y. Sawai, D. Chiszar, and H. Tanaka, pp. 417–422. Ithaca: Cornell University Press.

Savarie, P. J., R. L. Bruggers, and W. S. Wood. 1995. Fumigation of brown tree snakes with methyl bromide, sulfuryl fluoride, and phosphine on Guam. Unpublished report on file with the US Department of Agriculture, Fort Collins, Colorado: Animal and Plant Health Inspection Service, National Wildlife Research Center.

Savarie, P. J. and L. Clark. 2006. Evaluation of bait matrices and chemical lure attractants for brown tree snakes. *Proceedings of the Vertebrate Pest Conference* 22:483–488.

Savarie, P. J., T. C. Mathies, and K. A. Fagerstone. 2007. Flotation materials for aerial delivery of acetaminophen toxic baits to brown tree snakes. In *Managing Vertebrate Invasive Species: Proceedings of an International Symposium*, eds. G. W. Witmer, W. C. Pitt, and K. A. Fagerstone, pp. 218–223. Fort Collins: USDA/APHIS Wildlife Services, National Wildlife Research Center.

Savarie, P. J., J. A. Shivik, G. C. White, J. C. Hurley, and L. Clark. 2001. Use of acetaminophen for large-scale control of brown tree snakes. *Journal Wildlife Management* 65:356–365.

Savarie, P. J., W. S. Wood, G. H. Rodda, R. L. Bruggers, and R. M. Engeman. 2005. Effectiveness of methyl bromide as a cargo fumigant for brown tree snakes. *International Biodeterioration & Biodegradation* 56:40–44.

Savarie, P. J., D. L. York, J. C. Hurley, S. Volz, and J. E. Brooks. 2000. Testing the dermal and oral toxicity of selected chemicals to brown tree snakes. *Proceedings of the Vertebrate Pest Conference* 19:139–145.

Savidge, J. A. 1987. Extinction of an island forest avifauna by an introduced snake. *Ecology* 68:660–668.

Savidge, J. A. 1988. Food habits of *Boiga irregularis*, an introduced predator on Guam. *Journal of Herpetology* 22:275–282.

Savidge, J. A. 1991. Population characteristics of the introduced brown tree snake (*Boiga irregularis*) on Guam. *Biotropica* 23:294–300.

Savidge, J. A., F. J. Qualls, and G. H. Rodda. 2007. Reproductive biology of the brown tree snake, *Boiga irregularis* (Reptilia: Colubridae), during colonization of Guam and comparison with that in their native range. *Pacific Science* 61:191–199.

Savidge, J. A., J. W. Stanford, R. N. Reed, G. R. Haddock, and A. A. Yackel Adams. 2011. Canine detection of free-ranging brown tree snakes on Guam. *New Zealand Journal of Ecology* 35:174–181.

Schwab, B. W., E. P. Hayes, J. M. Fiori et al. 2005. Human pharmaceuticals in US surface waters: A human health risk assessment. *Regulatory Toxicology and Pharmacology* 42:296–312.

Schwenk, K. 1995. Of tongues and noses: Chemoreception in lizards and snakes. *Trends in Ecology & Evolution* 10:7–12.

Sharp, T. and G. Saunders. 2011. *A Model for Assessing the Relative Humaneness of Pest Animal Control Methods.* Canberra: Department of Agriculture, Fisheries and Forestry, Australia.

Shine, R. 1991. Strangers in a strange land: Ecology of the Australian colubrid snakes. *Copeia* 1991:120–131.

Shivik, J. A. 1998. Brown tree snake response to visual and olfactory cues. *Journal of Wildlife Management* 62:105–111.

Shivik, J. A. and L. Clark. 1997. Carrion seeking in brown tree snakes: Importance of olfactory and visual cues. *Journal of Experimental Zoology* 279:549–553.

Shivik, J. A. and L. Clark. 1999a. Ontogenetic shifts in carrion attractiveness to brown tree snakes (*Boiga irregularis*). *Journal of Herpetology* 33:334–336.

Shivik, J. A. and L. Clark. 1999b. The development of chemosensory attractants for brown tree snakes. In *Advances in Chemical Signals in Vertebrates*, eds. R. E. Johnson, D. Mueller-Schwarze, and P. W. Sorenson, pp. 649–654. Ithaca: Springer.

Shivik, J. A., P. J. Savarie, and L. Clark. 2002. Aerial delivery of baits to brown tree snakes. *Wildlife Society Bulletin* 30:1062–1067.

Shivik, J. A., W. G. Wright, and L. Clark. 2000. Seasonal variability in brown tree snake (*Boiga irregularis*) response to lures. *Canadian Journal of Zoology* 78:79–84.

Shwiff, S. A., K. Gebhardt, K. N. Kirkpatrick, and S. A. Shwiff. 2010. Potential economic damage from introduction of brown tree snakes, *Boiga irregularis* (Reptilia: Colubridae), to the Islands of Hawai'i. *Pacific Science* 64:1–10.

Siers, S. R. 2015. Microgeographic and ontogenetic variability in the ecology of invasive Brown Treesnakes on Guam, and effects of roads on their landscape-scale movements. PhD diss., Colorado State Univ.

Siers, S. R., R. N. Reed, and J. A. Savidge. 2016. To cross or not to cross: Modeling wildlife road crossings as a binary response variable with contextual predictors. *Ecosphere* 7:e01292.

Siers, S. R., J. A. Savidge, and R. N. Reed. 2014. Invasive brown treesnake movements at road edges indicate road-crossing avoidance. *Journal of Herpetology* 48:500–505.

Siers, S. R., G. A. Swayze, and S. P. Mackessey. 2013. Spectral analysis reveals limited potential for enhanced-wavelength detection of invasive snakes. *Herpetological Review* 44:56–58.

Stark, C. P., D. Chiszar, K. E. Stiles, and H. M. Smith. 2002. A laboratory situation for studying the effects of chemical and visual cues on prey trailing in brown tree snakes (*Boiga irregularis*). *Journal of Herpetology* 36:57–62.

Stevens, G. R. and L. Clark. 1998. Bird repellents: Development of avian-specific tear gases for resolution of human–wildlife conflicts. *International Biodeterioration & Biodegradation* 42:153–160.

Takashi, C., S. Seiichi, and T. Tatsuhiro. 1992. Animal repellent made of resin. Japanese Patent 04 046 107A.

Tobin, M. E., R. T. Sugihara, P. A. Pochop, and M. A. Linnell. 1999. Nightly and seasonal movements of *Boiga irregularis* on Guam. *Journal of Herpetology* 33:281–291.

Trombulak, S. C. and C. A. Frissell. 2000. Review of ecological effects of roads on terrestrial and aquatic communities. *Conservation Biology* 14:18–30.

Tyrrell, C. L., M. T. Christy, G. H. Rodda et al. 2009. Evaluation of trap capture in a geographically closed population of brown tree snakes on Guam. *Journal of Applied Ecology* 46:128–135.

USDA. 2003. Snake Repellents. USDA APHIS Wildlife Services Tech Note. https://www.aphis.usda.gov/wildlife_damage/nwrc/publications/Tech_Notes/TN_SnakeRepellents.pdf.

Vice, D. S. and R. M. Engeman. 2000. Brown tree snake discoveries during detector dog inspections following Supertyphoon Paka. *Micronesia* 33:105–110.

Vice, D. S., R. M. Engeman, and D. L. Vice. 2005. A comparison of three trap designs for capturing brown tree snakes on Guam. *Wildlife Research* 32:355–359.

Whittier, J., C. Macrokanis, and R. T. Mason. 2000. Morphology of the brown tree snake, *Boiga irregularis*, with a comparison of native and extralimital populations. *Australian Journal of Zoology* 48:357–367.

Wiewel, A. S., A. A. Yackel Adams, and G. H. Rodda. 2009. Distribution, density, and biomass of introduced small mammals in the southern Mariana Islands. *Pacific Science* 63:205–222.

Wiles, G. J., J. Bart, R. E. Beck, and C. F. Aguon. 2003. Impacts of the brown tree snake: Patterns of decline and species persistence in Guam's avifauna. *Conservation Biology* 17:1350–1360.

Witmer, G. W., P. W. Burke, W. C. Pitt, and M. L. Avery. 2007. Management of invasive vertebrates in the United States: An overview. In *Managing Vertebrate Invasive Species: Proceedings of an International Symposium*, eds. G. W. Witmer, W. C. Pitt, and K. A. Fagerstone, pp. 127–137. Fort Collins: USDA/APHIS Wildlife Services, National Wildlife Research Center.

Work, T. T., D. G. McCullough, J. F. Cavey, and R. Komsa. 2005. Arrival rate of nonindigenous insect species into the United States through foreign trade. *Biological Invasions* 7:323–332.

Zabel, A. and B. Roe. 2009. Optimal design of pro-conservation incentives. *Ecological Economics* 69:126–134.

8 Burmese Pythons

Michael E. Dorcas, Shannon E.
Pittman, and John D. Willson

CONTENTS

INTRODUCTION

Recent years have seen a substantial increase in the number of snake invasions. Snakes represent a particularly alarming emerging threat as damaging, but poorly understood, invasive predators (Kraus 2009). The most well-studied snake invasion is that of the brown tree snake (*Boiga irregularis*) in Guam (see Chapter 7). Accidental introduction of brown tree snakes to Guam shortly after World War II resulted in extirpation of over 50% of the island's native breeding birds and several lizards (Savidge 1987), with cascading effects on invertebrate communities and vegetation dynamics (Mortensen et al. 2008; Caves et al. 2013). Several other snake invaders have been introduced elsewhere, most as a result of the reptile pet trade. Burmese pythons (*Python molurus bivittatus*), northern African pythons (*Python sebae*), and boa constrictors (*Boa constrictor*) are now locally established in Florida (Snow et al. 2007). Boa constrictors are also established on the islands of Cozumel, Aruba, and Puerto Rico (Martínez-Morales and Cuaron 1999; Rogers et al. 2012; Romero et al. 2009). Finally, three species of water snakes (*Nerodia*) are established in California (Rose and Todd 2014), and most recently, California kingsnakes (*Lampropeltis getula*) on the Canary Islands (R. Reed, pers. comm.).

Perhaps the snake invasion that has received the most attention over the past decade is that of the Burmese python in Florida. The Burmese python, which is native to southern Asia, is one of the largest snakes in the world and is now found throughout much of southern Florida. Although a few pythons were found in the southern part of Everglades National Park (ENP) in the 1980s and 1990s, it was not until the turn of the century that pythons were recognized as an established reproducing population (Meshaka et al. 2000). Since that time, their numbers and geographic range in Florida have increased substantially, and it has become apparent that the potential impact of pythons on native ecosystems could be severe (Dorcas and Willson 2013). Our goals in this chapter are to: (1) provide background information on the history of the Burmese python invasion and spread, (2) describe the impacts pythons have had on native ecosystems and societal issues related to those impacts, (3) discuss current and potential management approaches and collection techniques, and (4) discuss challenges related to understanding and predicting future impacts and to managing and controlling python populations.

HISTORY OF THE INVASION

Pythons have been found in the South Florida Everglades since at least the 1980s (Snow et al. 2007). Many of these early records were assumed to be recently released pets; however, finding more pythons (including juveniles) in the late 1990s and early 2000s resulted in the conclusion that they were likely established as a reproducing population within ENP (Meshaka 2011). In the 1980s and 1990s, most pythons were found in wilderness areas of the southern part of ENP, especially along the Main Park Road just north of Flamingo. This area is covered by dense mangrove forest, often flooded by brackish water. A few pythons were also found on Long Pine Key in ENP, and one individual was found along the northeastern border of the park in the area known as Chekika (Snow et al. 2007). In the early 2000s, efforts to find pythons increased dramatically, as did the area known to be occupied by the species. Numerous pythons, including hatchlings, were found in the mangroves of the southern Everglades, both along the Main Park Road and even in remote areas far away from areas frequented by visitors. Shortly thereafter, the number of records increased dramatically in the Long Pine Key area and along the eastern border of the park (Snow et al. 2007). Pythons also began turning up in large numbers on the northern park boundary along Tamiami Trail (US 41), the Shark Valley Loop Road, and along nearby canals bordering and adjacent to Tamiami Trail. During this time, a few individuals were found in western portions of the park in Big Cypress National Preserve and in Collier-Seminole State Park where they have since become relatively common (Andreadis 2011; Dorcas and Willson 2011, 2013). During the last decade, the number of pythons found each year continues to be high but somewhat variable (Figure 8.1). Variable numbers of python records likely reflect a combination of variation in effort to find them and actual changes in python density. For example, the low numbers of pythons captured during 2011 and 2012 may be partly a result of a severe freeze that killed many pythons in ENP (Mazzotti et al. 2010). As of 2015, pythons appear to be ubiquitous in most areas south of Alligator Alley (I-75), and the number of pythons occurring north of I-75 appears to be increasing as well (Figure 8.2). Because of their extremely low detectability, it is often difficult to determine whether observations of pythons represent new releases or signs of newly established

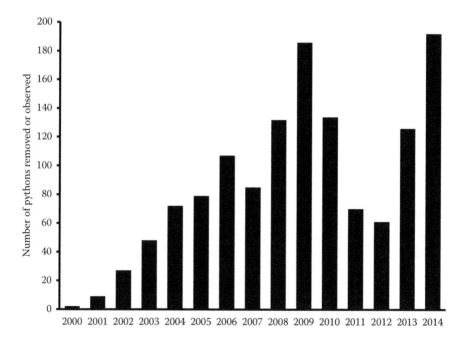

FIGURE 8.1 Number of pythons captured within ENP from 2000–2014. Note, data are not corrected for sampling effort. The low numbers of pythons captured during 2011 and 2012 may be partly a result of a severe freeze that killed many pythons in ENP. (Mazzotti et al. 2010; Unpublished data, Bryan Falk, USGS, Everglades National Park.)

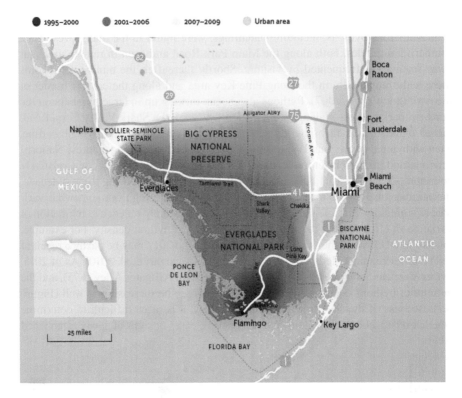

FIGURE 8.2 Approximate distribution of invasive Burmese pythons in south Florida through 2009. (Map reprinted with permission from Dorcas, M. E., and J. D. Willson, 2011, *Invasive Pythons in the United States: Ecology of an Introduced Predator*, Athens, GA, University of Georgia Press.)

or expanding populations (Dorcas and Willson 2013). Generally, by the time a few individuals are discovered in an area, it is likely that an established reproducing population is already present.

ESTABLISHMENT

Burmese pythons have been a mainstay of the reptile pet industry since the 1970s. They have been imported to the United States and bred by the tens of thousands over the last few decades. The docile nature of Burmese pythons in captivity combined with the fact that they are relatively easy to care for makes them favorite pet snakes for many reptile enthusiasts. Perhaps the primary reason many people obtain Burmese pythons as pets is because they can quickly grow to become very large snakes— something many people find fascinating. It is ironic then that their potentially large size is the very reason people often wish to get rid of them, thus increasing the likelihood that misguided pet owners might release unwanted animals in the wild.

Media attention focused on the python invasion has been intense, and several scenarios for python introduction have been proposed, generally with little or no support or scientific evaluation. Some of the more outlandish stories proposed include the idea that pythons were intentionally released by government scientists solely to prompt bans by the federal government on keeping snakes as pets. Other stories include pythons being blown, like Dorothy and Toto in *The Wizard of Oz*, through the air by Hurricane Andrew over many miles until they landed safely in the southern parts of ENP (Barker and Barker 2011). There are even stories of secret government laboratories deep in the heart of the Everglades where scientists trained pythons to crawl into buildings and tunnels after being fed pig carcasses laden with explosives, and when funding for the secret program got cut, the pythons were released. More realistically, it is likely that misguided pet owners released their unwanted pet pythons into ENP. Although it is difficult or even impossible to know for sure, two facts are nearly incontrovertible relating to the establishment of pythons in ENP. First, the vast majority of the pythons captured since 2000 are wild individuals (i.e., born in the wild and not released pets), and second, the source for the python population in Florida is undeniably the reptile pet trade.

The first evidence of wild reproduction by pythons included one juvenile python collected in 1995 and four hatchlings collected in 2002, all near West Lake in the far southern portion of ENP (Snow et al. 2007). This area is at least 32 km from the nearest agricultural or urban areas, and pythons were not regularly found in areas closer to Homestead and Miami until well after the year 2000 (Figure 8.2). Additionally, studies that have modeled python population growth based on life history characteristics indicate that it would take at least 30 years for the python population to grow from a few individuals to the thousands of pythons currently present in southern Florida (Willson et al. 2011a). Thus, in order for Hurricane Andrew (1992) to have been the source of the initial python population in ENP, as many have proposed, a large number of young snakes or a fewer number of large adults would have had to traverse at least 32 km to southern portions of ENP, without colonizing intervening habitats, where pythons have since become common (Willson et al. 2011a). Considering the potential population growth rate and the spatial patterns of the invasion, the most likely scenario for the establishment of pythons involves the introduction of a relatively small number of snakes to the southern portions of ENP around or before 1985 (Willson et al. 2011a). It is certainly possible, or even probable, that additional snakes kept as pets subsequently escaped or were released by owners, bolstering the wild population or perhaps speeding their spread across south Florida.

WHY ARE BURMESE PYTHONS SUCCESSFUL INVADERS?

The rapid expansion of the python population, both in terms of numbers and geographic range, is significant. After an apparent lag period, characteristic of many invasive species (Simberloff 2009), the python population expanded rapidly, eventually covering several thousand square km (Figure 8.2). Numerous ecological factors may be responsible for the success and spread of pythons, and these are discussed at length in Reed et al. (2011). Although the thought of one of the world's largest species of snakes thriving as an invasive species is difficult for many lay people and scientists alike to believe, a careful retrospective look can identify numerous factors that likely have contributed to its success as an invasive species. Such factors include the python's large body size, generalist diet, and high reproductive potential. Additionally, the adaptability of Burmese pythons (i.e., they occur in a variety of habitats—from deserts to tropical rainforests—over a broad geographic range in their native range) also likely contributes to their success as an invasive species.

Adult pythons are considerably larger than any native snake in south Florida, reaching lengths of nearly 6 m (Reed et al. 2012). Even as juveniles, pythons are larger than the adult size of many native snakes. Large size clearly confers an advantage by reducing the number of predators that can potentially prey on pythons, and as pythons grow, the number of potential predators only decreases. Apart from humans, large alligators, crocodiles, black bears, and perhaps Florida panthers are the only animals likely to prey on large adult pythons. A large body size also confers energetic advantages. Heavy-bodied snakes have the ability to store large energy reserves (i.e., fat), thus allowing them to survive long periods without food. Finally, the typically large body size of female pythons allows them to produce large clutches (potentially over 100) of relatively large eggs, when compared to native snakes, thus allowing for rapid population increase. The large size of female pythons also confers advantages after egg-laying because, unlike most snakes, female pythons remain with the eggs until hatching. Female pythons coil around their eggs, protecting them and helping to buffer them from environmental temperature fluctuations. Shivering thermogenesis, a behavior in which female Burmese pythons generate body heat (i.e., become endothermic) by contracting muscles in the body wall, actually warms the eggs, apparently helping to ensure warm and more uniform developmental temperatures (Dorcas and Willson 2011).

Burmese pythons can also grow very rapidly, likely reaching reproductive size within two to three years (Willson et al. 2014), and potentially allowing them to thrive in environments where longer-lived or slower-reproducing species cannot. In their native range, large constrictors such as Burmese pythons are frequently found in urbanized habitats, and sizeable Burmese pythons are known to inhabit large cities in southern Asia.

We are often asked "What controls the python population in its native range?" and "Why do they not have such severe impacts there?" Although there are numerous potential answers to these questions, the fact is, we know considerably more about Burmese pythons in southern Florida than we do in their native range. Only a few ecological studies of pythons have been conducted in Asia, and no study has explored the trophic roles they play in any detail. Difficulties with logistics and permitting severely limit research that could provide considerable insight into factors related to their population biology and trophic relationships. Thus, most of what we know about pythons in their native range is anecdotal.

SCOPE OF ISSUES: IMPACTS

Burmese pythons represent a novel apex predator substantially different from any others naturally occurring in the United States. Pythons are considerably larger than any native snake species; they grow rapidly, have higher fecundity, can likely occur in high densities, and can prey on a greater variety of prey. Mammals documented as prey of pythons include numerous species of rodents, rabbits, raccoons, opossum, bobcats, and deer (Dorcas and Willson 2011). Pythons have been documented preying on many species of birds, especially wading birds such as herons, egrets, and limpkin. Although they apparently do not regularly prey on fish, amphibians, or other snakes, pythons have also been documented on numerous occasions preying on alligators. Thus, a large python is a potential predator of nearly any species of large terrestrial vertebrate occurring in southern United States. The threat to native wildlife is the primary impact of invasive Burmese pythons in the United States.

Recently, we documented severe declines in several once-common mammals in ENP that correspond temporally and spatially with the proliferation of pythons in the region (Dorcas et al. 2012) (Figure 8.3). When we compared mammal relative abundances documented during road surveys from the time shortly before pythons

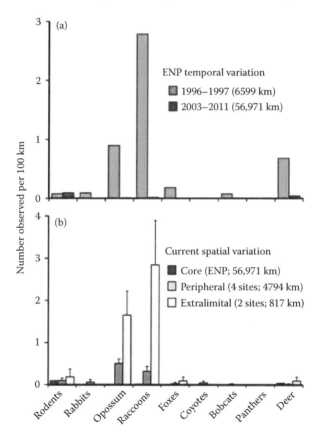

FIGURE 8.3 Variation in mammal abundances in south Florida in relation to the Burmese python invasion. (a) Temporal variation in mammal encounter rates in ENP, as reflected in distance-corrected road survey counts before (1996–1997) and after (2003–2011) pythons became common in the region. (b) Current spatial variation in mammal encounter rates in core (ENP), peripheral, and extralimital regions of python range in Florida. Pythons have been recorded for over a decade in the core region, only more recently in peripheral locations, and not at all in extralimital locations. (Adapted from Dorcas, M. E. et al. 2012, *Proceedings of the National Academy of Sciences of the United States of America,* 109 (7), 2418–22.)

were recognized as established (1996–1997) to a period after python proliferation (2003–2011), we found 99% declines of raccoons and opossums, 94% declines in white-tailed deer, and 87% declines in bobcats. Of particular note has been a complete absence of rabbits in the core of the python's introduced range since at least 2005 (Dorcas et al. 2012). A subsequent study on translocated marsh rabbits substantiates our research by directly linking pythons to the extirpation of marsh rabbits in ENP. McCleery et al. (2015) translocated marsh rabbits to two sites in ENP where they had been extirpated, as well as to a control site at the periphery of the python invasion. Translocations failed to re-establish marsh rabbit populations in ENP, and the authors found that python predation accounted for 77% of rabbit mortality within ENP, whereas mammals were the primary predators (accounting for 68% of mortality) at the control site (McCleery et al. 2015). The decline of midsized mammals in areas invaded by pythons is particularly alarming and raises many unanswered questions, including: (1) What characteristics have allowed pythons to have such dramatic impacts on once-common species of mammals? (2) Are pythons having similar impacts on species less tractable to sampling via road transects (e.g., birds, rare species, and aquatic species)? (3) Will native prey populations recover and/or adapt to the presence of pythons as predators? (4) What impacts will such severe reductions in mammals have on other components of the southern Florida ecosystems (i.e., indirect effects or trophic cascades)?

The most immediate concern related to python impacts is their potential impacts on rare or endangered species, particularly federally protected species like the wood stork and Key Largo woodrat, both of which have been recorded in the diets of pythons. Although not federally listed as endangered or threatened, numerous other bird species characteristic of the Everglades are likely threatened by pythons. These include rails of various species, herons, egrets, limpkins, roseate spoonbills, anhingas, purple gallinules, and white-crowned pigeons (Dove et al. 2011). Numerous American alligators have been documented as prey of pythons; the impacts on this iconic member of the Everglades food web, or its endangered cousin the American crocodile, remain unknown (Dorcas and Willson 2011). Large pythons have been reported to consume leopards in their native range, and thus, potential impacts of python predation on the highly endangered Florida panther cannot be discounted. Indirect effects of pythons on some endangered species could be even greater than the direct effects of predation. For example, the decline in the bobcat population in ENP may be related to direct predation, but may also be related to a severe decline in rabbits resulting from python predation. Likewise, python-related reductions in deer populations may substantially reduce the primary prey of Florida panthers (Maehr et al. 1990).

Although the declines in mammals attributable to pythons have received considerable attention, of even more concern are the potential long-term ecosystem-level changes that could result from the python invasion. Numerous plant and animal species may be indirectly affected by reductions in mammal and bird densities in areas where pythons are abundant. For example, the reduction in bobcat numbers (87% decline) in ENP could be a result of direct predation by pythons, but could also be related to apparent local extirpation of rabbits, an important prey species for bobcats. Ecosystem-level effects can be very complex, and some species may actually benefit from python predation on other species. For example, the reduction in numbers of foxes, raccoons, and opossum, all of which are major nest predators, may improve the nesting success of numerous species of birds and turtles (Dorcas and Willson 2011). Brown tree snakes have indirectly affected arthropod diversity and vegetation dynamics in Guam (Mortensen et al. 2008; Rogers et al. 2012), suggesting that predation by invasive snakes can alter entire ecosystems.

WHY SUCH SEVERE IMPACTS?

Snakes exhibit several key attributes that make them fundamentally different from other terrestrial predators (Reed et al. 2012). Most notably, snakes have very high assimilation efficiencies compared to endothermic predators and epitomize the "low energy" lifestyle characteristic of ectothermic vertebrates (Pough 1980; Bonnet et al. 1998). As ectotherms, snakes convert assimilated prey biomass into reproductive output and energy storage very efficiently, which allows them to take advantage of brief pulses of high prey abundance (e.g., bird migrations, rodent eruptions) and survive long periods of prey scarcity (Reed et al. 2012). Additionally, pythons are not territorial, and thus, python densities are not limited by intraspecific interactions (e.g., territoriality). Finally, because pythons are generalist predators, they can survive declines or extirpation of preferred prey (e.g., mammals) by switching to alternative prey (e.g., wading birds and alligators). These characteristics contribute to the ability of invasive snakes to literally eat prey to extinction, thus making pythons particularly damaging as invasive species.

RISK TO DOMESTIC ANIMALS

Although the full impact pythons are having on native ecosystems has yet to be determined, domestic animals have been subjected to python predation in southern Florida. In their native range, Burmese pythons are known to consume pets and other domestic animals, particularly poultry and small livestock such as goats. Burmese pythons in Florida have consumed domestic chickens, geese, turkeys, and house cats. In 2013, a wild northern African python killed a full-grown pet Siberian husky on the outskirts of Miami. The risks posed by pythons to domestic animals are likely similar to the risk already posed by American alligators. However, because pythons are so secretive, they may be able to persist and grow large in close proximity to people and their domestic animals without being noticed and, unlike alligators, are not necessarily tied to aquatic habitats. It is quite possible that pythons taking up residence in vacant lots or along the banks of canals—which are ubiquitous throughout southern Florida— might enter suburban yards while foraging and consume cats, dogs, and other pets. Such incidents will likely become a high-profile problem as pythons increase in abundance in suburban areas of southern Florida.

RISK TO HUMANS

Although remote, the possibility of a wild python in Florida attacking and killing a human is real, and the danger posed by large pythons should not be discounted. Unfortunately, media attention tends to greatly inflate the actual danger posed by pythons. Large adult Burmese pythons are certainly capable of killing, and even consuming, a child or small adult human. In Asia, wild Burmese pythons have been known to kill people. Such events are extremely rare, and Burmese pythons apparently attack people far less frequently than their longer relative, the reticulated python (*Broghammerus reticulatus*; Reed and Rodda 2009).

At least 13 people have been killed by pet pythons in the United States since 1980 according to the Humane Society of the United States (Dorcas and Willson 2011). Whether pythons view humans as potential food items on a regular basis is equivocal, but one particularly tragic case indicates that pythons will at least occasionally do so. In 2009, an eight-foot long pet albino Burmese python that escaped from its cage in a Florida home subsequently sought out, constricted, and killed the homeowner's two-year-old daughter. As of 2015, no deaths have been attributed to wild pythons in Florida. However, at least five strikes (and two bites) attributable to predatory, not defensive, attacks on biologists studying pythons in ENP have been documented (Reed and Snow 2014).

HISTORIC AND CURRENT MANAGEMENT

As the scope of the Burmese python invasion of south Florida has progressed, the need for management to suppress or eradicate populations has increased. Growing evidence that pythons are negatively impacting native species and threatening the ecological integrity of some of the most treasured and unique natural areas in the United States has further heightened the need for viable, science-based management options. Unfortunately, we currently lack methods that could be used to suppress or eradicate python populations, even in small areas. The slow progress in developing methods for python management stems from an almost dizzying array of challenges, some of which relate to the biology of the species and some of which are a result of the characteristics of the invaded region. In this section, we discuss the major challenges related to python management, and then specifically discuss strengths and weaknesses of techniques that have been used to remove pythons and might be used to manage their populations.

CHALLENGES POSED BY PYTHON BIOLOGY

Understanding individual detection probability is critical to management of nearly any invasive species. Individual detection probability is the probability of detecting (i.e., capturing or observing) a particular individual of a species, but is not directly affected by density. It is important not to confuse detection probability with overall capture rate, which is a product of individual detection probability and density. High capture rates can be an indication of an effective technique, but ultimately, a thorough understanding of individual detection probability is needed to determine the effort needed to substantially suppress or eradicate a population. Individual detection probability can vary strongly among capture techniques, habitats, demographic groups, environmental conditions, and resource environments (Gragg et al. 2007; Tyrrell et al. 2009; Christy et al. 2010; Willson et al. 2011b). Thus, a thorough understanding of factors affecting detection probability is necessary to assess effectiveness of management activities in different areas or using different techniques

(Willson et al. 2011b). Finally, in the context of invasive species management, a simple estimate of average detectability for a species is inadequate. Because eradication requires detection and removal of every individual, managers must also understand variation in detection probability among individuals (Rodda 2012). The few studies that have comprehensively evaluated detectability of snakes have often found high levels of heterogeneity in detection probability, with some individuals being highly refractory to capture (Dorcas and Willson 2009; Tyrrell et al. 2009). As the science of invasive snake management progresses, it is likely that the presence of a few essentially "uncatchable" individuals will be the primary impediment to complete or local eradication of species.

Snakes are among the most difficult species to study, primarily as a result of their low detectability (Dorcas and Willson 2009). Despite the fact that snakes can reach high densities in suitable habitat, their secretive habits, low levels of activity, cryptic coloration, and use of habitats that are not easily sampled (e.g., thick vegetation, subterranean habitats, and wetlands) make snakes difficult to detect under the best of circumstances. Although large, Burmese pythons are particularly secretive snakes. They have excellent camouflage, spend most of their time in habitats that are difficult to search, do not seem to be particularly attracted to baited traps or specific microhabitats, and as ambush predators, probably spend much of their time inactive or in well-hidden ambush locations (Dorcas and Willson 2013). Only one study to date has estimated individual detection probability of Burmese pythons. Dorcas and Willson (2013) conducted experimental searches for 10 male Burmese pythons that were collected from Florida and kept in a 31 × 25-m seminatural outdoor enclosure in South Carolina as part of another study (Figure 8.4a; Dorcas et al. 2011). The authors arranged 19, 30-minute searches of the enclosure by observers, many of which were experienced herpetologists but who had no prior knowledge of snake locations. Out of 190 possible opportunities for snake detection (19 searches × 10 snakes), only two snakes were located, yielding an individual detection probability of approximately 1%. Low detection was often a result of a snake's use of aquatic or subterranean habitats, where they were not visible, but even highly visible snakes were often missed by observers (Figure 8.4b; Dorcas and Willson 2013). Although a detection probability of 1% is extremely low, it is important to remember that this is based on an unusually high density of large snakes contained in a small enclosure with a relatively searchable habitat (Figure 8.4a). Undoubtedly, the detection probability of pythons would be substantially lower in many regions of south Florida where the habitat is much more difficult to search. However, no studies have rigorously estimated detectability of free-ranging pythons in the field. If the 1% value determined by Dorcas and Willson (2013) is a good benchmark, it indicates that for every python collected in Florida, 99 pythons are missed by observers in the area being searched.

Extremely low detectability impairs nearly every aspect of python research and management. Obviously, low detectability suggests that densities must be high in areas where pythons are encountered regularly, and that incredibly high efforts would be needed to substantially suppress or locally eradicate populations. However, low detectability complicates or precludes even estimation of density, which is a necessary first step in assessing the effectiveness of capture methods or the development of

(a)

(b)

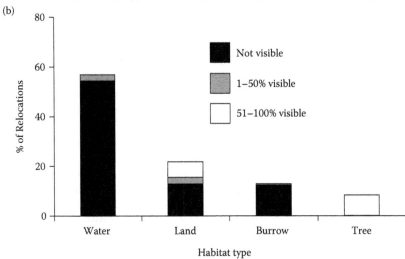

FIGURE 8.4 Habitat use and visibility of ten male Burmese pythons maintained in a seminatural outdoor enclosure in South Carolina and monitored by radiotelemetry. (Dorcas, M. E., et al. 2011, *Biological Invasions*, 13 (4), 793–802; Dorcas, M. E., et al. 2013, *Reptiles in Research: Investigations of Ecology, Physiology, and Behavior from Desert to Sea*, pp. 367–85, New York, Nova Biomedical.) All snakes were relocated approximately three times per week and were considered visible if any part of their body could be observed without altering the habitat. (a) Photograph of the enclosure used to evaluate habitat use and detectability of pythons. (b) Python visibility in relation to habitat use; bars represent frequency of relocations in various habitats, coded by visibility. Note that pythons used water most frequently and were seldom visible, even when relocated using telemetry. However, observers, many of which were experienced herpetologists, failed to detect even highly visible snakes during systematic searches of the enclosure.

effective management plans (Dorcas and Willson 2013). Clearly, estimating python detection probability and density in the field will be difficult, but it is a critical first step in any management plan. Years of basic research on factors influencing detection probability of brown tree snakes on Guam have yielded invaluable information that has formed the foundation of effective science-based management strategies (Rodda et al. 1999, 2007; Gragg et al. 2007; Tyrrell et al. 2009; Christy et al. 2010). A similar approach is sorely needed if we are to make headway toward solutions to the Burmese python problem in south Florida.

CHALLENGES POSED BY THE INVADED REGION

Curbing invasion of a species as secretive as the Burmese python would be challenging under any circumstance, but their management in South Florida is further impeded by characteristics of the invaded region. Paramount among these challenges is the sheer scale of the invasion. By the time pythons were recognized as established in the Everglades, they were already spread over an area greater than 260 km^2 (100 mi^2). As of 2015, established python populations have spread over an area of at least 10,000 km^2 (4000 mi^2) extending from Key Largo and Cape Sable north past I-75 (Alligator Alley) and west to Naples. To put the scale of the invasion in perspective, this area is approximately 20 times the land area of Guam, where after decades of research, managers have only recently implemented large-scale suppression efforts for brown tree snakes (Clark and Savarie 2012). The habitats inhabited by pythons in Florida are also some of the most rough and inaccessible in the country. Across the python's current introduced range, only a few areas are accessible on foot or by vehicle. Apart from a handful of roads and man-made canals, most of this region is a vast subtropical wetland accessible only by airboat or plane/helicopter, and consisting of shallow water, sawgrass, tangled mangrove forests, and impenetrable tropical hardwood hammocks. It is difficult to imagine any type of management plan capable of even small-scale control or eradication in such challenging habitat.

An additional challenge to python control in south Florida is the fact that the invasion now encompasses lands owned or managed by numerous different stakeholders, including the National Park Service (ENP and Big Cypress National Preserve), the State of Florida (e.g., the South Florida Water Management District, Florida Fish and Wildlife Commission), Native American Tribes, and private landowners. Although recent efforts have been spearheaded to bring together stakeholders to address the python issue, coordinating efforts across multiple jurisdictions with different priorities can be difficult. This is particularly true for those areas, such as ENP, that contain sensitive ecological and cultural resources that must be considered and protected as much as possible during python control efforts.

CAPTURE TECHNIQUES

Visual Searching

Soon after it became apparent that Burmese pythons were established and breeding in the Everglades, python biologists and recreational snake hunters realized that one of the most reliable ways to find pythons was to walk canal banks, road edges, or other accessible habitats in the winter and look for pythons basking in the sun or moving on the surface during the day. Like all snakes, pythons are able to use basking and other thermoregulatory behaviors to maintain body temperatures substantially higher than ambient air temperature. In fact, pythons maintained in an outdoor enclosure in South Carolina were often able to achieve body temperatures in excess of 25°C on days when the minimum temperature dropped below 10°C (Dorcas et al. 2011). During the coldest parts of winter in south Florida, when nightly temperatures can drop below 10°C, pythons can be found basking along canal banks, and hundreds of pythons have been collected during the last decade by experienced searchers using this technique. Pythons captured during visual searches are also larger, on average, than pythons collected using most other techniques.

Despite the large number of pythons that have been captured using visual searches, this technique has several major limitations that make it unlikely to be useful as a large-scale management technique. Of primary concern is that visual searches are only effective in limited areas and under specific conditions. Virtually the only situation where pythons are frequently located by visual searching is along canals in the winter. In warmer weather, pythons are seldom visible during the day (Dorcas and Willson 2013), and they do not appear to bask conspicuously except during the coldest weather (Mazzotti et al. 2010). Further, even in cool weather, few pythons have been captured using visual searches in more natural habitats such as sawgrass marshes, mangroves, or pine uplands. It is unclear whether the frequency of snakes seen along canals reflects a higher number of snakes in those habitats, greater visibility (detectability) of snakes there, or simply that these habitats are the most accessible and other habitats are not being searched (Dorcas and Willson 2009). Road surveys and radiotelemetry have revealed that pythons do use other habitats regularly. Thus, the apparent abundance of pythons along canals is likely at least partially a result of increased detectability and high search efforts in those habitats because of easier accessibility to humans. Although no rigorous scientific study has compared the efficacy of visual searches along canals to those conducted in more natural habitats, it is likely that this technique will only be useful for small areas of accessible habitat, located primarily along canals, and that it will be useless as a management technique for most of the vast acreage of natural habitats in the Everglades. Further, lacking data on detectability or density of pythons inhabiting canals, it is nearly impossible to evaluate whether this technique will even be viable for small-scale suppression in those habitats. In fact, high efforts and removal of dozens to hundreds of pythons from several small (<10 km) sections of publically accessible canals over the past 10 years does not appear to have had a substantial effect on the population; individuals, including large adults, are still found regularly in those areas. Whether this is a simple reflection of high density and extremely low detectability of resident animals or whether these habitats see a regular influx of individuals from more remote wilderness areas remains a critical question in evaluating the utility of this technique.

Bounty Systems: The "Python Challenge"

Bounty programs or participation in management by the general public has been effective for the management of many species, including eradication of many large predators. To date, the only large-scale, organized attempt at python removal, the

"Python Challenge," has focused primarily on visual searching. The first Python Challenge was a five-week long event during the winter of 2013 that was organized and implemented by the Florida Fish and Wildlife Conservation Commission (Harvey et al. 2015). The event offered prizes for the largest and most pythons collected and allowed open participation by the public following an entrance fee and training session. The event gained substantial media attention, and nearly 1600 participants registered. In total, only 68 pythons were removed, suggesting that this approach shows little promise as a way to significantly reduce python populations. However, the event did yield several useful insights. First, it showed that it may be possible to attract significant numbers of volunteers for python control initiatives—although the vast majority of those registered found no pythons. Second, it suggested that training might greatly improve the success of individual python collectors. Most of the 68 pythons captured were found by a few experienced collectors (Harvey et al. 2015). If the knowledge of these few could be imparted to other participants, the overall success of the program could potentially be increased. Finally, the Python Challenge demonstrated that public events could yield meaningful scientific information. Of the snakes captured during the challenge, several were from areas that are not regularly searched by python researchers and, thus, had few python records. Also, the event helped ensure that pythons collected by members of the general public were returned to researchers for future study and undoubtedly helped to raise overall public awareness of the problem. Another Python Challenge was held in early 2016 with similar results.

Detector Dogs

Use of trained detector dogs to find pythons has been investigated thoroughly by researchers from Auburn University led by Christina Romagosa and Melissa Miller (Romagosa et al. 2011). Using dogs specifically selected and trained for tracking wildlife (Auburn University EcoDogs), the team conducted an extensive series of searches with and without dogs along canal banks and other accessible habitats in south Florida. The dogs were able to reliably locate snakes in terrestrial habitats, locating 19 pythons during the six-month study. In experimental searches of terrestrial habitat, dogs located 92% of experimentally placed pythons, which was 27% better and 2.5 times faster than human searchers (Romagosa et al. 2011). However, dogs were generally unable to locate pythons in the water (at times signaling that a snake has been in the area, but unable to pinpoint the current location) or in particularly rough terrain such as impenetrable mangrove habitats. Thus, detector dogs show promise as a way to improve the efficacy of visual searches to suppress python populations in small, high-priority areas (e.g., nesting bird rookeries or neighborhoods where pythons have been seen), but are still not a viable technique for management of python population in more remote areas of south Florida. The high cost of training and maintaining detector dogs further limits their widespread use. An additional issue is that dogs that might stray from their trainer could become prey of pythons or alligators.

Road Collecting

Road collecting, slowly driving on roads at night and watching for animals crossing the road, is an effective technique for capturing many secretive snake species (Dorcas and Willson 2009; Willson 2016) and is the primary way that pythons have been captured during the warmer months of the year. During April through November, when pythons are primarily nocturnal, the vast majority of python captures are on roads at night. ENP has implemented a program of "authorized agents," volunteers that are permitted to collect pythons within the national park. These agents have been responsible for the removal of many pythons, primarily via road collecting. All pythons collected by authorized agents are provided to scientists for study. Road collecting is probably the single most cost-effective technique for capturing pythons, and does not require the experience and expertise needed to successfully find pythons during visual searches. However, road collecting suffers from the same fundamental flaw as visual searches in terms of its utility for large-scale python management: it can only be used in a few small areas (i.e., along roads). As with visual searches, road collection might be useful for suppression in the limited number of areas where it can be implemented (e.g., along roads in ENP), but this possibility cannot be assessed without a thorough understanding of density and detectability of snakes in these areas.

Despite the limitations of road collecting as a large-scale management technique, the method does have several strengths. First, road collecting is likely the most repeatable existing method for long-term monitoring of python relative abundance, because it suffers less from observer bias and sensitivity to environmental conditions than visual surveys (Willson 2016). Even lacking a density estimate, it is likely that synthesis of large quantities of standardized road-collection data will yield valuable insights on the dynamics of the invasive population, especially within ENP. Additionally, road collecting appears to sample a broader demographic spectrum of individuals than other methods, and thus far is the only method currently available for reliably finding juvenile pythons.

Traps

Although relatively expensive to construct and monitor, traps have often proven to be an effective technique for extensive sampling of secretive snakes across a variety of habitats (Dorcas and Willson 2009; Tyrrell et al. 2009; Willson 2016). Traps suffer minimally from observer bias typically associated with visual searching and show the most promise of any currently used collection technique that could be used away from roads or canal banks in South Florida. Most traps used for snakes are some variant of a funnel trap, with inward-facing funnels that direct the animal into a holding chamber, from which they have difficulty escaping. Trapping success is usually improved with either bait placed within the trap, a barrier (fence) that guides animals toward the funnel openings, or both. Although funnel traps have seldom been used to capture snakes the size of Burmese pythons, they have been used successfully in a wide array of habitats, ranging from aquatic to arboreal (Willson 2016). Despite the potential advantages of traps as a method for python management, there has only been limited field-testing of traps designed for pythons. Reed et al. (2011) tested large funnel traps baited with live rats over a four-month period in a fallow agricultural field outside of ENP. Over 6053 "trap-nights" (i.e., number of traps × number of nights deployed) resulted in a captured total of three pythons; 11 were found during subsequent mowing/disking of the study area (Reed et al. 2011). It is tempting to view this result as a failure, and indeed capture rates were very low. However, it is unclear if high rodent abundances in the field may have suppressed attraction of pythons to the baited traps or if python densities in the field were representative of those in natural habitats nearby. Clearly, we are far from having a python trap that is effective enough for widespread use, but there is reason to be optimistic. Development of an effective trap for an atypical species takes time. Years of research went into improving trap design for brown tree snakes on Guam, eventually resulting in a highly effective trap suitable for suppression at small to moderate spatial scales (Rodda et al. 1999; Tyrrell et al. 2009). One potential but yet unexplored possibility is to use pheromones as attractants to lure pythons into traps. Male pythons presumably find reproductive females via pheromones, and these pheromones could be used as a way to attract males. However, eliminating large, reproductive females would be a key to any eradication effort, and the effectiveness of any pheromonal attractant for females is currently unknown. As we learn more about python behavior, movement, activity, and diet, we may be able to modify traps to increase capture rates and target individuals that have the greatest effect on population growth.

"Judas Snakes"
Perhaps the least traditional method for python collection that has been used extensively to date is the "Judas Snakes" technique; that is, using radio-tagged pythons to lead researchers to other individuals, thereby "betraying" their location. The "Judas" approach has been used effectively for other invasive species, especially those that live in groups (e.g., goats, schooling fish; Campbell and Donlan 2005; Bajer et al. 2011). Although primarily solitary, Burmese pythons do engage in mating aggregations, typically consisting of a receptive female and one or more males (Dorcas and Willson 2013). Python researchers tracking snakes to understand movement and habitat use have found that tracked snakes occasionally lead them to other individuals. This technique does have one major advantage over most other techniques—it can be used to target reproductive females. Unfortunately, the cost of tracking pythons is high, and snakes often move into areas that are difficult or impossible to access on foot, thereby making it impossible to look for any associating untagged individuals.

Toxicants or Biological Controls
The only methods that currently show any promise for rangewide suppression or eradication are methods that do not require detection of individual pythons, such as toxicants or biological controls (biocontrols). Toxicants (acetaminophen in baits) have been developed and tested for brown tree snakes (Lardner et al. 2013), and this method is currently being considered as a method for large-scale suppression of that species on Guam. However, the use of toxicants may not be possible in Florida because of the risks posed to many native animals (Clark et al. 2012). Biocontrol methods such as viruses or other pathogens that target Burmese pythons but no other animal species may be possible, but development of these control agents is difficult and the risk of incidental environmental damage is high and cannot be eliminated (Simberloff and Stiling 1996; Fagan et al. 2002).

Other Methods

Despite the desperate need for rigorous testing, comparison, and improvement of existing methods for python collection mentioned above, there has been significant interest in evaluating novel techniques in hopes of finding a "silver bullet" that could be used to control the expanding python population. Drones equipped with infrared cameras were considered, but were found to be ineffective for locating snakes that were typically under water or in dense vegetation. If such a technique were found for easily locating pythons, it is unclear if it could then be used to actually capture snakes far from ground access. Python researchers regularly receive well-meaning but outlandish ideas for python control. Examples submitted to us and our colleagues include giant fishing lures and polyvinyl chloride (PVC) tubes equipped with razor blades. Although we should continue to seek new and innovative potential solutions to the python problem, the most obvious current direction for improving our ability to manage these atypical invaders is rigorous hypothesis-driven testing and innovation of existing techniques. With improvement, these techniques could likely be used to suppress python populations at small spatial scales.

FUTURE CHALLENGES AND ERADICATION/CONTROL

CHALLENGES TO RESEARCH

Efficient use of resources to manage the Burmese python invasion depends on solid scientific, hypothesis-driven research. However, numerous factors, especially the low detectability of pythons, limit our ability to study the invasion effectively (Dorcas and Willson 2013). Many observations of Burmese pythons occur along roadsides or within areas frequented by humans, potentially biasing estimates of abundance and habitat associations (Dorcas and Willson 2011). Determining whether individuals captured in new areas represent spreading of the current population, a new successful introduction, or an isolated nonreproducing individual is difficult. Low detectability—and an inability to determine whether captured pythons are wild or released—may have played a role in the initial lag time before the species was recognized as established (Willson et al. 2011a).

The most commonly used methods to estimate demography and density in animal populations are mark-recapture methods—animals are captured, given an individual or cohort mark, and released, generally at the site of capture. Scientists calculate the proportion of marked to unmarked animals captured during subsequent sampling events to estimate abundance and important demographic variables such as survivorship or recruitment (Williams et al. 2002a,b). Unfortunately, these methods are not feasible for studying Burmese pythons in Florida because releasing captured pythons is undesirable (they are euthanized). Additionally, mark-recapture studies are difficult to conduct on species with low detectability, requiring an intensive sampling effort over long periods of time and/or high population density to obtain the recapture rates necessary to generate precise estimates of demographic variables. Therefore, the traditional methods used to estimate demography and density of species are not available for the Burmese python in Florida, and alternative methods must be developed (Dorcas and Willson 2011).

DENSITY

An estimate of population density is a fundamental component of understanding and managing invasive species. Currently, we do not know what proportion of the python population any particular management strategy removes, and therefore, we have no method by which to evaluate effectiveness. Despite media reports, there are currently no reliable estimates of Burmese python density in Florida. The most critical first step toward managing the invasion therefore lies in developing methods by which to estimate density that do not require the rerelease of large numbers of animals after capture. Instead, we must develop methods to estimate density using data that are available—such as rate of python sightings from standardized road surveys. An estimation of density will allow scientists to evaluate the usefulness of management strategies.

DEMOGRAPHY

Understanding the life history characteristics of Burmese pythons in south Florida is critical for predicting population dynamics. While reproductive potential and growth are fairly well studied (Reed and Rodda 2009; Reed et al. 2012), age-specific survivorship is not well known, and is an important area for future research. Survivorship is often estimated using mark-recapture methods, which are not feasible for Burmese pythons. Future studies could estimate these important demographic variables without mark-recapture by monitoring the fates of snakes in different age classes using radiotelemetry (Currylow et al. 2011). The long life span and presumably high adult survivorship of Burmese pythons necessitate long-term radiotelemetry studies to arrive at estimates of survivorship of the older age classes. Known-fate analysis of radiotelemetry data may also provide insight into how habitat affects both survivorship and fecundity. One limitation to these studies is the considerable logistical challenge associated with radiotracking pythons in the wetland ecosystems of southern Florida (Pittman et al. 2014; Hart et al. 2015; Walters et al. 2016). GPS tracking, which can be used to follow snakes' movements using satellites, may prove feasible for monitoring larger snakes, but currently, juvenile pythons are too small for this technique.

SPREAD

Prediction of the spatial spread of invasive species is critical to the implementation of effective control or containment measures (Mack et al. 2000). There are two important aspects of invasive spread: (1) the possible geographic extent of invasion in the introduced range, and (2) the dynamics of the spread itself, such as rate of spread and important invasion corridors (Kot et al. 1996; With 2002). Understanding both will provide insight into ways to manage spread and mitigate negative impacts of the expanding python population.

Geographic Extent of Invasion

Several studies have attempted to predict the possible geographic extent of Burmese pythons in the United States (Pyron et al. 2008; Rodda et al. 2009, 2011; van Wilgen et al. 2009). These studies identify potential biotic and abiotic conditions associated with the geographic distribution extent of the species in its native range and project those conditions onto the introduced range to predict the possible geographic extent of an invasion in the introduced range. The studies applying these methods to Burmese pythons have varied in terms of the factors identified as important to describing the native geographic range and the analytical methods used to project those factors on the introduced range. Despite the differences in methodology, most studies agreed that the possible geographic range of the Burmese python in the United States includes the majority of Florida and potentially much of the southern United States (Figure 8.5; Rodda et al. 2009, 2011; van Wilgen et al. 2009). The drawback to these methods concerns their underlying assumptions: that the environmental variables used in models are actually the factors that limit the distribution of the species in the native range, and that these factors are the same as those that will limit the invasive range (Rodda et al. 2011). These are nontrivial assumptions upon which the interpretation of model projections is highly dependent. One additional concern is that these models are only as good as the data used to inform them. We currently have limited information on the boundaries of different population segments in the native range or the geographic origins of the pythons introduced to Florida. Pyron et al. (2008) predicted that pythons would be limited to extreme southern Florida based

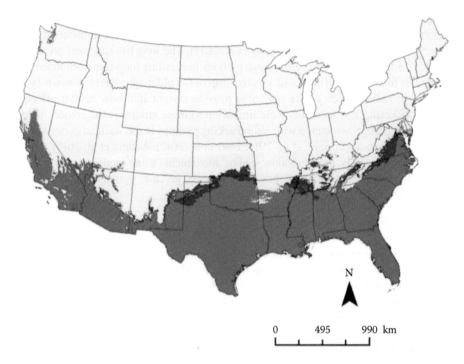

FIGURE 8.5 Area of the United States matching the climate envelope of *Python molurus* (dark green) based on climate matching. Other colors represent areas that are either too cold or too dry. (Used with permission from Rodda, G. H. et al. 2009, *Biological Invasions* 11 (2), 241–52.)

on climate, but a subsequent study showed that the study was confounded by the use of data from snakes other than Burmese pythons that, when removed, expanded the predicted region of suitable climate to essentially all of Florida and much of the U.S. Gulf Coast (Rodda et al. 2011).

Future attempts to identify a potential geographic range of Burmese pythons in the United States could improve predictions by explicitly studying and modeling the mechanisms that could limit the range of the species. Mechanistic models that identify the biophysical processes behind range limitation may have the potential to predict the potential geographic extent of Burmese pythons in the United States without relying exclusively on correlations to the environmental conditions of the native range (Tingley et al. 2014). However, the development of mechanistic models will require extensive research into the behavior and physiology of Burmese pythons, both in their native and introduced ranges.

Invasion Dynamics

Management of the current Burmese python invasion necessitates an understanding of how the population will spread northward. Understanding the dynamics of spread will allow managers to enact preemptive control strategies to limit negative impacts on susceptible ecosystems and to target potential invasion corridors for management. Predicting the dynamics of invasive spread requires understanding the dispersal behavior of the species, particularly the movement of individuals at the front of the invasion (Clark et al. 2001, 2003). Dispersal, or the movement of an individual from a birth site to a new breeding site, often occurs at a prebreeding life stage (Clobert et al. 2001). Unfortunately, little is known about the movement or behavior of juvenile Burmese pythons. Dispersal behavior, and long-distance dispersal in particular, is often critically linked to local population dynamics; individuals are often more likely to disperse if the local population density is high (Travis et al. 2009). The need to develop methods by which to estimate population density is therefore critical not only for controlling populations and predicting impacts, but also for understanding how and where population spread might occur.

Habitat throughout Florida, both natural and anthropogenically modified, varies considerably, and movement behavior of Burmese pythons is likely to vary by habitat. A greater understanding of how juvenile snakes disperse and make movement decisions within different habitat types is critical to developing any realistic prediction of the dynamics of invasive spread in heterogeneous landscapes (With 2002). For example, certain types of land use may impede or facilitate invasion, especially land use associated with the creation of straight-line features such as roads or canals (Parendes and Jones 2000; Gelbard and Belnap 2003). Recent research has indicated that dispersing pythons may preferentially use agricultural levees and canals to make dispersal movements (Pittman et al. unpub data). Therefore, straight-line features such as canals, roads, and embankments could facilitate the rapid spread of pythons through the Florida peninsula.

One important future research direction will be to understand how snakes move through habitats in areas north of the area they currently occupy. For example, Burmese pythons may use gopher tortoise or armadillo burrows for mating and nesting (Metzger 2013). Although armadillos are absent from much of the Everglades, they are abundant in Florida north of the Everglades and their burrows may provide habitat that facilitates northward invasion by providing overwintering sites well below ground. Experimental releases of juvenile Burmese pythons within enclosures composed of habitat types more common in the northern portion of the invaded range may provide insight into how snakes disperse through these areas.

Another important, but rarely studied, factor that could influence the dynamics of population spread is the evolution of movement behavior at the invasion front. Previous studies of other invasive species have found that individuals at the invasion front move differently than individuals at the core of the invasion (e.g., cane toads, Argentine ants; Suarez et al. 2001; Llewelyn et al. 2009). These differences may be the result of "spatial selection," which occurs as a result of spatial sorting of phenotypes (Shine 2012). Individuals at the front of an invasion are more likely to be individuals that are long-distance dispersers and move more frequently. These individuals tend to mate with other individuals that are also long-distance dispersers because those are the individuals present at the invasion front, and if dispersal behavior is heritable, the leading edge of the population may evolve enhanced dispersal behaviors. Spatial selection can thus result in an accelerating invasion front and make prediction of spread rate difficult. One way to test for the existence of spatial selection would be to sample individuals at different locations along the invasion front and measure their activity levels or their propensity to leave cover/take risks (Brown et al. 2014, 2015) with behavioral assays that correlate with dispersal behavior (Rehage and Sih 2004; Sih et al. 2004). Consistent differences in movement behavior among individuals at different locations along the invasion front may be indicative of spatial selection.

Evolution of behavioral traits important to survival at the front of the invasion may also impact future spread northward. For example, Burmese pythons may be limited by cold temperatures (Avery et al. 2010; Mazzotti et al. 2010). A study of the survival of adult Burmese pythons released into an enclosure in South Carolina found that no individuals survived the (unusually harsh) winter. Several snakes exhibited maladaptive behaviors during winter months—such as failing to locate areas that would have provided refuge from freezing temperatures (Dorcas et al. 2011). We currently know little about the Burmese python's ability to withstand subfreezing winter temperatures, but the ability to overwinter could be accomplished through behavioral traits rather than adjustments of physiological processes. Consequently, if aspects of the thermoregulatory behavior of pythons are heritable, there may be selection for behavioral traits that allow survival of harsh winter temperatures, even in the absence of physiological traits associated with cold tolerance.

DÉJÀ VU

In 2013, the authors attended a technical meeting focused on the problem of invasive pythons. At this meeting, there was considerable discussion about the risk pythons pose to U.S. ecosystems and potential methods for management of their populations. However, despite data presented by multiple scientists showing convincing evidence of the impacts of pythons on native fauna, it was clear that many attendees were still skeptical about the basic idea that an invasive snake could have such dramatic impacts over such a short period of time. Thus, there was little consensus on what steps should be taken to address the threat pythons pose to native ecosystems in the United States.

Earl Campbell (U.S. Fish and Wildlife Service), who has worked extensively on the problem of invasive brown tree snakes, was in attendance at the meeting and provided valuable perspective on the issue toward the end of the two-day meeting. Earl announced that he was experiencing déjà vu—that is, the discussions about whether pythons posed significant problems and what to do about them were reminiscent of the meetings he participated in decades ago regarding the brown tree snake invasion on Guam. His main take-home message was that the python problem appears to be very real, and that we should act rapidly while there is still time, rather than being overly skeptical of the growing body of scientific evidence that implicates pythons as damaging invaders. He emphasized that figuring out ways to manage pythons will require coordination of the numerous agencies and other entities involved, extensive hypothesis-driven research, and substantial funding. In reality, several factors make invasive pythons potentially more difficult to manage than brown tree snakes. First, pythons are not contained on an island, and thus, a natural geographic boundary (e.g., a large ocean surrounding Guam) will not limit expansion of their range in the United States. Second, pythons are already found over a much larger geographic area than invasive brown tree snakes, and much of that area is extremely difficult for humans to access. Third, thus far, trapping appears to be a much less effective method for capturing pythons than for capturing brown tree snakes. Fourth, because of the presence of numerous other species that might be harmed, management of pythons cannot use methods, such as toxicants, that are designed to kill all snakes (or reptiles)—a method currently being developed and tested on Guam. In conclusion, the challenges in understanding and managing invasive pythons in the United States are daunting and can clearly only be addressed via extensive, well-coordinated research that seeks to better understand factors affecting the density and detectability of pythons, the impacts of invasive pythons on our natural ecosystems, the risks they pose to other parts of the United States, and innovative approaches that show promise for effective management of this species.

ACKNOWLEDGMENTS

We have benefited from discussing issues related to pythons and working with many volunteers, technicians, and researchers, and many have helped with python collection and/or logistics of our research. These include Paul Andreadis, Ian Bartosek, Rick Bauer, Matt Brien, Kristen Cecala, Michael Cherkiss, Justin Davis, Brett DeGregorio, Evan Eskew, Bryan Falk, Anthony Flanagan, J. Whitfield Gibbons, Chris Gillette, Judy Greene, Wellington Guzman, Cris Hagen, Kristen Hart, Rebecca Harvey, Bobby Hill, Toren Hill, Josh Holbrook, Harry Greene, Trey Kieckhefer, Kenneth Krysko, Frank Mazzotti, Melissa Miller, David Millican, Tony Mills, Lori Oberhofer, Scott Pfaff, Melissa Pilgrim, Sean Poppy, Steven Price, Tom Rayhill, Robert Reed, Kenneth Rice, Michael Rochford, Gordon Rodda, LeRoy Rodgers, Christina Romagosa, Ron Rozar, Brian Smith, Dustin Smith, Skip Snow, Charlotte

Steelman, Bryan Stuart, Tracey Tuberville, Lynea Witczak, Alex Wolf, and Pat Zollner. J. Whitfield Gibbons, Katie Greene, Natalie Haydt, Emma Rose Parker, Kyle Dembart, Xzavier Killings, Kristen Gillespy, Brielle Bowerman, and Annalee Tutterow, who provided comments that improved the manuscript.

Funding that helped support work on this chapter was provided by Davidson College, the University of Arkansas, the J.E. and Majorie Pittman Foundation, the Associated Colleges of the South, the Duke Energy Foundation, and the Department of Defense Legacy Resource Management Program.

REFERENCES

Andreadis, P. T. 2011. *Python molurus bivittatus* (Burmese python). Reproducing population. *Herpetological Review* 42 (2): 302–3.

Avery, M. L., R. M. Engeman, K. L. Keacher, J. S. Humphrey, W. E. Bruce, T. C. Mathies, and R. E. Mauldin. 2010. Cold weather and the potential range of invasive Burmese pythons. *Biological Invasions* 12 (11): 3649–52.

Bajer, P. G., C. J. Chizinski, and P. W. Sorensen. 2011. Using the Judas technique to locate and remove wintertime aggregations of invasive common carp. *Fisheries Management and Ecology* 18 (6): 497–505. doi:10.1111/j.1365–2400.2011.00805.x.

Barker, D. G., and T. M. Barker. 2011. A review of: Willson, J. D., M. E. Dorcas and R. W. Snow. 2010. Identifying plausible scenarios for the establishment of invasive Burmese pythons (*Python molurus*) in southern Florida. Biological Invasions. *Bull. Chicago Herp. Soc* 46 (4): 41–3. doi 10.1007/s10530-010-9908-3.

Bonnet, X., D. Bradshaw, and R. Shine. 1998. Capital versus income breeding: An ectothermic perspective. *Oikos* 83 (2): 333. doi:10.2307/3546846.

Brown, G. P., B. L. Phillips, and R. Shine. 2014. The straight and narrow path: The evolution of straight-line dispersal at a cane toad invasion front. *Proceedings of the Royal Society B: Biological Sciences* 281 (1795): 20141385–5. doi:10.1098/rspb.2014.1385.

Brown, G. P., B. L. Phillips, and R. Shine. 2015. Directional dispersal has not evolved during the cane toad invasion. *Functional Ecology* 29 (6): 830–38. doi:10.1111/1365-2435.12397.

Campbell, K., and C. J. Donlan. 2005. Feral goat eradications on islands. *Conservation Biology* 19 (5): 1362–74. doi:10.1111/j.1523-1739.2005.00228.x.

Caves E. M., S. B. Jennings, J. HilleRisLambers, J. J. Tewksbury, and H. S. Rogers. 2013. Natural experiment demonstrates that bird loss leads to cessation of dispersal of native seeds from intact to degraded forests. *PLoS ONE* 8 (5): e65618. doi:10.1371/journal.pone.0065618.

Christy, M. T., A. A. Yackel Adams, G. H. Rodda, J. A. Savidge, and C. L. Tyrrell. 2010. Modelling Detection probabilities to evaluate management and control tools for an invasive species. *Journal of Applied Ecology* 47 (1): 106–13. doi: 10.1111/j.1365-2664.2009.01753.x.

Clark, J. S., M. Lewis, and L. Horvath. 2001. Invasion by extremes: Population spread with variation in dispersal and reproduction. *American Naturalist* 157 (5): 537–54.

Clark, J. S., M. Lewis, J. S. McLachlan, and J. HilleRisLambers. 2003. Estimating population spread: What can we forecast and how well? *Ecology* 84 (8): 1979–88.

Clark, L., and P. J. Savarie. 2012. Efficacy of aerial broadcast baiting in reducing. *Human–Wildlife Interactions* 6 (2): 212–21.

Clark, L., P. J. Savarie, J. A. Shivik, S. W. Breck, and B. S. Dorr. 2012. Efficacy, effort, and cost comparisons of trapping and acetaminophen-baiting for control of brown treesnakes on Guam. *Human–Wildlife Interactions* 6 (2): 222–36.

Clobert, J., E. Danchin, A. Dhondt, and J. Nichols. 2001. *Dispersal*. New York, NY: Oxford University Press.

Currylow, A. F., P. A. Zollner, B. J. MacGowan, and R. N. Williams. 2011. A survival esti-mate of midwestern adult eastern box turtles using radiotelemetry. *American Midland Naturalist* 165 (1): 143–49.

Dorcas, M. E., and J. D. Willson. 2009. Innovative methods for studies of snake ecology and conservation. In *Applied Ecology and Conservation*, eds. S. J. Mullin and R. A. Seigel, pp. 5–37. Ithaca, NY: Cornell University Press.

Dorcas, M. E., and J. D. Willson. 2011. *Invasive Pythons in the United States: Ecology of an Introduced Predator*. Athens, GA: University of Georgia Press.

Dorcas, M. E., and J. D. Willson. 2013. Hidden giants: Problems associated with study-ing secretive invasive pythons. In *Reptiles in Research: Investigations of Ecology, Physiology, and Behavior from Desert to Sea*, ed. W. I. Lutterschmidt, pp. 367–85. New York: Nova Biomedical.

Dorcas, M. E., J. D. Willson, and J. W. Gibbons. 2011. Can invasive Burmese pythons inhabit temperate regions of the southeastern United States? *Biological Invasions* 13 (4): 793–802.

Dorcas, M. E., J. D. Willson, R. N. Reed et al. 2012. Severe mammal declines coincide with proliferation of invasive Burmese pythons in Everglades National Park. *Proceedings of the National Academy of Sciences of the United States of America* 109 (7): 2418–22.

Dove, C. J., R. W. Snow, M. R. Rochford, and F. J. Mazzotti. 2011. Birds consumed by the inva-sive Burmese python (*Python molurus bivittatus*) in Everglades National Park, Florida, USA. *The Wilson Journal of Ornithology* 123 (1): 126–31. doi:10.1676/10-092.1.

Fagan, W. F., M. A. Lewis, M. G. Neubert, and P. Van Den Driessche. 2002. Invasion theory and biological control. *Ecology Letters* 5 (1): 148–57.

Gelbard, J. L., and J. Belnap. 2003. Roads as conduits for exotic plant invasions in a semiarid landscape. *Conservation Biology* 17 (2): 420–32.

Gragg, J. E., G. H. Rodda, J. A. Savidge, G. C. White, K. Dean-Bradley, and A. R. Ellingson. 2007. Response of brown treesnakes to reduction of their rodent prey. *Journal of Wildlife Management* 71 (7): 2311. doi:10.2193/2006-444.

Hart, K. M., M. S. Cherkiss, B. J. Smith et al. 2015. Home range, habitat use, and movement patterns of non-native Burmese pythons in Everglades National Park, Florida, USA. *Animal Biotelemetry* 3: 8. doi:10.1186/s40317-015-0022-2.

Harvey, R. G., L. Perez, and F. J. Mazzotti. 2015. Not seeing is not believing: Volunteer beliefs about Burmese Pythons in Florida and implications for public participation in invasive species removal. *Journal of Environmental Planning and Management* 2015: 1–19.

Kot, M., M. A. Lewis, and P. Van Den Driessche. 1996. Dispersal data and the spread of invading organisms. *Ecology* 77 (7): 2027–42.

Kraus, F. 2009. *Alien Reptiles and Amphibians: A Scientific Compendium and Analysis*. New York: Springer.

Lardner, B., A. A. Yackel Adams, J. A. Savidge, G. H. Rodda, R. N. Reed, and C. S. Clark. 2013. Effectiveness of bait tubes for brown treesnake control on Guam: Baiting for brown treesnakes. *Wildlife Society Bulletin*, June, n/a – n/a. doi:10.1002/wsb.297.

Llewelyn, J., B. L. Phillips, R. A. Alford, L. Schwarzkopf, and R. Shine. 2009. Locomotor performance in an invasive species: Cane toads from the invasion front have greater endurance, but not speed, compared to conspecifics from a long-colonised area. *Oecologia* 162 (2): 343–48. doi:10.1007/s00442-009-1471-1.

Mack, R. N., D. Simberloff, W. M. Lonsdale, H. Evans, M. Clout, and F. A. Bazzaz. 2000. Biotic invasions: Causes, epidemiology, global consequences, and control. *Ecological Applications* 10 (3): 689–710.

Maehr, D. S., R. C. Belden, E. D. Land, and L. Wilkins. 1990. Food habits of panthers in southwest Florida. *Journal of Wildlife Management* 54: 420–3.

Martínez-Morales, M. A., and A. D. Cuaron. 1999. Boa constrictor, an introduced predator threatening the endemic fauna on Cozumel Island, Mexico. *Biodiversity & Conservation* 8 (7): 957–63.

Mazzotti, F. J., M. S. Cherkiss, K. M. Hart et al. 2010. Cold-induced mortality of invasive Burmese pythons in South Florida. *Biological Invasions* 13 (1): 143–51.

McCleery, R. A., A. Sovie, R. N. Reed, M. W. Cunningham, M. E. Hunter, and K. M. Hart. 2015. Marsh rabbit mortalities tie pythons to the precipitous decline of mammals in the Everglades. *Proceedings of the Royal Society B: Biological Sciences* 282 (1805): 20150120. doi:10.1098/rspb.2015.0120.

Meshaka, W. E. Jr. 2011. A runaway train in the making: The exotic amphibians, reptiles, turtles, and crocodilians of Florida. *Herpetological Conservation & Biology* 6 (monograph 1).

Meshaka W. E. Jr., B. P. Butterfield, and J. B. Hauge. 2004. *The Exotic Amphibians and Reptiles of Florida*. Malabar, FL: Krieger Publishing.

Metzger, C. J. 2013. *Python molurus bivittatus* (Burmese Python): Habitat use. *Herpetological Review* 44: 333–34.

Mortensen, H. S., Y. L. Dupont, and J. M. Olesen. 2008. A snake in paradise: Disturbance of plant reproduction following extirpation of bird flower—visitors on Guam. *Biological Conservation* 141 (8): 2146–54. doi:10.1016/j.biocon.2008.06.014.

Parendes, L. A., and J. A. Jones. 2000. Role of light availability and dispersal in exotic plant invasion along roads and streams in the H. J. Andrews Experimental Forest, Oregon. *Conservation Biology* 14: 64–75.

Pittman, S. E., K. M. Hart, M. S. Cherkiss, R. W. Snow, I. Fujisaki, B. J. Smith, F. J. Mazzotti, et al. 2014. Homing of invasive Burmese pythons in south Florida: Evidence for map and compass senses in snakes. *Biology Letters* 10 (3): 20140040. doi:10.1098/rsbl.2014.0040.

Pough, F. H. 1980. Advantages of ectothermy for tetrapods. *American Naturalist* 115: 92–112.

Pyron, R. A., F. T. Burbrink, and T. J. Guiher. 2008. Claims of potential expansion throughout the U.S. by invasive python species are contradicted by ecological niche models. ed. A. Hector. *PLoS ONE* 3 (8): e2931. doi:10.1371/journal.pone.0002931.

Reed, R. N., K. M. Hart, G. H. Rodda et al. 2011. A field test of attractant traps for invasive Burmese pythons (*Python molurus bivittatus*) in southern Florida. *Wildlife Research* 38 (2): 114. doi:10.1071/WR10202.

Reed, R. N., and G. H. Rodda. 2009. *Giant constrictors: Biological and management profiles and an establishment risk assessment for nine large species of pythons, anacondas, and the boa constrictor.* U.S. Geological Survey Open-File Report.

Reed, R. N., and R. W. Snow. 2014. Assessing risks to humans from invasive Burmese pythons in Everglades National Park, Florida, USA. *Wildlife Society Bulletin* 38 (2): 366–69. doi:10.1002/wsb.413.

Reed, R. N., J. D. Willson, G. H. Rodda, and M. E. Dorcas. 2012. Ecological correlates of invasion impact for Burmese pythons in Florida. *Integrative Zoology* 7 (3): 254–70. doi:10.1111/j.1749-4877.2012.00304.x.

Rehage, J. S., and A. Sih. 2004. Dispersal behavior, boldness, and the link to invasiveness: A comparison of four Gambusia species. *Biological Invasions* 6 (3): 379–91.

Rodda, G. H. 2012. Population size and demographics. In *Reptile Biodiversity: Standard Methods for Inventory and Monitoring*, eds. R. W. McDiarmid, M. S. Foster, C. Guyer, J. W. Gibbons, and N. Chernoff, pp. 283–322. Berkeley, CA: University of California Press.

Rodda, G. H., T. H. Fritts, C. S. Clark, S. W. Gotte, and D. Chiszar. 1999. A state-of-the-art trap for the brown treesnake. In *Problem Snake Management: The Habu and the Brown Treesnake*, eds. G. H. Rodda, Y. Sawai, D. Chiszar, and H. Tanaka, pp. 285–305. Ithaca, NY: Cornell University Press.

Rodda, G. H., C. S. Jarnevich, and R. N. Reed. 2009. What parts of the U.S. mainland are climatically suitable for invasive alien pythons spreading from Everglades National Park? *Biological Invasions* 11 (2): 241–52. doi:10.1007/s10530-008-9228-z.

Rodda, G. H., C. S. Jarnevich, and R. N. Reed. 2011. Challenges in identifying sites climatically matched to the native ranges of animal invaders. ed A. Hector. *PLoS ONE* 6 (2): e14670. doi:10.1371/journal.pone.0014670.

Rodda, G. H., J. A. Savidge, C. L. Tyrrell, M. T. Christy, and A. R. Ellingson. 2007. Size bias in visual searches and trapping of brown treesnakes on Guam. *Journal of Wildlife Management* 71 (2): 656–61. doi:10.2193/2005-742.

Rogers, H., J. H. R. Lambers, R. Miller, and J. J. Tewksbury. 2012. Natural experiment demonstrates top-down control of spiders by birds on a landscape level. *PLoS One* 7: e43446.

Romagosa, C. M., C. Guyer, M. A. Miller, B. Rogers, T. C. Angle, and R. L. Gillette. 2011. Assessment of detection dogs as a potential tool for python detection efforts. National Park Service and South Florida Water Management District.

Romero, S., J. F. Campbell, J. R. Nechols, and K. A. With. 2009. Movement behavior in response to landscape structure: The role of functional grain. *Landscape Ecology* 24 (1): 39–51.

Rose, J. P., and B. D. Todd. 2014. Projecting invasion risk of non-native watersnakes (*Nerodia fasciata* and *Nerodia sipedon*) in the western United States. ed. B. R. Schmidt. *PLoS ONE* 9 (6): e100277. doi:10.1371/journal.pone.0100277.

Savidge, J. A. 1987. Extinction of an island forest avifauna by an introduced snake. *Ecology* 68 (3): 660. doi:10.2307/1938471.

Shine, R. 2012. Invasive species as drivers of evolutionary change: Cane toads in tropical Australia. *Evolutionary Applications* 5 (2): 107–16.

Sih, A., A. M. Bell, J. C. Johnson, and R. E. Ziemba. 2004. Behavioral syndromes: An integrative overview. *Quarterly Review of Biology* 79 (3): 241–77.

Simberloff, D. 2009. The role of propagule pressure in biological invasions. *Annual Review of Ecology, Evolution, and Systematics* 40 (1): 81–102. doi:10.1146/annurev.ecolsys.110308.120304.

Simberloff, D., and P. Stiling. 1996. How risky Is biological control? *Ecology* 77 (7): 1965. doi:10.2307/2265693.

Snow, R. W., M. L. Brien, M. S. Cherkiss, L. Wilkins, and F. J. Mazzotti. 2007. Dietary habits of the Burmese python, *Python molurus bivittatus*, in Everglades National Park, Florida. *Herpetological Bulletin* 101: 5–7.

Suarez, A. V., D. A. Holway, and T. J. Case. 2001. Patterns of spread in biological invasions dominated by long-distance jump dispersal: Insights from Argentine ants. *Proceedings of the National Academy of Sciences of the United States of America* 98 (3): 1095–100.

Tingley, R., M. Vallinoto, F. Sequeira, and M. R. Kearney. 2014. Realized niche shift during a global biological invasion. *Proceedings of the National Academy of Sciences* 111 (28): 10233–38. doi:10.1073/pnas.1405766111.

Travis, J. M. J., K. Mustin, T. G. Benton, and C. Dytham. 2009. Accelerating invasion rates result from the evolution of density-dependent dispersal. *Journal of Theoretical Biology* 259 (1): 151–58. doi:10.1016/j.jtbi.2009.03.008.

Tyrrell, C. L., M. T. Christy, G. H. Rodda et al. 2009. Evaluation of trap capture in a geographically closed population of brown treesnakes on Guam. *Journal of Applied Ecology* 46 (1): 128–35. doi:10.1111/j.1365-2664.2008.01591.x.

van Wilgen, N. J., N. Roura-Pascual, and D. M. Richardson. 2009. A quantitative climate-match score for risk-assessment screening of reptile and amphibian introductions. *Environmental Management* 44 (3): 590–607. doi:10.1007/s00267-009-9311-y.

Walters, T. M., F. J. Mazzotti, and H. Carl Fitz. 2016. Habitat selection by the invasive species Burmese python in southern Florida. *Journal of Herpetology.* 50 (1): 50–6. http://www.journalofherpetology.org/doi/abs/10.1670/14-098.

Williams, B. K., J. D. Nichols, and M. J. Conroy. 2002a. Estimating abundance for closed populations with mark-recapture methods. In *Analysis and Management of Animal Populations*, pp. 290–331. San Diego, CA: Academic Press.

Williams, B. K., J. D. Nichols, and M. J. Conroy. 2002b. Estimation of demographic parameters. In *Analysis and Management of Animal Populations*, pp. 334–62. San Diego, CA: Academic Press.

Willson, J. D. 2016. Surface-dwelling reptiles: Coverboards, drift fences, and arrays. In *Reptile Ecology and Conservation: A Handbook of Techniques*, ed. C. K. Dodd, Jr., pp. 125–138. Oxford, UK: Oxford University Press.

Willson, J. D., M. E. Dorcas, and R. W. Snow. 2011a. Identifying plausible scenarios for the establishment of invasive Burmese pythons (*Python molurus*) in Southern Florida. *Biological Invasions* 13 (7): 1493–504.

Willson, J. D., R. W. Snow, R. N. Reed, and M. E. Dorcas. 2014. *Python molurus bivittatus* (Burmese Python). Minimum size at maturity. *Herpetological Review* 45: 2.

Willson, J. D., C. T. Winne, and B. D. Todd. 2011b. Ecological and methodological factors affecting detectability and population estimation in elusive species. *The Journal of Wildlife Management* 75 (1): 36–45. doi:10.1002/jwmg.15.

With, K. A. 2002. The landscape ecology of invasive spread. *Conservation Biology* 16 (5): 1192–203.

9 Frogs (Coqui Frogs, Greenhouse Frogs, Cuban Tree Frogs, and Cane Toads)

Karen H. Beard, Steve A. Johnson, and Aaron B. Shiels

CONTENTS

INTRODUCTION

Amphibians are perhaps most well known for their highly threatened status, which often masks appreciation for the great numbers of species that are widespread global invaders (Kraus 2009). Both purposeful and accidental introductions of amphibians have occurred worldwide. Motivations for purposeful amphibian introductions include their use as biocontrol agents and culinary ambitions (Storer 1925; Kraus 2009). However, there are an increasing number of amphibians that are being accidentally introduced and becoming widespread (Kraus 2009). These introductions are in some ways more disconcerting because they may be the most difficult to prevent in the future.

There are 19 nonnative amphibians that have become successfully established in 28 of the 50 U.S. states (Figure 9.1; Kraus 2009). The most successful nonnative amphibian is the bullfrog (*Lithobates catesbeianus*), which has become established in 19 states outside of its native range on the eastern side of the United States, followed by the Cuban greenhouse frog (*Eleutherodactylus planirostris*), which has established itself in six states, and five frog species, including the Puerto Rican coqui (*E. coqui*), which are now established in three states outside of their native range (Figure 9.1; Kraus 2009). The state with the most nonnative frogs is California with eight species, followed by Hawaii with six, and Florida and Arizona with four (Table 9.1; Kraus 2009). Many nonnative amphibians in the United States, particularly in the western United States, are from other parts of the United States, namely, east of the Mississippi River. However, there are also many nonnative amphibians with tropical or subtropical origins that are primarily successful in tropical and subtropical states, such as Florida and Hawaii, and territories, such as Guam.

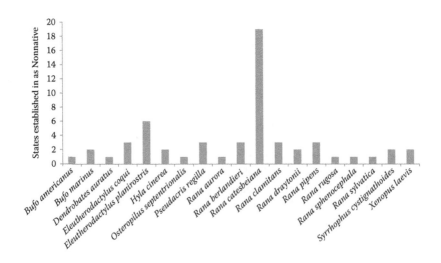

FIGURE 9.1 Number of U.S. states that have established nonnative frog species. (From Kraus, F., 2009, *Alien Reptiles and Amphibians: A Scientific Compendium and Analysis Series, Invading Nature—Springer Series in Invasion Ecology*, Dordrecht, Springer.)

TABLE 9.1
Number of Established Nonnative Frog Species by U.S. State

State/Territories	Number of Nonnative Frog Species
Alabama	1
Alaska	2
Arizona	4
California	8
Colorado	1
Florida	4
Georgia	1
Hawaii	6
Idaho	1
Illinois	1
Iowa	2
Kansas	1
Louisiana	2
Massachusetts	2
Minnesota	1
Mississippi	1
Missouri	1
Montana	1
Nebraska	1
Nevada	3
New Mexico	1
Oklahoma	1
Oregon	1
South Dakota	1
Texas	3
Utah	3
Washington	2
Wyoming	1
Territories	
American Samoa	1
Guam	5
Puerto Rico and islands	9
U.S. Virgin Islands	4

Source: Kraus, F., 2009, *Alien Reptiles and Amphibians: A Scientific Compendium and Analysis Series, Invading Nature—Springer Series in Invasion Ecology*, Dordrecht, Springer.

From an ecological perspective, the major concern with nonnative amphibians is a reduction in native species through competition or predation (Meshaka 2001; Beard and Pitt 2005) and the spread of chytrid fungus (Beard and O'Neill 2005), which has been devastating to amphibian populations around the world, including populations in the western United States (e.g., the boreal toad in Colorado and the

FIGURE 9.2   (a) Photograph of a coqui frog taken in Hilo, Hawaii. (Photo by Steve Johnson.) (b) Photograph of a greenhouse frog taken in Orlando, Florida. (Photo by Bob Fewster.) (c) Photograph of a Cuban tree frog taken in Lakeland, Florida. (Photo by Steve Johnson.) (d) Photograph of a cane toad taken in Lakeland, Florida. (Photo by Steve Johnson.)

mountain yellow-legged frog in California; Muths et al. 2003; Briggs et al. 2005). From an economic perspective, nonnative amphibians in the United States have lowered homeowner property values (Kaiser and Burnett 2006), cost the horticulture industry in terms of loss revenue and treating infestations (Beard et al. 2009), caused power outages (Johnson et al. 2010), and cost government agencies millions of dollars for management (Anonymous 2010). From a social perspective, nonnative amphibians have been blamed for noise pollution (Kalnicky et al. 2014) and producing toxic chemicals that harm humans and occasionally kill pets (Krakauer 1968; Reeves 2004).

Because this book is focused on terrestrial species, this chapter will review nonnative amphibians that are primarily terrestrial. For example, while the bullfrog is a notorious invader in the western United States (Kraus 2009), it will not be covered here because of its more aquatic lifestyle. Although salamanders, caecilians, and frogs are classified as amphibians, nearly 90% of all amphibian species are frogs, and indeed the most invasive amphibians are frogs. We will focus this chapter on two purely terrestrial species (Puerto Rican coqui and Cuban greenhouse frog; Figure 9.2a,b), meaning they do not require water for breeding, as well as two species that are primarily terrestrial but that use water bodies for breeding (Cuban tree frog and the cane toad; Figure 9.2c,d).

CASE STUDIES

Coqui Frogs (*Eleutherodactylus coqui*)

History of the Introduction and Spread

The coqui is endemic to the island of Puerto Rico but has been introduced to several areas in the United States. It was introduced to Florida in the early 1970s, likely via nursery plants (Austin and Schwartz 1975; Wilson and Porras 1983). It became established on the Puerto Rican islands of Culebra and Vieques and two U.S. Virgin Islands in the late 1970s and early 1980s (Rivero 1978; MacLean 1982). The coqui established in Hawaii in the late 1980s; it was brought over in nursery plants either from Puerto Rico or via Florida, which had populations in nurseries at the time (Kraus et al. 1999). The last reported population in Florida around the year 2000 (Meshaka et al. 2004). Florida populations may have died off because of cold winters. By 2001, the coqui had been collected from over 275 distinct locations throughout the islands of Hawaii, Maui, Oahu, and Kauai (Kraus and Campbell 2002). The coqui has been introduced to Guam and California from Hawaii (Campbell and Kraus 2002; Christy et al. 2007). In Guam, the few individuals introduced were quickly eradicated; in California, populations established outside of nurseries have not been confirmed.

While the coqui was once present on the four main Hawaiian Islands (Hawaii, Maui, Oahu, and Kauai), it currently only has established populations on the Big Island (Hawaii) and Maui. Genetic and morphological analyses indicate that populations on these islands started as two separate introductions; past populations on Oahu and Kauai came from the Big Island introduction (Peacock et al. 2009). Very diligent control operations were able to eradicate the coqui frog from Oahu and Kauai. Cooler climates, colder winters, and more manageable terrain might have played a role in these successful eradications, although there are continuing reports of individual calling frogs on these islands that are managed as incipient populations (Beachy et al. 2011; Pitt et al. 2012).

Currently, the coqui is widespread across the Big Island, particularly in the forested lowlands and on the windward side on the island, occupying over 30% of sites along major roads across the island (Anonymous 2010; Olson et al. 2012a). While there have been 36 different incipient populations reported on Maui (Kraus and Campbell 2002), after persistent control efforts, the coqui has been largely restricted to one last stronghold in Maliko Gulch on the north side of the island; however, calling individuals are occasionally reported in different parts of the island. The persistent Maliko Gulch population in Maui, as well as those on the Big Island, makes continuous monitoring, management, and control of incipient populations a reality on Maui as well as Oahu and Kauai.

Once established in Hawaii, interisland and within-island movement of nursery plants played a strong role in their spread, although hitchhiking on vehicles also likely contributed (Peacock et al. 2009; Everman and Klawinski 2013). Many new coqui populations begin adjacent to nurseries, such as the previously established Kauai population (K. Beard, pers. obs.). On the Big Island and Oahu, many landowners state that they started hearing coquis calling after they placed recently purchased nursery plants in their yards (Kalnicky et al. 2014). There were also intentional introductions. In the early 2000s, people on the Big Island moved coquis to state parks where they deposited individuals near the parking areas in an effort to make the coqui too widespread to eradicate and as a misguided insect control effort (Beard and Pitt 2012).

Traits That Help Them Establish

The coqui is a small frog, with mean snout-vent length (SVL) around 34 mm for males and 40 mm for females (Beard 2007), which probably helps it establish because it may go undetected. It has direct development, meaning there is no free-living tadpole phase and metamorphosis occurs within the egg to produce froglets (Townsend and Stewart 1985, 1994). Coqui typically lay their eggs in leaf litter or rock crevices, but also can reproduce in man-made objects, such as nursery pots, as long as there is sufficient moisture (Beard et al. 2003). Direct development, along with year-round breeding, means that unlike some other frog species, the coqui does not require a landscape feature (such as a water body) or many other individuals, such as breeding chorus, to reproduce. Each clutch has, on average, 28 eggs (Townsend and Stewart 1994).

The coqui frog has male parental care, such that the male frog sits and guards the eggs before and a few days after hatch, which prevents desiccation and predation (Townsend et al. 1984). Because of this, if a male protecting a clutch is introduced, the eggs will be protected during transport, and tens of individuals could be introduced from a single introduction event. Genetic diversity of coquis was greatly reduced when they invaded the Big Island and Maui, yet they continued to establish successfully and reach extremely high densities (Peacock et al. 2009). Furthermore, findings from Peacock et al. (2009) suggest that a single clutch may be sufficient to establish a population.

The coqui is a generalist sit-and-wait predator (Woolbright and Stewart 1987; Beard 2007). In Hawaii, it consumes primarily leaf litter invertebrates, such as ants, amphipods, spiders, beetles, roaches, springtails, and mites, but shifts its diet at different sites based on availability (Beard 2007). The coqui is found in Hawaii and Puerto from 0 m to 1200 m above sea level (as reviewed in Beard et al. 2009), and within this elevation range it is generally found anywhere that has adequate humidity and cover (Schwartz and Henderson 1991). In Hawaii, it is mostly found in lowland forests on the eastern side and in private residences. The elevation limit and associated minimal survival temperature of the coqui in Hawaii is unknown, but it has not established populations above 1200 m in the over 25 years since its introduction.

Scope of the Issues/Damage Caused by the Species

The coqui frog has one of the highest densities of any terrestrial frog species in the world. In its native Puerto Rico, densities are typically around 20,000 frogs/ha, but in some areas in Hawaii it has been documented to reach densities up to 90,000 frogs/ha and consume 690,000 invertebrates/ha/night (Woolbright et al. 2006; Beard et al. 2008). Because of its high densities and generalist feeding behavior, it was hypothesized to reduce invertebrates and change ecosystem functions (Beard and Pitt 2005). The coqui has been found to change invertebrate communities across the Big Island (Choi and Beard 2012). Leaf litter insects such as mites and ants, in particular, are reduced where coquis invade, whereas flies increase (Choi and Beard 2012). Furthermore, coquis have been associated with increases in leaf litter decomposition rates, increased nutrient cycling rates, and faster growth rates of nonnative plants (Sin et al. 2008). While it has been hypothesized that coquis may compete with native insectivorous birds, recent research does not support this (Smith et al. 2017, Smith 2016). They have been hypothesized to bolster introduced mammal populations (Kraus et al. 1999), but this hypothesis has not yet been fully explored. Finally, because coquis do not appear to be affected by, but can be carriers of, chytrid fungus, their introductions could impact native amphibians by carrying this and other diseases (Beard and O'Neill 2005). This is not an issue in Hawaii, where there are no native amphibians, but could be important in other invaded areas.

From an economic perspective, the coqui has negatively influenced the floriculture industry and homeowners. The floriculture and nursery industry in Hawaii produces over $100 million annually (Hara et al. 2010). For the floriculture industry, because coquis are mostly transported via plants, economic impacts include decreased sales, destruction of plant shipments, increased costs to control infestations, and increased quarantine procedures (Beard et al. 2009). Plant shipments from Hawaii to Guam, the continental United States, and other countries require a phytosanitary certificate that certifies shipments are pest-free. Interisland and international plant shipments from the Big Island, in particular, are supposed to be inspected and treated prior to shipment. This is often done by bathing plants using a hot water treatment (Hara et al. 2010). While some hotels and resorts have complained of potential loss of revenue, economic loss has only been documented for private landowners (Kaiser and Burnett 2006). Kaiser and Burnett (2006) found that complaints about the loud frog calls were related to housing prices, and that the closer the complaint to the marketed property, the greater the reduction in the housing price. While the Big Island housing unit drop in price associated with coqui establishment was almost always less than 1%, property values were estimated to drop $7.6 million islandwide as a result of coquis. Furthermore, businesses, private landowners, and local, state, and federal government covered the cost of managing coquis. At its highest point in the late 2000s, public agencies were spending $4 million per year to control the frogs (Anonymous 2010; Beard and Pitt 2012).

The primary public concern over the coqui has to do with its extremely loud mating call. The coqui produces a two-note mating call, which sounds like "ko-kee" (80–90 dB at 0.5 m) and exceeds the noise level set to minimize impacts for enjoyment of life (70 dB, Department of Health, Hawai'i Revised Statutes Section 324F-1; Beard and Pitt 2005). In Hawaii, there has been a lot of support by the general public in the form of coqui control groups to eradicate the coqui. These groups raised funds to rent or purchase control equipment, and invested endless hours of volunteer time monitoring and controlling populations (Anonymous 2010; Beard and Pitt 2012). Despite much local support in Hawaii for coqui suppression or eradication, there were also people that enjoyed the call of the coqui while others became accustomed to it (Kalnicky et al. 2014). More specifically, Kalnicky et al. (2014) found that people with more frogs on their property and those who owned property tended to have less-negative attitudes toward the frog. If tolerance for the species is in fact growing, that would hinder attempts to engage the general public in control efforts.

Historic and Current Management

Control efforts for coquis have primarily been performed in Hawaii, despite Guam and California placing restrictions for live plant importation from Hawaii as well as unlawful possession of coqui without permit (Hara et al. 2010). Since the frog's introduction to Hawaii, many control measures have been considered (Pitt et al. 2012). Some of the most effective measures identified for killing frogs were not approved for use because they were not deemed safe for nontargets or humans, or accepted by the general public (Pitt et al. 2012). Furthermore, most control measures were only found to work in limited situations, not across large areas with varying environmental conditions (Pitt et al. 2012).

Chemical control has been the most effective means of controlling coqui frogs. The U.S. Department of Agriculture (USDA) tested 90 chemical agents (agricultural pesticides and pharmaceutical and household products) and 170 chemical formulations as potential frog pesticides (reviewed in Pitt et al. 2012). Eight chemicals killed frogs, but only three were ever approved for control (Pitt et al. 2012). From 2001 to 2002, a 2% caffeine solution was approved for limited use and was very effective, but was not approved for widespread use because of human health concerns and a lack of public support (Pitt et al. 2012). From 2005 to 2008, a 6% hydrated lime solution was approved to control coquis but later discontinued because of caustic effects; it killed vegetation. Homeowners liked using hydrated lime because it was inexpensive (Pitt et al. 2012). In 2002, 16% citric acid, a food additive, was approved for widespread use and it is the only chemical currently approved for controlling frogs in Hawaii (Pitt et al. 2012). Citric acid has been used by landowners, government agencies, and nurseries to control coquis (Hara et al. 2010). For example, in 2005 alone, over 24,000 kg of citric acid was used to control coquis on Maui (L. Strohecker, pers. comm.). The successful eradication of hundreds of calling frogs on Oahu can be attributed to years of spraying citric acid using ground operations (Beachy et al. 2011).

Mechanical methods have also been evaluated for controlling coqui frogs, including hot water treatments, traps, vegetation management, hand capture, and barriers (Pitt et al. 2012). Frogs and their eggs are killed when exposed to hot water or steam applied for three minutes at 45°C (Hara et al. 2010). These methods are particularly important to prevent movement of coqui frogs via plant shipments because they often do not harm plants, unlike citric acid, which has phytotoxic effects. Traps have been developed that capture frogs, and thus they can be removed if diligently checked; however, frogs can use some traps to breed in, such as polyvinyl chloride (PVC) pipes (Stewart and Pough 1983), and they have not been effective at reducing populations (reviewed in Pitt et al. 2012). Vegetation management can reduce and help eradicate coqui frogs (Kalnicky et al. 2013). Experiments removing 100% of leaf litter and 100% understory vegetation showed marked reduction in coqui populations (Kalnicky et al. 2013). Removing tree canopies to create gaps in tree stands also reduced the coqui population (Klawinski et al. 2014). Hand capturing and erecting barriers can be effective in small areas, such as in and around greenhouses, on private yards, and with other incipient populations (reviewed in Pitt et al. 2012). For example, hand capturing prevented the coqui from establishing in Guam when they came over in a plant shipment from Hawaii (Beard et al. 2009). Investigating shipments for frogs as they come in may be the most critical step in stopping some introductions.

Biological controls have probably been the least explored. Chytrid is already established in coqui frog populations in Hawaii, and coquis are relatively resistant to the disease (Beard and O'Neill 2005). Research to identify parasites has not found one that reduces growth or survival (Marr et al. 2010). However, the high density of coqui frogs in their native range suggests that even if predators or parasites from Puerto Rico could be identified, they are unlikely to eradicate populations and biological controls bring associated risk (Pitt et al. 2012).

Combinations of management approaches can be important in the successful eradication of coqui frogs. The Kauai eradication, in particular, involved a great amount of vegetation removal, covering about 6 ha in addition to continual citric acid spraying (Pitt et al. 2012). However, once large populations are established in difficult terrain and in remote areas, even the most intensive efforts can make eradication impractical. For example, attempts to eradicate coquis from Manuka Natural Area Reserve on the Big Island and Maliko Gulch on Maui using large-scale citric acid helicopter drops, semipermanent spray systems, and ground operations have not been successful despite measurable reductions in coquis for periods of time (Tuttle et al. 2008; Anonymous 2010; Beard and Pitt 2012; Pitt et al. 2012). The terrain in these locations, including rock crevices and steep slopes, makes complete eradication of coquis from these areas unlikely because it is hard to spray all areas, and because coquis can hide from chemical spray in abundant crevices, particular during daytime spraying (K. Beard, pers. obs.). In the case of both the Oahu and Kauai eradications, the populations were isolated, the terrain was less difficult to maneuver through, and there were fewer rock crevices than the previously mentioned Big Island and Maui sites (K. Beard, pers. obs.). If the coqui invades other areas where the terrain is not as difficult and the sites are not as remote, methods developed to control coqui in Hawaii may work. Vegetation removal with citric acid spraying appears particularly effective at reducing populations.

Greenhouse Frogs (*Eleutherodactylus planirostris*)

History of the Introduction and Spread

The greenhouse frog is native to Cuba and the Bahamas (four islands). It is one of the most widespread frog species in the world. Within the United States, it has become established in Hawaii (on five islands), Florida (widespread on the peninsula), Alabama (one county), Georgia (five counties), Louisiana (10 parishes), Mississippi (one city), and on Guam (widespread) (Olson et al. 2012b). Internationally its spread is beginning to be appreciated in Mexico, the Philippines, and on many Caribbean Islands (Kraus et al. 1999; Olson et al. 2014; Rogelio Cedeno-Vazquez et al. 2014). The first report of this frog in the United States occurred in Florida in the late 1800s; it is possible they established naturally, perhaps via driftwood (Goin 1947; Meshaka et al. 2004; Heinicke et al. 2011). The greenhouse frog is thought to have established itself in Hawaii in 1994 (Kraus and Campbell 2002). It first appeared in Hawaii in plants imported from Florida, and in Guam in 2003 in plants imported from Hawaii. Its spread is likely almost exclusively attributable to accidental introductions, primarily through the floriculture trade but also probably through cargo (Kraus et al. 1999).

In general, its introductions and spread have not been well studied (Olson et al. 2012b). Part of the reason for this may be its cryptic nature and relatively quiet mating call (Kraus and Campbell 2002). For example, *Eleutherodactylus planirostris* and *E. coqui* invaded Hawaii around the same time, and while there were many complaints about the coqui and a large effort was launched to control its spread, few funds were directly spent on controlling the greenhouse frog. This might explain why the greenhouse frog became more widespread on the Big Island than the coqui frog (Olson et al. 2012a). In a study designed to determine their distributions on the Big Island, greenhouse frogs were found to occupy 35% of sites along the major road systems around the island compared to 31% for the coqui (Olson et al. 2012a). While no systematic surveys have confirmed the distribution of greenhouse frog across the other Hawaiian Islands, they are thought to be widespread on Oahu, Kauai, and Maui (Olson et al. 2012b). Similarly, the greenhouse frog has become widespread in Florida and Guam. In Florida, greenhouse frogs are found throughout the peninsula and at a growing number of locations in the state's panhandle, predominantly in coastal areas (Krysko et al. 2011). Recent studies show that they are also spreading in Alabama (Alix et al. 2014) and Mississippi (Mann et al. 2015).

Traits That Help Them Establish

The frog is small (SVL on average 24 mm) and cryptic. It is brown in color and blends in with leaf litter and soil, where it is mostly found. Like the coqui frog, it has direct development; therefore, it does not need a water body to breed in; can lay eggs leaf litter, rock crevices, or soil; and unlike other members of the genus, there is no guarding of the eggs (reviewed in Olson et al. 2012b), which may increase the chances of an inadvertent introduction. Each clutch has, on average, 16 eggs (Goin 1947). Eggs require 100% humidity to hatch and can be submerged in water for a period of up to 25 days and still remain viable (Goin 1947). Eggs hatch 13–20 days after deposition (Goin 1947).

While the greenhouse frog consumes a diversity of invertebrates including spiders, mites, springtails, and beetles in the leaf litter, the large majority of its diet is typically ants (Goin 1947; Stewart 1977; Olson and Beard 2012; Ferreria et al. 2015). This specialization on ants may assist their establishment into previously uninvaded areas, considering that ants comprise the majority of invertebrate biomass in tropical areas (Hölldobler and Wilson 1990). The greenhouse frog has invaded areas like Guam and Florida, with high potential predator densities (as reviewed in Olson et al. 2012b). However, this does not appear to have controlled frog establishment and spread.

The greenhouse frog has a high tolerance for warm and dry conditions compared to other *Eleutherodactylus* species; for example, it has established itself in Florida, whereas the coqui did not (Olson et al. 2012b). Furthermore, on the Big Island, the coqui and greenhouse frog overlap in about a third of their occupied sites; yet, the greenhouse frog is more often found in drier sites on the western side of the island (Olson et al. 2012a). In Florida and Hawaii, it is common in wet and dry forests, open grasslands and pastures, coastal areas, scrub habitats, nurseries, residential gardens, and resort areas (Meshaka et al. 2004; Olson et al. 2012a). Its distribution in its nonnative range appears to reflect warmest-month temperatures in Cuba, but in the southeastern United States, it lives in areas as cool as 4°C (Tuberville et al. 2005; Rödder and Lötters 2010). In Hawaii, it has been detected from sea level up to 1115 m (Olson et al. 2012a).

The soft call of the greenhouse frog, with sound pressure levels around 35–45 dB at 0.5 m (K. Beard, unpublished data), prevents reporting of new infestations. A clear example of how the quieter call and more cryptic nature of the greenhouse frog, compared to the coqui frog, likely facilitated their range expansion is their comparative histories in Guam. When the coqui was introduced to Guam, its establishment was prevented very shortly after it came out of shipments. The greenhouse frog, on the other hand, established and rapidly spread (as reviewed in Christy et al. 2007; Olson et al. 2012b). In addition, through conversations with private landowners in Hawaii, most residents were aware when coquis were on their properties but many did not recognize the call of a greenhouse frog, and when they were informed that it was indeed a frog they said they did not mind the sound (K. Beard, unpublished data). Public opinion influences invasive species management in Hawaii; citizens have been very involved in control coqui frogs. The lack of public concern or awareness about the greenhouse frog likely contributed to its spread.

Scope of the Issues/Damage Caused by the Species

The greenhouse frog is most likely to impact leaf litter invertebrates where it invades because it only moves vertically several centimeters from the ground and has an insectivorous diet (Olson and Beard 2012). Assessments of ecological impacts of greenhouse frogs are most common in Hawaii, although their impacts are probably transferrable to other locations where they have invaded. Their densities have been estimated to reach up to 13,000 frogs/ha in Hawaii, and they have been estimated to consume 129,000 invertebrates/ha/night (Olson and Beard 2012). Also, in Hawaii, the greatest concern may be their potential to reduce rare or threatened populations of invertebrates. In other areas, where there are native frogs or other native species that depend on the leaf litter community, the greenhouse frog may compete with them for prey. The greenhouse frog may alter nutrient cycling, like the coqui frog (Sin et al. 2008), or serve as a food source for nonnative predators. Brown tree snakes in Guam are thought to consume greenhouse frogs (Mathies et al. 2012). Although they are likely consumed by a diversity of small carnivorous animals, documented predators of greenhouse frogs in the southeastern United States are exceedingly few (Meshaka et al. 2004; Jensen 2008; Dodd 2013).

Similar to the coqui frog, the greenhouse frog has impacted the floriculture industry in Hawaii. There is no summary of the amount of funds expended to control greenhouse frogs, but because intentional transport of frogs into the State of Hawaii is illegal (Kraus and Campbell 2002), it is very likely that nursery owners expend funds to keep their nurseries pest-free when they have infestations as well as treat shipments going off the island. Unfortunately, because greenhouse frogs are less obvious than coquis, greenhouse frogs and frog eggs are probably not detected as frequently in shipments as are the coqui. In addition, some resorts in Hawaii have attempted to reduce greenhouse frogs on their properties with moderate success, as described in the next section (Olson et al. 2012b).

Historic and Current Management

Unlike the coqui frog, the greenhouse frog has not been the target of large-scale control or eradication efforts in Hawaii or elsewhere that we know of. However, many of the same chemicals that were found to kill coquis are equally effective on greenhouse frogs (government documents supporting this are reviewed in Olson et al. 2012b). Specifically, 16% citric acid solution is 100% effective at killing greenhouse frogs in laboratory conditions (reviewed in Olson et al. 2012b). In addition, hydrated lime and caffeine are effective but currently not permitted for *Eleutherodactylus* control in Hawaii (Kraus and Campbell 2002; Pitt et al. 2012). In locations where coquis are sympatric with greenhouse frogs, both species were likely reduced by efforts targeting coquis. Because there have been almost no large-scale efforts to control greenhouse frogs, they have spread unabated in most places where they have been introduced. However, the chemical control measures developed for coqui frogs would be effective in controlling or reducing them (reviewed in Olson et al. 2012b) and should be considered as a management option.

Many of the mechanical controls described for coqui frogs would also work on greenhouse frogs. For example, hot water treatments are effective at killing *Eleutherodactylus* species (Hara et al. 2010) and should be required in any area shipping plant material where there are infestations. Traps may work on greenhouse frogs, and one of the most common locations to find greenhouse frogs at any site in Hawaii is an irrigation box (Ferreria et al. 2015). While hand captures of greenhouse frogs may be less effective than for coqui frogs because the greenhouse frog is smaller and more cryptic, hand capturing from irrigation boxes at several resorts in Hawaii did result in lower populations over time (as reviewed in Olson et al. 2012b). Leaf litter removal may be particularly effective for greenhouse frog control because they are often found close to the forest floor (Olson et al. 2012b; Kalnicky et al. 2013). While not tested on greenhouse frogs, barriers may also be effective in small areas (as reviewed in Olson et al. 2012b).

Cuban Tree Frogs (*Osteopilus septentrionalis*)

History of the Introduction and Spread

The Cuban tree frog, as its common name implies, is native to Cuba (including Isle of Pines) as well as the Cayman and Bahama islands (Schwartz and Henderson 1991). Individuals have been introduced to numerous islands of the West Indies and populations are established on several of these islands, including Puerto Rico and the U.S. Virgin Islands (Lever 2003; Kraus 2009). In the United States, Cuban tree frogs are only established in peninsular Florida; however, there are numerous isolated records of the species from many counties in the state's panhandle (Johnson 2007). Additional records of single frogs have been reported from other states, especially in the southeastern United States.

The first Cuban tree frogs from Florida were observed in the late 1920s in Key West, and they most likely were transported there inadvertently as stowaways in cargo on boats from Cuba (Barbour 1931). However, some authorities suggest Cuban tree frogs may have also colonized the Florida Keys naturally (Meshaka 2001). Based on records maintained by the Florida Museum of Natural History at the University of Florida, by the early 1930s the frogs had made their way to the southern extreme mainland of Florida. Records into the 1950s are also confined to the southern tip of the peninsula, but 20 years later the frogs had been documented throughout much of the southern third of the peninsula. By the 1990s, Cuban tree frogs could be found as far north as Orlando, with records from several counties even further up the Atlantic Coast of Florida. By 2010, they had been recorded from every county in the peninsula and were established as far north as Cedar Key on Florida's Gulf Coast, Gainesville in north-central Florida, and Jacksonville on the Atlantic Coast.

Traits That Help Them Establish

Cuban tree frogs are the largest species of tree frog in the United States (Dodd 2013), and this facilitates their ability to consume large prey. They show pronounced sexual size dimorphism, and the largest male reported in Florida is 85 mm SVL and the largest female is 122 mm SVL (McGarrity and Johnson 2009). They are generalist predators that consume a great variety of invertebrates dominated by spiders, roaches, and beetles, as well as small vertebrates including native tree frogs (Meshaka 2001; Glorioso et al. 2012; Johnson, unpublished).

Females are quite fecund, and Meshaka (2001) reported clutch sizes from 1177 to 16,371 eggs based on a sample of 153 females from the Florida Everglades. They breed in a diversity of water bodies ranging from retention ponds to shallow cypress swamps (Dodd 2013). They also breed in ornamental fishponds, rain barrels, and swimming pools that are not well chlorinated. In northern and central Florida, they reproduce during the spring and summer, but may be reproductively active any time of the year in extreme southern Florida. Larval development (tadpole stage) to adult transformation can occur in the field in three to four weeks (Meshaka 2001).

In natural areas, Cuban tree frogs prefer closed canopy, forested habitats and tend to avoid more open habitats (McGarrity and Johnson 2010). They are common in moist tropical hardwood hammocks, bottomland forests, cypress swamps, and mangrove swamps, but also occur in pine rocklands, pine flatwoods, sandhills, and xeric hammocks (Meshaka 2001; Johnson 2007; Campbell et al. 2009; Rice et al. 2011). They also inhabit human-modified environments, such as Brazilian pepper stands, orange groves, agricultural landscapes, and urban/suburban neighborhoods. In urban and suburban settings, they are found on buildings and homes, and among landscape plants.

Cuban tree frogs have exceptionally large, sticky toe pads and are excellent climbers. Their affinity for human-modified environments (e.g., suburban settings), tendency to hide in confined spaces, and ability to climb well certainly facilitates their invasion of new habitats via human transport. The main invasion pathway for Cuban tree frogs within the state of Florida and beyond is via hitchhiking on vehicles and trailers, as well as on ornamental plants, especially palms (Meshaka 1996). They also have the ability to move rapidly from open, inhospitable areas to preferred habitats (McGarrity and Johnson 2010).

Tadpoles and adults tolerate a broad range of water temperatures. Meshaka (2001) reported finding Cuban tree frog tadpoles in water ranging from 12°C to 41°C. As expected, tadpoles in warm water reached metamorphosis much faster (three to four weeks) than those in colder water (five to six months). Although prolonged exposure to below freezing temperatures in the wild is lethal to adults and can lead to local declines of Cuban tree frog populations (S. Johnson, unpub. data), they are surprisingly tolerant of short-term exposure to cold (Simpson 2013). Although several species of native snakes and birds are documented predators of Cuban tree frogs, Cuban tree frogs exhibit several antipredator behaviors (e.g., crypsis, large body size, and sheltering in confined spaces) and emit noxious skin secretions (Meshaka 2001) that likely make them less palatable than native tree frogs to potential predators.

Scope of the Issues/Damage Caused by the Species

Cuban tree frogs in Florida are a quintessential invasive species. They cause detrimental ecological and economic impacts, and they negatively affect quality of life for some Floridians. They readily consume small vertebrates and their large gape allows adult Cuban tree frogs of both sexes to capture and swallow native tree frogs (Wyatt and Forys 2004; Glorioso et al. 2012). In fact, they are now the most commonly encountered tree frog in urban and suburban neighborhoods in much of peninsular Florida, likely due to their predation of native green and squirrel tree frogs (S. Johnson, pers. obs.). They may also depress populations of native tree frogs in natural areas. Cuban tree frogs reduced capture probability of native tree frogs and dominated captures in PVC pipe refuges at sites in south and central Florida during three different studies in which hundreds of frogs were encountered (Campbell et al. 2010; Waddle et al. 2010; Rice et al. 2011). These findings are even more astounding considering the fact that Cuban tree frogs prefer a more natural hiding place over a PVC pipe refuge, at least in an experimental setting (Hoffmann et al. 2009). Although there are no empirical data available, it seems possible, given their density in some natural areas and their broad diet, that Cuban tree frogs might compete for food with native species. Additionally, Cuban tree frog tadpoles were found to be superior competitors against two native species of tadpoles in controlled experiments (Knight et al. 2009), and Cuban tree frog tadpoles are documented predators of native *Hyla squirella* tadpoles (Smith 2005). Cuban tree frogs will also invade bird nest boxes erected to benefit native wildlife. However, the prevalence of this behavior and its potential effect on use of nest boxes by native birds has not been studied to our knowledge. In addition, how Cuban tree frogs use natural tree cavities is unknown.

The close association of Cuban tree frogs with human-dominated landscapes and the frogs' propensity to seek tight, enclosed spaces during the day has led to negative economic impacts. Although poorly documented and in need of additional research, Cuban tree frogs have been responsible for short-circuiting electrical equipment. In one instance, a Cuban tree frog was deemed responsible for invading a switchgear box and causing a short circuit leading to a $20,000 repair for the Lakeland Electric company (Johnson et al. 2010). An engineer with Lakeland Electric reported that Cuban tree frog invasion of other equipment has also led to power outages, but no estimates of monetary damages were given (S. Perkins, pers. comm.). Cuban tree frogs are also known to have caused damage to air conditioner compressor units and water pumps, which had to be repaired at cost to the homeowner (S.A. Johnson, pers. obs.). An additional economic burden, which has yet to be quantified, is the cost to

Floridians resulting from hospital visits by children and pets as a result of Cuban tree frog poisoning. When handled or harassed, the frogs secrete a sticky substance that is extremely irritating to mucous membranes (S.A. Johnson, pers. obs.). If the frog's secretion gets into a child's eyes, it causes an intense burning sensation, and we know of at least one instance when a young boy had to be taken to an emergency room for treatment resulting from the child's handling of a Cuban tree frog. Although we do not know of any deaths of pets, we have communicated with people who report their cat or small dog had health issues after encounters with Cuban tree frogs; significant costs were incurred for veterinarian care that resulted from such exposure. It seems probable that cases of Cuban tree frog poisoning in pets and children go unreported and undiagnosed.

Cuban tree frogs often inadvertently invade homes and become a serious nuisance to people. They do so by jumping through an open door or window, catching a ride on a potted plant brought inside, and often via plumbing vent stacks located on the roofs of homes (Johnson 2007). Once in the plumbing system, they can make their way to toilets and sinks. It is not uncommon for Florida residents to contact the University of Florida's (UF) Cooperative Extension Service with an image of a Cuban tree frog in a toilet requesting advice on identification and removal of the interloper. At night, Cuban tree frogs perch on windows and walls of homes and buildings to feed on insects attracted to lights. As a result, the frogs leave unsightly feces on the sides of people's homes.

Though nowhere near the magnitude of the problem caused by coqui frogs in Hawaii, an additional annoyance caused by Cuban tree frogs that has been consistently reported to UF Extension is the racket from their breeding choruses. The call of male Cuban tree frogs is a squeaking raspy sound and, compared to native species, does not sound very loud. Nonetheless, the density of these pests around homes, and the fact that they breed in ornamental ponds and poorly maintained swimming pools near bedroom windows, has resulted in people losing sleep (S.A. Johnson, unpublished data). However, we do not know of any studies that have measured the volume of their mating calls.

Historic and Current Management

There is no agency-led, statewide plan to manage Cuban tree frogs on public lands. Instead, most Cuban tree frog control efforts occur at local scales on private property and are conducted by citizens. These people most often initiate hand-removal of Cuban tree frogs after viewing educational materials or receiving guidance from the UF Cooperative Extension Service, including their website (http://ufwildlife.ifas.ufl. edu). Goals of these localized removal efforts are often twofold: (1) people remove frogs and take action on their property to help mitigate the nuisances caused by this pest, and (2) they seek to reduce the number of frogs with the hope that native frogs will benefit.

Hand-capture and removal of Cuban tree frogs via PVC pipes are the two methods recommended by the UF Cooperative Extension Service to people who wish to remove the frogs from their property (Johnson 2007). After capture, citizens are directed to apply liberally a benzocaine-containing ointment to the frogs until they are anesthetized and then secure the frogs in a plastic bag and place them in the freezer for 24 hours (http://ufwildlife.ifas.ufl.edu/cuban_treefrog_inFL.shtml). Another humane method of euthanasia for Cuban tree frogs and other small ecto-therms, including cane toads, is to place individual frogs in a standard refrigerator for several hours then transfer them to a freezer overnight to ensure death (Shine et al. 2015). Although the efficacy of PVC pipe refuges has been questioned (Wyatt and Forys 2004) and there is bias among tree frog species in their propensity to shel-ter in PVC pipe "traps" (Hoffmann et al. 2009), Rice et al. (2011) found no evidence that Cuban tree frogs were behaviorally excluding native species from the pipes. Therefore, this passive sampling method remains an efficient means for Cuban tree frog removal. Additional strategies used as part of an integrated pest management plan for Cuban tree frogs on local scales include eliminating breeding opportunities (e.g., keep pools well chlorinated, dump out standing water, exclude frogs from rain barrels/cisterns), management of eggs and tadpoles (e.g., net egg masses from the surface of ponds, dump out containers with tadpoles), and excluding frogs from hid-ing places around homes.

The most common pathway for invasion of human dwellings by Cuban tree frogs is via plumbing vent pipes located on the roofs of buildings. Cuban tree frogs appar-ently enter the pipe seeking shelter from desiccation and eventually find their way to toilets and sinks. Commercially available covers can be placed on the vent pipes or hardware cloth can be cut and attached with cable ties to the pipes to exclude frogs. Homeowners can also purchase and apply a proven wildlife chemical deterrent that has been shown to be effective for dissuading Cuban tree frogs from enclosed spaces. In response to problems that Cuban tree frogs were causing for electric generat-ing utilities, Johnson et al. (2010) tested the ability of the product "Sniff 'n' Stop" (ICORP-IFOAM Specialty Products Corporation, https://sniffnstop.com) to repel Cuban tree frogs. They tested several versions of the product (foam, gel, tape, epoxy) in controlled laboratory experiments. Although none of the applications of "Sniff 'n' Stop" were 100% effective at excluding frogs, they all deterred frogs from using treated hiding places.

CANE TOADS (*RHINELLA MARINA*)

History of the Introduction and Spread

Cane toads are native to Central and South America and extreme south Texas (Conant and Collins 1991). They are thought to be the most widespread amphibian in the world, and they are established in at least 40 countries (Lever 2001). Starting in the 1920s, cane toads were introduced to many areas to control pests in sugar-cane plantations (Krakauer 1968). Internationally, they are established on islands throughout the Caribbean and Oceania. Cane toads were introduced to Puerto Rico in 1920 with a second group in 1923 (Schwartz and Henderson 1991). The intro-duction of the cane toad to Puerto Rico was deemed a success at controlling the white-grub on sugarcane (Anonymous 1934), but later its role in reducing the pest was questioned. However, in many ways, the damage was done once the Puerto Rico cane toad population was deemed successful, and they were quickly introduced elsewhere. By 1932, they were well established in Puerto Rico and introduced from there to Oahu. They established themselves quickly in Oahu, and were moved to the other Hawaiian Islands (Easteal 1981). Cane toads were introduced to Guam in 1937 (Christy et al. 2007) and have also been introduced into the Northern Marianas and American Samoa in the Pacific, as well as in the U.S. Virgin Islands in the Caribbean (Lever 2001).

Initial releases in the 1930s and 1940s in Florida to control sugarcane pests were not thought to result in establishment. It appears that later accidental releases in the airport and by pet dealers, in the 1950s and 1960s, established the populations that continue to persist today (Krakauer 1968). Introductions to Louisiana were unsuccessful (Easteal 1981). Cane toads were introduced from Hawaii to Australia, where they are particularly problematic (Krakauer 1968). They are cold intolerant and therefore are unlikely to become widespread in most parts of the continental United States. However, recent research on cane toads in New South Wales, Australia, demonstrated that toads have the ability to acclimate rapidly to low temperatures (McCann et al. 2014). It has been noted that while their populations expand and increase quickly, they also experience precipitous drops in population size (Simberloff 2004). This was noticed in Puerto Rico and in Australia with no clear indication of why these population drops occurred; the populations in Puerto Rico never recovered (Freeland 1986).

Traits That Help Them Establish

Cane toads are large frogs, with lengths and sizes varying across populations, but reaching up to around 100–150 mm SVL in Florida (Marshall and Meshaka 2006) and 80 mm SVL in Hawaii (Beard and Pitt 2006). Their large size may help them avoid predation and outcompete other anurans. Their skin is resistant to desiccation, which helps them survive in arid conditions, such as the dry seasons of tropical environments (Duellman and Trueb 1986). Similar to the Cuban tree frog, they have minimal requirements for breeding other than presence of sufficient water (saline or fresh) for their eggs to hatch and mature, which can include small ditches (Ely 1944; De León and Castillo 2015). Salt water does not seem to deter cane toads, and adults are able to survive in salinities of up to 40% sea water (Liggins and Grigg 1985); they have been observed swimming in the sea and crossing 600 m of salt water between two islands (Lever 2001). Furthermore, Lever (2001) reports several observations of cane toads spawning in brackish water and calling on tidal mudflats.

Like most invasive species, cane toads are highly fecund, and females produce clutches of approximately 20,000 eggs (Hagman and Shine 2008). The time required for eggs to mature varies with climate, but they typically hatch within 48 hours, and it takes about 30 days for tadpoles to mature. Sexual maturity in cane toads can be reached within one year, depending on the environment (Shanmuganathan et al. 2010). They breed year round (Doody et al. 2014). Their generalist diet, which includes both vertebrates and invertebrates, has almost certainly helped them attain high population densities (over 2000 frogs/ha) in their introduced ranges (Freeland 1986). Finally, eggs, tadpoles, and adults are poisonous to many predators as they contain bufadienolides (alkaloid substances toxic to vertebrates), which make cane toad control via natural predators in the environment challenging (Shanmuganathan et al. 2010).

Habitat modeling in Australia has predicted that cane toads will greatly expand their current distribution of northern and eastern Australia because they can tolerate temperatures of less than 5°C (e.g., southeast Australia) and temperatures of greater than 37°C (e.g., parts of the Northern Territory) (Urban et al. 2007). Similar limits may exist in the United States. Cane toads also appear to favor open and human-disturbed habitats, such as roadside, urban, and suburban areas (Urban et al. 2007); diurnal shelters are important for maintaining body temperature and moisture year round but especially outside of the wet season (Seebacher and Alford 2002).

Scope of the Issues/Damage Caused by the Species

The cane toad can be devastating from an ecological perspective. The species has primarily been researched in Australia, where studies of its impacts have shed light on the invasion process in general. Because every part of their life cycle can be poisonous to predators, cane toads can have severe effects on potential predators and most specifically on native reptiles and mammals in Australia. They have been predicted to affect a large number of species in Australia (Phillips et al. 2003; Beckmann and Shine 2012); declines in snakes and crocodiles, in particular, as well as native mammals have been attributed to the cane toad invasion; however, impacts from cane toad invasions often show both population-level differences (Doody et al. 2009; Somaweera et al. 2013) and, in some cases, minimal impacts (Kamper et al. 2013).

Perhaps one of the most serious negative impacts of cane toads in Australia was their widespread, local extinction of a marsupial carnivore, the northern quoll (*Dasyurus hallucatus*); shortly after cane toads colonized natural areas containing the quolls, they went extinct (O'Donnell et al. 2010). Phillips et al. (2003) determined that many native Australian snakes die from consuming a single cane toad, and that approximately 30% of the populations of native snakes are threatened by cane toads. Boland (2004) discovered that cane toads significantly reduce fledging success of ground-nesting rainbow bee-eater birds (*Merops ornatus*) by usurping their burrows and consuming eggs and chicks. Some vertebrates, including at least seven native birds and three native rodents, can consume cane toads successfully either because they only eat the nontoxic parts of the toad or because they have developed some resistance to the toxicant (Covacevich and Archer 1975; Lever 2001).

Its impacts on native frogs are particularly equivocal as it might benefit them in areas where it reduces their predators (Shine 2014), although they do cause mass mortality of native frog larvae through direct consumption of the cane toad eggs (Crossland et al. 2008). There has been concern that cane toad egg and larvae presence alone would kill native organisms sharing the same water body, but data supporting this concern are lacking, and several native Australian frogs have been tested and survive in water bodies infected with cane toad toxicants (Hagman and Shine 2009b). Despite the findings that all native frogs consuming cane toad eggs die, mortality varies by species because not all native anurans' breeding coincides with that of cane toads (Crossland et al. 2008), and reductions in some native frog species may release others from competition and therefore increase their survival (Crossland 2000). Effects of cane toads on competitive interactions have also been demonstrated by Doody et al. (2015), where cane toad suppression of monitor lizards, which are predators of native birds, resulted in 55%–81% increase in fledging success of a native finch.

Stomach content analyses reveal prey vulnerable to cane toad invasions. In their native range of Venezuela, 269 stomachs contained at least 18 invertebrate orders; beetles dominated on a mass basis (27%), followed by ants (13%), larvae of dragon-flies/damselflies (8%), grasshoppers/crickets (4%), and butterflies/moths (3%) (Evans and Lampo 1996). In forested areas of northern Queensland, 81 stomachs contained spiders, cockroaches, earwigs, true bugs, flies, grasshoppers/crickets, centipedes, and millipedes (Heise-Pavlov and Longway 2011). In Papua New Guinea, snails, including the small *Sublina octina* and the large *Achatina fulica*, composed 42% of their diet in cacao plantations (Bailey 1976). Greenlees et al. (2006) used outdoor exclosure trials in northern Australia to determine cane toad effects on invertebrates. Their findings revealed that, although cane toads reduced invertebrate abundance and species richness, the level of reduction was equivalent to native frogs tested. Greenlees et al. (2006) point out that in the field, the massive amount of cane toad biomass relative to that of native frogs will result in a large nutrient sink in cane toads resulting from their feeding on invertebrates.

Whereas the impact of the cane toad in Australia is of great concern because of the naivety of the endemic fauna to this species; in Hawaii, where there are no native terrestrial reptiles or ground-dwelling mammals, its effect on potential predators has not been as well studied. In a study to determine if cane toads consume other nonnative frogs (*Eleutherodactylus* spp.) in wet forests in Hawaii, they were only found to consume roughly equal proportions of plants and invertebrates (36% snails, 23% millipedes, 17% beetles, and 10% butterflies/moths) (Beard and Pitt 2006). In Florida, in the same study referenced above where Cuban tree frogs were found to affect native frog larvae, cane toads were not found to affect larval growth or development of two native species (Smith 2005). There is valid concern for the native ecosystems of the United States where cane toads may establish because of the wide range of largely negative effects of these invaders in Australian ecosystem.

Similar to other frogs, cane toads carry and transmit diseases (Speare 1990). Large numbers of potentially pathogenic bacteria (e.g., *Aeromonas hydrophila*, *Mycobacterium* spp.), including *Salmonella* spp., fungi (*Fonsecaeapedrosoi* and *Candida* sp.), an amoeba, and helminths (*Spirametra mansoni* and *Rhabdias sphe-rocephala*), have been isolated from cane toads, but the level to which these cause disease is apparently small (Speare 1990; Drake et al. 2013). The extent to which these pathogens are transferred to native species or humans is not well known.

Most of the cost associated with cane toad management in Australia comes from control measures. Australia has spent about $1 million per year supporting research and community-based control measures to reduce their impacts and developing a national plan (Shine et al. 2006; Shine 2010). While the United States has not focused any funds directly to control cane toads and no formal economic studies have covered their invasion in the United States, economic costs will range from losses in biodiversity to human and domestic animal health and safety (e.g., exposure to cane toad toxicants or carried pathogens). Given the wide biodiversity and human establishment along coastlines where cane toads are particularly common (Urban et al. 2007), the economic costs associated with cane toad invasion and spread could be significant and will likely become more of a burden in the future with the potential for range expansion with global temperature changes.

From a societal perspective, the cane toad primarily is considered a pest because of the defensive secretions emitted from their parotoid as well as other glands that can cause illness to humans and death in domestic animals if ingested, and cause pain if it enters the eyes (Krakauer 1968). The poison is often released directly outside the gland, and while they cannot eject toxin at will, parotoid glands under pressure can squirt the toxin close to a meter (S. Johnson, pers. obs.). A review of 90 cases of cane toad intoxication in dogs conducted over 2.5 years in Brisbane showed that while most pet dogs became sick, survival rates were very high at 96% (Reeves 2004). In a similar study in southern Florida, Roberts et al. (2000) examined 94 cases of cane toad intoxication in dogs conducted over one year, and while in all dogs received some type of treatment, treatment varied considerably among patients and four dogs died as a result of toad exposure. Other societal impacts from cane toads include *Salmonella* contamination of water sources (including catchments), consumable vegetation, and dog food bowls (Drake et al. 2013).

Historic and Current Management

Traps, attractants, and barriers have been trialed in the past for cane toad control, mainly in Australia. Cage traps have been historically suggested as the most effective and least labor intensive of the current cane toad control techniques (Lampo and De Leo 1998), yet trapping cane toads using pit traps or cage traps, and manual removal of frogs in general, is difficult and only likely to be successful with small, isolated, incipient populations (Schwarzkopf and Alford 2007). However, use of attractants in combination with trapping can help increase trap effectiveness (Schwarzkopf and Alford 2007; Yeager et al. 2014). Existing cane toad trap designs in Australia use lights to lure insects to traps, and toads enter the traps to feed on the insects (Schwarzkopf and Alford 2007). Schwarzkopf and Alford (2007) used large pens and field trials to demonstrate that cane toads are attracted to conspecific mating vocalizations, and trap success increased threefold when playback vocalizations were present in traps compared to traps without playbacks. Furthermore, both male and female toads were attracted to quiet (47 dB at 1 m) playbacks, whereas only males responded to loud (67 dB at 1 m) playbacks (Schwarzkopf and Alford 2007). Yeager et al. (2014) demonstrated that using sound attractant inside a trap with ultraviolet light was particularly effective at capturing female cane toads. Schwarzkopf and Alford (2007) suggested that outfitting cage traps containing a light source with an inexpensive media player, speaker, and associated batteries would be about $50.

Arranging trapping near water sources, particularly at the end of the dry season, in combination with other attractants, such as a light source and auditory attractant, has been a favorable method in Australia (Taylor and Edwards 2005). In theory, fencing could be used to keep cane toads out of a desired habitat, yet no trials investigating the minimum aperture size of the fencing mesh or the integrity of the fencing in gulches and water crossings have been carried out (Taylor and Edwards 2005). In Hawaii, the aperture size of fencing used to exclude invasive ungulates is too large to exclude cane toads (A.B. Shiels, pers. obs.). If logistically feasible, draining a breeding habitat may prevent breeding, but investigations of the efficacy of such a control method have not occurred. Moist microsites for shelter and food may also limit cane toad establishment or densities. Heise-Pavlov and Longway (2011) found that restoration measures (e.g., providing rock piles or logs) appeared to benefit cane toad presence. Removal of moist microsites during the dry season may be particularly effective. Taking advantage of local dispersal pathways and favorable vegetation cover conditions may be another technique to improve effectiveness of cane toad control efforts. Road corridors and open landscapes that lack dense vegetation are areas of high cane toad activity and dispersal within Queensland, Australia (Brown et al. 2006).

The Australian government provided funds to support the National Cane Toad Taskforce, which also identified four approaches to long-term and widespread control of cane toads that seem trial-worthy in the future, and these were: (1) identification and release of a *Bufo*-specific pathogen, (2) development and release of a sterility agent or sterile males, (3) identification and use of a cane toad-specific toxicant, and (4) development of a genetically modified organism (GMO) that would interfere with the development of cane toads (as reviewed in Taylor and Edwards 2005). Although some of these approaches have ongoing research efforts (e.g., GMO), all require secured funding and commitment for at least five to ten years to conduct research trials (Taylor and Edwards 2005; Schwarzkopf and Alford 2007). For example, establishment of new toxicants or biocontrols would have to ensure minimal impacts to nontarget organisms, and in Australia, this would mean ensuring minimal risk for over 200 native frogs (Tyler and Knight 2009).

Unlike installing and using various types of barriers and trap designs, implementing biocontrol methods typically takes years or decades and often is unsuccessful (Yeager et al. 2014). There is no known pathogen that is specifically lethal to cane toads, which is why a GMO to serve this purpose has been explored (Shanmuganathan et al. 2010). Research on a virus (*Toddia* sp.) showed promise for controlling cane toads because all experimental infections caused cane toad death (Speare 1990). However, Shanmuganathan et al. (2010) reviewed the biocontrol efforts attempted over an eight-year period in Australia, and concluded that, so far, research efforts have failed to produce a tool for large-scale control of cane toad populations.

Chemical addition to water bodies may represent a future technique for suppressing the number and sizes of cane toads developing to the tadpole and adult stages (Crossland and Shine 2012). Recent research by Crossland and Shine (2012) has shown that cane toad eggs placed in water that already had a cohort of conspecific tadpoles had 45% greater mortality and 40% smaller tadpole body sizes than tadpoles grown in fresh water. Therefore, cane toad tadpoles produce waterborne chemical cues that suppress conspecific embryoic development (Crossland and Shine 2012). This apparent adaptation, to presumably reduce intraspecific competition or avoid development in predator-active environments (Hagman and Shine 2009a), may be applied as a management technique to reduce cane toads in water bodies where breeding occurs. Unlike the chemical cues exuded by tadpoles, many other chemical trials, such as adding scents of foods, native-range fish predators (cichlids— *Cichlasoma labiatum*), native-range axolotls predators (*Ambystoma mexicanum*), as well as thermal stimuli, did not illicit any response in cane toad tadpoles (Hagman and Shine 2008). Furthermore, examination of the negative effects of chemical cues from cane toad tadpoles was tested against nine native frogs in Australia and was found to have no negative effects on the native frogs, which is encouraging for using these low-level cane toad alarm pheromones as a management technique to control this invasive species (Hagman and Shine 2009b).

Use of baits to attract cane toad predators or cause aversion to cane toads has also been trialed in Australia. Cat food bait was successfully used to recruit predatory ants (*Iridomyrmex reburrus*) to water edges where newly morphed cane toads had emerged (Ward-Fear et al. 2010). Additionally, taste aversion to cane toads has been trialed successfully in Australia and has potential as a management technique to benefit cane toad-threatened fauna. O'Donnell et al. (2010) successfully induced a taste aversion to live toads in juvenile quolls. The taste aversion (thiabendazole) was placed in dead toads to induce vomiting when eaten by quolls. From these results, O'Donnell et al. (2010) suggest that wildlife managers could aerially deploy taste aversion baits ahead of the invasion front of cane toads to benefit native wildlife that commonly consume cane toads. To our knowledge, such aerial application of bait to help avert native fauna from predation of cane toads on a landscape scale has not yet occurred. The breadth of management techniques that are being researched and implemented in Australia is impressive, and such techniques are certainly applicable to cane toad management in the United States.

In addition to future cane toad reduction approaches, biosecurity efforts to help prevent cane toad range expansion and otherwise prevent their colonization of new areas have been a priority for Australia but have been generally absent from the United States. Offshore islands are apparently colonized by cane toads from flood waters and cargo transport in Australia; significant educational campaigns have helped to inform the public about cane toad impacts and control efforts needed, and public involvement has been successful in some areas in northern and eastern Australia (e.g., community groups involved in "toad buster" or "tad round-up" activities; Schwarzkopf and Alford 2007).

FUTURE CHALLENGES

Preventing the introduction and establishment of nonnative species is the most efficient means of controlling their spread and negative impact. Most terrestrial, invasive amphibians are tropical or subtropical in origin, and as a result they are primarily establishing and spreading in tropical and subtropical U.S. states, such as Florida and Hawaii. Terrestrial breeders, such as the coqui and greenhouse frog, can invade with one egg mass (Peacock et al. 2009). For many of these species, the mode of transport has been via the horticulture trade (Kraus 2003, 2009). While an extremely important source of introduced amphibians is the pet trade (Kraus 2003, 2009), most amphibians introduced this way are aquatic (*Lithobates catebeiana*, *L. berlandieri*, and *Xenopus laevis*) (Herrel and van der Meijden 2014). The Cuban tree frog and cane toad are examples of invasive water-body breeders that were not introduced via the pet trade. These species are likely successful because they invade locations with plenty of water, do not have strict breeding requirements, breed year round, and produce a large number of offspring; therefore, in appropriate conditions, their populations grow quickly. While we focused on four terrestrial species that are the most problematic invaders in the United States, there are several others that are probable candidates for invasion and negative impacts in the United States, and any potential introductions should be monitored closely: *Scinax ruber* (red-snouted treefrog), *Scinax x-signatus* (Venezuela snouted treefrog), and *Eleutherodactylus johnstonei* (Johnstone's Whistling Frog) (Global Invasive Species Database 2008a,b).

There are several areas of research and management in the United States that need to be addressed concerning the four species covered in this chapter. First, our knowledge of the location and spread of many of these species is still minimal. For example, we have not yet determined how widespread the greenhouse frog is on islands in Hawaii other than the Big Island (Olson et al. 2012a). Monitoring the potential spread of all these species is critical and citizen scientists can play a role. Second, there is great need for more research on the ecological impacts of all four of these species. As their ranges expand, these problematic species are likely to cause even more harm to wildlife, economies, and citizens. Third, we need greater educational campaigns and detailed risk assessments to curtail new invasions. Finally, development of more effective methods for population control of all four species is needed. Kalnicky et al. (2014) found that in the case of the coqui frog, there was a growing tolerance toward the species from people with greater exposure to them. If this is true for other nonnative amphibians, it suggests that managers may have a short window of public support for control. This may need to be kept in mind for all nonnative species.

REFERENCES

Alix, D.M., C. Guyer, and C.J. Anderson. 2014. Expansion of the range of the introduced greenhouse frog, *Eleutherodactylus planirostris*, in Coastal Alabama. *Southeastern Naturalist* 13:N59–N62.

Anonymous. 1934. Toads save sugar crop. *Nature* 134:877–877.

Anonymous. 2010. Hawai'i's Coqui frog management, research and education plan. *State of Hawaii* 877–877.

Austin, D.F., and A. Schwartz. 1975. Another exotic amphibian in Florida, *Eleutherodactylus coqui*. *Copeia* 1975:188.

Bailey, P. 1976. Food of the marine toad, *Bufo marinus*, and six species of skink in cacao planation in New Britain, Papua New Guinea. *Australian Wildlife Research* 3:185–188.

Barbour, T. 1931. Another introduced frog in North America. *Copeia* 1931:140.

Beachy, J.R., R. Neville, and C. Arnott. 2011. Successful control of an incipient invasive amphibian: *Eleutherodactylus coqui* on O'ahu, Hawai'i. In *Island Invasives: Eradication and Management. Proceedings of the International Conference on Island Invasives*, eds. C.R. Veitch, M.N. Clout, and D.R. Towns, pp. 140–147. Gland, Switzerland: IUCN.

Beard, K.H. 2007. Diet of the invasive frog, *Eleutherodactylus coqui*, in Hawaii. *Copeia* 2007:281–291.

Beard, K.H., R. Al-Chokhachy, N.C. Tuttle, and E.M. O'Neill. 2008. Population density estimates and growth rates of *Eleutherodactylus coqui* in Hawaii. *Journal of Herpetology* 42:626–636.

Beard, K.H., S. McCullough, and A.K. Eschtruth. 2003. Quantitative assessment of habitat preferences for the Puerto Rican terrestrial frog, *Eleutherodactylus coqui*. *Journal of Herpetology* 37:10–17.

Beard, K.H., and E.M. O'Neill. 2005. Infection of an invasive frog *Eleutherodactylus coqui* by the chytrid fungus *Batrachochytrium dendrobatidis* in Hawaii. *Biological Conservation* 126:591–595.

Beard, K.H., and W.C. Pitt. 2005. Potential consequences of the coqui frog invasion in Hawaii. *Diversity and Distributions* 11:427–433.

Beard, K.H., and W.C. Pitt. 2006. Potential predators of an invasive frog (*Eleutherodactylus coqui*) in Hawaiian forests. *Journal of Tropical Ecology* 22:345–347.

Beard, K.H., and W.C. Pitt. 2012. Chapter 26: Caribbean tree frog (*Eleutherodactylus coqui*). In *Handbook of Global Freshwater Invasive Species*, ed. R.A. Francis, pp. 311–319. London: Earthscan.

Beard, K.H., E.A. Price, and W.C. Pitt. 2009. Biology and impacts of Pacific Island invasive species: *Eleutherodactylus coqui*, the Coqui frog (Anura: Leptodactylidae). *Pacific Science* 63:297–316.

Beckmann, C., and R. Shine. 2012. How many of Australia's ground-nesting birds are likely to be at risk from the invasive Cane Toad (*Rhinella marina*)? *Emu* 112:83–89.

Boland, C.R.J. 2004. Introduced cane toads *Bufo marinus* are active nest predators and competitors of rainbow bee-eaters *Merops ornatus*: Observational and experimental evidence. *Biological Conservation* 120:53–62.

Briggs, C.J., V.T. Vredenburg, R.A. Knapp, and L.J. Rachowicz. 2005. Investigating the population-level effects of chytridiomycosis: An emerging infectious disease of amphibians. *Ecology* 86:3149–3159. doi: 10.1890/04–1428.

Brown, G.P., B.L. Phillips, J.K. Webb, and R. Shine. 2006. Toad on the road: Use of roads as dispersal corridors by cane toads (*Bufo marinus*) at an invasion front in tropical Australia. *Biological Conservation* 133:88–94.

Campbell, E.W., and F. Kraus. 2002. Neotropical frogs in Hawaii: Status and management options for an unusual introduced pest. In *Proceeding of the 20th Vertebrate Pest Conference*, University of California Davis, eds R.M. Timm and R.H. Schmidt, pp. 316–318.

Campbell, K.R., T.S. Campbell, and S.A. Johnson. 2010. The use of PVC pipe refugia to evaluate spatial and temporal distributions of native and introduced treefrogs. *Florida Scientist* 73:78–88.

Campbell, T.S., P. Irvin, K.R. Campbell, K. Hoffmann, M.E. Dykes, A.J. Harding, and S.A. Johnson. 2009. Evaluation of a new technique for marking anurans. *Applied Herpetology* 6:247–256.

Choi, R.T., and K.H. Beard. 2012. Coqui frog invasions change invertebrate communities in Hawaii. *Biological Invasions* 14:938–948.

Christy, M.T., C.S. Clark, D.E. Gee, D. Vice, D.S. Vice, M.P. Warner, C.L. Tyrrell, et al. 2007. Recent records of alien anurans on the Pacific Island of Guam. *Pacific Science* 61:469–483.

Conant, R., and J. Collins. 1991. *Reptiles and Amphibians: Eastern/Central North America*. New York: Houghton Mifflin Company.

Covacevich, J., and M. Archer. 1975. The distribution of the cane toad, *Bufo marinus*, in Australia and its effects on indigenous vertebrates. *Memoirs of the Queensland Museum* 17:305–310.

Crossland, M., and R. Shine. 2012. Embryonic exposure to conspecific chemicals suppress cane toad growth and survival. *Biology Letters* 8:226–229.

Crossland, M.R. 2000. Direct and indirect effects of the introduced toad *Bufo marinus* (Anura: Bufonidae) on populations of native anuran larvae in Australia. *Ecography* 23:283–290.

Crossland, M.R., G.P. Brown, M. Anstis, C.M. Shilton, and R. Shine. 2008. Mass mortality of native anuran tadpoles in tropical Australia due to the invasive cane toad (*Bufo marinus*). *Biological Conservation* 141:2387–2394.

De León, L.F., and A. Castillo. 2015. *Rhinella marina* (Cane Toad). Salinity tolerance. *Herpetological Review* 46:237–238.

Dodd, C.K., Jr. 2013. *Frogs of the United States and Canada.* Baltimore: Johns Hopkins University Press.

Doody, J.S., B. Green, D. Rhind, C.M. Castellano, R. Sims, and T. Robinson. 2009. Population-level declines in Australian predators caused by an invasive species. *Animal Conservation* 12:46–53.

Doody, J.S., P. Mayes, S. Clulow, D. Rhind, B. Green, C.M. Catellano, D. D'Amore et al. 2014. Impacts of the highly invasive cane toad on aquatic reptiles in a highly modified ecosystem: The importance of replicating impact studies. *Biological Invasions* 16:2303–2309.

Doody, J.S., R. Soanes, C.M. Catellano, D. Rhind, B. Green, C.R. Mchenry, and S. Clulow. 2015. Invasive toads shift predator-prey densities in animal communities by removing top predators. *Ecology* 96:2544–2554.

Drake, M., V. Amandi, U. Zieger, R. Johnson, and H. Hariharan. 2013. Prevalence of Salmonella spp. in cane toads (*Bufo marinus*) from Grenada, West Indies, and their antimicrobial susceptibility. *Zoonose and Public Health* 60:437–441.

Duellman, W.E., and L. Trueb. 1986. *Biology of Amphibians.* New York: McGraw-Hill.

Easteal, S. 1981. The history of introductions of *Bufo marinus* (Amphibia: Anura); a natural experiment in evolution. *Biological Journal of the Linnean Society* 16:93–113.

Ely, C.A. 1944. Development of *Bufo marinus* larvae in dilute sea water. *Copeia* 56:256.

Evans, M., and M. Lampo. 1996. Diet of *Bufo marinus* in Venezuela. *Journal of Herpetology* 30:73–76.

Everman, E., and P. Klawinski. 2013. Human-facilitated jump dispersal of a non-native frog species on Hawai'i Island. *Journal of Biogeography* 40:1961–1970.

Ferreria, R.B., K.H. Beard, R.T. Choi, and W.C. Pitt. 2015. Diet of the non-native greenhouse frog in Maui. *Journal of Herpetology* 49:586–593.

Freeland, W.J. 1986. Populations of cane toad, *Bufo marinus*, in relation to time since colonization. *Australian Wildlife Research* 13:321–329.

Global Invasive Species Database, GISD. 2008a. *Scinax Ruber. Invasive Species Specialist Group (ISSG) of the IUCN Species Survival Commission.* www.iucngisd.org. Accessed Retrieved June 29, 2015.

Global Invasive Species Database, GISD. 2008b. *Scinax x-Signatus. Invasive Species Specialist Group (ISSG) of the IUCN Species Survival Commission.* Accessed Retrieved June 29, 2015.

Glorioso, B.M., J.H. Waddle, M.E. Crockett, K.G. Rice, and H.F. Percival. 2012. Diet of the invasive Cuban Treefrog (*Osteopilus septentrionalis*) in pine rockland and mangrove habitats in South Florida. *Caribbean Journal of Science* 46:346–355.

Goin, C.J. 1947. *Studies on the Life History of Eleutherodactylus ricordii planirostris (Cope) in Florida: With Special Reference to the Local Distribution of an Allelomorphic Color Pattern.* Vol. IV, Biological Science Series. Gainesville: University of Florida Press.

Greenlees, M.J., G.P. Brown, J.K. Webb, B.L. Phillips, and R. Shine. 2006. Effects of an invasive anuran [the cane toad (*Bufo marinus*)] on the invertebrate fauna of a tropical Australian flood plain. *Animal Conservation* 9:431–438.

Hagman, M., and R. Shine. 2008. Understanding the toad code: Behavioural responses of cane toad (*Chaunus marina*) larvae and metamorphs to chemical cues. *Austral Ecology* 33:37–44.

Hagman, M., and R. Shine. 2009a. Factors influencing responses to alarm pheromone by larvae of invasive cane toads, *Bufo marinus. Journal of Chemical Ecology* 35:265–271.

Hagman, M., and R. Shine. 2009b. Species-specific communication in an introduced toad compared with native frogs in Australia. *Aquatic Conservation: Marine and Freshwater Ecosystem* 19:724–728.

Hara, A.H., C.M. Jacobsen, S.R. Marr, and R.Y. Niino-DuPont. 2010. Hot water as a potential disinfestation treatement for an invasive anuran amphibian, the coqui frog, *Eleutherodactylus coqui* Thomas (Leptodactylidae), on potted plants. *International Journal of Pest Management* 56:255–263.

Heinicke, M.P., L.M. Diaz, and S.B. Hedges. 2011. Origin of invasive Florida frogs traced to Cuba. *Biology Letters* on-line:1–5.

Heise-Pavlov, S.R., and L.J. Longway. 2011. Diet and dietary selectivity of cane toads (*Rhinella marina*) in restoration sites: A case study in far north Queensland, Australia. *Ecological Management and Restoration* 12:230–233.

Herrel, A., and A. van der Meijden. 2014. An analysis of the live reptile and amphibian trade in the USA compared to the global trade in endangered species. *Herpetological Journal* 24:103–110.

Hoffmann, K.E., S.A. Johnson, and M.E. McGarrity. 2009. Intraspecific variation in use of polyvinyl chloride (PVC) refuges by hylid treefrogs: A potential source of capture bias. *Herpetological Review* 40:423–426.

Hölldobler, B., and E.O. Wilson. 1990. *The Ants.* Boston: Harvard University Press.

Jensen, J.B. 2008. Greenhouse Frog (*Eleutherodactylus planirostris*). In *Amphibians and Reptiles of Georgia*, eds. J.B. Jensen, C.D. Camp, W. Gibbons, and M.J. Elliot, pp. 92–93. University of Georgia Press: Athens, Georgia.

Johnson, S.A. 2007. *The Cuban treefrog in Florida.* Vol. Available from: http://edis.ifas.ufl.edu/UW259., Gainesville, FL: University of Florida Cooperative Extension Service. Available: http://edis.ifas.ufl.edu/UW259 Accessed: October 20, 2015. Gainesville: University of Florida.

Johnson, S.A., M.E. McGarrity, and C.L. Staudhammer. 2010. An effective chemical deterrent for invasive Cuban Treefrogs. *Human–Wildlife Interactions* 4:112–117.

Kaiser, B., and K. Burnett. 2006. Economic impacts of *E. coqui* frogs in Hawaii. *Interdisciplinary Environmental Review* 8:1–11.

Kalnicky, E.A., K.H. Beard, and M.W. Brunson. 2013. Community-level response to habitat structure manipulations: An experimental case study in a tropical ecosystem. *Forest Ecology and Management* 307:313–321.

Kalnicky, E.A, M.W. Brunson, and K.H. Beard. 2014. A social–ecological systems approach to non-native species: Habituation and its effect on management of coqui frogs in Hawaii. *Biological Conservation* 180:187–195.

Kamper, W., J.K. Webb, M.S. Crowther, M.J. Greenlees, and R. Shine. 2013. Behaviour and survivorship of a dasyurid predator (*Antechinus flavipes*) in response to encounters with the toxic and invasive cane toad (*Rhinella marina*). *Australian Mammalogy* 35:136–143.

Klawinski, P.D., B. Dalton, and A.B. Shiels. 2014. Coqui frog populations are negatively affected by canopy opening but not detritus deposition following an experimental hurricane in a tropical rainforest. *Forest Ecology & Management* 332:118–123.

Knight, C.M., M.J. Parris, and W.H.N. Gutzke. 2009. Influence of priority effects and pond location on invaded larval amphibian communities. *Biological Invasions* 11:1033–1044.

Krakauer, T. 1968. The ecology of the neotropical toad, *Bufo marinus*, in south Florida. *Herpetologica* 214–221.

Kraus, F. 2003. Invasion pathways for terrestrial vertebrates. In *Invasive Species: Vectors Andmanagement Strategies*, eds. G.M. Ruiz, and J.T. Carlton, pp. 68–92. Washington, DC: Island Press.

Kraus, F. 2009. *Alien Reptiles and Amphibians: A Scientific Compendium and Analysis Series, Invading Nature—Springer Series in Invasion Ecology.* Dordrecht: Springer.

Kraus, F., and E.W. Campbell. 2002. Human-mediated escalation of a formerly eradicable problem: The invasion of Caribbean frogs in the Hawaiian Islands. *Biological Invasions* 4:327–332.

Kraus, F., E.W. Campbell, A. Allison, and T. Pratt. 1999. *Eleutherodactylus* frog introductions to Hawaii. *Herpetological Review* 30:21–25.

Krysko, K.L., K.M. Enge, and P.E. Moler. 2011. Atlas of amphibians and reptiles in Florida. In *Final Report, Project Agreement 08013, Florida Fish and Wildlife Conservation Commission*. Tallahassee, Florida, 524 pp.

Lampo, M., and G. De Leo. 1998. The invasive ecology of the toad Bufo marinus: From South America to Australia. *Ecological Applications* 8:388–396.

Lever, C. 2001. *The Cane Toad. The History and Ecology of a Successful Colonist*. Otley, UK: Westbury Academic and Scientific Publishing.

Lever, C. 2003. *Naturalized Reptiles and Amphibians of the World*. Oxford, UK: Oxford University Press.

Liggins, G.W., and G.C. Grigg. 1985. Osmoregulation of the cane toad *Bufo marinus* in salt water. *Comparative Biochemistry and Physiology* 82A:613–619.

MacLean, W.P. 1982. *Reptiles and Amphibians of the Virgin Islands*. London, Basingstoke: Macmillan Caribbean.

Mann, D.L., T. Mann, W. Lu, and N. Winstead. 2015. *Eleutherodactylys planirostris* (Greenhouse Frog). Geographic Distribution. *Herpetological Review* 46:377.

Marr, S.R., S.A. Johnson, A.H. Hara, and M.E. McGarrity. 2010. Preliminary evaluation of the potential of the helminth parasite *Rhabdias elegans* as a biological control agent for invasive Puerto Rican coquis (*Eleutherodactylus coqui*) in Hawaii. *Biological Control* 54:69–74.

Marshall, S.E., and W.E. Meshaka. 2006 An island of cane toads (*Bufo marinus*) in an ocean of xeric uplands in South-Central Florida. *Florida Scientist* 69:169–179.

Mathies, T., W.C. Pitt, and J.A. Rabon. 2012. *Boiga irregularis (Brown Treesnake) Diet*. *Herpetological Review* 43:143–144.

McCann, S., M.J. Greeless, D. Newell, and R. Shine. 2014. Rapid acclimation to cold allows the cane toad to invade montane areas within its Australian range. *Functional Ecology* 28:1166–1174.

McGarrity, M.E., and S.A. Johnson. 2009. Geographic trend in sexual size dimorphism and body size of *Osteopilus septentrionalis*: Implications for invasion of the southeastern United States. *Biological Invasions* 11:1411–1420.

McGarrity, M.E., and S.A. Johnson. 2010. A radio telemetry study of invasive Cuban treefrogs. *Florida Scientist* 73:225–235.

Meshaka, W.E. Jr. 1996. Vagility and the Florida distribution of the Cuban treefrog (*Osteopilus septentrionalis*). *Herpetological Review* 27:37–39.

Meshaka, W.E., Jr. 2001. *The Cuban Treefrog in Florida: Life History of a Successful Colonizing Species*. Gainesville, Florida: Univ. Press of Florida.

Meshaka, W.E., B.P. Butterfield, and J. Brian Hauge. 2004. *The Exotic Amphibians and Reptiles of Florida*. Malabar: Krieger Publishing Company.

Muths, E., P.S. Corn, A.P. Pessier, and D.E. Green. 2003. Evidence for disease-related amphibian decline in Colorado. *Biological Conservation* 110:357–365. doi: Pii s0006-3207(02)00239-2 10.1016/s0006-3207(02)00239-2.

O'Donnell, S., J.K. Webb, and R. Shine. 2010. Conditioned taste aversion enhances the survival of an endangered predator imperilled by a toxic invader. *Journal of Applied Ecology* 47:558–568.

Olson, C.A., and K.H. Beard. 2012. Diet of the invasive greenhouse frog in Hawaii. *Copeia* 2012:121–129.

Olson, C.A., K.H. Beard, D.N. Koons, and W.C. Pitt. 2012a. Detection probabilities of two introduced frogs in Hawaii: Implications for assessing non-native species distributions. *Biological Invasions* 14:889–900.

Olson, C.A., K.H. Beard, and W.C. Pitt. 2012b. Pacific Island invasive species: 8. *Eleutherodactylus planirostris*, the greenhouse frog (Anura: Leptodactylidae). *Pacific Science* 66:255–270.

Olson, C.A., A. Diesmos, and K.H. Beard. 2014. Geographical distribution: *Eleutherodactylus planirostris* (greenhouse frog). *Herpetological Review* 45:652–653.

Peacock, M.M., K.H. Beard, E.M. O'Neill, V.S. Kirchoff, and M.B. Peters. 2009. Strong founder effects and low genetic diversity in introduced populations of coqui frogs. *Molecular Ecology* 18:3603–3615.

Phillips, B.L., G.P. Brown, and R. Shine. 2003. Assessing the potential impact of cane toads on Australian snakes. *Conservation Biology* 17:1738–1747.

Pitt, W.C., K.H. Beard, and R.E. Doratt. 2012. Management of invasive coqui frog populations in Hawaii. *Outlooks on Pest Management* 23:166–169. doi: 10.1564/23aug05.

Reeves, M.P. 2004. A retrospective report of 90 dogs with suspected cane toad (*Bufo marinus*) toxicity. *Australian Veterinary Journal* 82:608–611.

Rice, K.G., J.H. Waddle, M.W. Miller, M.E. Crockett, F.J. Mazzotti, and H.F. Percival. 2011. Recovery of native treefrogs after removal of nonindigenous Cuban treefrogs *Osteopilus septentrionalis*. *Herpetologica* 67:105–117.

Rivero, J.A. 1978. *Los Anfibios y Reptiles de Puerto Rico*, ed. Editorial Universitaria. San Juan, Puerto Rico: Universidad de Puerto Rico.

Roberts, B.K., M.G. Aronsohn, B.L. Moses, R.L. Burk, J. Toll, and F.R. Weeren. 2000. *Bufo marinus* intoxication in dogs: 94 cases (1997–1998). *Journal of the American Veterinary Medical Association* 216:1941–1944.

Rödder, D., and S. Lötters. 2010. Explanative power of variables used in species distribution modelling: An issue of general model transferability or niche shift in the invasive greenhouse frog (*Eleutherodactylus planirostris*). *Naturwissenschaften* 97:781–796.

Rogelio Cedeno-Vazquez, J., J. Gonzalez-Vazquez, A. Martinez-Arce, and L. Canseco-Marquez. 2014. First record of the invasive greenhouse frog (*Eleutherodactylus planirostris*) in the Mexican Caribbean. *Revista Mexicana De Biodiversidad* 85:650–653. doi: 10.7550/rmb.43234.

Schwartz, A., and R.W. Henderson. 1991. *Amphibians and Reptiles of the West Indies: Descriptions, Distributions, and Natural History*. Gainesville: University Press of Florida.

Schwarzkopf, L., and R.A. Alford. 2007. Acoustic attractants enhance trapping success for cane toads. *Wildlife Research* 34:366–370.

Seebacher, F., and R.A. Alford. 2002. Shelter microhabitats determine body temperature and dehydration rates of a terrestrial amphibian (*Bufo marinus*). *Journal of Herpetology* 36:69–75.

Shanmuganathan, T., J. Pallister, S. Doody, H. McCallum, T. Robinson, A. Sheppard, C. Hardy et al. 2010. Biological control of the cane toad in Australia: A review. *Animal Conservation* 13:16–23.

Shine, R. 2010. The ecological impact of invasive cane toads (*Bufo marinus*) in Australia. *Quarterly Review of Biology* 85:253–291.

Shine, R. 2014. A review of ecological interactions between native frogs and invasive cane toads in Australia. *Austral Ecology* 39:1–16.

Shine, R., J. Amiel, A.J. Munn, M. Stewart, A.L. Vyssotski, and J.A. Lesku. 2015. Is "cooling then freezing" a humane way to kill amphibians and reptiles? *Biology Open* doi: 10.1242/bio.012179.

Shine, R., G.P. Brown, B.L. Phillips, J.K. Webb, and M. Hagman. 2006. The biology, impact and control of cane toads: An overview of the University of Sydney's research program. In *Proceedings of the Cane Toad Workshop*. Canberra (Australia): University of Canberra, Invasive Animals Cooperative Research Centre.

Simberloff, D. 2004. Now you see them, now you don't!—Population crashes of established introduced species. *Biological Invasions* 6:161–172.

Simpson, S.E. 2013. Assessing critical thermal minima to determine the thermal limits of the invasive Cuban treefrog (*Osteopilus septentrionalis*). *MS thesis*, University of Florida.

Sin, H., K.H. Beard, and W.C. Pitt. 2008. An invasive frog, *Eleutherodactylus coqui*, increases new leaf production and leaf litter decomposition rates through nutrient cycling in Hawaii. *Biological Invasions* 10:335–345.

Smith, K.G. 2005. Effects of nonindigenous tadpoles on native tadpoles in Florida: Evidence of competition. *Biological Conservation* 123:433–441.

Smith, R.L. 2016. Invasive coqui frogs serve as novel prey for birds in Hawaii, and not as competitors. M.S. Thesis, Utah State University, Logan, Utah.

Smith, R.L., K.H. Beard, and A.B. Shiels. 2017. Different prey resources suggest little competition between non-native frogs and insectivorous birds despite isotopic niche overlap. *Biological Invasions* 19:1001–1013.

Somaweera, R., R. Shine, J. Webb, T. Dempster, and M. Letnic. 2013. Why does vulnerability to toxic invasive cane toads vary among populations of Australian freshwater crocodiles? *Animal Conservation* 16:86–96.

Speare, R. 1990. A review of the diseases of the cane toad, *Bufo marinus*, with comments on biological control. *Australian Wildlife Research* 17:287–410.

Stewart, M.M. 1977. "The role of introduced species in a Jamaican frog community." *Actas del IV Simposium Internacional de Ecologica Tropical*, Panama City, pp. 113–146.

Stewart, M.M., and F.H. Pough. 1983. Population density of tropical forest frogs: Relation to retreat sites. *Nature* 221:570–572.

Storer, T.I. 1925. A synopsis of the amphibian of California. *University of California Publications in Zoology* 27:1–342.

Taylor, R., and G. Edwards. 2005. A review of the impacts and control of cane toads in Australia with recommendations for future research and management approaches. The National Cane Toad Task Force Report.

Townsend, D.S., and M.M. Stewart. 1985. Direct development in *Eleutherodactylus coqui* (Anura: Leptodactylidae): A staging table. *Copeia* 1985:423–436.

Townsend, D.S., and M.M. Stewart. 1994. Reproductive ecology of the Puerto Rican Frog *Eleutherodactylus coqui*. *Journal of Herpetology* 28:34–40.

Townsend, D.S., M.M. Stewart, and F.H. Pough. 1984. Male parental care and its adaptive significance in a Neotropical frog. *Animal Behavior* 32:421–431.

Tuberville, T.D., J.D. Willson, M.E. Dorcas, and J. Whitfield Gibbons. 2005. Herpetofaunal Species Richness of Southeastern National Parks. *Southeastern Naturalist* 4:537–569.

Tuttle, N.C., K.H. Beard, and R. Al-Chokhachy. 2008. Aerially applied citric acid reduces an invasive frog. *Wildlife Research* 35:676–683.

Tyler, M.J., and F. Knight. 2009. *Field guide to Australian Frogs*. Australia: CSIRO Publishing.

Urban, M.C., B.L. Phillips, D.K. Skelly, and R. Shine. 2007. The cane toad's (*Chaunus (Bufo) marinus*) increasing ability to invade Australia is revealed by a dynamically updated range model. *Proceedings of the Royal Society B: Biological Sciences* 274:1413–1419.

Waddle, J.H., R.M. Dorazio, S.C. Walls, K.G. Rice, J. Beauchamp, M.J. Schuman, and F.J. Mazzotti. 2010. A new parameterization for estimating co-occurrence of interaction species. *Ecological Applications* 20:1467–1475.

Ward-Fear, G., G.P. Brown, and R. Shine. 2010. Using a native predator (the meat ant, *Iridomyrmex reburrus*) to reduce the abundance of an invasive species (the cane toad, *Bufo marinus*) in tropical Australia. *Journal of Applied Ecology* 47:273–280.

Wilson, L.D., and L. Porras. 1983. *The Ecological Impact of Man on the South Florida Herpetofauna*, ed. University of Kansas Museum of Natural History, Special Publication No. 9. Lawrence: University of Kansas.

Woolbright, L.L., A.H. Hara, C.M. Jacobsen, W.J. Mautz, and F.L. Benevides. 2006. Population densities of the Coqui, *Eleutherodactylus coqui* (Anura: Leptodactylidae) in newly invaded Hawaii and in native Puerto Rico. *Journal of Herpetology* 40:122–126.

Woolbright, L.L., and M.M. Stewart. 1987. Foraging success of the tropical frog, *Eleutherodactylus coqui*: The cost of calling. *Copeia* 1987:6975.

Wyatt, J.L., and F.A. Forys. 2004. Conservation implications of predation by Cuban tree-frogs (*Osteopilus septentrionalis*) on native hylids in Florida. *Southeastern Naturlaist* 3:695–700.

Yeager, A., J. Commito, A. Wilson, D. Bower, and L. Schwarzkopf. 2014. Sex, light, and sound: Location and combination of multiple attractants affect probability of cane toad (*Rhinella marina*) capture. *Journal of Pest Science* 87:323–329.

10 Ecology, Impacts, and Management of Invasive Rodents in the United States

Gary W. Witmer and Aaron B. Shiels

CONTENTS

INTRODUCTION TO INVASIVE RODENTS

Approximately 42% of all mammalian species in the world are rodents, amounting to about 2277 species (Wilson and Reeder 2005). Rodents have adapted to all lifestyles: terrestrial, aquatic, arboreal, and fossorial (underground). Most species are small, secretive, nocturnal, adaptable, and have keen senses of touch, taste, and smell. For most species of rodents, the incisors continually grow throughout their life span, requiring constant gnawing to keep the incisors sharp and at an appropriate length. This can result in extensive damage to seeds, fruits, field crops, structures, wires, and insulation. Rodents are known for their high reproductive potential; however, there is much variability between species as to the age at first reproduction, size of litters, and the number of litters per year. All these characteristics make many rodent species ideal invaders.

Rodents have ecological, scientific, social, and economic values (Witmer et al. 1995; Dickman 1999). Rodents are important in seed and spore dispersal, pollination, seed predation, energy and nutrient cycling, the modification of plant succession and species composition, and as a food source for many predators. Additionally, some species provide food and fur for human uses. Hence, the indiscriminate removal of native rodents from ecosystems, including agroecosystems, is not the best management option in many cases (Villa-Cornejo et al. 1998; Aplin and Singleton 2003; Brakes and Smith 2005).

FIGURE 10.1   Introduced rodents, such as this Norway rat (*Rattus norvegicus*), can cause extensive damage to island flora and fauna and to agricultural production. (Photo by Jack Jeffrey.)

Introduced rodents, whether purposefully or accidently introduced, have caused serious impacts to native flora and fauna, agriculture, property, and other resources (Capizzi et al. 2014). Long (2003) reviewed the many rodent introductions around the world and briefly discussed the resultant damage. As invasive species, rodents are particularly problematic because they have many characteristics that make them effective invaders (e.g., Pitt et al. 2011a), and as a result, numerous invasive rodents have become established in parts of the United States and its territories (Figure 10.1). By far the prominent invasive rodents in the United States are species of *Rattus* (*R. exulans, R. norvegicus,* and *R. rattus*) and the house mouse (*Mus musculus*); the Gambian giant pouched rat (*Cricetomys gambianus*) is also a formidable concern as they have established and become invasive in Florida. Nutria (*Myocastor coypus*) are among the most-damaging invasive rodents in the United States (Witmer et al. 2012a), but because of their aquatic lifestyle they are not covered in this terrestrially focused book. While this chapter focuses on *Rattus*, house mice, and Gambian giant pouched rats, there are several additional introduced terrestrial rodents that have become locally or regionally invasive in the United States, and these include hoary marmots (*Marmota caligata*), arctic ground squirrels (*Spermophilus parryii*), eastern fox squirrels (*Sciurus niger*), and eastern gray squirrels (*Sciurus carolinensis*). While fox squirrels and eastern gray squirrels were presumably introduced accidently into states and regions outside their native range, arctic ground squirrels (*Spermophilus parryii*) and hoary marmots (*Marmota caligata*) were introduced to Alaskan islands for food or fur. Additional native species of rodents (voles, *Microtus* spp. and deer mice, *Peromyscus* spp.) have been placed on some islands of the United

States, at least on a temporary basis, to study interactions among rodent species (e.g., Crowell and Pimm 1976; Crowell 1983).

In this chapter, we review the most prominent rodent species introduced into terrestrial ecosystems in the United States, and discuss their impacts to humans and ecosystems, management strategies, and the methods used to reduce invasive populations and their impacts. We also review invasive rodent eradication projects and methods used in the United States.

ISSUES AND DAMAGE CAUSED BY INVASIVE RODENTS

Several types of damage have been caused by introduced rodents in the United States (Hygnstrom et al. 1994; Witmer and Singleton 2010). The substantial and worldwide loss of human food, both standing crops and stored foodstuffs, has been documented in several reviews (Meerburg et al. 2009b; Witmer and Singleton 2010). In addition to consuming human foods, rodents also contaminate much more stored food through defecation and urination. Rodents also transmit many diseases to humans, companion animals, and livestock (Meerburg et al. 2009a). For example, the plague bacteria, *Yersinia pestis*—causal agent of the Black Death which killed millions of humans worldwide in several pandemics—reached North America in the late 1800s via infected rats on ships arriving in California ports (Witmer 2004).

Rodents can be prolific on islands where they have few or no predators. Their high reproductive potential, omnivorous foraging strategy, and aggressive predatory behavior have led to the endangerment or extinction of numerous native island species, especially birds (Moors and Atkinson 1984; Witmer et al. 1998; Veitch and Clout 2002; Engeman et al. 2006; Towns et al. 2011). While their impacts to seabirds have been long known, invasive rodents also impact seeds and seedlings, invertebrates, sea turtle eggs and hatchings, and other parts of the ecosystem (Witmer et al. 2007a; Caut et al. 2008; Angel et al. 2009; Towns et al. 2009; St Clair 2011; Drake et al. 2011). Most seabirds and many endemic land birds that nest on islands have not evolved to deal with mammalian predators and can be highly vulnerable to predation from introduced rodents and other nonnative predators. In addition to direct effects, rodents can have many indirect effects on island resources through competition and trophic cascade effects (Russell 2011). Invasive rodents have reached over 80% of the world's island groups where they have caused the demise of hundreds of endemic species (Atkinson 1985). As a result, there has been a concerted worldwide effort to eradicate introduced rodents from islands, with numerous successes (Howald et al. 2007; Witmer et al. 2011). These efforts have used a combination of techniques but relied heavily on the use of rodenticides (Howald et al. 2007; Witmer et al. 2007b).

BIOLOGY, ECOLOGY, AND DAMAGE OF SOME INVASIVE RODENTS

Norway rats (*Rattus norvegicus*; Figure 10.1) are native to a large part of Asia, but now occur worldwide with the exception of the polar regions (Long 2003). They were most likely first introduced to North America via transatlantic shipping beginning in the 1700s (Brooks 1973; Meehan 1984) but are now well established in both rural and urban areas throughout the United States, including Alaska, Hawaii, and all territories. This species is one of the most successful invasive vertebrates in the United States and is responsible for a variety of types of damage to crops and stored commodities (Jackson 1977; Timm 1994a). Norway rats likely spread rapidly and systematically across the country in conjunction with shipping of commodities and along wagon, riverboat, and rail routes. One of the three common commensal rodent species on the North American continent, the Norway rat is closely tied to human settlements.

Norway rat pelage is typically brown above, with the ventral region lighter brown/ yellow or gray. The tail is sparsely haired and scaly and typically about the same length as the body. Adult body weights range from 200 to 500 g. Breeding may occur throughout the year, depending on conditions. Females produce litters of six to 12 young and are known to bear four to six litters per year (Timm 1994a). Gestation is about three weeks, and animals reach sexual maturity approximately three weeks after birth (Timm 1994a). Given their reproductive potential, populations can expand rapidly when food, water, and habitat are available.

In farm settings, damage to stored food and grains, damage to garden crops, and predation on eggs and baby chickens are common. Grain consumption and fecal contamination are common problems in commercial grain storage facilities (Jackson 1977). Damage to roads, bridges, railroad track beds, and hydraulic structures may result from burrowing activities and the associated soil loosening or flooding (Timm 1994a). Structural damage in buildings results from gnawing and burrowing and may include damage to doors, window sills, and walls as well as to pipes and wiring. Insulation may be damaged or removed in the course of nest building. In urban areas, Norway rat populations are commonly associated with poor sanitation or accumulation of trash and food refuse in inner-city areas, although outdoor feeding of pets and wildlife often supports suburban populations as well. Norway rats serve as reservoirs of a number of diseases that may affect humans and domestic animals, most commonly salmonellosis, leptospirosis, and trichinosis (Meehan 1984). In areas with high rat populations in close association with humans, rat bites may occur, particularly to babies or young children.

Management of Norway rats appears to be best achieved with habitat manipulations. Davis (1953) demonstrated that wild, free-ranging urban populations could be completely managed by environmental control and sanitation. However, Fall and Jackson (1998) contended that the political impossibility of maintaining diligence by urban residents and sustained support by public and private sectors has allowed Norway rat problems to continue unabated. Numerous products are available commercially to property owners for Norway rat control, and extensive professional rodent control services are available through the pest control industry (Timm 1994a; Corrigan 2001).

Roof rats (*Rattus rattus*; Figure 10.2), also known as black rats or ship rats, are native to a large portion of Asia, probably throughout the Indo-Malayan region and through southern China (Long 2003). They also are now widespread worldwide and are the dominant *Rattus* species found on tropical islands. Roof rats are the most successful of the three commensal *Rattus* species. In the United States, they occur in and along port and shore areas in southeastern and western North America and throughout Hawaii and tropical and temperate Atlantic and Pacific Ocean islands. Although known most commonly as a commensal species closely tied to people

FIGURE 10.2   Roof rat in native Hawaiian forest. Note the metal ear tag on the characteristically large ears. (Photo by Aaron Shiels.)

and their movements, roof rats, particularly in warmer areas, readily establish in undeveloped areas, including native forests in Hawaii and on oceanic islands (Shiels et al. 2014). According to Brooks (1973), roof rats were well established in Virginia in the early 1600s and in North America's east coast areas by the 1800s. They occur sporadically in warmer inland areas but rarely persist. However, a recent infestation discovered in urban Phoenix, Arizona raised concerns that the species could permanently establish in patches of suitable habitat and subsequently threaten crops and orchards (Nolte et al. 2003). In cooler temperate areas, roof rats compete poorly with the larger and more aggressive Norway rat, and occur mostly in port areas and generally indoors (Meehan 1984). However, in island natural areas, particularly forests, roof rats have been identified as the most destructive rodent to native species and ecosystems (Ruffino et al. 2009; Traveset et al. 2009; Banks and Hughes 2012; Shiels et al. 2014).

The roof rat pelage is reddish-brown, brown, gray, or black with the ventral area being lighter or white. The tail is generally about 27 mm longer than the body (Shiels et al. 2014). Adult roof rats weigh 150–250 g on continents, but some adults on islands in the Pacific (Shiels et al. 2014), including Hawaii (Shiels 2010), weigh just 90 g. As in Norway rats, breeding may occur throughout the year if resources are available; the pattern of breeding and the reproductive potential are similar between roof rats and Norway rats. Roof rat females typically bear three or more litters with five to eight young per litter each year (Marsh 1994). They are sexually mature by one to two months and may have a life expectancy of around one year. The typical life span for a roof rat in the wild is one year or less (Shiels 2010). Like many rodent species, they are primarily nocturnal, although when densities are high, such as on some tropical islands, they also are often active during the day.

Recently, a variant of *Rattus rattus*, the Asian house rat, has been separated taxonomically as *Rattus tanezumi* (Musser and Carleton 2005). Animals of both species are generally similar in appearance; however, *R. tanezumi* appears more variable and has a somewhat shorter tail. A chief distinguishing feature is a differing number of chromosomes between the two species, but this is of course not evident without use of special laboratory techniques, and some authorities have not accepted the name change. *Rattus tanezumi* has recently been reported as a new invasive species in North America based on collections in California (James 2006). However, species in this complex are difficult to separate morphologically. Additional molecular evidence shows that what has historically been identified as *Rattus rattus* is actually a complex of approximately five to seven species (Robins et al. 2007; Pages et al. 2010).

Like the Norway rat, the roof rat invades homes and structures, causing damage and contamination of stored food and commodities (Marsh 1994). However, it survives well in field and forest habitats in tropical and subtropical areas (Shiels et al. 2014) and causes damage to orchard, grain, and sugarcane crops in such states as California and Hawaii (Kami 1966; Baldwin et al. 2014). Because of their arboreal nature, roof rats can prey on adult birds, nestlings, and eggs. Furthermore, they are recognized worldwide as the likely cause of rare bird extinctions in many island areas, including Hawaii (Munro 1945; Atkinson 1977; Pitt and Witmer 2007). Roof rats are well-known predators of seabirds, especially those that are ground- and burrow-nesting (Jones et al. 2008) and especially small-egged species. Latorre et al. (2013) demonstrated that although roof rats were able to consume all sized eggs offered (12–68 g), larger eggs were 13 times more likely to survive roof rat interaction than the smallest eggs. Roof rats also eat native and nonnative snails, and in Hawaii they depredate the introduced predatory snail, *Euglandina rosea*, which has complicated management strategies to protect native tree snails (Meyer and Shiels 2009). Roof rats also pose substantial threats to native and endangered plants through seed predation (Pender et al. 2013; Shiels and Drake 2015), as well as potentially aiding in the spread of nonnative seeds via dispersal (Shiels 2011; Shiels and Drake 2011). Although the majority of the roof rat diet in island natural areas is plant material (Shiels and Pitt 2014; Shiels et al. 2014), insects comprise the second most common food item (Shiels et al. 2013), which reveals their link to potential impacts on additional ecosystem services such as pollination and decomposition (St Clair 2011).

Roof rats are a reservoir for a number of diseases of humans and animals, but are most notorious for their role in bringing bubonic plague, the "Black Death," to 14th century Europe. The occurrence of bubonic plague in Hawaii during 1899 to 1958 was associated with this species (Tomich 1986), as were the initial outbreaks in California in the early 1900s (Witmer 2004). While roof rats rarely transmit plague today, they are known to transmit other bacteria that negatively affect humans, such as leptospirosis, and transmit several types of harmful nematodes, including *Capillaria hepatica* (Berentsen et al. 2015) and *Angiostrongylus cantonensis* (Wang et al. 2008). Rat lung-worm disease, which is caused by *A. cantonensis* nematodes infecting the human brain and causing symptoms ranging from severe headaches to coma and death (Wang et al. 2008), is particularly known in the United States in wet regions of Hawaii (Jarvi et al. 2015), yet additional cases have originated in Florida, Louisiana, California, and the Caribbean.

Roof rat control methods are the same or similar to those used for Norway rats (e.g., toxicants, traps, barriers, deterrents). However, roof rats have been a particular target of recent efforts, both in the United States and in many other countries, to eradicate them from islands where seabirds or other desirable native species are threatened by rat predation (Howald et al. 2007; Witmer et al. 2007a). In Hawaii, Pitt et al. (2011b) recently developed and tested a nest box for endangered Hawaiian cavity-nesting birds that prevents access by roof rats (Figure 10.3).

Polynesian rats (*Rattus exulans*; Figure 10.4), also known as the Pacific rat, or Kiore in New Zealand, is a small tropical rat native to the Southeast Asia mainland that has spread throughout islands in the Pacific in conjunction with human settlement of the region (Matisoo-Smith and Robins 2004). Although they do not occur on the United States mainland, they are well established on most tropical and subtropical islands (less than about 30° latitude) throughout the Pacific, including the Hawaiian Islands (Roberts 1991). Polynesian rats are the smallest species (110–150 mm adult body length) in the genus *Rattus* and are slender (40–100 g for adults) with relatively small feet. Their pelage is reddish-brown to gray-brown on the dorsal surface and light gray or white on the ventral area. Polynesian rats may breed throughout the year and have up to four litters annually with three to six young in each (Tobin 1994). They are sexually mature by one to two months and may have a life expectancy of around one year. Like many rodent species, they are primarily nocturnal, although when densities are high, such as on some tropical islands, they are diurnal and nocturnal.

Polynesian rats may be common to a wide range of habitats from forests to grasslands, and in agricultural croplands, such as sugarcane (Kami 1966). They are good climbers but poor swimmers, so their dispersal to new islands is generally limited by human movement via ships and cargo (McCartney 1970; Spenneman

FIGURE 10.3 Rat-proof bird nest box designed to protect endangered Hawaiian cavity-nesting birds. (Photo by William Pitt.)

FIGURE 10.4 Adult Polynesian rat (approximately 11 cm body length) that is anesthetized to fit a radio-collar for tracking movement patterns in a montane forest on Oahu, Hawaii. (Photo by Aaron Shiels.)

1997; Matisoo-Smith and Robins 2004). Like the other invasive rodents discussed, Polynesian rats are opportunistic omnivores, and their diets vary greatly by what is available according to season and location so as to exploit locally abundant food sources (Kami 1966; Kepler 1967; Fall et al. 1971; Crook 1973; Tobin and Sugihara 1992; Sugihara 1997; Rufaut and Gibbs 2003). In general, their diet is nearly equally split between plant material and arthropods (Shiels et al. 2013; Shiels and Pitt 2014). Predators of Polynesian rats include mongooses, cats, other larger rodents, and birds (Marshall 1962). In addition, many Polynesian cultures consider this rat to be a valuable food resource, and this species and other rodents may have been introduced into new areas intentionally for food (Spenneman 1997).

Polynesian rats are a significant agricultural pest throughout the Pacific region, as they damage a variety of crops including rice, corn, macadamia nuts, sugarcane, coconut, cacao, pineapple, soybeans, and root crops (Strecker 1962; Kami 1966; Tobin and Sugihara 1992). Previous research documented the extensive effects of rat damage on sugarcane, but sugarcane production has largely been replaced by diversified agriculture in Hawaii (Pitt and Witmer 2007). Rat damage has now shifted to high-value seed crops (corn, soybean), tropical fruits, and native plants. Because Polynesian rats were spread through the western Pacific Basin several thousand years ago, and the eastern Pacific at least 600 to 800 years ago (Wilmshurst et al. 2011), modern population and community-level impacts of this rat on the native flora and fauna are not always apparent (Kepler 1967; Crook 1973; Rufaut and Gibbs 2003; Meyer and Butaud 2009). Furthermore, the more recent introductions of the roof and Norway rats, as well as house mice, have potentially masked some of the negative impacts of Polynesian rats on native ecosystems (Shiels 2010; Shiels et al. 2013). Polynesian rats are effective predators of seabirds, lizards, insects, and sensitive plant species that did not evolve with mammalian predators. Recent eradication efforts of Polynesian rats on islands have revealed the extent of their negative impacts as species recovery has occurred, including invertebrates and vertebrates (Gibbs 2009; Newton et al. 2016).

A variety of methods have been employed to reduce the effects of Polynesian rats on agriculture and natural resources (Jackson 1977). The most successful attempts have integrated rodenticides, alteration of cultural practices, and trapping (Sugihara 1977). Rodenticides have been effectively used to reduce agricultural damage, protect forest birds, and protect seabird colonies. Previous attempts to control rat damage using biological methods (e.g., predator introductions) have been unsuccessful and deleterious for other species. The most frequently cited failure is the introduction of the mongoose (*Herpestes auropunctatus*) to Hawaii in 1883, which has had cascading impacts on ecosystem function (Pitt and Witmer 2007).

Gambian giant pouched rats (*Cricetomys gambianus*; Figure 10.5), referred to hereafter as the Gambian rat, are native to a large area of central and southern Africa. They had become popular in the pet industry and likely were released by a pet breeder and subsequently became established on Grassy Key in the Florida Keys in 1999 (Engeman et al. 2006; Perry et al. 2006). Despite a prolonged eradication effort, a free-ranging and breeding population remained on the island (Engeman et al. 2006, 2007; Witmer and Hall 2011). There is a concern that if this species reaches the mainland, there could be damage to the Florida fruit industry because Gambian rats are known to damage numerous types of agricultural crops in Africa (Fiedler 1994). Imported Gambian rats also may serve as reservoirs of monkey pox and other diseases. An outbreak of monkeypox occurred in the Midwestern United States in 2003 as a result of infected Gambian rats imported from Africa for the pet industry (Enserink 2003). A climate-habitat modeling study suggested that their new range in North America could expand substantially if they were to become established on the United States mainland (Peterson et al. 2006).

FIGURE 10.5 Gambian giant pouched rat captured in a racoon-sized cage trap on Grassy Key, Florida. (Photo by Gary Witmer.)

Gambian rat pelage is gray-brown in color, and they can reach a considerable size: about 2.8 kg in weight and about 1 m in length, body and tail combined (Kingdon 1974). Females produce four young per litter and can bear eight or more litters per year (Ajayi 1975). Because of their reproductive potential and large size, Gambian rats have been raised in captivity as a source of protein in Africa (Ajayi 1975). Since free-ranging Gambian rats are a relatively recent addition to the rodent fauna of North America, relatively little is known about their biology, habitat use, impacts, and interactions with native species or about the most effective means to capture or control these rodents. Hence, current efforts are concentrating on use of traditional live trap capture methods (Figure 10.5) and rodenticides in bait stations (Engeman et al. 2007; Witmer and Hall 2011). Eradicating Gambian rats from Grassy Key has proven problematic because of the large number of private properties on the island, some of whose owners will not allow government employees or the use of rodenticides on their property (Witmer and Hall 2011). Therefore, it will be important to develop additional tools to manage or eradicate this species and other rodent invaders in the United States (Witmer et al. 2010a,b; Witmer and Hall 2011).

House mice (*Mus musculus* and *M. domesticus*; Figure 10.6) are native to southern Europe, northern Africa, and Asia (Long 2003). They now occur worldwide and are probably the most numerous and widespread mammalian species in the world next to humans (Witmer and Jojola 2006). While house mice are thought to have first originated in the grasslands of Central Asia, they have been transported by humans to most parts of the world, largely as stowaways on ships and in cargo. House mice have remarkable abilities that have allowed them to be highly successful in many habitats around the world; chief among these are their reproductive potential and their adaptability in different environments (Timm 1994b; Witmer and Jojola 2006).

House mice are small, slender rodents with a pelage that is grayish-brown on the dorsal surface and gray to buff on the ventral area. This small (maximum mass in United States is about 20 g for adults) and highly prolific animal is a continuous breeder in many situations; a female can produce five to 10 litters—each with five to six young—per year (Timm 1994b). The young mature within about three weeks and soon become reproductively active. House mice are short-lived (generally less than one year) and have high population turnover. In one study, 20 mice placed in an outdoor enclosure with abundant food, water, and cover became a population of 2000 in eight months (Corrigan 2001).

House mice cause many types of damage (Timm 1994b; Witmer and Jojola 2006). A major concern is their role in the consumption and contamination of stored foods; it has been estimated that substantial amounts of stored foods are lost each year in this manner. Mice may damage many types of crops in the field, especially corn, cereal grains, and legumes. Mice also consume and contaminate large amounts of livestock feed at animal production facilities. While mice generally live in close proximity to humans (Corrigan 2001), sometimes remote populations occur such as in many natural areas in Hawaii from sea level to nearly 4000 m elevation (Shiels 2010). Australia has mouse "plagues" periodically, resulting in enormous losses to stored crops and crops in fields (Brown et al. 2004). In buildings, a mouse infestation can be a considerable nuisance because of the noise, odors, and droppings. More

FIGURE 10.6   House mice have amazing abilities which allow them to access almost any available area or resource. (Photo source unknown.)

importantly, they damage insulation and wiring (Hygnstrom 1995). House fires have been caused by mice gnawing electrical wires; likewise, communication systems have been shut down for periods of time, resulting in economic losses (Timm 1994b). Additionally, house mice are susceptible to a large number of disease agents and endoparasites. Consequently, they serve as reservoirs and vectors of disease transmission to humans, pets, and livestock (Gratz 1994). Important among these diseases are leptospirosis, plague, salmonella, lymphocytic choriomeningitis, and toxoplasmosis.

When introduced to islands, house mice can cause significant damage to natural resources, including both flora and fauna. For example, on Gough Island in the South Atlantic, house mice fed on nestling albatross chicks (Cuthbert and Hilton 2004). Additionally, Witmer et al. (2012b) documented seedling damage by house mice in a pen study. House mice are omnivores, yet their diet is largely dominated by insects, some of which are likely plant pollinators (Shiels et al. 2013; Shiels and Pitt 2014). House mice are subordinate to introduced rats, so the impacts of mice may go unnoticed when rats are also present on the island (Angel et al. 2009). This phenomenon was demonstrated by the large increase in mice abundance on Buck Island, U.S. Virgin Islands, after invasive roof rats were eradicated (Witmer et al. 2007a). In very dry habitats on islands, house mice may numerically dominate over introduced rats.

A large number of methods and materials have been developed to help solve house mouse problems. In general, the use of multiple approaches and materials—integrated pest management (IPM)—is more likely to reduce a mouse problem to a tolerable level (Witmer 2007). For example, sanitation and blocking of small access openings can be combined with some use of traps (kill traps and/or live traps) and/or toxicants. The tools available and their proper use have been extensively reviewed (Brooks 1973; Prakash 1988; Timm 1994b; Corrigan 2001).

INVASIVE RODENT MANAGEMENT TOOLS AND RESEARCH NEEDS

Many methods and tools have been developed and used to control rodent populations or to reduce the damage they cause (Table 10.1). The methods that are commonly used vary greatly from region to region around the world, as well as between developed and undeveloped countries. Methods used also vary with regard to the type of management. When long-term population suppression is the management goal (such as in agricultural and urban/suburban settings), a variety of approaches are used, generally through an IPM strategy (Witmer 2007). While traps and rodenticides are the mainstays of rodent population management, IPM also employs habitat management, exclusion, and sanitation (Hygnstrom et al. 1994). On the other hand, if eradication of the invasive rodent species is the management goal, rodenticides are heavily relied upon, although traps may be used to some extent in combination with rodenticides. Some of the methods are highly regulated, and regulations vary across political jurisdictions. The many methods used to manage rodent populations and damage have been described at length by Prakash (1988), Corrigan (2001), Buckle and Smith (2015), Hygnstrom et al. (1994), and Caughley et al. (1998). One novel technique that was recently tested and may deserve further research was the use of live laboratory rats as lures to trap invasive wild Norway rats; the live rats, regardless of gender, were more efficient than food baits for catching invasive rats (Shapira et al. 2013b). Similar testing using laboratory mice to attract wild house mice has also been investigated (Shapira et al. 2013a). In this chapter, we will only address in detail traps and rodenticides as invasive rodent population management techniques.

A wide array of traps have been developed and used to manage rodents, and many types are commercially available (Hygnstrom et al. 1994; Proulx 1999). Trap types are subdivided into live traps and kill traps. With live traps, the rodent becomes contained in a box or cage trap after tripping a treadle. Another type of live trap is

TABLE 10.1
Methods and Techniques for Rodent Control That Have Been Suggested, Tested, or Used to Reduce Rodent Populations and Damage around the World

Physical	Chemical	Biological	Other
Rodent proof construction	Baits/baiting systems	Virally vectored control	Appeasement
Passive barriers	Glues	Immunogens	Insurance
Electric barriers	Poison sprays	Habitat modification	Bounties
Drift fences	Poison moats	Cultural practices	Harvest
Trapping	Tracking powder	Crop timing	Compensation
Flooding burrows	Tracking greases, gel	Crop diversification, and species selection	
Drives	Repellents	Buffer crops	
Hunting	Attractants	Parasites	
Clubbing	Aversive agents	Diseases	
Frightening devices	Plant systematics	Predators	
Flame throwers	Sterilants	Ultrasonics	
Burrow destruction	Fumigation	Biosonics	
Habitat destruction	Psychotropic drugs	Resistant plants	
Harborage removal	Herbicides	Lethal genes	
Supplemental feeding		Endophytic grasses	
Digging		Unpalatable plants	
Dogs together with flooding or digging			

Source: Modified from Witmer, G. and Singleton, G., *Agricultural Production*, New York, New York, Nova Science Publishers, Inc., 1–38, 2010.

the leg-hold trap which, when tripped by the rodents paw, springs the jaws of the trap to close tightly around the leg and hold the animal until the trapper arrives. Leg-hold traps are generally only used for larger rodent species such as nutria, muskrats (*Ondatra zibethicus*), and beaver (*Castor canadensis*). Invasive rodents captured in live traps can be used in research or euthanized. An advantage of live traps is that nontarget animals captured can often be released unharmed.

Kill traps cause the rapid death of the rodent by body constriction when the animal trips the trap's trigger mechanism. The most common type of rodent kill trap is the snap trap. Alternatively, Conibear kill traps can be used for larger rodent species. Automatically self-resetting rat kill traps have been recently developed in New Zealand by Goodnature® (e.g., Peters et al. 2014). Known as the Goodnature® A24 rat traps (or A24s), they can fire up to 24 times before the CO_2 cartridge must be replaced. The A24 traps are relatively expensive compared to nonautomatic traps, and currently cost about $112 per unit. Although A24 traps have been used widely over the past few years, including in Hawaii, California, and Puerto Rico, mixed results have been observed in their reliability and effectiveness. In fact, the New Zealand government spent NZ$4 million to test the Goodnature® traps, and the A24s were determined to be ineffective for rat control due to mechanical failure and bait palatability issues (Gillies et al. 2012). Improved reliability and effectiveness of such automated traps as the Goodnature® A24s will certainly help expand the number of options to efficiently suppress invasive rodents (Campbell et al. 2015).

Hygnstrom et al. (1994) provided good illustrations of various types of traps and directions for their proper and effective use. Effective trapping requires skill and practice, and use of the proper type of trap for the situation, proper placement, and appropriate bait is important to achieve a high level of trap success (i.e., a high capture rate). A disadvantage of kill traps is they can injure or kill nontarget animals, including birds. Various types of traps are also used to monitor rodent populations. Rodent population monitoring is essential so that necessary management action can be taken before populations get too large, at which point extensive damage to resources cannot be avoided (Witmer 2005). Rodent population monitoring before and during rodent suppression is also important to justify the use and effectiveness of the rodent control technique.

Rodenticides are widely used in the United States as well other parts of the world. Because of the risk of harm to people, pets, and livestock, rodenticides are carefully regulated by the United States Environmental Protection Agency (EPA), as well as by state agencies. There are many types of rodenticides, and these vary by active ingredient as well as formulation (Table 10.2). These materials vary widely in their mode of action and in toxicity. The types and uses of rodenticides in the United States were reviewed by Witmer and Eisemann (2007), and their specific use for conservation purposes (i.e., the control or eradication of invasive rodents) was reviewed by Witmer et al. (2007b).

Proper training and careful use are required to safely use rodenticides so that they are effective in reducing rodent populations while minimizing the hazard to nontarget animals. An EPA-approved product label provides considerable information on the product and its use, including: the registrant and EPA registration number(s), active ingredient and concentration, target species and settings in which it can be used, directions for use, storage and disposal requirements, precautionary statements, safety and environmental hazards, and threatened and endangered species considerations (Figure 10.7).

Both primary (direct consumption) and secondary hazards (consuming a poisoned rodent or poisoned nontarget animal) can occur to nontarget animals when rodenticides are used (Masuda and Jamieson 2013; Pitt et al. 2015). Rodenticides such as brodifacoum (a second generation anticoagulant) are highly toxic, but also result in persistent residues in body tissues of animals that consume lethal or sublethal doses or consume rodents that have been poisoned (Witmer and Eisemann 2007). There are growing concerns about persistence, chronic use, and the subsequent chronic secondary effects of these residues in nontarget and predatory animals (Thomas et al. 2011; Masuda and Jamieson 2013; Pitt et al. 2015; Elliott et al. 2016). The main safeguard for the safe use of rodenticides in the United States is carefully following the EPA label instructions for the product. Other considerations include: the product used; when, where, and how it is applied; cleaning up spills promptly; and not using rodenticides where highly valued or protected wildlife occur (determined by scouting and surveying the area before use).

TABLE 10.2
The Main Rodenticides Used in the United States by Category and Percent Active Ingredient

Acute rodenticides
- Cholecalciferol (0.075%)
- Strychnine (0.5%)
- Zinc phosphide (2%)
- Bromethalin (0.01%)

Fumigants
- Aluminum phosphide (56%)
- Magnesium phosphide (56%)
- Acrolein (95%)
- Gas cartridges (variable)

First-Generation Anticoagulants
- Chlorophacinone (0.005%)
- Diphacinone (0.005%)
- Warfarin (0.025%)
- Pindone (0.025%)

Second-Generation Anticoagulants
- Bromadiolone (0.005%)
- Brodifacoum (0.005%)
- Difethialone (0.0025%)

Source: Witmer, G. and Eisemann, J. D., *Proceedings of the 12th Wildlife Damage Management Conference*, Corpus Christi, Texas, 12th Wildlife Damage Management Conference, 114–118, 2007.

Additional research is needed to improve existing methods and to develop new methods for invasive rodent detection and control, including both lethal and nonlethal means of resolving rodent damage situations (Witmer et al. 1995; Eason et al. 2010; Witmer and Singleton 2010; Blackie et al. 2014; Campbell et al. 2015). Future research should include, but not be limited to, developing new rodenticides, more effective repellents and barriers, improved biological control, fertility control, and habitat manipulation. Some exciting areas of new research on invasive rodent control methods include:

- Resetting, long-life toxin delivery systems
- Increasing knowledge of pest behavior
- New active ingredient toxicants
- Long-life bait coatings
- Crab deterrents for bait
- Transgenic rodents (genetic manipulation)
- More stakeholder engagement and increased understanding of social processes

PRECAUTIONARY STATEMENTS

HAZARDS TO HUMANS AND DOMESTIC ANIMALS

Keep away from humans, domestic animals and pets. If swallowed, this material may reduce the clotting ability of the blood and cause bleeding. Wear protective gloves when applying or loading bait. With detergent and hot water, wash all implements used for applying bait. Do not use these implements for mixing, holding, or transferring food or feed.

ENVIRONMENTAL HAZARDS

This pesticide is toxic to birds, mammals and aquatic organisms. Predatory and scavenging mammals and birds might be poisoned if they feed upon animals that have eaten bait.

PERSONAL PROTECTIVE EQUIPMENT (PPE)

Applicators and other handlers must wear:
-long sleeved shirt and long pants
-gloves
-shoes plus socks

For aerial application, in addition to the above PPE, loaders must wear protective eyewear or a face shield and a dust/mist filtering respirator (MSHA/NIOSH TC-21C).

USE RESTRICTIONS

It is a violation of Federal law to use this product in a manner inconsistent with its labeling. A copy of this label must be in the possession of the user at the time that the product is applied.

READ THIS LABEL: Read this entire label and follow all use directions and precautions.

IMPORTANT: Do not expose children, pets or other non-target animals to rodenticides. To help prevent accidents:
1) Keep children out of areas where this product is used or deny them access to bait by use of tamper resistant bait stations.
2) Store this product in locations out of reach of children, pets, and other nontarget animals.
3) Apply bait only according to the directions authorized.
4) Dispose of product container and unused, spoiled, or unconsumed bait as specified in the "STORAGE AND DISPOSAL" section.

(SEE RIGHT PANEL FOR ADDITIONAL USE RESTRICTIONS)

RESTRICTED USE PESTICIDE

DUE TO HAZARDS TO NON-TARGET SPECIES

For retail sale to and use only by Certified Applicators or persons under their direct supervision and only for those uses covered by the Certified Applicators certification.

For use by or in cooperation with government conservation agencies.

BRODIFACOUM-25D CONSERVATION

PELLETED RODENTICIDE BAIT FOR CONSERVATION PURPOSES

For control or eradication of invasive rodents in dry climates on islands or vessels for conservation purposes

ACTIVE INGREDIENT
Brodifacoum (CAS No. 56073-10-0) 0.0025%
INERT INGREDIENTS 99.9975%
TOTAL 100.0000%

KEEP OUT OF REACH OF CHILDREN

CAUTION

First Aid

If swallowed	-Call a physician or poison control center immediately for treatment advice. -Have person sip a glass of water if able to swallow. -Do not induce vomiting unless told to do so by a poison control center or doctor. -Do not give anything by mouth to an unconscious person.
If on skin or clothing	-Take off contaminated clothing. -Rinse skin immediately with plenty of water for 15-20 minutes. -Call a poison control center or doctor for treatment advice.
If inhaled	-Move person to fresh air. -If person is not breathing, call 911 or an ambulance, then give artificial respiration, preferably mouth-to-mouth if possible. -Call a poison control center or doctor for further treatment advice.
If in eyes	-Hold eye open and rinse slowly and gently with water for 15-20 minutes. Remove contact lenses, if present, after the first 5 minutes, then continue rinsing eye. -Call a poison control center or doctor for treatment advice.

Have the product container or label with you when calling a poison control center or doctor, or when going for treatment.

For a medical emergency involving this product, call (877) 854-2494

NOTE TO PHYSICIAN: If swallowed, this material may reduce the clotting ability of blood and cause bleeding. If ingested, administer Vitamin K₁ intramuscularly or orally, as indicated in bishydroxycoumarin overdose. Repeat as necessary based on monitoring of prothrombin times.

USE RESTRICTIONS, (CONT)

This product may be used to control or eradicate Norway rats (*Rattus norvegicus*), roof rats (*Rattus rattus*), Polynesian rats (*Rattus exulans*), house mice (*Mus musculus*) or other types of invasive rodents on islands for conservation purposes, or on grounded vessels or vessels in peril of grounding.

This product may be applied using bait stations, burrow baiting, canopy baiting or by aerial and ground broadcast application techniques.

This product is to be used for the protection of State or Federally-listed Threatened or Endangered Species or other species determined to require special protection.

Do not apply this product to food or feed.

Treated areas must be posted with warning signs appropriate to the current rodent control project.

This product is for use in dry climates.

DIRECTIONS FOR USE

BAIT STATIONS: Tamper-resistant bait stations must be used when applying this product to grounded vessels or vessels in peril of grounding, or when used in areas of human habitation. Bait must be applied in locations out of reach of children, non-target wildlife, or domestic animals, or in tamper-resistant bait stations.
TO BAIT RATS: Apply 4 to 16 ounces (113 to 454 grams) of bait per placement. Space placements at intervals of 16 to 160 ft (about 5 to 50 meters). Placements should be made in a grid over the area for which rodent control is desired.
TO BAIT MICE: Apply 0.25 to 0.5 ounces (7 to 14 grams) of bait per placement. Space placements at intervals of 6 to 12 ft (about 2 to 4 meters). Larger placements, up to 2 ounces (57 grams) may be needed at points of very high mouse activity. Placements should be made in a grid over the area for which rodent control is desired.
FOR BOTH RAT AND MOUSE BAITING: Maintain an uninterrupted supply of fresh bait for at least 15 days or until signs of rodent activity cease. Where a continuous source of infestation is present, permanent bait stations may be established and bait replenished as needed.

Page 1 of 2
EPA Approved 03/18/08
EPA Reg. No. 56229-37

FIGURE 10.7 EPA-approved label for a rodenticide designed for invasive rodent eradication on islands.

We believe there is a need to identify effective, commercially available rodenticide formulations for the various invasive rodents in each region of the country, as Pitt et al. (2011c) have done for rats and mice in Hawaii, and Witmer and Moulton (2014) did for Norway rats and mice in the North American mainland. Another important research need is greater evaluation of the effectiveness of combinations of techniques, given that combinations could potentially be much more effective in the reduction of damage and may be more acceptable to the public.

INVASIVE RODENT ERADICATION

Since the early 1990s, federal and state agencies, along with conservation organizations, have been eradicating rodents from various islands in the United States, primarily for conservation purposes. To date, 644 successful eradications of rats on islands worldwide have been reported (DIISE Partners 2016), and the great majority of these have involved roof, Norway, and Polynesian rats. In general, successful rat eradications from tropical islands (89%) have been somewhat lower than successes in temperate islands (96%; Keitt et al. 2015). Some suggestions on why the lower success rate on tropical islands include (1) increased crab and insect densities resulting in competition for bait, (2) year-round and unpredictable breeding by rats, and (3) increased or unpredictable availability of alternative, natural foods (Keitt et al. 2015; Holmes et al. 2015b). Witmer et al. (2011) documented the attempted eradications of introduced rodents in the United States and its territories. Of about 40 island eradication attempts, 22 (55%) or more appear to have succeeded. For several islands, however, it is too early to determine if the attempted eradication has been successful. Additionally, experimental rat eradication trials on 12 small islands in The Bay of Islands, Adak, Alaska, failed or rapid reinvasion occurred, and those perhaps should not be included in the list of more concerted eradication efforts as eradication methods were being investigated (Witmer et al. 2011). In some cases, what appeared to be failed eradications may have resulted from rapid reinvasion by rats from nearby islands, suggesting the need to eradicate rats from groups of islands as an eradication unit (Savidge et al. 2012). Genetic analyses of rats before and after eradications are often necessary in helping sort out the issue of reinvasion versus failed eradication (Keitt et al. 2015). Numerous islandwide rodent eradications are underway or being planned. Most rodent eradications around the world have used the second-generation anticoagulant brodifacoum (Howald et al. 2007). In the United States, however, most eradications have used the first-generation anticoagulant diphacinone (Witmer et al. 2011). Early rodent eradications used hand-broadcast and bait stations of rodenticides, but in recent years, aerial broadcast via helicopter has become common. Aerial broadcast of rodenticide bait allows rodent eradications on much larger and more rugged islands, such as Rat Island, Alaska (2700 ha; Witmer et al. 2011). There are large cost differences between aerial and ground applications of rodenticides for island eradications; most aerial-based operations cost between $1 and $3 million (Holmes et al. 2015a), whereas ground-based operations are mere thousands, and the most expensive listed in the Holmes et al. (2015a) review was about $300,000 (Buck Island; Witmer et al. 2007a). Aerial application, however, has allowed very large islands to be treated, islands so large they could not be treated by hand broadcast (e.g., Rat Island in the Aleutian Islands) where rats were successfully eradicted, is 2780 ha). In the United States and most other countries in the world, aerial application of rodenticide is typically used for invasive rodent eradication purposes rather than invasive rodent population suppression; however, in New Zealand, aerial broadcast of toxic baits (e.g., 1080) has been used for decades to suppress, rather than eradicate, invasive possums (*Trichosurus vulpecula*) and rats (*Rattus* spp.; Morgan et al. 2015). Currently, the U.S. Department of Agriculture (USDA) Animal and Plant Health Inspection Service (APHIS) has two rodenticides registered with the EPA for island conservation purposes: one formulation of diphacinone pellets and two formulations of brodifacoum pellets (Witmer et al. 2007b; Figure 10.7).

When using rodenticides, a variety of considerations and mitigation measures are employed to avoid, reduce, or minimize nontarget hazards and environmental impacts; some include the rodenticide type, amount (Pott et al. 2015), formulation, method and timing of baiting (Keitt et al. 2015), captive holding and later release of some nontarget wildlife species until after the baiting operation, removal of rodent carcasses to prevent secondary poisoning, and avoidance of bait placement in aquatic systems (Witmer et al. 2007b). In general, impacts to nontarget species during invasive rodent eradications should be considered in terms of population-level effects, rather than the effects to individuals, and the benefits should outweigh the costs of the project implementation. While there will probably always be some losses of nontarget animals, proper precautions should minimize such risk and allow for the rapid recovery of affected populations (Howald et al. 2005). For example, Croll et al. (2016) determined that despite initial nontarget mortalities (including bald eagle reduction from 24 individuals to two individuals), within five years posteradication of rats, the bald eagles, terrestrial birds, and shorebirds had colonized, recolonized, or otherwise increased in abundance to levels near to or above those of pre-eradication on an Aleutian Island in Alaska. Those involved with successful invasive rodent eradications on islands are often surprised at how rapidly the island's flora and fauna recover or change after rodents are removed (Witmer et al. 2007a; Croll et al. 2016).

Planning and conducting a successful invasive rodent eradication from an island poses many challenges and should not be undertaken without a thorough commitment and adequate resources. The basic tenets of a successful eradication are: all individuals (i.e., target rodent species) must be put at risk; rodents must be removed faster than they can reproduce; and the risk of immigration must be zero (Parkes and Murphy 2003). An eradication attempt that is 99% successful can ultimately result in 100% failure due to the high reproductive potential of the remaining rodent population. Because of the large commitment of resources and usually public funds in eradication efforts, the potential for failure should be minimized, and there have been a number of factors that have been identified that commonly influence whether a rodent eradication attempt is successful (e.g., see Keitt et al. 2015; Holmes et al. 2015b). Planning and implementation components include:

- Preliminary monitoring and research
- Feasibility of eradication
- Regulatory compliance
- Public information and communications media
- Public support
- Technical assistance and operations
- Planning
- Logistics
- Procurement of equipment and other services
- Monitoring and research
- Staff recruitment and training
- Implementation
- Contingency planning
- Follow-up monitoring
- Implementation of a biosecurity plan

REMAINING AND FUTURE CHALLENGES

A number of challenges remain with invasive rodent management and eradication in the United States. Some of the challenges faced include public and agency concerns about animal welfare issues, uncertainties with pesticide use, the necessary investments to achieve conservation goals, the use of certain toxicants and traps, land access (especially to private lands), public attitudes, resource availability, and detection and monitoring difficulties (Witmer and Hall 2011; Witmer et al. 2011).

Nonetheless, we will hopefully continue to relieve the burdens on insular and mainland ecosystems caused by rodent introductions. This is also essential to ensure adequate food resources for humans and livestock worldwide. The flora and fauna of islands generally respond rapidly, and conservation goals are achieved after invasive rodents are removed and often without much additional input by people (e.g., Witmer et al. 2007a; Witmer et al. 2011; Croll et al. 2016; Newton et al. 2016). Endemic, threatened, or endangered species can be, and have been, reintroduced after successful rodent eradications. For example, the endangered St. Croix ground lizard (*Ameiva polops*) was recently reintroduced to Buck Island in the U.S. Virgin Islands after the successful eradication of roof rats (Witmer et al. 2007a). The recent eradication of Polynesian rats and house mice from Cocos Island (a small island off of Guam) set the stage for the reintroduction of the endangered Guam rail, *Gallirallus owstoni* (Lujan et al. 2010).

CONCLUSIONS

Invasive rodents will continue to pose challenges to land and resource managers, commodity producers, and homeowners. Although preventing rodent invasion beyond their native range is the most idealized way of preventing their negative impacts as an invasive species, it is not always possible, and the rodents described in this chapter have clearly invaded much of the United States. Many tools are available to reduce rodent populations and associated damage, and these tools should be used in a well thought out IPM approach. Rodenticides will continue to be an important tool against rodents and their damage, but care must be exercised in their use. It is probably safe to assume that much of the public will continue to scrutinize certain tools, such as toxicants. Hence, public outreach and education will be important to ensure continued availability of adequate tools, such as rodenticides, until such time as new and emerging technologies can be developed, tested, registered and regulated, and adopted widely (Witmer et al. 2009; Eason et al. 2010; Witmer and Singleton 2010; Blackie et al. 2014; Campbell et al. 2015). Continued technology development and transfer to agencies, companies, and the public are essential to improve the effectiveness and safety of rodenticides and other methods used to control or eradicate invasive rodents. With proper planning, nontarget losses from invasive rodent suppression or eradication will be minimal, and these nontarget populations, along with other island and mainland resources, will be ensured and often recover quickly after the invasive rodents have been removed.

REFERENCES

Ajayi, S. 1975. Observations on the biology, domestication, and reproductive performance of the African giant rat *Cricetomys gambianus* waterhouse in Nigeria. *Mammalia* 39, 343–364.
Angel, A., Wanless, R.M., and Cooper, J. 2009. Review of impacts of the introduced house mouse on islands in the Southern Ocean: Are mice equivalent to rats? *Biological Invasions* 11, 1743–1754.
Aplin, K. and Singleton, G. 2003. Balancing rodent management and small mammal conservation in agricultural landscapes: Challenges for the present and the future. In: G. Singleton, L. Hinds, C. Krebs, and D. Spratt (Eds.), *Rats, Mice and People: Rodent*

Biology and Management. (Pp. 80–88), Canberra, Australia: Australian Centre for International Agricultural Research.

Atkinson, I. 1985. The spread of commensal species of *Rattus* to oceanic islands and their effects on island avifaunas. In: P. Moors (Ed.). *Conservation of Island Birds.* (Pp. 35–81), Cambridge, United Kingdom: International Council of Bird Preservation, Technical Bulletin No. 3.

Atkinson, I.A.E. 1977. A reassessment of factors, particularly *Rattus rattus* L. that influenced the decline of endemic forest birds in the Hawaiian Islands. *Pacific Science* 31, 109–133.

Baldwin, R.A., Quinn, N., Davis, D.H., and Engeman, R.M. 2014. Effectiveness of rodenticides for managing invasive roof rats and native deer mice in orchards. *Environmental Science and Pollution Research International* 21, 5795–5802.

Banks, P.B., and Hughes, N.K. 2012. A review of the evidence for potential impacts of black rats (*Rattus rattus*) on wildlife and humans in Australia. *Wildlife Research* 39, 78–88.

Berentsen, A.R., Vogt, S., Guzman, A.N., Vice, D.S., Pitt, W.C., Shiels, A.B., and Spraker, T.R. 2015. *Capillaria hepatica* infection in black rats (*Rattus rattus*) on Diego Garcia, British Indian Territory. *Journal of Veterinary Diagnostic Investigation* 27, 241–244.

Blackie, H.M., MacKay, J.W.B., Allen, W.J., Smith, D.H.V., Barrett, B., Whyte, B.I., Murphy, E.C. et al. 2014. Innovative developments for long-term mammalian pest control: Innovative developments for mammalian pest control. *Pest Management Science* 70, 345–351.

Brakes, C., and Smith, R. 2005. Exposure of nontarget small mammals to rodenticides: Short-term effects, recovery and implications for secondary poisoning. *Journal of Applied Ecology* 42, 118–128.

Brooks, J. E. 1973. A review of commensal rodents and their control. *CRC Critical Reviews in Environmental Control* 3(4), 405–453.

Brown, P.R., Davies, M., Singleton, G., and Croft, J. 2004. Can farm-management practices reduce the impact of house mouse populations on crops in an irrigated farming system? *Wildlife Research* 31, 597–604.

Buckle, A.P., and Smith, R.H. 2015. *Rodent Pests and Their Control.* 2nd Edn. Wallingford, U.K.: CAB International. 422 pp.

Campbell, K.J., Beek, J., Easton, C.T., Glen, A.S., Godwin, J., Gould, F., Holmes, N.D. et al. 2015. The next generation of rodent eradications: Innovative technologies and tools to improve species specificity and increase their feasibility on islands. *Biological Conservation* 185, 47–58.

Capizzi, D., Bertolino, S., and Mortelliti, A. 2014. Rating the rat: Global patterns and research priorities in impacts and management of rodent pests. *Mammal Review* 44, 148–162.

Caughley, J., Bomford, M., Parker, B., Sinclair, R., Griffiths, J. and Kelley, D. 1998. *Managing Vertebrate Pests: Rodents.* Canberra, Australia: Bureau of Resource Sciences and Grains Research and Development Corporation. 130 pp.

Caut, S., Angulo, E., and Courchamp, F. 2008. Dietary shift of an invasive predator: Rats, seabirds and sea turtles. *Journal of Applied Ecology* 45, 428–437.

Corrigan, R.M. 2001. *Rodent Control: A Practical Guide for Pest Management Professionals.* Richfield, Ohio: GIE, Inc. 355 pp.

Croll, D.A., Newton, K.M., McKown, M., Holmes, N., Williams, J.C., Young, H.S., Buckelew, S. et al. 2016. Passive recovery of an island bird community after rodent eradication. *Biological Invasions* 18, 703–715.

Crook, I. G. 1973. The tuatara, *Sphenodon punctatus* Gray, on islands with and without populations of the Polynesian rat, *Rattus exulans* (Peale). *Proceedings of the New Zealand Ecological Society* 20, 115–120.

Crowell, K.L. 1983. Islands—Insight or artifact? Population dynamics and habitat utilization in insular rodents. *Oikos* 27, 442–454.

Crowell, K.L., and Pimm, S.L. 1976. Competition and niche shifts of mice introduced onto small islands. *Oikos* 27, 251–258.

Cuthbert, R., and Hilton, G. 2004. Introduced house mice: A significant predator of threatened and endemic birds on Gough Island, South Atlantic Ocean? *Biological Conservation* 117, 483–489.

Davis, D. E. 1953. The characteristics of rat populations, *Quarterly Review of Biology* 28(4), 373–401.

Dickman, C. 1999. Rodent-ecosystem relationships: A review. In: G. Singleton, L. Hinds, H. Leirs, and Z. Zhang (Eds.), *Ecologically-Based Management of Rodent Pests.* (Pp. 113–135), Canberra, Australia: Australian Centre for International Agricultural Research.

DIISE Partners. 2016. The Database of Island Invasive Species Eradications, developed by Island Conservation, Coastal Conservation Action Laboratory UCSC, IUCN SSC Invasive Species Specialist Group, University of Auckland and Landcare Research, New Zealand. http://diise.islandconservation.org. Accessed January 8, 2016.

Drake, D.R., Bodey, T.W., Russell, J.C., Towns, D.R., Nogales, M., and Ruffino, L. 2011. Direct Impacts of Seabird Predators in Island Biota other than Seabirds. In: C. P. H. Mulder, W. B. Anderson, D. R. Towns, and P. J. Bellingham (Eds.), *Seabird Islands: Ecology, Invasion, and Restoration.* (Pp. 91–132), New York, New York: Oxford University Press.

Eason, C., Fagerstone, K., Eisemann, J., Humphrys, S., O'Hare, J., and Lapidge, S. 2010. A review of existing and potential New World and Australian vertebrate pesticides with a rationale for linking use patterns to registration requirements. *International Journal of Pest Management* 56, 109–125.

Elliott, J., Rattner, B., Shore, R., and Van Den Brink, N. 2016. Paying the piper: Mitigating the impact of anticoagulant rodenticides on predators and scavengers. *BioScience* 66, 401–407.

Engeman, R.M, Witmer, G.W., Bourassa, J.B., Woolard, J.W., Constantin, B., Hall, P.T., Hardin, S., and Perry, N.D. 2007. The path to eradication of the Gambian giant pouched rat in Florida. In G. W. Witmer, W. C. Pitt, and K. A. Fagerstone (Eds.), *Managing Vertebrate Invasive Species: Proceedings of an International Symposium.* (Pp. 305–311), Fort Collins, CO: USDA/APHIS/Wildlife Services, National Wildlife Research Center.

Engeman, R., Woolard, J., Perry, N., Witmer, G., Hardin, S., Brashears, L., Smith, H., Muiznieks, B., and Constantin, B. 2006. Rapid assessment for a new invasive species threat: The case of the Gambian giant pouched rat in Florida. *Wildlife Research* 33, 439–448.

Enserink, M. 2003. U.S. monkey pox outbreak traced to Wisconsin pet dealer. *Science* 300, 1639.

Fall, M.W., and Jackson, W. B. 1998. A new era of vertebrate pest control? An introduction. *International Biodeterioration and Biodegradation* 42, 85–91.

Fall, M.W., Medina, A.B., and Jackson, W. B. 1971. Feeding patterns of *Rattus rattus* and *Rattus exulans* on Eniwetok Atoll, Marshall Islands. *Journal of Mammalogy* 52, 69–76.

Fiedler, L.A. 1994. Rodent Pest Management in Eastern Africa. *FAO Plant Production and Protection Paper 123.* Rome, Italy.

Gibbs, G. W. 2009. The end of an 80-million year experiment: A review of evidence describing the impact of introduced rodents on New Zealand's "mammal-free" invertebrate fauna. *Biological Invasions* 11, 1587–1593.

Gillies, C., Gorman, N., Crossan, I., Harawira, R., Hawaikirangi, R., Long, J., and McCool, E. 2012. *A second progress report of DOC S&C Investigation 4276 "Operational scale trials of self-resetting traps for ground-based pest control for conservation in NZ forests." Department of Conservation Science and Capability Group Report*, Hamilton, New Zealand. 24 Pp.

Gratz, N.G. 1994. Rodents as carriers of disease. In A. P. Buckle and R. H. Smith (Eds.), *Rodent Pests and Their Control.* (Pp. 85–108), Wallingford, United Kingdom: CAB International.

Holmes, N.D., Campbell, K.J., Keitt, B.S., Griffiths, R., Beek, J., Donlan, C.J., and Broome, K.G. 2015a. Reporting costs for invasive vertebrate eradications. *Biological Invasions* 17, 2913–2925.

Holmes, N.D., Griffiths, R., Pott, M., Alifano, A., Will, D., Wegmann, A.S., and Russell, J.C. 2015b. Factors associated with rodent eradication failure. *Biological Conservation* 185, 8–16.

Howald, G., Donlan, C.J., Galvan, J.P., Russell, J.C., Parkes, J., Samaniego, A., Wang, Y. et al. 2007. Invasive rodent eradication on islands. *Conservation Biology* 21(5), 1258–1268.

Howald, G., Faulkner, K., Tershy, B., Keitt, B., Gellerman, H., Creel, E., Grinnell, M. et al. 2005. Eradication of black rats from Anacapa Island: Biological and social considerations. In D. Garcelon and C. Schwemm (Eds.), *Proceedings of the 6th California Islands Symposium.* (Pp. 299–312), Arcata, California: National Park Service Technical Publication CHIS-05–01, Institute for Wildlife Studies.

Hygnstrom, S. E. 1995. House mouse damage to insulation. *International Biodeterioration and Biodegradation* 36, 143–150.

Hygnstrom, S., Timm, R., and Larson, G. 1994. *Prevention and Control of Wildlife Damage.* Lincoln, Nebraska: University of Nebraska Cooperative Extension Service.

Jackson, W. B. 1977. Evaluation of rodent depredations to crops and stored products. *European Plant Protection Organization Bulletin* 7, 439–458.

James, D.K. 2006. New rat species in North America. *Vector Ecology Newsletter* 37(1), 5.

Jarvi, S.L., Pitt, W.C., Farias, M., Shiels, L., Severino, M., Howe, K., Jacquier, S. et al. 2015. Detection of *Angiostrongylus cantonensis* in the blood and peripheral tissue of wild Hawaiian rats (*Rattus rattus*) by a quantitative PCR (qPCR) assay. *PLoS ONE* 10: 1–12.

Jones, H.P., Tershy, B.R., Zavaleta, E.S., Croll, D.A., Keitt, B.S., Finkelstein, M.E., and Howald, G.R. 2008. Severity of the effects of invasive rats on seabirds: A global review. *Conservation Biology* 22, 16–26.

Kami, H. T. 1966. Foods of rodents in the Hamakua District, Hawaii. *Pacific Science* 20, 367–373.

Keitt, B., Griffiths, R., Boudjelas, S., Broome, K., Cranwell, S., Millett, J., Pitt, W., et al. 2015. Best practice guidelines for rat eradication on tropical islands. *Biological Conservation* 185, 17–26.

Kepler, C. B. 1967. Polynesian rat predation on nesting Laysan albatrosses and other Pacific seabirds. *Auk* 84, 426–430.

Kingdon, J. 1974. *East African Mammals. An Atlas of Evolution in Africa Volume 2, Part B (Hares and Rodents).* Chicago, Illinois: University of Chicago Press.

Latorre, L., Larrinaga, A.R., and Santamaria, L. 2013. Rats and seabirds: Effects of egg size on predation risk and the potential of conditioned taste aversion as a mitigation method. *PLoS ONE* 8(9), e76138.

Long, J. 2003. *Introduced Mammals of the World.* Collingwood, Australia: CSIRO Publishing. 590 pp.

Lujan, D.T., Vice, D.S., Guerrero, J.P., and Candaso, C.C. 2010. Rodent eradication on Cocos Island, Guam: Integrating wildlife damage management, resort operations, and non-target concerns. In R.M. Timm and K.A. Fagerstone (Eds.), *Proceedings of the 24nd Vertebrate Pest Conference.* (Pp.9–13), Davis, California: University of California.

Marsh, R.E. 1994. Roof rats. In S. E. Hygnstrom, R. M. Timm, and G. E. Larson, (Eds.), *Prevention and Control of Wildlife Damage.* (Pp. B-125–B-132), Lincoln, Nebraska: University of Nebraska Cooperative Extension.

Marshall, Jr., J. T. 1962. Predation and natural selection. In T. I. Storer (Ed.), *Pacific Island Rat Ecology: Report on a Study Made on Ponape and Adjacent Islands 1955–1958.* (Pp. 177–187), Honolulu, Hawaii: Bulletin 225, Bernice P. Bishop Museum.

Masuda, B.M. and Jamieson, I.G. 2013. Response of a reintroduced bird population to a rat reinvasion and eradication. *New Zealand Journal of Ecology* 37, 224–231.

Matisoo-Smith, E. and Robins, J.H. 2004. Origins and dispersals of Pacific peoples: Evidence from mtDNA phylogenies of the Pacific rat. *Proceedings of the National Academy of Science* 101, 9167–9172.

McCartney, W.C. 1970. Arboreal behavior of the Polynesian rat (*Rattus exulans*). *BioScience* 20, 1061–1062.

Meehan, A.P. 1984. *Rats and Mice.* East Grinstead, UK: Rentokil Ltd. 383 pp.

Meerburg, B., Singleton, G., and Kijlstra, A. 2009a. Rodent-borne diseases and their risks for public health. *Critical Reviews in Microbiology* 35, 221–270.

Meerburg, B.G., Singleton, G.R., and Leirs, H. 2009b. The Year of the Rat ends—time to fight hunger. *Pest Management Science* 65, 351–352.

Meyer, J.Y. and Butaud, J.F. 2009. The impacts of rats on the endangered native flora of French Polynesia (Pacific Islands): Drivers of plant extinction or coup de grâce species? *Biological Invasions* 11(7), 1569–1585.

Meyer, W.M. and Shiels, A.B. 2009. Black rat (*Rattus rattus*) predation on nonindigenous snails in Hawaii: Complex management implications. *Pacific Science* 63, 339–347.

Moors, P.J. and Atkinson, I.A.E. 1984. Predation on seabirds by introduced animals, and factors affecting its severity. In P. J. Moors (Ed.), *Conservation of Island Birds: Case Studies for the Management of Threatened Island Birds.* (Pp. 667–690), Cambridge, United Kingdom: International Council for Bird Preservation Technical Publication No. 2.

Morgan, D., Warburton, B., and Nugent, G. 2015. Aerial prefeeding followed by ground based toxic baiting for more efficient and acceptable poisoning of invasive small mammalian pest. *PLoS ONE* 10(7), e0134032.

Munro, G. C. 1945. Tragedy in bird life. *Elepaio* 5, 48–51.

Musser, G.G., and Carleton, M.D. 2005. Superfamily Muroidea. In D. E. Wilson and D. M. Reeder (Eds.), *Mammal Species of the World: A Taxonomic and Geographic Reference (3rd Ed).* (Pp. 894–1531), Baltimore and London: Johns Hopkins University Press.

Newton, K., McKown, M., Wolf, C., Gellerman, H., Coonan, T., Richards, D., Harvey, A. et al. 2016. Response of native species 10 years after rat eradication on Anacapa Island, California. *Journal of Fish and Wildlife Management* 7, 72–85.

Nolte, D.L., Bergman, D., and Townsend, J. 2003. Roof rat invasion of an urban desert island. In G. R. Singleton, L. A. Hinds, C. J. Krebs, and D. M. Spratt (Eds.), *Rats, Mice, and People: Rodent Biology and Management.* (Pp. 481–484), Canberra, Australia: Australian Centre for International Agriculture Research.

Pages, M., Chaval, Y., Herbreteau, V., Waengsothorn, S., Cosson, J.-F., Hugot, J.-P., Morand, S. et al. 2010. Revisiting the taxonomy of the Rattini tribe: A phylogeny-based delimitation of species boundaries. *BMC Evolutionary Biology* 10, 184.

Parkes, J., and Murphy, E. 2003. Management of introduced mammals in New Zealand. *New Zealand Journal of Zoology* 30, 335–359.

Pender, R.J., Shiels, A.B., Bialic-Murphy, L., and Mosher, S.M. 2013. Large-scale rodent control reduces pre—and post-dispersal seed predation of the endangered Hawaiian lobeliad, *Cyanea superba* subsp. *superba* (Campanulaceae). *Biological Invasions* 15, 213–223.

Perry, N.D., Hanson, B., Hobgood, W., Lopez, R.L., Okraska, C.R., Karem, K., Damon, I.K. et al. 2006. New invasive species in southern Florida: Gambian rat. *Journal of Mammalogy* 87, 262–264.

Peters, D., Schumacher, K., Schumaker, R., and Baigent, D. 2014. Goodnature automatic traps for vertebrate pest control: Field trials using new kill traps targeting animal pests in New Zealand. In R. Timm and J. O'Brien. (Eds.), *Proceedings of the 26th Vertebrate Pest Conference.* (Pp. 405–410), Davis, California: University of California.

Peterson, A.T., Papeş, M., Reynolds, M.G., Perry, N.D., Hanson, B., Regnery, R.L., Hutson, C.L., et al. 2006. Native-range ecology and invasive potential of *Cricetomys* in North America. *Journal of Mammalogy* 87, 427–432.

Pitt, W.C., Berentsen, A.R., Shiels, A.B., Volker, S.F., Eisemann, J.D., Wegmann, A., and Howald, G. 2015. Nontarget species mortality and the measurement of brodifacoum rodenticide residues after a rat (*Rattus rattus*) eradication on Palmyra Atoll, tropical Pacific. *Biological Conservation* 185, 36–46.

Pitt, W.C., Driscoll, L.C., and Sugihara, R.T. 2011c. Efficacy of Rodenticide Baits for the Control of Three Invasive Rodent Species in Hawaii. *Archives of Environmental Contamination and Toxicology* 60, 533–542.

Pitt, W.C., Driscoll, L.C., and VanderWerf, E.A. 2011b. A rat-resistant artificial nest box for cavity-nesting birds. *Human–Wildlife Interactions* 5, 100–105.

Pitt, W.C., Sugihara, R.T., Driscoll, L.C., and Vice, D.S. 2011a. Physical and behavioural abilities of commensal rodents related to the design of selective rodenticide bait stations. *International Journal of Pest Management* 57, 189–193.

Pitt, W.C. and Witmer, G.W. 2007. Invasive predators: A synthesis of the past, present, and future. In A. M. T. Elewa (Ed.), *Predation in Organisms—a Distinct Phenomenon.* (Pp. 265–293), Heidelberg, Germany: Springer Verlag.

Pott, M., Wegmann, A.S., Griffiths, R., Samaniego-Herrera, A., Cuthbert, R.J., Brooke, M.D.L., Pitt, W.C. et al. 2015. Improving the odds: Assessing bait availability before rodent eradication to aid in selecting bait application rates. *Biological Conservation* 185, 27–35.

Prakash, I. 1988. *Rodent Pest Management.* Boca Raton, Florida: CRC Press. 496 pp.

Proulx, G. 1999. *Mammal Trapping.* Sherwood Park, Alberta, Canada: Alphs Wildlife Research & Management, Ltd. 231 pp.

Roberts, M. 1991. Origin, dispersal routes, and geographic distribution of *Rattus exulans*, with special reference to New Zealand. *Pacific Science* 45, 123–130.

Robins, J.H., Hingston M., Matisoo-Smith, E. and Ross, H.A. 2007. Identifying *Rattus* species using mitochondrial DNA. *Molecular Ecology Notes* 7, 717–729.

Rufaut, C.G., and Gibbs, G.W. 2003. Response of a tree weta population (*Hemideina crassidens*) after eradication of the Polynesian rat from a New Zealand island. *Restoration Ecology* 11, 13–19.

Ruffino, L., Bourgeois, K., Vidal, E., Duhem, C., Paracuellos, M., Escribano, F., Sposimo, P., et al. 2009. Invasive rats and seabirds after 2,000 years of an unwanted coexistence on Mediterranean islands. *Biological Invasions* 11, 1631–1651.

Russell, J.C. 2011. Indirect effects of introduced predators on Seabird Islands. In C. P. H. Mulder, W. B. Anderson, D. R. Towns, and P. J. Bellingham (Eds.), *Seabird Islands: Ecology, Invasion, and Restoration.* (Pp. 261–279), New York, New York: Oxford University Press.

Savidge, J., Witmer, G., Jojola, S., Pierce, J., and Burke, P. 2012. Genetic evaluation of an attempted *Rattus rattus* eradication on Congo Cay, U.S. Virgin Islands, identifies importance of eradication units. *Biological Invasions* 14, 2343–2354.

Shapira, I., Brunton, D., Shanas, U., and Raubenheimer, D. 2013a. Conspecific attraction in invasive wild house mice: Effects of strain, sex and diet. *Animal Behaviour Science* 147, Shiels, A. B. 2010. Ecology and impacts of introduced rodents (*Rattus* spp. and *Mus musculus*) in the Hawaiian Islands. Ph.D. Diss. University of Hawai'i at Mānoa. 186–193.

Shapira, I., Shanas, U., Raubenheimer, D., Knapp, C., Alberts, S., and Brunton, D. 2013b. Laboratory rats as conspecific biocontrol agents for invasive Norway rats *R. novegicus*. *Biological Control* 66, 83–91.

Shiels, A.B. 2010. Ecology and impacts of introduced rodents (Rattus spp. and Mus musculus) in the Hawaiian Islands. Ph.D. Diss. University of Hawai'i at Mānoa, pp. 186–193.

Shiels, A.B. 2011. Frugivory by introduced black rats (*Rattus rattus*) promotes dispersal of invasive plant seeds. *Biological Invasions* 13, 781–792.

Shiels, A.B. and Drake, D.R. 2011. Are introduced rats (*Rattus rattus*) both seed predators and dispersers in Hawaii? *Biological Invasions* 13, 883–894.

Shiels, A.B. and Drake, D.R. 2015. Barriers to seed and seedling survival of once-common Hawaiian palms: The role of invasive rats and ungulates. *AoB PLANTS* 7, plv057, 1–10.

Shiels, A.B., Flores, C.A., Khamsing, A., Krushelnycky, P.D., Mosher, S.M., and Drake, D.R. 2013. Dietary niche differentiation among three species of invasive rodents (*Rattus rattus*, *R. exulans*, *Mus musculus*). *Biological Invasions* 15, 1037–1048.

Shiels, A.B. and Pitt, W.C. 2014 A review of invasive rodent diets (*Rattus* spp. and *Mus musculus*) on Pacific islands. *Proceedings of the 26th Vertebrate Pest Conference*, March 3–6, 2014, Waikaloa, Hawaii.

Shiels, A.B., Pitt, W.C., Sugihara, R.T., and Witmer, G.W. 2014. Biology and impacts of Pacific island invasive species. 11. *Rattus rattus*, the black rat (Rodentia: Muridae). *Pacific Science* 68, 145–184.

Spenneman, D.H.R. 1997. Distribution of rat species (*Rattus* spp.) on the atolls of the Marshall Islands: Past and present dispersal. *Atoll Research Bulletin* 446, 1–18.

St Clair, J.J.H. 2011. The impacts of invasive rodents on island invertebrates. *Biological Conservation* 144, 68–81.

Strecker, R.L. 1962. Economic relations. In T. I. Storer (Ed.), *Pacific Island Rat Ecology: Report on a Study Made on Ponape and Adjacent Islands 1955–1958, Bulletin 225.* (Pp. 200–218), Honolulu, Hawaii: Bernice P. Bishop Museum.

Sugihara, R.T. 1997. Abundance and diets of rats in two native Hawaiian forests. *Pacific Science* 51, 189–198.

Sugihara, R.T., Pank, L.F., Fellows, D.P., Hirata, D.N., Stott, R.S., Hilton, H.W., and Kaya, H. 1977. Non-crop habitat manipulation as a means of controlling rats and reducing damage to sugarcane. *Hawaiian Sugar Technical Report*, 83–90.

Thomas, P.J., Mineau, P., Shore, R.F., Champoux, L., Martin, P.A., Wilson, L.K., Fitzgerald, G., et al. 2011. Second generation anticoagulant rodenticides in predatory birds: Probabilistic characterization of toxic liver concentrations and implications for predatory bird populations in Canada. *Environment International* 37, 914–920.

Timm, R.M. 1994a. Norway rats. In S. E. Hygnstrom, R. M. Timm, and G. E. Larson, (Eds.), *Prevention and Control of Wildlife Damage*. (Pp. B-105–B-120), Lincoln, Nebraska: University of Nebraska Cooperative Extension.

Timm, R.M. 1994b. House mice. In S. E. Hygnstrom, R. M. Timm, and G. E. Larson (Eds.), *Prevention and Control of Wildlife Damage*. (Pp. B-31–B-46), Lincoln, Nebraska: University of Nebraska Cooperative Extension.

Tobin, M.E. 1994. Polynesian rats. In S. E. Hygnstrom, R. M. Timm, and G. E. Larson, (Eds.), *Prevention and Control of Wildlife Damage*. (Pp. B-121–B-124), Lincoln, Nebraska: University of Nebraska Cooperative Extension.

Tobin, M.E. & Sugihara, R.T. 1992. Abundance and habitat relationships of rats in Hawaiian sugarcane fields. *Journal of Wildlife Management* 56, 815–821.

Tomich, P.Q. 1986. *Mammals in Hawaii*, 2nd Ed. Honolulu, Hawaii: Bishop Museum Press. 375 pp.

Towns, D.R., Byrd, G.V., Jones, H.P., Rauzon, M.J., Russell, J.C. and Wilcox, C. 2011. Impacts of introduced predators on seabirds. In C. P. H. Mulder, W. B. Anderson, D. R. Towns, and P. J. Bellingham (Eds.), *Seabird Islands: Ecology, Invasion, and Restoration*. (Pp. 56–90), New York, New York: Oxford University Press.

Towns, D.R., Wardle, D.A., Mulder, C.P.H., Yeates, G.W., Fitzgerald, B.M., Parrish, G.R., Bellingham, P.J., et al. 2009. Predation of seabirds by invasive rats: Multiple indirect consequences for invertebrate communities. *Oikos* 118, 420–430.

Traveset, A., Nogales, M., Alcover, J.A., Delgado, J.D., Lopez-Darias, M., Godoy, D., Igual, J.M., et al. 2009. A review of the effects of alien rodents in the Balearic (western Mediterranean Sea) and Canary Islands (eastern Atlantic Ocean). *Biological Invasions* 11, 1653–1670.

Veitch, C. and Clout, M. (Eds.). 2002. *Turning the Tide: The Eradication of Invasive Species. IUCN SSC Invasive Species Specialist Group*. Gland, Switzerland and Cambridge, U.K.: IUCN.

Villa-Cornejo, B., Lopez-Forment, W., Cornejo, M.V., and Prescott, C.V. 1998. Not all Sigmodontine rodents in the sugarcane fields in coastal Veracruz, Mexico are pests. In R. O. Baker and A. C. Crabb (Eds.), *Proceedings of the 18th Vertebrate Pest Conference* (Pp. 236–241), Davis, California: University of California.

Wang, Q-P., Lai, D.H., Zhu, X.U., Chen, X-G., and Lun, Z-R. 2008. Human angiostrongyliasis. *Lancet Infectious Disease* 8, 621–630.

Wilmshurst, J.M., Hunt, T.L., Lipo, C.P., and Anderson, A.J. 2011. High-precision radiocarbon dating shows recent and rapid ignition human colonization of East Polynesia. *Proceedings of the National Academy of Science* 108, 1815–1820.

Wilson, D.E. and Reeder, D.M. 2005. *Mammal Species of the World*. 3rd Ed. Baltimore, Maryland: Johns Hopkins University Press. 2, 142 pp.

Witmer, G.W. 2004. Rodent ecology and plague in North America. In: *Proceedings of the 19th International Congress of Zoology*. (Pp. 154–156), Beijing, China: China Zoological Society.

Witmer, G. 2005. Wildlife population monitoring: Some practical considerations. *Wildlife Research* 32, 259–263.

Witmer, G. 2007. The ecology of vertebrate pests and integrated pest management (IPM). In M. Kogan and P. Jepson (Eds.), *Perspectives in Ecological Theory and Integrated Pest Management*. (Pp. 393–410), Cambridge, UK: Cambridge University Press.

Witmer, G., Boyd, F., and Campbell, E. 1998. Rat management for endangered species protection in the U.S. Virgin Islands. In R. Baker and C. Crabb (Eds.), *Proceedings of the 18th Vertebrate Pest Conference*. (Pp. 281–286), Davis, California: University of California.

Witmer, G.W., Boyd, F., and Hillis-Starr, Z. 2007a. The successful eradication of introduced roof rats (*Rattus rattus*) from Buck Island using diphacinone, followed by an irruption of house mice (*Mus musculus*). *Wildlife Research* 34, 108–115.

Witmer, G., and Eisemann, J.D. 2007. Rodenticide Use in Rodent Management in the United States: An Overview. In D. L. Nolte, W. M. Arjo, and D. H. Stalman (Eds.), *Proceedings of the 12th Wildlife Damage Management Conference* (Pp. 114–118), Corpus Christi, Texas: 12th Wildlife Damage Management Conference.

Witmer, G., Eisemann, J., and Howald, G. 2007b. The use of rodenticides for conservation efforts. In D. L. Nolte, W. M. Arjo, and D. H. Stalman (Eds.), *Proceedings of the 12th Wildlife Damage Management Conference*. (Pp. 160–166), Corpus Christi, Texas: 12th Wildlife Damage Management Conference.

Witmer, G.W., Fall, M., and Fiedler, L. 1995. Rodent control, research, and technology transfer. In J. Bissonette and P. Krausman (Eds.), *Integrating People and Wildlife for A Sustainable Future: Proceedings of the First International Wildlife Management Congress*. (Pp. 693–697), Bethesda, Maryland: The Wildlife Society.

Witmer, G.W., and Hall, P. 2011. Attempting to eradicate invasive Gambian giant pouched rats (*Cricetomys gambianus*) in the United States: Lessons learned. In C. Veitch, M. Clout and D. Towns (Eds.), *Island Invasives: Eradication and Management*. (Pp. 135–138), Gland, Switzerland: International Union for Conservation of Nature (IUCN).

Witmer, G., and Jojola, S. 2006. What's up with house mice?—A review. In R. M. Timm and J. M. O'Brien (Eds.), *Proceedings of the 22nd Vertebrate Pest Conference* (Pp.124–130), Davis, California: University of California.

Witmer, G.W., Keirn, G., Hawley, N., Martin, C., and Reaser, K. 2009. Human dimensions of invasive vertebrate species management. In J. Boulanger (Ed.), *Proceedings of the 13th Wildlife Damage Management Conference* 13, 100–105.

Witmer, G., and Moulton, R. 2014. Improving invasive house mice control and eradication strategies via more effective rodenticides. In R. Timm and J. O'Brien. (Eds.), *Proceedings of the 26th Vertebrate Pest Conference*. (Pp. 67–72), Davis, California: University of California.

Witmer, G.W., Pierce, J., and Pitt, W.C. 2011. Eradication of invasive rodents on islands of the United States. In C. Veitch, M. Clout and D. Towns (Eds.), *Island Invasives: Eradication and Management*. (Pp. 135–138), Gland, Switzerland: International Union for Conservation of Nature (IUCN).

Witmer, G., Sheffels, T., and Kendrot, S. 2012a. The introduction, impacts, and management of a large, invasive, aquatic rodent in the United States. In D. Abreau and S. deBorbon (Eds.). *Marshes: Ecology, Management, and Conservation*. (Pp. 49–89), New York, New York: Nova Science Publishers, Inc.

Witmer, G., and Singleton, G. 2010. Sustained Agriculture: The Need to Manage Rodent Damage. In C. Felix (Ed.), *Agricultural Production*. (Pp. 1–38), New York, New York: Nova Science Publishers, Inc.

Witmer, G., Snow, N., and Burke, P. 2010a. Potential attractants for detecting and removing invading Gambian giant pouched rats (*Cricetomys gambianus*). *Pest Management Science* 66, 412–416.

Witmer, G., Snow, N., and Burke, P. 2010b. Evaluating commercially available rodenticide baits for invasive Gambian giant pouched rats (*Cricetomys gambianus*). *Crop Protection* 29, 1011–1014.

Witmer, G., Snow, N., Moulton, R., and Swartz, J. 2012b. An assessment of seedling damage by wild house mice and wild deer mice. *Canadian Journal of Forest Research* 42, 1168–1172.

11 Wild Pigs

John J. Mayer and James C. Beasley

CONTENTS

INTRODUCTION

Nonnative wild pigs (*Sus scrofa*) are the most abundant and widespread medium- to large-sized invasive vertebrate species presently found in the United States (Mayer and Brisbin 2009). That fact combined with their destructive nature and pervasive disease risk potential makes these invasive animals a priority national concern for aggressive and proactive control in this country. In addition, wild pigs encompass a distinctive combination of issues and problems that are unique among invasive species found in the United States. As such, these animals represent a singularly complicated national challenge for which there is no viable solution at the moment.

Wild pigs have been present in both continental and insular portions of the United States for centuries; however, beginning in the early 1990s, populations began to dramatically increase in both numbers and distribution. This rapid and substantial increase was unexpected given generally stable population sizes for the rest of the twentieth century predating the 1990s and has created a critical management need to control expanding populations. As with the initial presence of these animals in the United States, most of the recent increases in wild pig populations are attributable to an anthropogenic origin, largely illegal translocations of animals (Gipson et al. 1998; Mayer and Brisbin 2009).

Concurrent with the aforementioned increase has been an upsurge in the impacts invasive pigs have on both natural and human environments. Not only has the scale and frequency of these impacts increased, but also the types of observed damage being caused by wild pigs. Such damage has been recently estimated to cost the United States billions of dollars annually (Pimentel 2007).

In spite of the generally accepted invasive categorization of wild pigs, the situation with this species in the United States is complicated by the fact that these animals are also a widely popular game species. Such strongly polarized views (i.e., unwanted destructive pest vs. desirable huntable game resource) are not typical for an invasive species issue. Because of this ongoing debate, even with the negatives greatly outweighing the positives, the circumstances associated with invasive wild pigs represent a complex dilemma that will be difficult to solve on a national scale.

HISTORY OF THE INTRODUCTION AND SPREAD

As with all of the Suidae, wild pigs belonging to the species *Sus scrofa* (Figure 11.1) are not native to the Western Hemisphere. Their presence in the United States and other countries in the Americas is attributable to both accidental (e.g., escaped domestic pigs that have gone wild) and intentional (e.g., Eurasian wild boar released to provide a new huntable game species) introductions by mankind (Figure 11.2). The initial introductions of pigs into the United States date as far back as the fourth century AD in the Hawaiian Islands and the early to mid-1500s on the mainland (Mayer and Brisbin 2008, 2009; Maly et al. 2013). These introductions occurred

FIGURE 11.1   Appearances of a typical invasive wild pig (*Sus scrofa*) that is currently found in the United States. Top image is of a mature male, and the bottom image is of a mature female with offspring.

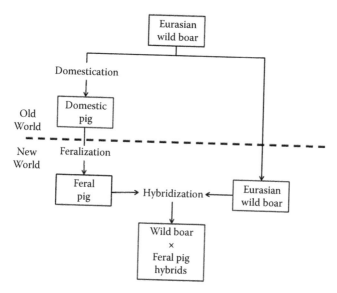

FIGURE 11.2 Simple taxonomic phylogeny of wild *Sus scrofa* being introduced to the New World. The three categories at the bottom represent the current types of wild pigs belonging to this species that are found in the United States.

concurrently with the exploration and early colonization of these areas that would later become parts of the United States. Thus, colonization occurred along two main importation routes, first from Southeast Asia through Oceania into the Hawaiian Archipelago in northern Polynesia, and second from western Europe across the Atlantic Ocean into North America (Mayer and Brisbin 2008, 2009).

The wild pigs found in the United States are typically assigned to the species *Sus scrofa*. However, based on anecdotal morphological evidence (i.e., reported facial warts) and the origin from Southeast Asia through Melanesia and finally into Polynesia, Groves (1983) posited that wild pigs on the Hawaiian Islands would likely be either pure Sulawesi warty pigs (*Sus celebensis*) or hybrids between *Sus scrofa* and *Sus celebensis* rather than pure *Sus scrofa*. In contrast to that theory, Larson et al. (2005, 2007) used mitochondrial DNA and morphological data to determine that the pigs translocated throughout Polynesia, including into the Hawaii Archipelago, were *Sus scrofa* and not *Sus celebensis* or hybrids, as suggested by Groves (1983). Irrespective of the source, those original Polynesian pigs in Hawaii have subsequently hybridized with the larger European domestic pigs that were later introduced throughout the Pacific basin (Allen et al. 2001; Maly et al. 2013).

Counter to the aforementioned importation routes, one theory (Quinn 1970) has posited that an archaic introduction occurred across the Bering land bridge during Altithermal time. This theory was based on the purported presence of Old World *Sus* skeletal material in pre-Columbia archaeological deposits in Arkansas and Texas. However, most archaeozoologists discount this theory, attributing such presence to the inadvertent incorporation of recent materials into older paleological assemblages (Mayer and Brisbin 2008, 2009).

As noted previously, the first introduction of *Sus scrofa* into the United States occurred with the Polynesian discovery and colonization of the Hawaii Islands. The reported timing of that initial colonization varies from 300 to 400 AD (Kirch 1982; Maly et al. 2013) up to 1219–1266 AD (Wilmshurst et al. 2011). Pigs were among the first animals introduced to these islands during the initial Polynesian settlement voyages, with pig skeletal remains being found at the two earliest known habitation sites in Hawaii (Kirch 1982). The pigs belonging to these ancient residents were loosely controlled and allowed to be free-roaming (Maly et al. 2013), and early free-ranging stocks formed the initial basis of the wild pig populations found in the Hawaiian Islands. Populations subsequently expanded, and wild pigs were abundant throughout the archipelago at the time of the first European contact. Aside from some evidence indicating a visit by Spanish sailors in the early 1600s, the first documented European arrival to the Hawaiian Islands and introduction of European domestic pigs is attributed to Captain James Cook's expedition in 1778. With the passage of time, more domestic pigs were introduced and free-ranged by European settlers to this archipelago (Mayer and Brisbin 2008, 2009). A recent study using mtDNA and MC1R sequencing, conducted on four of the islands within the archipelago, found that the current Hawaiian wild pigs are a mix of animals of both the Polynesian and European introductions, with 70% of those animals sampled being descendants from the original Polynesian introduction (Linderholm et al. 2016).

The first importation of pigs into North America occurred in the eleventh century with the Norse settlements in southern Greenland; however, the presence of those pigs declined and completely disappeared soon thereafter. In spite of evidence that these animals were free-ranged, no indications exist of any wild populations becoming established (Nelson et al. 2012). The first sustaining introduction of pigs to the New World occurred with Columbus' second voyage in 1493. His expedition obtained eight domestic pigs from the Canary Islands and transported these animals to the Caribbean. The Spanish supply ships that followed Columbus' second voyage brought more pigs from the Canaries. These animals were subsequently dispersed throughout the expanding Spanish colonies in the West Indies and released into open range to forage. This environment had abundant food, negligible competition, and almost no predators, which enabled the pigs to multiply rapidly. As early as 1500, wild pigs had begun depredating colonists' crops of maize and sugarcane; in a short time, these pigs became so destructive the Spanish crown issued a proclamation in 1506 directing colonists to reduce the numbers of pigs to control the damage they were causing (Zadik 2005; Mayer and Brisbin 2008, 2009).

When the Spanish began to mount expeditions to the mainland, they took with them herds of pigs (i.e., some purchased domestic stock and others live-captured feral animals) to supply their men with ambulatory larders of fresh meat. During these journeys through the New World, the pigs either escaped from or were released by the expedition, and in many instances went wild. These initial expeditions were followed by other European expeditions and colonists, who also brought domestic pigs with them to the North American mainland. By the mid-1500s, feral populations of pigs were already established in what would later become the present-day southwestern U.S. states and northeastern Florida (Zadik 2005; Mayer and Brisbin 2008, 2009).

Domestic pigs, including descendants of the aforementioned introductions as well as newly imported stock, were widely introduced in the European colonies throughout what would later become the eastern and southwestern United States. Following a long-standing husbandry practice for rearing domestic livestock in Europe, these pigs were turned loose into the forests and woodlands to forage for themselves. Manpower was in short supply in the colonies, being needed to clear land and build structures; so, unlike in the Old World, where such free-ranged domestic stock was tended by a herdsman, these pigs were also turned loose to fend for themselves. In the early exploration and colonization of the United States, this free-ranging practice led to an increasing number of feral populations of pigs (Figure 11.3). Typically, however, these free-ranging pigs were considered to be private property, and not wild animals, in most locations. This was the era of hog rights, where these animals were ear notched to identify their ownership. As a generally accepted piece of the Colonial landscape, there were only minimal reports of damage from pigs during this time period (Mayer and Brisbin 2009).

By the late 1800s, populations of wild pigs were widespread in the United States. Most of these were composed of free-ranged domestic pigs that had gone wild/feral. Beginning in 1890, pure Eurasian wild boar were brought to the United States and released into fenced hunting preserves in various locations (e.g., Corbin's Park, NH—1890; Litchfield Park, NY—1902; Hooper Bald, NC—1912; St. Charles Ranch, TX—1930; Powderhorn Ranch, TX—1939; Edward's Plateau, TX—1940s) to provide a new huntable big game animal. However, these animals escaped out of every one of those early enclosures and established populations in the wild. These now free-ranging wild boars expanded their range and hybridized with feral pigs that were already present in most of these areas (Mayer and Brisbin 2008, 2009).

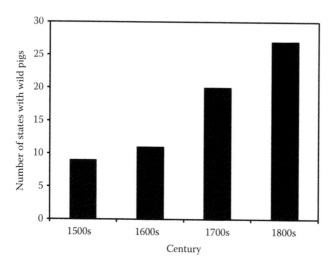

FIGURE 11.3 Collective number of what would later become U.S. states where wild pigs (*Sus scrofa*) had been reported to be present during each century from the 1500s through the 1800s.

From the 1940s through the 1980s, the populations of wild pigs in this country remained relatively unchanged. However, during that same time frame, several U.S. states began to actively promote wild pigs as a desirable game species. This included changes in hunting regulations as well as the active stocking of wild pigs in some states. The promotion of wild pigs as desirable game was rapidly spread by the media (e.g., via magazines and TV/cable) and private individuals and organizations also began active stocking of wild pigs into new areas.

For most of the twentieth century, the abundance and distribution of wild pigs in U.S. states remained fairly stable (Figure 11.4); however, after 1989, wild pigs underwent a dramatic increase in both range and population size. A small component of this growth was natural expansion, but the primary factors were anthropogenic in origin. In particular, the promotion of pigs as game animals played a substantial role in their expansion throughout the United States, with wild pigs becoming the second-most popular big game animal in United States, second only to white-tailed deer (*Odocoileus virginianus*) based on numbers harvested each year by the early 2000s (Kaufman et al. 2004). This popularity led to numerous clandestine illegal translocations and releases, especially in northern states where wild pigs had heretofore not existed in large numbers. Such illegal translocations continue today, despite a growing awareness of the negative impacts of invasive wild pigs. Hundreds of commercial fenced shooting operations opened up across the country in the latter half of the twentieth century, trying to further capitalize on this popularity. Every state that had high-fence operations had at least one incident of escaped animals, some dating back to the 1960s. Further, the expanding wild pig population in Canada even spread down into North Dakota in the early 2000s. The numbers of wild pigs also increased, even within their established range, going from 1 to 2 million in the 1980s up to an estimated 6.3 million in 2014 (Mayer 2014). By the second decade of

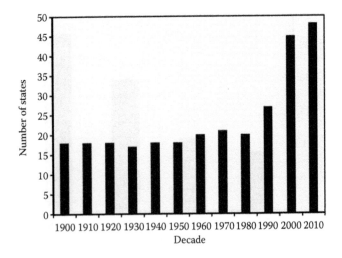

FIGURE 11.4 U.S. states reporting the presence of wild pigs (*Sus scrofa*) during each decade of the twentieth century. Such a "presence" does not imply an established or prolonged occurrence in some instances.

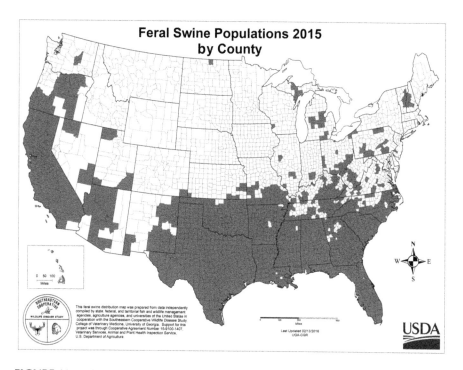

FIGURE 11.5 Recent range of wild pigs (*Sus scrofa*) in the United States. (Map courtesy of the Southeastern Cooperative Wildlife Disease Study at the University of Georgia.)

the twenty-first century, a total of 48 states had reported at least a temporary presence of wild pigs. By February 2015, wild pigs were reported to be established in 35 states (Figure 11.5) (J. L. Corn, SCWDS, pers. comm.).

The basis for the successful establishment of wild pigs in new areas where they have expanded naturally or been introduced by man is largely due to their highly adaptable biological makeup. Pigs are ecological generalists and opportunistic omnivores that can live in most nonpolar habitats. Wild pigs also have few effective non-human predators in most locations where they are found in the United States. Last, pigs have the highest reproductive potential for mammals their size or larger, producing several offspring up to two times a year under optimal conditions (Dzieciolowski et al. 1992; Mayer and Brisbin 2009). Given year-round access to food, water, shade, and escape cover (Mayer and Brisbin 2009), invasive pigs can live almost anywhere in the lower 48 states, Hawaii, and the temperate coastal/maritime regions of Alaska.

SCOPE OF THE IMPACTS

The recent population increase in invasive wild pigs throughout the United States resulted in a nationwide escalation in the damages caused by these animals. This increase has collectively included upsurges in the locations, extent, and diversity of these impacts. The potential for environmental and economic impacts from wild pigs in any given area is high (Coblentz and Bouska 2004; Massei and Genov 2004), and such damage is typically observed shortly after wild pigs are initially present in an area. Most of this damage is either directly or indirectly the result of their foraging activities; however, other impacts not specifically related to their feeding behavior do occur (e.g., wallowing/rubbing, scent marking, collision with vehicles, attacks on humans, some types of property damage) (Mayer and Brisbin 2009; Beasley et al. 2013).

Recent summaries of the impacts caused by invasive wild pigs were reported by Campbell and Long (2009), Mayer and Brisbin (2009), West et al. (2009), Barrios-Garcia and Ballari (2012), Bevins et al. (2014), USDA APHIS (2015), and Keiter and Beasley (2017). However, the extent and implications of many of these impacts are largely unknown and remain an important area of needed research. In general, wild pigs cause extensive damage to both the natural and human environments, with the magnitude being rivaled only by the diversity of such impacts. The following paragraphs provide summaries of the different types of damage that invasive wild pigs cause.

Rooting: Wild pigs root year round in search of forage resources at or below the surficial soil horizon; such rooting can be extensive (Figure 11.6). Rooting is the most common type of wild pig damage observed, affecting both the natural and the human environments. Excavation of the soil column by such foraging behavior breaks up and loosens the soil, which can lead to destabilization, erosion, and slumping of surface soils as well as sedimentation in downgradient habitats (e.g., streams or wetlands) (Mayer and Brisbin 2009). Mechanical disturbance and the reduction of ground vegetative cover and leaf litter created by rooting behavior can also cause alteration or severe destruction of habitats of small mammals in the ground-story layer (Singer et al. 1984) as well as impact archaeological/cultural resources through mechanical disruption of soil profiles (Engeman et al. 2013). Wild pig rooting also

FIGURE 11.6   Given a localized forage resource that wild pigs are targeting, the rooting in some areas can be extensive. The above image is from the Aransas National Wildlife Refuge in coastal Texas.

modifies the chemistry and nutrient cycling within the soil column, accelerating the leaching of Ca, P, Zn, Cu, and Mg from the leaf litter and upper soil horizons (Singer et al. 1984). The aggressive widespread predation/removal of earthworms (e.g., *Lumbricus terrestris*) by pigs, a preferred food item, can interfere with the decomposition cycle in soils (Tisdell 1982). Combined with wallowing behavior by pigs, rooting can negatively impact water quality in wetland habitats (Tate 1984). Subsurface foraging by pigs can cause significant damage to turf or grassed areas such as lawns, parks, sporting venues, and cemeteries (Mayer and Brisbin 2009). Rooting in forested habitats can damage tree roots and increase sprouting and root suckers in certain hardwood species (Huff 1977; Lacki and Lancia 1986). Severely rooted areas in croplands can cause damage to or impair agricultural and farm equipment and machinery being operated there (Nunley 1999).

Agricultural damage: In the United States as well as the rest of the species global range, wild pigs are well known for their depredation of agricultural crops. Estimated economic losses to crop damage by wild pigs are extensive, costing in the tens of millions of dollars annually in individual states where estimates exist (Table 11.1). This damage is the result of foraging activities by pigs, which entails impacts from direct feeding along with damage resulting from the related trampling and rooting associated with pig foraging behavior. Agricultural depredation caused by wild pigs in the United States entails a variety of crops, including grains/cereals (e.g., wheat, sorghum, barley, hay, rye, oats, millet, maize/corn, and rice), vegetables (e.g., various potatoes, yams, squash, turnips, rutabagas, beets, cassava, lettuce, cabbage, beans, peas, soybeans, and artichokes), fruits (e.g., pumpkins, grapes, blueberries, pineapples, avocadoes, bananas, apples, various citrus species, watermelons, cantaloupes, and coconuts), and other crops (e.g., cotton, linseed, sunflower, peanuts, almonds, oilseed/rapeseed, groundnuts, pecans, clover, thyme, basil, sugarcane, and turf/sod/pasturage) (Mayer and Brisbin 2009).

Forestry/timber damage: Wild pigs affect various forestry and timber resources in the United States, the most serious and well-known form of damage being depredation of planted pine seedlings. Documented initially in the early to mid-1900s in the southeast, damages typically consist of wild pigs pulling up recently planted pine seedlings and then chewing on the plant's rootstock and lower stem. That plant is then dropped to the ground and the pig moves on to the next seedling, systematically moving from one seedling to the next in one planted row after another in pine regeneration plots, ultimately causing extensive damage to the crop (Lucas 1977). The rate of such depredation has been reported as six seedlings per minute per animal, with sustained damage of as many as 400 to 1000 seedlings per day (Hopkins 1948; Wakely 1954). Depredation primarily has been documented for the grass stage of planted longleaf pine (*Pinus palustris*), but has been observed with other pine species as well. During the mid-1900s, in a period of intensive planting of longleaf and slash pine by the U.S. Forest Service, entire pine plantations in Alabama, Florida, Louisiana, and Mississippi were reported to be complete failures because of wild pig depredations (Lucas 1977; Lipscomb 1989). Wild pigs have also been found to dig up and consume the rootstock of planted hardwood seedlings (Mayer et al. 2000). Other forms of damage wild pigs cause to forestry/timber resources include girdling of mature pine trees through rubbing, damage to the lateral roots of mature pine

TABLE 11.1

Examples of Dollar Amounts of Damage Caused by Wild Pigs in the United States

Damage Amount (in US$)	Location or Area Covered	Time Frame	Impacted Resource(s)	References
$800 million	United States—nationwide	Annual	Environmental and crop damage	Pimentel et al. (2005)
$1.5 billion	United States—nationwide	Annual	Crop damage and control costs	Pimentel (2007)
$36 million	United States—nationwide	Annual	Wild pig–vehicle collisions/accidents	Mayer and Johns (2007)
$190 million	Alabama, Arkansas, Florida, Georgia, Louisiana, Mississippi, Missouri, North Carolina, South Carolina, and Texas	2014	Annual losses to corn, soybean, wheat, rice, peanut and sorghum crops	Anderson et al. (2016)
~$75 million	Alabama—statewide	2009	Damage to crops	Shi et al. (2012)
>$74 million	Louisiana—statewide	2013	Agricultural production losses and other nonproduction damage	Tanger et al. (2015)
~$12 million	Missouri—statewide	Single disease outbreak, 45 days in length	Modeled outbreak of foot-and-mouth disease being transmitted from wild pigs to Missouri's livestock	Cozzens et al. (2010)
$115.3 million	South Carolina—statewide	Annual	Total statewide losses to wild pig damage to agriculture, livestock, and timber resources	S. L. Rodriguez, G. Yarrow, and M. Bolick, Clemson University, unpublished data
$150.6 million	89 of the 159 counties in Georgia	2014	Crop damage and noncrop damage	Mengak (2016)
>$2 million	Central Florida	Annual	Pasture production losses in managed rangeland	Bankovich et al. (2016)
>$52 million	7 of the 8 ecological regions in Texas	Annual	Damage to Texas agriculture	Higginbotham et al. (2008)
$1 to 4 million	Single location in Florida	One incident	Rooting/foraging damage to 5.4 ha of exposed basin marsh	Engeman et al. (2003)
$16 million	Jacksonville International Airport, Florida	One incident	Single F-16 fighter jet collided with two wild pigs attempting to cross a runway causing the plane to crash; jet was a complete financial loss	Anon. (1988)

trees by rooting and chewing, and damage to the bark of trees by tusking (i.e., scent marking with the tusk glands) (Conley et al. 1972; Lucas 1977; Stegeman 1938). As many as 90% of the pine trees at a nature preserve in California were impacted by wild pig girdling (Chipping 1993). Wild pig damage to forestry resources has been documented to cause problems with the control of insect pests, invasive species, and pathogens (Kliejunas and Ko 1976; Diong 1982; Lease et al. 1996). Last, wild pigs can also cause both serious and long-lasting damage to forest understory and herbaceous communities (Bratton 1975; Howe et al. 1981).

Competition with native animals: As an invasive species, one of the concerns about wild pigs has been competition with native animal species, both game and nongame, for available resources. Primarily based on dietary overlap of the annual mast crop, wild pigs are thought to compete with white-tailed deer, wild turkeys (*Meleagris gallopavo*), squirrels, black bears (*Ursus americanus*), and other native species that feed on mast (Conley et al. 1972). The existence of general dietary overlap has also included the inferred or suggested forage resource competition between wild pigs and collared peccaries (*Pecari tajacu*), striped skunks (*Mephitis mephitis*), Virginia opossums (*Didelphis virginiana*), red (*Vulpes vulpes*) and gray (*Urocyon cinereoargenteus*) foxes, raccoons (*Procyon lotor*), bobcats (*Lynx rufus*), muskrats (*Ondatra zibethicus*), nutria (*Myocastor coypus*), eastern cottontail (*Sylvilagus floridanus*), swamp rabbits (*Sylvilagus aquaticus*), hawks, owls, sandhill cranes (*Grus canadensis*), and waterfowl (Mayer and Brisbin 2009). Although such dietary overlap is generally accepted (e.g., as with the annual mast crop), quantification of the assumed competition has been challenging. If such competitive impacts did exist, the effects would be worse during years of poor mast crops (Tate 1984). In fact, Yarrow (1987) suggested that in such times of low mast availability, competition from wild pigs for scarce resources may limit white-tailed deer populations. Finally, being scavengers, wild pigs will also compete with coyotes, vultures, and eagles for carrion (Zappacosta 2005; Turner et al. 2017).

Wild pigs also may competitively exclude species from preferred foraging areas. For example, based on observations of interspecific aggression, Hellgren (1993) reported intense competition for space between sympatric populations of wild pigs and collared peccaries. However, Ilse and Hellgren (1995) and Gabor et al. (2001) found that diet and habitat partitioning allowed for the coexistence of sympatric populations of these two species in south Texas. White-tailed deer reportedly will avoid feeders, food plots, or natural foraging areas that are utilized by wild pigs (Tolleson et al. 1995). In addition, wild pigs have even been reported to exclude deer from feeding on acorns (Taylor and Hellgren 1997). Keever (2014) found that deer avoid wild pigs and may be spatially and temporally partitioning forage resources. Last, wild pig use of salt/mineral blocks has been demonstrated to exclude some native game species (Mayer and Brisbin 2009).

Competition with domestic livestock: Similar to competition with native animals, wild pigs will also compete with domestic livestock (e.g., range cattle) for forage resources due to dietary overlap, which would include feed, pasturage, and salt/mineral blocks (Mayer and Brisbin 2009). Competition between wild pigs and cattle has also been observed at range feeders (Carswell 2011), representing a potential means of disease transmission between these two species (see below).

Depredation of native plants: Based on their dietary preferences, wild pigs are primarily herbivorous and will forage on a diversity of native vegetation. Such impacts vary by location and the seasonal availability of floral species. Of particular concern is depredation of protected or listed floral species. For example, wild pigs have been observed to either uproot or consume one-third of wildflowers listed as occurring in the Great Smoky Mountains National Park. Further, these invasive animals destroyed an entire population of Turk's cap lily (*Lilium superbum*) in the park (Bratton 1974). In Marin County, California, wild pig rooting severely impacted colonies of fairy slipper orchids (*Calypso bulbosa*) (Chipping 1993). Last, one of the major factors that led to the extinction of four species of Hawaiian flowering lobelioids (i.e., *Cyanea linearifolia, C. parrifolia, C. purpurellifolia*, and *C. pycnocarpa*) was extensive rooting/foraging by wild pigs (Beacham 1997).

Predation of native animals: In spite of being mostly herbivorous, wild pigs will also prey on a variety of native animal species. Most often, this predation is directed at either young individuals or less-mobile species. This can include the egg, larval/immature stage, and adult forms of both invertebrate and vertebrate animal taxa. The diverse invertebrate taxa that have been identified as being consumed by invasive pigs include various species of annelids, arthropods, arachnids, crustaceans, insects, myriapods, mollusks, and nematods. Among vertebrates, wild pig prey includes species of fish, amphibians, reptiles, birds, and mammals. Wild pigs will readily consume the eggs and hatchlings of sea turtles, crocodilians, and lizards, and will prey on eggs through adult-age classes of ground-nesting game birds (Mayer and Brisbin 2009). The declines of both the Hawaiian goose or Nēnē (*Branta sandvicensis*) and the dark-rumped petrel (*Pterodroma phaeopygia*) in Hawaii were attributed in part to nest predation by wild pigs (Bryan 1937; Smith 1952; Lever 1985). A variety of native mammal species have also been documented as prey of wild pigs, including shrews, moles, armadillos, rodents, and deer fawns. In addition to live prey, wild pigs will readily feed on carrion (Mayer and Brisbin 2009; Turner et al. 2017).

Predation of/attacks on domestic animals: Both predatory and defensive attacks on domestic animals (i.e., pets and livestock) by wild pigs have been reported in the United States. Similar to the impact on native vertebrates, wild pigs can be voracious predators of domestic livestock, including killing and eating goats/kids, sheep/lambs, cattle/calves, foals, and chickens (Mayer and Brisbin 2009). Annual impacts can reach thousands of head of stock predated in states like California and Texas (Beach 1993). Nationally, there have also been reports of pet dogs being attacked and even killed by wild pigs. This has included dogs that were running loose/off leash as well as dogs that were being walked on a leash (Mayer 2013; USDA APHIS 2015).

Disease transmission: Wild pigs are susceptible to a wide variety of pathogens, including some zoonotic bacterial and viral infections (Nettles 1989; Davis 1993; Davidson 2006; Hutton et al. 2006). As such, invasive wild pigs can serve as disease reservoirs/vectors, possibly impacting native wildlife, domestic animals, and humans alike. Some of these pathogens (e.g., *E. coli*) also have the potential to contaminate water supplies as well as agricultural crops (e.g., the outbreak in California spinach in 2006) (Jay et al. 2007). In particular, disease transmission represents a major threat to livestock industry nationally, primarily due to risks associated with brucellosis and the pseudorabies virus. Although these two pathogens have been eradicated in the domestic stock population nationally, both are endemic in wild pig populations in a number of states. Either direct or indirect (e.g., foraging in the same range feeder) contact between wild pigs and domestic livestock has been observed to occur in a number areas, so circumstances associated with the likely transmission of such pathogens do exist (Mayer and Brisbin 2009; USDA APHIS 2015). Last, foreign animal diseases, such as classical swine fever, African swine fever, and foot-and-mouth disease, represent a serious concern. All of these diseases are currently found in the native portion of the *S. scrofa* global range (e.g., Eurasia and Africa), and their inadvertent introduction into the United States (e.g., through the importation of infected food products) could cost this country billions of dollars in damages to the U.S. livestock industry (Paarlberg et al. 2002; Alexandrov et al. 2013; EFSA 2014).

Property damage: Wild pigs can cause a variety of damage to property within anthropogenic ecosystems. Many impacts result from foraging activities, but unquestionably the most frequently observed type of property damage is impacts to fences and gates caused when pigs try to breach these barriers. In livestock areas, such damage compromises the ability of fencing to contain domestic stock and to exclude predators. In suburban areas, wild pigs have been reported to root up lawn irrigation and sprinkling systems, and then break the excavated polyvinyl chloride (PVC) lines to access water for drinking. The natural curiosity of these animals while foraging leads to wild pigs accidentally knocking over and breaking personal property (e.g., lawn furniture, glass-top patio tables, flower pots/planters, scientific field instruments) (Mayer and Brisbin 2009; West et al. 2009; J. J. Mayer, unpublished data).

Wild pig–vehicle collisions: One of the impacts that has recently increased with the population and range of wild pigs in the United States has been collision with vehicles. Such accidents occur year round, being highest in the fall and winter months and at night. These collisions with vehicles involve both sexes and all age classes of wild pigs with the exception of neonates and very small piglets (<2 kg in body mass). Pig–vehicle collisions are poorly documented in the United States, although the number of wild pig roadkills in an area is positively correlated to local population numbers (Mayer and Johns 2007). Unlike collisions with most other ungulates, about 10% of collisions with pigs involve multiple animals and as many as 18 wild pigs have been involved in a single collision (Mayer and Brisbin 2009). Wild pig–vehicle collisions occur more frequently in areas of preferred habitat for this species, specifically in areas close to streams (Beasley et al. 2013). The consequences of such accidents can be serious, with the number of human fatalities increasing nationally (Mayer and Brisbin 2009). These accidents normally involve less than 5% of a wild pig population annually (Mayer and Johns 2007).

Attacks on humans: Wild pig attacks on humans are rare, but do occur. These attacks happen under both hunting and nonhunting circumstances, and can potentially have serious consequences, including fatalities among the human victims. Four human fatalities associated with an attack by a wild pig in the United States have been reported, the most recent of which occurred in 1996. Most human victims were adult males traveling on foot and alone. The wild pigs involved in these attacks were typically solitary, male, and large in size. Wild pig attacks on humans in developed areas (i.e., suburban and urban) have been increasing since the mid-1990s (Mayer 2013).

Contaminant uptake: The foraging by wild pigs in wetlands or other habitats downgradient from industrial operations or accident sites has the potential for the uptake of contaminants (i.e., metals, organics, and radionuclides). As an often-hunted species, wild pigs represent a potential vector for the introduction of contaminants into the human food chain as a result of the consumption of harvested animals through sport hunting (Mayer and Brisbin 2009). Such concerns have prompted the implementation of monitoring and studies looking at the uptake of such contaminants by wild pigs (e.g., on the Savannah River Site in South Carolina; Oldenkamp et al. 2017). As an example of the longevity of this type of impact, incidental ingestion of radiocesium by native wild boar in Europe through rooting and foraging activities remains a route of contamination in game meat in several countries 30 years after the Chernobyl nuclear accident (Vilic et al. 2005; Strebl and Tataruch 2007).

Economic impacts: The diversity and magnitude of damages caused by invasive wild pigs in the United States would suggest that such impacts would be economically costly. However, to date there has been no comprehensive nationwide estimate that encompasses all of the aforementioned types of damage. While some estimates do exist (e.g., agricultural damages, vehicle collisions), the impacts to the natural environment are particularly challenging to quantify (Pimentel et al. 2005). Some nationwide and state-specific damage estimates have been reported (Table 11.1), and based on that information, the annual cost of wild pig damage nationwide is in the hundreds of millions, if not billions, of dollars. Given the valuation of the U.S. livestock resources at risk in the event of a foreign animal disease outbreak, the cost of such damage could easily escalate into the hundreds of billions of dollars annually (J. J. Mayer, unpublished data).

HISTORIC AND CURRENT MANAGEMENT

Wild pigs present a unique management and/or control challenge as an invasive species in the United States. Due to their high reproductive potential (i.e., sexually mature at a young age, produce large litters, can have two litters per year, potential late onset of reproductive senescence), modeling has identified that 60%–80% of a wild pig population needs to be removed annually, year after year, to stabilize and begin reducing the numbers of these animals in a given area. Unfortunately, there are currently no logistically viable or cost-effective options for controlling populations of wild pigs over the extended time frame that would be necessary in most instances to either greatly reduce or completely eradicate populations at a large spatial scale. However, prolonged and extensive management can be effective at maintaining low densities of pigs at a highly localized scale, particularly in areas at the fringe of this species distribution. Moreover, management efforts to control or eliminate wild pigs have recently expanded at a national scale in the United States, prompting the pursuit of additional tools (e.g., toxicants, contraceptives, trap advancements, etc.) to aid in pig population control (Mayer and Brisbin 2009; USDA APHIS 2015).

Related to the history of wild pig management in the United States, the national management strategy has extensively transitioned over time from ensuring sustained free-ranging meat and big game resources to that of controlling a destructive invasive species. This evolution regarding the perception of wild pigs has been further complicated by the ongoing broad variation in how wild pigs are managed within individual states, as well as on federal versus state lands within the same state. Additional management differences are seen on public versus private lands.

Historic management (1800s to 1989): Prior to the late 1800s, wild pigs in the United States were considered to be free-ranging livestock that were subject to landowner management. Some recreational hunting did exist with the goal of procuring edible meat. With the exception of efforts to deal with the damage caused by individual or small groups of wild pigs in localized areas, no large-scale eradication or control efforts were put in place in this country during the nineteenth century.

In the early to mid-1900s, the perception, and therefore the management, of wild pigs in the United States began to change. For example, in the 1940s through 1960s, as the open range closed with the promulgation of fence laws, the legal ownership basis (e.g., "hog rights") for maintaining rural free-range livestock was eliminated. Specifically with wild pigs, the ownership rights typically defaulted to whoever land, public or private, on which the animals were standing at any point in time. The interest in this animal as a big game species with trophy value also began to evolve and increase during this same time frame (Mayer and Brisbin 2008, 2009).

Because of the damage being caused by wild pigs, some control efforts were conducted during this same period of time. For example, between 1910 and 1958, the Hawaii Territorial Board of Agriculture and Forestry conducted an eradication of wild pigs on all of the major islands in the archipelago, ultimately removing 169,592 animals (Diong 1982). Between 1938 and 1940, after Bull Island became part of the Cape Romain National Wildlife Refuge in South Carolina, the U.S. Fish and Wildlife Service implemented an eradication program that successfully removed all of the wild pigs off of that barrier island. In the 1940s and 1950s, hog cholera was introduced onto Santa Cruz and Santa Rosa Islands off of the coast of California by the islands' owners in an effort to reduce the numbers of wild pigs there. Each introduction of the virus was followed by large numbers of dead or dying pigs; however, these populations survived those efforts, and by the late 1980s, there was no longer any active virus found on these islands. In the 1960s, the U.S. Forest Service conducted trapping programs to remove wild pigs from National Forest lands in Arkansas, Florida, and Oklahoma (Mayer and Brisbin 2008).

In contrast to the aforementioned eradication efforts, starting in the 1940s, wild pigs began to gain popularity as a game species in several states. Such popularity led some states to promote the game status of these animals and/or manage them as a sustainable huntable resource. In several states (i.e., California, Florida, Hawaii, North Carolina, Tennessee, and West Virginia) these efforts led to regulations associated with the harvest of wild pigs (e.g., license and tag requirements, seasons, size/age restrictions, and bag limits). Six states (i.e., California, Florida, Mississippi, North Carolina, Tennessee, and West Virginia) even conducted active stocking of pigs from the 1950s through the 1970s. In a few instances, these stocking efforts were funded by private sport hunting organizations. In other cases, such stocking entailed the removal of nuisance wild pigs from state parks, which were then released onto public hunting lands. During this same time period, private individuals in California, Georgia, Florida, North Carolina, and Tennessee also conducted stocking efforts involving wild pigs. In several states, these public and private stocking efforts were intentionally undertaken to introduce or increase the Eurasian wild boar phenotype (i.e., in California, Florida, Mississippi, North Carolina, Tennessee, and West Virginia) (Mayer and Brisbin 2008, 2009).

Current management (1990 to present): Concurrent with both mounting interest in the control of invasive species and the initial expansion of wild pig populations in the United States in the early 1990s came a transition in most states toward the identification of nonnative wild pigs as a destructive invasive species. At the same time, this view was compromised with the popularity of pigs as a game species, especially in those states where they were still regarded as huntable game or had a listed game status.

As the numbers and range of these animals continued to increase during the last decade of the twentieth century, so did the damage that wild pigs were causing. Concurrent with the escalation of these impacts, control programs were initiated in a number of states at both the public and private levels. This was especially true in northern states, where populations of wild pigs had not existed during the twentieth and most of the nineteenth centuries. In addition to control efforts conducted by local private individuals and landowners, both state and federal agencies began to implement management programs to control the rising numbers of invasive wild pigs.

In 1999, President Clinton signed Executive Order 13112 on invasive species into effect. Between 2000 and 2010, there was a growing recognition among both state and federal agencies regarding the national scale of this problem with invasive wild pigs. As such, the need for a coordinated approach to successfully address this invasive species crisis was being discussed at regional and national conferences. Workshops at the local/county, state, regional, and national levels were held to spread the word to the public on this issue.

As the awareness of this nationwide issue grew after 2000, a national plan/program to deal with issues associated with wild pigs began to develop. The U.S. Department of Agriculture (USDA) began developing and implementing coordinated control/ eradication actions conducted by Animal and Plant Health Inspection Service (APHIS) Wildlife Services. Starting in 2004, similar efforts involving research to support such control efforts were undertaken with the National Wildlife Research Center. USDA also developed an environmental impact statement (EIS) regarding proposed actions to address this issue on a national scale. This EIS effort sought to coordinate efforts between federal and state agencies, as well as the neighboring countries of Canada and Mexico. Efforts at a national scale further expanded in 2014 when the U.S. Congress allocated funds to implement a collaborative national management program with the goal of stabilizing and eventually reducing the size of the wild pig population found in the United States (USDA APHIS 2015).

In an effort to control the illegal spread of these animals, spurred on by the popularity of wild pigs as game animals, several states (i.e., Delaware, Kansas, Nebraska, New York, North Dakota, and Tennessee) outlawed the sport hunting of these animals. The concept being, if one cannot legally hunt wild pigs as game animals, then there is no motivation to release these animals into new areas. Illinois and Pennsylvania instituted partial bans on sport hunting of wild pigs to control the public's interference with their state's control programs. Both North Carolina and

Tennessee, a long the epicenter of "wild boar" hunting in the southeast, removed the game status/listing for wild pigs. However, wild pigs remain a popular game species or commodity in several U.S. states (Bevins et al. 2014; USDA APHIS 2015).

MANAGEMENT/CONTROL OPTIONS

The various options employed for the management and control of wild pigs in the United States over the past couple of decades have included lethal removal, exclusion, repellents, supplemental feeding, damage abatement, and economic incentives. Control of wild pigs generally requires a multifaceted approach, and thus most efforts or programs dealing with wild pigs employ two or more of these general options.

Lethal removal: This option entails the direct reduction of individuals within a population, typically by shooting, trapping, dogging, Judas pig technique, snaring or sport hunting. Population modeling has shown that an annual removal rate of 60%–80% is necessary to stabilize or reduce a wild pig population over time. Such removal rates would have to be consistently conducted over several years to drive a population to extinction (Mayer and Brisbin 2009).

Shooting is one of the most widely applied control methods for wild pigs and encompasses both opportunist and strategic (i.e., using baited sites, shooting blinds, lighting systems, trail cameras to pattern wild pig presence at a site) formats, normally using either center-fire rifles or shotguns. Although shooting is generally inexpensive, success rates can be highly variable depending upon the specific application of this method and the abundance of the population being controlled. One advantage of shooting as a pig control method is the versatility of application, in that shooting can be conducted on foot, from surface vehicles, or out of an aircraft. In contrast, shooting can be difficult to use or unviable in suburban and urban areas, where recent increases in the presence of wild pigs have been observed. Recent technical advances in the control shooting of wild pigs have included muzzle-blast suppression (subsonic ammunition and muzzle suppressors), scopes and goggles for shooting in low light/ at night (night vision, thermal image, infrared), and light, remote-controlled camera drones for locating individuals (Mayer and Brisbin 2009; West et al. 2009).

Trapping of wild pigs entails live capture using a variety of trap designs (e.g., corral, box, cage, net, and drop traps) (Figures 11.7 and 11.8). Animals are attracted into the trap (typically with bait), captured, and then dispositioned (e.g., euthanized, removed live from the trap, or radio-collared and released). Trapping is a long-standing method for capturing wild pigs that can be reasonably straightforward and inexpensive to implement. Trapping is also the safest form of lethal removal for wild pigs, and can be readily deployed in developed areas. Because of the normally fixed nature of these traps, the individual trap cost, and the manpower required to operate a trap array, trapping is generally not practical for large-scale utilization. Limitations to trapping success for wild pigs include seasonal effectiveness (reduced during times of natural resource abundance; e.g., mast crops), trap-shy individuals, and trapping fatigue (i.e., the realized decrease of trapping success in an area over time). However, recent technical advances in wild pig trapping have improved the effectiveness and utility of trapping as a management tool for pigs, although trapping alone is insufficient to eliminate wild pigs in most areas. Recent advancements

FIGURE 11.7   Images of traps used to capture wild pigs. Styles include: Top image—Cage trap; Middle image—Narrow door corral trap; and Bottom image—Large door corral trap.

FIGURE 11.8   Images showing the function of drop traps used to capture wild pigs.

include corral trap innovations/designs, drop trap innovations/designs, bait evaluations and compositions, door innovations/designs, use of trail/game cameras, and trigger innovations/designs including wireless/remote triggering systems (Mayer and Brisbin 2009; West et al. 2009).

Dogging involves the use of trained hunting dogs to locate and catch wild pigs, which can then be euthanized. This method has the potential to remove large numbers of pigs. Further, in situations where one needs to quickly obtain a wild pig specimen for sample collection/testing (e.g., in an area with a suspected disease outbreak), dogging can produce such specimens quicker than almost any other method. However, acquiring and maintaining a good pack of "hog dogs" is expensive, can result in impacts to the dogs (e.g., physical and potentially fatal injuries caused by the pigs, exposure to pathogens like pseudorabies virus), and is opposed by numerous groups. Ultimately, the success of employing this method depends upon the skills of the dogs and their handler; many hunters claim to have excellent packs of "hog dogs," yet few actually do. Last, the use of dogging to control wild pigs in an area has been documented to reduce the success of other control techniques, such as shooting and trapping (Mayer and Brisbin 2009; West et al. 2009).

The Judas pig technique consists of radio-collaring a wild pig that is then released back into the source population. After a period of time (e.g., one or more weeks) to allow for that animal to re-establish itself within the social structure of the local population, the Judas animal is located and any other pigs found with the radio-collared animal are then shot and killed. At that point, the Judas pig is allowed to escape and the process is repeated at some periodic interval (Sharp and Saunders 2002). Although both females and males have been used as Judas pigs, sows have been shown to work best for this technique (McIlroy and Gifford 1997; Wilcox et al. 2004). It is also wise to sterilize mature females that are released as Judas pigs so that they do not continue to contribute to population recruitment (McCann and Garcelon 2008). Often the Judas pig technique requires a large amount of labor, which then only produces a low number of animals removed. Thus, this method may be most useful for eradicating small localized numbers of wild pigs or when used in conjunction with other proven lethal removal techniques. The Judas pig technique can also be useful in locating the last few remaining animals during eradication programs (Wilcox et al. 2004).

Snaring can be effective in removing individual wild pigs and is relatively low cost, both in terms of materials and labor. A snare is a loop of steel cable that is attached to a secure anchor point and placed in a location so that the loop catches the pig around the neck or base of the head as the animal passes through a small opening. After being caught by the snare, the animal then typically dies of strangulation or is euthanized when the trapper arrives. Due to both humaneness and nontarget species issues, snares are not legal to use in all states. The use of modified leg snares has also been reported for the live capture of wild pigs (Mayer and Brisbin 2009; West et al. 2009).

Recreational sport hunting also is used as a lethal removal method in some states. However, given the small annual mean percentage removal rate (i.e., 23%) that sport hunting has on a wild pig population (Mayer 2014), it is not considered to be an effective management tool for controlling this invasive species (West et al. 2009). Further, when states have implemented sport hunting opportunities with the intent of enlisting public assistance in controlling wild pig populations, this has often been followed by an increase in the illegal transport and release of these animals into areas to create hunting opportunities (Bevins et al. 2014).

Exclusion: Exclusion entails the prevention of access by wild pigs to an area through the use of pig-proof fencing. Although this option can be effective, fencing that is truly "pig-proof" can be expensive to both construct and maintain, and the most effective pig-proof fences are typically also the most expensive. The estimated cost of erecting pig-proof fencing is approximately $22,000 per kilometer, and cost can increase significantly in areas with uneven terrain (e.g., canyons, creeks, ditches, or steep slopes). Unfortunately, fencing seldom affords permanent control, since wild pigs are persistent and will eventually find a way through almost any type of fence over time. In addition, such fencing is subjected to breaches caused by both manmade and natural incidents (e.g., vandalism, trees falling, washouts due to flash flooding, and vehicle collisions with the structure) that would enable wild pigs to cross the barrier. Because of this, fences need to be maintained on a periodic basis, with an estimated annual cost of approximately $500 per kilometer. Due to the construction and maintenance costs, use of exclusion fencing is most often limited to small areas (Mayer and Brisbin 2009). However, in certain circumstances, it is practical to enclose a large area to facilitate wild pig eradication and the subsequent protection of lands surrounded by that barrier. Such extensive fencing projects for eradicating and subsequently excluding wild pigs have been undertaken at Pinnacles National Monument, Annadel State Park, and the Channel Islands in California, and at the Hawaii Volcanoes National Park in Hawaii (Barrett et al. 1988; Stone and Anderson 1988; Schuyler et al. 2002; McCann and Garcelon 2008). A novel design entailing low-level fencing (i.e., 86 cm in height) has been developed to exclude wild pigs from game feeders, but still allow access for white-tailed deer (Rattan et al. 2010).

Repellents: While repellents have been successful for use with some wildlife species, tests on wild pigs have had mostly mixed results. The various repellents that have been specifically tested on wild pigs can be divided into five categories as follows: olfactory, acoustic, visual, gustatory, and combinations of the four preceding groupings. Several commercial repellents are marketed for use against wild pigs (e.g., Schwegler©, Stop Jabali©, Wildschwein-Stopp®). A few have shown promise; however, most repellents tested on wild pigs have proven to be ineffective (Barrett and Birmingham 1994; West et al. 2009; USDA APHIS 2015).

Supplemental feeding: The use of supplemental feeding in the form of either planted food plots or established feeding stations attempts to divert wild pigs away from agricultural crops that could be damaged through depredation by these invasive animals. This management option is widely practiced in Europe (Wilson 2005). When natural foods (e.g., mast crop) are scarce, wild pig impacts to agricultural crops have been reported to increase (Mackin 1970). Using strategically located feeding stations, supplemental feeding has the potential to reduce such temporal crop damage (Goulding et al. 1998). In addition, carefully targeted supplemental feeding during specific periods of vulnerability of high-valued crops may be appropriate in some circumstances. In contrast to the said benefits, supplemental feeding also has the potential to increase reproductive output and size of a wild pig population, making control of that population more difficult (Wilson 2005).

Damage abatement: Recent efforts by government agencies and academic institutions have sought to reduce wild pig damage by providing abatement consultation services to affected landowners. These damage abatement efforts typically encompass programs that provide direct control support (i.e., wild pig lethal removal by agency personnel), subsidies to purchase or loan equipment (e.g., traps, game cameras), and damage prevention education (e.g., workshops, technical field events). For example, a two-year pilot project was undertaken by Texas A&M AgriLife Extension to provide on-site technical assistance to landowners in three areas of Texas, which were experiencing wild pig damage to their agricultural crops. This pilot project included both direct assistance/service to and the education of these landowners. At the conclusion of this effort, a total of 48 participating cooperators realized a total economic benefit of $4,459,312 as a direct result of this project (Higginbotham et al. 2008).

Economic incentives: Finally, programs have been implemented in several locations to economically incentivize the increased lethal removal of wild pigs. For example, bounty programs have been tried in areas of Alabama, Michigan, South Carolina, and Texas. However, these programs have done little to control the numbers of these animals. In addition, bounty programs are often susceptible to fraud, such as turning in domestic pigtails or ears acquired from slaughterhouses to claim bounty fees (Bevins et al. 2014; J. J. Mayer, unpublished data). Further, in addition to offering bounties, a couple of dozen counties in Texas operate a contest where financial grants are awarded to the three counties with the highest annual harvests of wild pigs. In addition, the Texas Animal Health Commission maintains a list of approved buying stations/exotic meat processors (i.e., 107 as of August 2016) located across the state that will purchase wild pigs from landowners, hunters, and trappers (TAHC 2016). These privately owned operations, which pay for live wild pigs based on the animal's weight, slaughter, package, and ship the processed meat to both national and international customers. From 2004 to 2009, a total of 460,911 wild pigs were processed in these Texas commercial operations. Sale of wild pig meat has become a source of nontraditional income for landowners trying to offset the cost of damage caused by these animals (Higginbotham 2010, 2013). Under § 35:15–34–5 of the Oklahoma Administrative Code, live wild pigs can be transported and sold to licensed sporting, handling, or slaughter facilities in that state. A total of 54 of these licensed facilities were listed as being present in Oklahoma in February 2016 (ODAFF 2016).

Because of the ability of wild pigs to rapidly recover from population-reduction activities (e.g., lethal removal), it is necessary that control programs for these animals be ongoing until success is achieved (e.g., eradication, target density or population size, or damage reduction). Postponing control activities (e.g., to save money) even for several months can result in substantial increases in the local wild pig numbers. The money saved initially by such delays is quickly lost by the increased efforts required to reduce population numbers back to where they were before the delay was implemented. The primary mistake that landowners and government agencies make with respect to newly established wild pig populations is to ignore the presence of these animals until their numbers increase to the level that problems and damage begin to occur. Once wild pig populations in an area have passed that threshold, the cost for controlling those animals goes up, and the chances of achieving success in managing them go down. To miss that small window of opportunity will become increasingly more costly in both dollars and damage as time passes (Mayer and Brisbin 2009).

Examples of successful eradication programs for long-established wild pig populations in the United States do exist; however, most of these took place either on islands (e.g., the California Channel Islands; Horn Island, Mississippi; Bull Island, South Carolina) or on mainland tracts of land that were enclosed by pig-proof fencing (Pinnacles National Monument, California; Annadel State Park, California). The isolated short-term presence of wild pigs has also resulted in successful eradications in a number of locations in the United States, especially in the northern tier of the lower 48 contiguous states (e.g., in Colorado, Idaho, Iowa, New York, Washington, and Wisconsin). Localized eradication programs in mainland states with widespread established wild pig populations are often only successful for short periods of time due to the likelihood of animals dispersing in from surrounding areas. In addition, such successes are confronted with the potential for clandestine illegal releases of pigs back onto those cleared properties.

Management/control options under development: In addition to the aforementioned technical advances of some lethal removal options (e.g., shooting, trapping), several other options for controlling wild pig populations are under development. These primarily include both pig-specific oral contraceptives and toxins, neither of which has a specific product approved for use in the United States at the moment.

The development of a pig-specific oral toxin/poison is a high priority for management with a clear goal to deliver a pig-specific oral toxicant in a bait to lethally reduce numbers of wild pigs in an area. Toxins have been used for years in Australia, with such poisoning campaigns being successful (Saunders et al. 1990; Choquenot et al. 1996). However, poisoning typically does not remove all of the pigs in an area; therefore, such campaigns must be followed up by the use of other lethal removal methods (e.g., shooting, trapping, dogging) to eradicate the residual animals. In addition, as with any baiting program, potential nontarget impacts must be mitigated (either through a pig-specific toxicant or a feeder only accessible by pigs) prior to the implementation of using toxicants as a control tool in the United States (Beasley et al. 2015). At this time, research groups in the United States are developing pig-specific toxicants using either sodium nitrite or warfarin as the active ingredient. These projects are attempting to identify a toxicant that can humanely kill pigs while having a benign effect on nontarget animals and the environment at large. Additionally, a pig-specific feeder will be needed for these toxin delivery programs (Mayer and Brisbin 2009; West et al. 2009; USDA APHIS 2015; J. J. Mayer, unpublished data).

Developmental research on a pig-specific oral contraceptive or reproductive inhibitors in the United States focuses on immunocontraception, which uses an animal's immune system to prevent it from either fertilizing offspring (i.e., females and males) or implanting embryos (i.e., females). The oral contraceptive would be delivered in a bait with the goal of reducing recruitment within that population. The development of injectable contraceptives has been successful, but an oral delivery would be preferable (e.g., logistically more feasible and cost-effective). A handful of laboratories (e.g., at Auburn University, Texas A&M University, and USDA's National Wildlife Research Center) are currently working on one or more of the two types of immunocontraceptives noted previously. Similar to the delivery of oral toxins, the deployment of this option would also require a pig-specific feeder to reduce any potential impacts to nontarget species (Mayer and Brisbin 2009; West et al. 2009; USDA APHIS 2015; J. J. Mayer, unpublished data).

In addition to active research into advancements in new and improved methods for controlling wild pigs, there is a surprising paucity of basic research on wild pig ecology/biology in the United States, and thus this remains an important area of research. Indeed, knowledge of key population parameters is essential to the initiation and evaluation of successful management efforts, yet to date, key attributes essential to the development of population and spatial models remain unknown or insufficiently quantified. For example, even basic parameters such as age-specific survival and fecundity and methods for reliably quantifying population size across large spatial scales have not been adequately evaluated in the United States for wild pigs. Given that wild *Sus scrofa* are distributed across the globe, such parameters have been established in other regions of the species range; however, climate, genetic composition, and many key population parameters have been demonstrated to differ for pig populations in the United States compared to other portions of their range, and thus there remains a need to develop models specific to U.S. pig populations. Moreover, pig reproductive rates, genetic makeup, etc. vary widely even within the United States (Mayer and Brisbin 2009), and thus there is a growing need for broad-scale research across the invasive range of this highly destructive species to maximize effects of control and eradication efforts.

FUTURE CONTROL CHALLENGES

The ultimate resolution of the invasive wild pig issue in the United States depends upon a number of factors. First and foremost, effective management/control options for this species must be developed and made available to both public and private individuals to employ. Even with the development of such options, there is no single all-inclusive and effective technique for controlling wild pig populations; so, successful control programs will need to involve a coordinated multifaceted approach using several management/control options. Second, there is a need to address the clandestine, grassroots support for this species by a component of the American sport hunting community, which continues to exist as a unique conundrum with respect to the control of this particular invasive species. Again, the recognized potential of wild pigs as game animals was the primary cause of the recent national increase in these animals. If that component of this issue is not addressed, then this invasive species crisis cannot be resolved. Next, a unified national approach will be needed to solve this problem among the 50 states, with common goals and objectives especially among those states where these animals exist. Assuming that the national goal is eradication, a situation where a state that lists these animals as a game species is adjacent to a state where wild pigs are considered to be an unwanted invasive species is not a plan for success. Fourth, additional research is needed to more clearly understand the movement behavior, social structure, and disease transmission dynamics of this species, as well as enhanced methods for estimating abundance. The more that one knows about the biology of an invasive species, the better one's chances of being able to control or eliminate that organism. Last, successful accomplishment of all of the above factors will also require adequate and committed funding.

The invasive wild pig crisis in the United States has been described as the greatest emerging wildlife management challenge that this country faces in the twenty-first century. Given the current inventory of legal management/control options, there is little possibility that eradication can be achieved on a large scale in the near future. The most realistic goal would be to eradicate populations in the states where these animals have appeared and become established since 1990. Eradication in states with widespread populations that number in the thousands of animals or more (e.g., Alabama, Arkansas, California, Florida, Georgia, Louisiana, Mississippi, Oklahoma, South Carolina, and Texas) is unlikely in the foreseeable future. Further, eradication in Hawaii is challenged due to important cultural traditional uses of the resident wild pigs. Given these circumstances, wild pigs will likely be present in this country for many years to come. As wild pigs continue to be present in this country, the threats that they pose to both the natural and human environments will persist.

ACKNOWLEDGMENTS

We thank T. A. Campbell for providing constructive comments on an early draft of the chapter and J. C. Kilgo for providing valuable support in identifying some of the reference materials used in this chapter. This work was supported by the U.S. Department of Energy Office of Environmental Management under Contract DE-AC09-08SR22470 to Savannah River Nuclear Solutions LLC (John J. Mayer) and Award Number DE-FC09-07SR22506 to the University of Georgia Research Foundation (James C. Beasley).

REFERENCES

Alexandrov, T., D. Stefanov, P. Kamenov, A. Miteva, S. Khomenko, K. Sumption, H. Meyer-Gerbaulet, et al. 2013. Surveillance of foot-and-mouth disease (FMD) in susceptible wildlife and domestic ungulates in Southeast of Bulgaria following a FMD case in wild boar. *Veterinary Microbiology* 166:84–90.

Allen, M. S., E. A. Matisoo-Smith, and A. Horsburgh. 2001. Pacific "Babes": Issues in the origins and dispersal of Pacific pigs and the potentials of mitochondrial DNA analysis. *International Journal of Osteoarchaeology* 11:4–13.

Anderson, A. C. Slootmaker, E. Harper, J. Holderieath, and S. A. Shwiff. 2016. Economic estimates of feral swine damage and control in 11 U.S. states. *Crop Protection* 89:89–94.

Anonymous. 1988. Wild pigs cause jet to crash. *The News & Courier*, Charleston, South Carolina, June 10:2A.

Bankovich, B., E. Boughton, R. Boughton, M. L. Avery, and S. M. Wisely, 2016. Plant community shifts caused by feral swine rooting devalue Florida rangeland. *Agriculture, Ecosystems and Environment* 220:45–54.

Barrett, R. H., and G. H. Birmingham. 1994. Wild pigs. Pp. D65–D70. In S. E. Hygnstrom, R. M. Timm, and G. E. Larson (eds.), *Prevention and Control of Wildlife Damage. 2 Volumes; Great Plains Agricultural Council*, Univ. of Nebraska, Lincoln, Nebraska, USA.

Barrett, R. H., B. L. Goatcher, P. J. Gogan, and E. L. Fitzhugh. 1988. Removing feral pigs from Annadel State Park. *Transactions of the California–Nevada Section of the Wildlife Society* 24:47–52.

Barrios-Garcia, M. N., and S. A. Ballari. 2012. Impact of wild boar (*Sus scrofa*) in its introduced and native range: A review. *Biological Invasions* 14:2283–2300.

Beach, R. 1993. Depredation problems involving feral hogs. Pp. 67–75. In C. W. Hanselka and J. F. Cadenhead (eds.), *Feral Swine: A Compendium for Resource Managers*. Texas Agricultural Extension Service, Kerrville, Texas, USA.

Beacham, W. 1997. *The World Wildlife Fund Guide to Extinct Species of Modern Times*. Beacham Publishing Corp., Osprey, Florida, USA.

Beasley, J. C., T. E. Grazia, P. E. Johns, and J. J. Mayer. 2013. Habitats associated with vehicle collisions with wild pigs. *Wildlife Research* 40(8):654–660.

Beasley, J. C., S. C. Webster, O. E. Rhodes, Jr., and F. L. Cunningham. 2015. Evaluation of Rhodamine B as a biomarker for assessing bait acceptance in wild pigs. *Wildlife Society Bulletin* 39:188–192.

Bevins, S. N., K. Pedersen, M. W. Lutman, T. Gidlewski, and T. J. Deliberto. 2014. Consequences associated with the recent range expansions of nonnative feral swine. *BioScience* 64:291–299.

Bratton, S. P. 1974. The effect of the European wild boar (*Sus scrofa*) on the high elevation vernal flora in Great Smoky Mountains National Park. *Bulletin of the Torrey Botanical Club* 101(4):198–206.

Bratton, S. P. 1975. The effect of the European wild boar (*Sus scrofa*) on grey beech forest in the Great Smoky Mountains. *Ecology* 56:1356–1366.

Bryan, L. W. 1937. Wild pigs in Hawaii. *Paradise Pacific* 49(12):31–32.

Campbell T. A., and D. B. Long. 2009. Feral swine damage and damage management in forested ecosystems. *Forest Ecology and Management* 257:2319–2326.

Carswell, C. 2011. Invasion of the feral pigs. *High Country News.* http://www.hcn.org/issues/43.14/invasion-of-the-feral-pigs. Accessed on August 25, 2011.

Chipping, D. H. 1993. Impacts of wild pig on native vegetation. Pp. 4. In W. Tietje and R. Barrett (eds.), *The Wild Pig in California Oak Woodland: Ecology and Economics.* University of California, Berkeley, Berkeley, California, USA.

Choquenot, D., J. McIlroy, and T. Korn. 1996. *Managing Vertebrate Pests: Feral Pigs.* Bureau of Rural Sciences, Australian Government Publishing Service, Canberra, Australia.

Coblentz, B., and C. Bouska. 2004. *Pest Risk assessment forFferal Pigs in Oregon.* Department of Fisheries and Wildlife, Oregon State University, Corvallis, Oregon, USA.

Conley, R. H., V. G. Henry, and G. H. Matschke. 1972. *Final report for the European Hog Research Project W-34.* Tennessee Game and Fish Commission, Nashville, Tennessee, USA.

Cozzens, T., K. Gebhardt, S. Shwiff, M. Lutman, K. Pedersen, and S. Swafford. 2010. Modeling the economic impact of feral swine-transmitted foot-and-mouth disease: A case study from Missouri. *Proceedings of the Vertebrate Pest Conference* 24:308–311.

Davidson, W. R. 2006. Wild swine. Pp 105–134. In *Field Manual of Wildlife Diseases in the Southeastern United States.* Third edition. Southeastern Cooperative Wildlife Disease Study, Athens, Georgia, USA.

Davis, D. S. 1993. Feral hogs and disease: Implications for humans and livestock. Pp. 84–87. In C. W. Hanselka and J. F. Cadenhead (eds.), *Feral Swine: A Compendium for Resource Managers.* Texas Agricultural Extension Service, Kerrville, Texas, USA.

Diong, C. H. 1982. Population biology and management of the feral pig (*Sus scrofa* L.) in Kipahula Valley, Maui. *PhD Dissertation*, University of Hawaii, Honolulu, Hawaii, USA.

Dzieciolowski, R. M., C. M. H. Clarke, and C. M. Frampton. 1992. Reproductive characteristics of feral pigs in New Zealand. *Acta Theriologica* 37:259–270.

EFSA (European Food Safety Authority). 2014. Evaluation of possible mitigation measures to prevent introduction and spread of African swine fever virus through wild boar. *EFSA Journal* 12(3):1–23.

Engeman, R. M., K. J. Couturier, R. K. Felix Jr., and M. L. Avery. 2013. Feral swine disturbance at important archaeological sites. *Environmental Science and Pollution Research* 20:4093–4098.

Engeman, R. M., H. T. Smith, R. G. Severson, M. A. M. Severson, S. A. Shwiff, B. Constantin, and D. Griffin. 2003. *Amount and Economic Valuation of Feral Hog Damage to a Unique Basin Marsh Wetland in Florida. Florida Park Service "Parknership" Technical Report*, Florida Department of Environmental Protection, Florida Park Service, Hobe Sound, Florida, USA.

Gabor, T. M., E. C. Hellgren, and N. J. Silvy. 2001. Multiscale habitat partitioning in sympatric suiforms. *Journal of Wildlife Management* 65:99–110.

Gipson P. S., B. Hlavachick, and T. Berger. 1998. Range expansion by wild hogs across the central United States. *Wildlife Society Bulletin* 26:279–286.

Goulding, M. J., G. Smith, and S. J. Baker. 1998. *Current Status and Potential Impact of Wild Boar (Sus scrofa) in the English Countryside: A Risk Assessment.* Central Science Laboratory, Ministry of Agriculture, Fisheries and Food, London, England.

Groves, C. P. 1983. Pigs east of the Wallace Line. *Journal de la Société des Océanistes* 77(39):105–119.

Hellgren, E. C. 1993. Biology of feral hogs (*Sus scrofa*) in Texas. Pp. 50–58. In C. W. Hanselka and J. F. Cadenhead, (eds.), *Feral Swine: A Compendium for Resource Managers: Proceedings of a Conference.* Texas Agricultural Extension Service, Kerrville, Texas, USA.

Higginbotham, B. 2010. *Outreach Education Effort Summary: Feral Hogs (2006–09).* Texas AgriLife Extension Service. Texas A&M University, Overton, Texas, USA.

Higginbotham, B. 2013. *Wild Pig Damage Abatement Education and Applied Research Activities.* Texas A&M AgriLife Research and Extension Center. Texas A&M University, Overton, Texas, USA.

Higginbotham B., G. Clary, L. Hysmith, and M. Bodenchuk. 2008. *Final Report—Statewide Feral Hog Abatement Pilot Project, 2006–2007.* Texas AgriLife Extension Service, Texas A&M University, Overton, Texas, USA.

Hopkins, W. 1948. Hogs or logs—Longleaf pine seedlings and range hogs won't grow together. *Naval Stores Review* 57:12–13.

Howe, T. D., F. J. Singer, and B. B. Ackerman. 1981. Forage relationships of European wild boar invading northern hardwood forests. *Journal of Wildlife Management,* 45:748–754.

Huff, M. H. 1977. *The effect of the European wild boar (Sus scrofa) on the woody vegetation of the gray beech forest in the Great Smoky Mountains.* Res/Resource Manage. Rpt. 18. Uplands Field Research Laboratory, Great Smoky Mountains National Park, Gatlinburg, Tennessee, USA.

Hutton, T., T. DeLiberto, S. Owen, and B. Morrison. 2006. *Disease Risks Associated With Increasing Feral Swine Numbers and Distribution in the United States.* Midwest Association of Fish and Wildlife Agencies, Wildlife and Fish Health Committee. Madison, Wisconsin, USA.

Ilse, L. M., and E. C. Hellgren. 1995. Resource partitioning by sympatric populations of collared peccaries and feral hogs in southern Texas. *Journal of Mammalogy* 76:784–799.

Jay, M. T., M. Cooley, D. Carychao, G. W. Wiscomb, R. A. Sweitzer, L. Crawford-Miksza, J. A. Farrar, et al. 2007. *Escherichia coli* O157:H7 in feral swine near spinach fields and cattle, central California coast. *Emerging Infectious Diseases* 13(12):1908–1911.

Kaufman, K., R. Bowers, and N. Bowers. 2004. *Kaufman Focus Guide to Mammals of North America.* Houghton Mifflin, New York, USA.

Keever, A. C. 2014. Use of N-mixture models for estimating white-tailed deer populations and impacts of predator removal and interspecific competition. *M. S. Thesis.* Auburn University, Auburn, Alabama, USA.

Keiter, D. K., and J. C. Beasley. 2017. Hog Heaven? Challenges of managing introduced wild pigs in natural areas. *Natural Areas Journal* 37:6–16.

Kirch, P. V. 1982. The impact of the prehistoric Polynesians on the Hawaiian ecosystem. *Pacific Science* 36:1–14.

Kliejunas, J. T., and W. H. Ko. 1976. Dispersal of *Phytophthora cinnamomi* on the island of Hawaii. *Phytopathology* 66:457–460.

Lacki, M. J., and R. A. Lancia. 1986. Effects of wild pigs on beech growth in Great Smoky Mountains National Park. *Journal of Wildlife Management,* 50:655–659.

Larson, G., T. Cucchi, M. Fujita, E. Matisoo-Smith, J. Robins, A. Anderson, B. Rolett et al. 2007. Phylogeny and ancient DNA of Sus provides insights into neolithic expansion in Island Southeast Asia and Oceania. *Proceedings of the National Academy of Sciences of the United States of America* 104(12):4834–4839.

Larson, G., K. Dobney, U. Albarella, M. Fang, E. Matisoo-Smith, J. Robins, S. Lowden et al. 2005. Worldwide phylogeography of wild boar reveals multiple centers of pig domestication. *Science* 307(5715):1618–1620.

Lease, J. K., R. J. Dusek, and C. T. Atkinson. 1996. Feral pig control is effective in reducing mosquito populations (abstract). *The 1996 Hawaii Conservation Conference*, July 18–19, Honolulu, Hawaii, USA.

Lever, C. 1985. *Naturalized Mammals of the World*. Longman, London, United Kingdom.

Linderholm, A., D. Spencer, V. Battista, L. Frantz, R. Barnett, R. C. Fleischer, H. F. James et al. 2016. A novel MC1R allele for black coat colour reveals the Polynesian ancestry and hybridization patterns of Hawaiian feral pigs. *Royal Society Open Science* 3:160304. http://dx.doi.org/10.1098/rsos.160304

Lipscomb, D. J. 1989. Impacts of feral hogs on longleaf pine regeneration. *Southern Journal of Applied Forestry* 13(4):177–181.

Lucas, E. G. 1977. Feral hogs—Problems and control on national forest lands. Pp. 17–21. In G. W. Wood (ed.), *Research and Management of Wild Hog Populations*. Belle Baruch Forest Science Institute of Clemson University, Georgetown, South Carolina, USA.

Mackin, R. 1970. Dynamics of damage caused by wild boar to different agricultural crops. *Acta Theriologica* 15:447–458.

Maly, K., B. K. Pang, and C. P. Burrows. 2013. *Pigs in Hawai'i, From Traditional to Modern*. East Maui Watershed Partnership, Makawao, Hawaii, USA.

Massei, G., and P. V. Genov. 2004. The environmental impact of wild boar. *Galemys* 16(no. especial):135–145.

Mayer, J. J. 2013. Wild pig attacks on humans. *Proceedings of the Wildlife Damage Management Conference* 15:17–25.

Mayer, J. J. 2014. *Estimation of the Number of Wild Pigs Found in the United States*. SRNS–STI–2014–00292, Savannah River Nuclear Solutions, LLC, Savannah River Site, Aiken, South Carolina, USA.

Mayer, J. J., and I. L. Brisbin, Jr. 2008. *Wild Pigs in the United States: Their History, Comparative Morphology, and Current Sstatus*. 2nd Edition. The University of Georgia Press, Athens, Georgia, USA.

Mayer, J. J., and I. L. Brisbin, Jr. (eds.). 2009. *Wild Pigs: Biology, Damage, Control Techniques and Management*. SRNL-RP-2009–00869. Savannah River National Laboratory, Aiken, South Carolina, USA.

Mayer, J. J., and P. E. Johns. 2007. Characterization of wild pig–vehicle collisions. *Proceedings of the Wildlife Damage Management Conference* 12:175–187.

Mayer, J. J., E. A. Nelson, and L. D. Wike. 2000. Selective depredation of planted hardwood seedlings by wild pigs in a wetland restoration area. *Ecological Engineering* 15(Supplement 1):S79–S85.

McCann, B. E., and D. K. Garcelon. 2008. Eradication of feral pigs from Pinnacles National Monument. *Journal of Wildlife Management* 72(6):1287–1295.

McIlroy, J. C., and E. J. Gifford. 1997. The "Judas" pig technique: A method that could enhance control programmes against feral pigs, *Sus scrofa*. *Wildlife Research* 24:483–491.

Mengak, M. T. 2016. *2015 Georgia Wild Pig Survey—Final Report*. Publ. No. 16–23. Warnell School of Forestry and Natural Resources, University of Georgia, Athens, USA.

Nelson, E., J. Heinemeier, J. Møhl, and J. Arneborg. 2012. Isotopic analysis of the domestic animals of Norse Greenland. *Journal of the North Atlantic Special Volume* 3:77–92.

Nettles, V. F. 1989. Disease of wild swine. Pp. 16–18. In N. Black (ed.), *Proceedings: Feral Pig Symposium*. April 27–29, Orlando, Florida. Livestock Conservation Institute, Madison, Wisconsin, USA.

Nunley, G. L. 1999. The Cooperative Texas Wildlife Damage Management Program and feral swine damage management. Pp. 27–30. In *Proceedings of the Feral Swine Symposium.* June 2–3. Texas Animal Health Commission, Fort Worth, Texas, USA.

ODAFF (Oklahoma Department of Agriculture, Food, and Forestry). 2016. Licensed Feral Swine Facilities. http://www.oda.state.ok.us/ais/feralswinefac.pdf. Accessed on March 24, 2016.

Oldenkamp, R. E., A. L. Bryan, Jr., R. A. Kennamer, J. C. Leaphart, S. C. Webster, and J. C. Beasley. 2017. Trace elements and radiocesium in game species near contaminated sites. *Journal of Wildlife Management* 81, in press.

Paarlberg, P., J. G. Lee, and A. H. Seitzinger. 2002. Potential revenue impact of an outbreak of foot and mouth disease in the United States. *Journal of the American Veterinary Medical Association* 220:988–992.

Pimentel, D. 2007. Environmental and economic costs of vertebrate species invasions into the United States. Pp. 2–8. In G. W. Witmer, W. C. Pitt, and K. A. Fagerstone (eds). *Managing Vertebrate Invasive Species: Proceedings of an International Symposium.* USDA/APHIS Wildlife Services, National Wildlife Research Center, Fort Collins, Colorado, USA.

Pimentel, D., R. Zuniga, and D. Morrison. 2005. Update on the environmental and economic costs associated with alien-invasive species in the United States. *Ecological Economics* 52:273–288.

Quinn, J. H. 1970. Occurrence of *Sus* in North America. *Geological Society of America, Abstracts* 2(4):298.

Rattan, J. M., B. J. Higginbotham, D. B. Long, and T. A. Campbell. 2010. Exclusion fencing for feral hogs at white-tailed deer feeders. *The Texas Journal of Agriculture and Natural Resource* 23:83–89.

Saunders, G., B. Kay, and R. Parker. 1990. Evaluation of a warfarin poisoning programme for feral pigs (*Sus scrofa*). *Australian Wildlife Research* 17(5):525–533.

Schuyler, P. T., D. K. Garcelon, and S. Escover. 2002. Eradication of feral pigs (*Sus scrofa*) on Santa Catalina Island, California, USA. Pp. 274–286. In C. R. Veitch and M. N. Clout (eds.), *Turning the Tide: The Eradication of Invasive Species.* IUCN SSC Invasive Species Specialist Group, International Union for the Conservation of Nature and Natural Resources, Cambridge, United Kingdom.

Sharp, T. and G. Saunders. 2002. *Use of Judas Pigs.* SOP PIG004. Natural Heritage Trust, New South Wales, Sydney, Australia.

Shi, W., B. Zheng, Y. Zhang, and S. Ditchkoff. 2010. Economic estimate of wild pig damage to farm land owners in Alabama. *Proceedings of the 2010 International Wild Pig Symposium.* April 11–13, 2010, Pensacola, Florida, USA.

Singer, F. J., W. T. Swank, and E. E. C. Clebsch. 1984. The effects of wild pig rooting in a deciduous forest. *Journal of Wildlife Management* 48:464–473.

Smith, J. D. 1952. The Hawaiian goose (nēnē) restoration program. *Journal of Wildlife Management* 16:1–9.

Stegeman, L. J. 1938. The European wild boar in the Cherokee National Forest, Tennessee. *Journal of Mammalogy,* 19(3):279–290.

Stone, C., and S. Anderson. 1988. Introduced animals in Hawaii's natural areas. *Proceedings of the Vertebrate Pest Conference* 13:134–140.

Strebl, F., and F. Tataruch. 2007. Time trends (1986–2003) of radiocesium transfer to roe deer and wild boar in two Austrian forest regions. *Journal of Environmental Radioactivity* 98:137–152.

TAHC (Texas Animal Health Commission). 2016. Texas Approved Feral Swine Holding Facilities. http://www.tahc.state.tx.us/animal_health/swine/FeralSwineFacilities.pdf Accessed on August 25, 2016.

Tanger, S. M., K. Guidry, H. Nui, C. Richard and M. Abreu. 2015. *Dollar Estimates of Feral Hog Damage to Agriculture in Louisiana*. Research Information Sheet 113, Agricultural Economics and Agribusiness, Louisiana State University, Baton Rouge, Louisiana, USA.

Tate, J. 1984. Techniques for controlling wild hogs in the Great Smoky Mountains National Park. *Proceedings of a workshop*, November 29–30, Research/Resources Mgmt. Rpt. SRE-72. U.S. Department of the Interior, National Park Service, Southeast Regional Office, Atlanta, Georgia, USA.

Taylor, R. B., and E. C. Hellgren. 1997. Diet of feral hogs in the western south Texas plains. *The Southwestern Naturalist* 42:33–39.

Tisdell, C. A. 1982. *Wild Pigs: Environmental Pest or Economic Resource?* Pergamon Press, New York, New York, USA.

Tolleson, D. R., W. E. Pinchak, D. Rollins, and L. J. Hunt. 1995. Feral hogs in the Rolling Plains of Texas: Perspectives, problems, and potential. *Great Plains Wildlife Damage Control Workshop Proceedings* 12:124–128.

Turner, K. L., E. F. Abernethy, L. M. Conner, O. E. Rhodes, Jr., and J. C. Beasley. 2017. Abiotic and biotic factors modulate carrion fate and scavenging community dynamics. *Ecology* 98, in press.

USDA APHIS (U.S. Department of Agriculture Animal and Plant Health Inspection Service). 2015. *Final Environmental Impact Statement—Feral Swine Damage Management: A National Approach*. U.S. Department of Agriculture, Washington, DC, USA.

Vilic, M., D. Barisic, P. Kraljevic, and S. Lulic. 2005. ^{137}Cs concentration in meat of wild boars in Croatia a decade and a half after the Chernobyl accident. *Journal of Environmental Radioactivity* 81:55–62.

Wakely, P. C. 1954. Planting the Southern pine. *Forest Service Agricultural Monograph* 18:1–233.

West, B. C., A. L. Cooper, and J. B. Armstrong. 2009. Managing wild pigs: A technical guide. *Human–Wildlife Interactions Monograph* 1:1–54.

Wilcox, J. T., E. T. Ashehoug, C. A. Scott, and D. H. Van Vuren. 2004. A test of the Judas technique as a method for eradicating feral pigs. *Transactions of the Western Section of the Wildlife Society* 40:120–126.

Wilmshurst, J. M., T. L. Hunt, C. P. Lipo, and A. J. Anderson. 2011. High-precision radiocarbon dating shows recent and rapid initial human colonization of east Polynesia. *Proceedings of the National Academy of Sciences* 108(5):1815–1820.

Wilson, C. J. 2005. *Feral Wild Boar in England: Status, Impact and Management*. DEFRA, RDS National Wildlife Management Team, Exeter, United Kingdom.

Yarrow, G. K. 1987. The potential for interspecific resource competition between white-tailed deer and feral hogs in the post oak savannah region of Texas. *PhD Dissertation*, Stephen F. Austin University, Nacogdoches, Texas, USA.

Zadik, B. J. 2005. The Iberian pig in Spain and the Americas in the time of Columbus. *M.A. Thesis*. University of California, Berkeley, USA.

Zappacosta, O. 2005. Wild pigs and plants. http://www.stanford.edu/~rawlings/pigs.htm Accessed on December 30, 2008.

12 Ecology of the Small Indian Mongoose (*Herpestes auropunctatus*) in North America

Are R. Berentsen, William C. Pitt, and Robert T. Sugihara

CONTENTS

GENERAL ECOLOGY AND DISTRIBUTION

The small Indian mongoose (*Herpestes auropunctatus*) is a diurnal opportunistic omnivore native to parts of the Middle East, India, and Asia (Corbet and Hill 1992; Lekagul and McNeely 1977; Veron et al. 2007). Much of what is known about the species comes from records of populations where they were introduced to control rodents on sugarcane plantations (predominantly the Caribbean Islands and Hawaii) rather than their native range (Horst et al. 2001). In published research, the introduced mongoose is alternately, and often synonymously, identified as *H. auropunctatus* or *H. javanicus*. However, research by Veron et al. (2007) suggests that *H. auropunctatus* and *H. javanicus* are distinct taxa with unique biogeographic ranges: *H. auropunctatus* from the Middle East to Myanmar and *H. javanicus* from Myanmar and east, throughout Southeast Asia. Myanmar represents the eastern and western limits of *H. auropunctatus* and *H. javanicus*, respectively (Veron et al. 2007). Given documentation by Espeut (1882) that the mongoose's introduced to the Caribbean, and later Hawaii, originated from Calcutta, India, it is now generally accepted that the mongoose species introduced to North America is *H. auropunctatus*.

Small Indian mongooses were introduced to several regions worldwide to reduce rodent damage to sugar plantations and in some cases for viper control (e.g., habu

[*Trimeresurus flavovirids*] and horned viper [*Vipera ammodytes*] [Barun et al. 2011]; fer-de- lance [*Bothrops lanceolatus, B. atrox*] [Barun et al. 2011; Myers 1931]; St. Lucia Lancehead pit viper [*B. caribbaeus*] [Des Vœux 1903]). These introductions took place in Croatia (Barun et al. 2008, 2011), Fiji (Gorman 1975), Okinawa (Yamada 2002), Australia (failed attempt, reviewed by Peacock and Abbott 2010), Trinidad (Urich 1914), portions of coastal South America (Nellis 1989), throughout the Caribbean (Hoagland et al. 1989), and Hawaii (Doty 1945; Baldwin et al. 1952).

Mongooses are slender with short legs, an elongated muzzle, and short ears (Nellis 1989). The dental formula is similar to the basic formula for carnivores: I3/3, C1/1, P4/4, M2/2. Mongooses exhibit sexual dimorphism: females reach sexual maturity at approximately 305 g with an average adult weight of 434 g, whereas males typically reach sexual maturity at approximately 395 g and average 650 g as adults (Nellis and Everard 1983). However, males weighing over 1000 g have been recorded in food-rich habitats (Berentsen, unpublished data). The term "*auropunctatus*" means "golden points" and refers to the speckled, gold-colored tips of the pelage. Mongooses are terrestrial and diurnal (Baldwin et al. 1952) and occupy terrestrial burrows, including spaces beneath large boulders and likely spaces around roots and logs as well as natural cracks in lava substrates in Hawaii (Pimentel 1955a; Hinton and Dunn 1967). They are opportunistic generalists and feed on a variety of prey items, including insects, lizards, rats, crustaceans, human refuse, seeds, birds, bird eggs, vegetable matter, and carrion (Williams 1918; Spencer 1950; Baldwin et al. 1952; Wolcott 1953; Kami 1964; Vilella and Zwank 1993; Vilella 1998; Horst et al. 2001; Pitt et al. 2015). Mongooses are largely considered a solitary species but will congregate around locally abundant food resources and discarded animal carcasses (Pitt et al. 2015). Throughout their native range, mongooses are found in open, forested, and scrub habitats (Shekhar 2003). Pimentel (1955a) suggested that because mongooses are poor climbers they avoid forest habitats, but in many Caribbean regions they have expanded their range to include broad expanses of subtropical dry forest and subtropical rainforest (Vilella 1998; Horst et al. 2001). In addition, their introduced range includes agricultural land and urban areas (Spencer 1950; Baldwin et al. 1952). In Asia, the species can be found in elevations up to 2100 m (Simberloff et al. 2000) and has been recorded at elevations of up to 3000 m on Hawaii (Baldwin et al. 1952).

Mongooses are capable of breeding year round, although two to three birth peaks tend to occur throughout the year in an apparent correlation with day length (Nellis and Everard 1983), as has been suggested in Fiji (Gorman 1976). Gestation is approximately seven weeks (Asdell 1964), and young remain with the mother for four to six months (Hays and Conant 2007). Typical litter size is two to four pups (Asdell 1964; Nellis and Everard 1983; Coblentz and Coblentz 1985a). Precise life expectancy is unknown as traditional aging techniques relying on tooth wear are inconsistent as a result of differing regional diets and a paucity of known-age reference specimens. Pearson and Baldwin (1953) established relative age criteria based on tooth irruption in mongooses on Hawaii and documented mongooses up to 40 months of age, but did not have known-age samples as a reference. Personnel from the U.S. Department of Agriculture (USDA)/Wildlife Services captured and uniquely numbered mongooses via subcutaneous injection of a passive integrated transponder (PIT tag) on Hawaii that were recaptured four and a half years later during a subsequent study

(R. Sugihara, pers. obs.). In Puerto Rico, Horst et al. (2001) used PIT tagging of known-aged animals and documented ages of at least 81 months.

Population density and home range estimates throughout the Caribbean and Hawaii are highly variable and can differ among studies depending on experimental design and analysis methods. Mongoose population density in the Caribbean ranges from 0.19 to 9.0 mongooses/ha (Pimentel 1955a,b; Hoagland et al. 1989; Corn and Conroy 1998; Vilella 1998; Horst et al. 2001; Quinn and Whisson 2005; Hudson 2010; Johnson et al. 2016). Horst et al. (2001) suggested that mongoose population density in Puerto Rico was lower in grasslands than semiwooded regions, Vilella (1998) found relatively low densities in montane humid and rainforest regions, but Guzmán-Colón and Roloff (2014) found no correlation between habitat type and population density. Population densities in Hawaii are estimated at 0.04 mongoose/ha in lowland wetlands to 3.0/ha in moist forests (Stone et al. 1994). Population densities on Hawaii are reported to be higher at elevations between 1515 and 2060 m above sea level (Duffy 2007).

Mongoose home range estimates in the Caribbean range from 1.0 to over 50 ha (Berentsen, unpublished data; Nellis 1989; Quinn and Whisson 2005; Edwards 2006) and may vary with season. Males tend to have larger home ranges than females (Berentsen, unpublished data; Quinn and Whisson 2005). Mongoose home range estimates in Hawaii are comparable to those found in the Caribbean and range from 6.0 to over 70 ha (Hays 1999; Pitt et al. 2015). Male mongooses tend to have larger home ranges and longer point-to-point travel distances than females (Hays and Conant 2003; Pitt et al. 2015).

INTENTIONAL INTRODUCTIONS TO THE CARIBBEAN AND HAWAII

Small Indian mongooses have been introduced to over 64 islands worldwide (Barun et al. 2011). However, in this section, we restrict our discussion to sugarcane-producing islands of the Caribbean and Hawaii.

As early as 1814, it was suggested that the mongoose be introduced to control rats (Lunan 1814). The first recorded introduction of mongooses to the Caribbean region took place around 1870 when an unknown number of mongooses originating from India were introduced to the island of Trinidad (Urich 1914). Little documentation exists regarding its outcome, but Urich (1914) suggested a second introduction took place in 1898, which implies the initial attempt in 1870 was unsuccessful. By 1870, approximately one-fifth of some sugar plantation harvests in Jamaica were lost to rat damage, resulting in significant financial losses to plantation owners (Lewis 1953). Numerous attempts at importing mongooses into Jamaica for rodent control were made circa 1872, but all failed as the mongooses were obtained from captive colonies in the United Kingdom (Lewis 1953). However, on February 13, 1872, nine mongooses (four males and five females) were introduced to the Spring Garden Estate, Jamaica, directly from Calcutta, India, by W. B. Espeut (Espeut 1882). Initial success in suppressing rat populations in Jamaica resulted in the export of mongooses to several other Caribbean Islands, including Cuba and Puerto Rico (Espeut 1882). To date, the small Indian mongoose has become established on at least 29 islands throughout the Caribbean, with most introductions having occurred prior to 1900 (Horst et al. 2001).

Following the initial success of rat control in the Caribbean (West Indies), the small Indian mongoose was introduced several times to Hawaii in 1883 (Doty 1945; Baldwin et al. 1952), and mongooses currently are found on the Hawaiian Islands of Hawaii, Maui, Molokai, and Oahu (Baldwin et al. 1952; Hays and Conant 2007). Anecdotal reports describe early introduction attempts onto sugar plantations on Kauai, likely during the late 1800s. These reports describe crate(s) of mongooses shipped from Hawaii to Kauai where, upon being bitten by a mongoose in one of the crates, a dockworker allegedly disposed of the crates by throwing them off the pier. While these reports have circulated among families who worked on the Kauai sugar plantations, they have not been substantiated. Sporadic sightings occurred on Kauai in 1976 when a lactating female on Kauai was killed by an automobile (Tomich 1986). More recent reports suggest mongooses persist on Kauai, with two being trapped (USDA, unpublished data), but it is unclear whether an extensive breeding population has become established. Credible sightings have been reported (Menard et al. 2013) but large-scale trapping is prohibited, making population estimation problematic.

Interisland travel within the Hawaiian Islands, islands in the Caribbean, as well as between the Caribbean and the Florida Keys has raised concern about unintentional introduction of the mongoose into mongoose-free islands as well as into the continental United States. The mongoose has not yet reached the North American continent, but in November 1976, mongoose sightings were reported on Dodge Island off the coast of Florida (Nellis et al. 1978). Intensive trapping was conducted in February 1977 and a young female was captured. This followed reports of two other mongooses being killed by local residents in the preceding months. In September 2016 a small Indian mongoose was captured after escaping from a food delivery truck in Orange County, Florida (Florida Fish and Wildlife Conservation Commission, pers. comm.). The origin of this animal was unknown, but was of the same species found in the Caribbean and highlights the potential for inadvertent translocation of invasive species through interstate or international commerce. No other mongooses have been reported from this area, and mongooses have not been reported since on the U.S. mainland.

DAMAGE TO NATIVE FAUNA

By 1882, 10 years after their introduction, mongooses had spread throughout Jamaica and scientists were noting a sharp decrease in the number of ground-nesting birds, fowl, and reptiles (Lewis 1953). Eight years later, in 1890, initial attempts to reduce mongoose populations were being recommended. Hoagland et al. (1989) found Norway rat (*R. norvegicus*) densities were indirectly correlated with mongoose densities in Jamaica, whereas mouse (*Mus musculus*) densities and black rat densities were directly correlated with mongoose densities. Hoagland et al. (1989) go on to suggest that mongooses are ineffective predators of black rats and house mice. Thus, the initial success of mongoose suppression of rodent populations may have been restricted to Norway rats, with little effect on other crop-damaging rodent species. In addition, Lewis (1953) suggested that initial reduction in rat populations also caused rats to become more arboreal and thus relatively safe from mongooses, which are notably poor climbers. Furthermore, the diurnal nature of mongooses versus the nocturnal habits of rats likely doomed the introductions for rodent control to failure from the beginning.

Westermann (1953) provides an early summary of the overall effects of introduced fauna, including mongooses, to local flora and fauna throughout the West Indies. Although extinctions attributed directly to mongooses are likely few (Henderson 1992), there are several cases in which population declines have been exacerbated by mongoose introduction. In Puerto Rico, the introduced mongoose has been implicated in the decline of the Puerto Rican parrot (*Amazonia vittata*; Engeman et al. 2006) and Puerto Rican nightjar (*Antrostomus noctitherus*; Vilella and Zwank 1993). Mongooses are also known to prey upon the Puerto Rican giant anole (*Anolis cuvieri*; Schwartz and Henderson 1991), coqui frogs (*Eleutherodactylus* spp.; Wolcott 1953; Pimentel 1955a,b), and eggs of the Puerto Rican slider (*Trachemys stejnegeri stejnegeri*; León and Joglar 2005). Wetmore (1927) suggested the mongoose was responsible for the decline of the West Indian nighthawk (*Chordeiles gundlachii*), short-eared owl (*Asio flameus*), and two species of quail doves. Espeut (1882) also mentions a decline in a variety of species on Jamaica that he attributed to the mongoose, including snakes, lizards, and crabs, among others. Collar et al. (1992) attributed the decline of the Jamaican petrel (*Pterodroma caribbaea*) to the mongoose. Also in Jamaica, Lewis et al. (2011) suggested the mongoose was responsible for a reduction in blue-tailed galliwasp (*Celestus duquesneyi*) and endangered Jamaican iguana (*Cyclura collei*) populations. The mongoose has been documented damaging nests of the endangered hawksbill turtle (*Eretmochelys imbricate*), green turtle (*Chelonia mydas*), and loggerhead sea turtle (*Caretta caretta*) throughout the U.S. Virgin Islands (Seaman and Randall 1962; Nellis and Small 1983; Coblentz and Coblentz 1985b).

Mongoose introduction has also been suggested as a contributing factor in the decline of the endangered Hispaniolan solenodon (*Solenodon paradoxus*) in Haiti (Turvey et al. 2008; Thomas 2009). Westermann (1953) stated that the Cuban solenodon (*S. cubanus*) went extinct around 1910 and suggested mongooses were a significant factor in the solenodon's demise. However, three live Cuban solenodons were captured in 1974–1975 and remnant populations survive in isolated pockets, and the species is now listed as endangered (Soy and Mancina 2008). The impact, or lack thereof, of mongooses on populations of the Cuban solenodon remains unclear. While Westermann (1953) suggested the mongoose was likely responsible for its extermination, Borroto-Páez (2009) suggested domestic dogs and cats were more likely predators of solenodons than mongooses, as the two species are not considered sympatric on Cuba. However, given the adaptability of mongooses to a variety of habitats and evidence of mongoose predation of solenodons in Haiti (Turvey et al. 2008), potential impacts should not be ruled out without further investigation.

Nellis and Everard (1983) discuss the disappearance of various snakes, including *Alsophis sancte-crucis*, *A. ater*, and *A. rufiventris*, from St. Croix, Jamaica, and St. Kitts and Nevis and attribute their demise to mongooses. Damage to domestic poultry is also a concern, although mongooses are more likely to affect subsistence producers rather than commercial poultry operations in the Caribbean. There is anecdotal information on damage to domestic poultry farms (hens, chicks, and eggs) in Hawaii and to sweet potatoes on Okinawa, although published reports are scant.

Overall, it is estimated that the annual economic impact of mongooses associated with public health, poultry loss, extinctions of various reptiles and amphibians, as well as destruction of native bird species is approximately U.S. $50 million in the Hawaiian Islands and Puerto Rico alone (Pimentel et al. 2005).

Hawaii has not been excluded from damage by mongoose introduction. The mongoose's impacts on bird populations were noted as early as 1900 (Anonymous 1900). Since its introduction to Hawaii, the mongoose is suspected to have contributed to the decline or extinction of various species of ground-nesting birds, including several species of honeycreepers (Smith and Remington 1996) and the Hawaiian goose (*Nesochen [Branta] sandvicensis*; Loope et al. 1988; Banko 1982, 1992). Hodges and Nagata (2001) attributed 38% of known-cause Hawaiian petrel (*Petrodoma phaeopygia sandwichensis*) mortality to mongooses or domestic cats (*Felis catus sylvestri*) and 41% to rats (*Rattus* spp.). Exclusion of mongooses has also been suggested as a strategy to assist in the recovery of the Hawaiian petrel and Newell's shearwaters (*Puffins newelli*; Cowan et al. 2014).

The mongoose is not only a contributing factor to the extirpation of some species, but also potentially influences behavior. For example, the bridled quail dove (*Geotrygon mystacea*), a ground-nesting bird on St. Croix, was thought to be extinct, but Nellis and Everard (1983) suggested it has become an arboreal nester in response to nest predation by mongooses. Also, on St. Croix, the snake *Alsophis sancticrucis* and two species of lizard (*Amevia polops* and *A. exsul*) have declined presumably due to mongoose predation (Nellis 1982). Follow-on effects of mongoose predation may increase populations of various species in trophic cascades, such as in Japan, where Watari et al. (2008) suggested decreases in native predators, such as frogs and lizards, may result in an increase in the abundance of smaller animals. Nellis (1982) also reported an increase in populations of *Spaerodactylus* and *Hemidactylus* lizards following mongoose introduction and, presumably as a result of reduced predation by *Ameiva* spp., an increase in populations of the moth *Cactoblastis cactorum*, with a resulting estimated 90% reduction in the cactus *Opuntia* sp. Townsend (2006) suggested that deforestation may have forced nesting of the threatened golden swallow (*Tachycineta euchrysea*) in an unsuitable habitat in the Dominican Republic, resulting in nest predation by mongooses, and Allen (1911) describes a reduction in *Ameiva* lizards on Grenada likely due to mongoose predation.

DISEASE

Cases of leptospirosis have been reported as a result of mongooses in Puerto Rico (Pimentel 1955a,b), Trinidad, and Grenada (Everard et al. 1976). In addition, leptospirosis is highly prevalent in Hawaii (Alicata and Breaks 1943; Alicata 1958; Minette 1964; Higa and Fujinaka 1976), with dozens of cases reported statewide, some requiring hospitalization (Wong et al. 2012). Cases of *Salmonella* spp. and *Campylobacter* spp. have been reported in free-ranging mongooses on Barbados (Rhynd et al. 2014; Matthias and Levett 2002), and although rabies is not present in Hawaii, the mongoose is the primary rabies reservoir on Puerto Rico, Cuba, Grenada, the Dominican Republic, and most likely Haiti (Everard and Everard 1992; Zieger et al. 2014; Berentsen et al. 2015).

References to rabies in Puerto Rico date back to the 1840s, prior to the introduction of the mongoose; the first clinically documented case of rabies in mongooses did not occur until 1950 (Tierkel et al. 1952). Seroprevalence in mongooses ranges from 11.7% up to 40% on some islands (Zieger et al. 2014; Berentsen et al. 2015), and in Puerto Rico, 70% of animals testing positive for rabies virus in 2013 were mongooses (Dyer et al. 2014). Rabies virus strains from Grenada (Zieger et al. 2014), Cuba (Nadin-Davis et al. 2006), and Puerto Rico (Nadin-Davis et al. 2008) suggest independent introduction of rabies to different Caribbean islands. No oral rabies vaccination program exists for mongooses, although research into potential oral rabies vaccine baits in the Caribbean has been conducted (Linhart et al. 1993; Creekmore et al. 1994; Berentsen et al. 2014). With large populations of feral dogs and cats, along with no compulsory vaccination for kept domestic animals, risks of rabies transmission between mongooses, domestic animals, and, ultimately, humans remain high.

MANAGEMENT STRATEGIES

As the perception that mongooses are useful in rodent control waned, they rapidly became a pest species in areas where they have been introduced, and currently the mongoose is listed as one of 100 of the world's worst invasive species (Lowe et al. 2000). Globally, there have been six successful attempts to eradicate mongooses from islands (Barun et al. 2011). All of these islands were less than or at most 115 ha, and populations were eradicated using a variety of methods, including trap/remove and primary or secondary poisoning using rodenticides such as brodifacoum, bromadiolone, or thallium sulfate. Two large-scale ongoing mongoose eradication attempts are being conducted in Japan (Abe 2013). The eradication efforts on 71,200-ha Amami-Oshima Island began in 1993, starting with support from local villages and then taken over by the Japanese Ministry of the Environment in 2000. The mongoose eradication effort in northern Okinawa also began in 2000 and encompasses much of the island (227,130 ha). Because these areas have native mammals, control methods are limited to kill traps or live traps requiring euthanasia. Where mongoose populations have been reduced to low levels, native species are recovering (Watari et al. 2008). The challenge remains, however, to develop methods for detecting and controlling mongooses at low densities.

In the Caribbean, early management strategies for mongooses took place in Trinidad through a bounty system which ultimately proved ineffective and costly (Urich 1931). In an effort to protect populations of endangered Puerto Rican parrots as well as human health and safety, limited mongoose control is performed on Puerto Rico (Engeman et al. 2006; Quinn et al. 2006). Rather than large-scale eradication efforts, local control of mongooses has been conducted much more frequently, with live box traps the most common method employed (Barun et al. 2011). Morton (2005) describes trap and removal efforts to protect native iguana (*Iguana iguana*) nesting sites on St. Lucia. Kill-trapping and acute toxicants such as thallium sulfate, strychnine, sodium monofluoroacetate (Compound 1080), and even ground glass were used historically in the Caribbean (Pimentel 1955b; Everard and Everard 1992; Barun et al. 2011). Currently there is no islandwide control program for mongooses on any islands in the Caribbean.

The history of organized mongoose control in Hawaii began in 1915 when the Territorial Legislature had appropriated funds for a bounty on the mongoose (Public Health Reports: 1896–1970). The Territorial Fish and Game Commission and individual counties eventually led extermination campaigns that offered cash prizes for the greatest number of mongoose heads (Anonymous 1921). Mongoose bounties eventually proved ineffective and were discarded. Early research into mongoose control included the use of warfarin and zinc phosphide. Although zinc phosphide was found to be effective in killing mongooses, because of its extreme toxicity and risks to domestic livestock, its use was not recommended (Woodworth and Woodside 1953). Subsequent eradication efforts included the use of toxicants such as thallium sulfate (Kridler 1965), and sodium monofluoroacetate injected into chunks of meat that were deposited in crevices near Hawaiian goose nests and throughout sanctuaries (Walker 1974). Later, the anticoagulant diphacinone was also employed (Stone et al. 1995; Smith et al. 2000).

For a wide-ranging carnivore like the mongoose, the efficacy of a control method depends upon the ability of the target species to detect a bait or lure from a distance, follow it to its source, and be attracted enough to it to enter a trap or bait station. The bait or lure must also be evaluated for its stability and longevity: how long does it remain both detectable and attractive? For a toxic bait, the toxicity to the target species must be quantitatively evaluated to ensure that the toxicant is lethal to the majority of individuals in consumable doses. The palatability of the matrix of toxic bait is critical to ensure that the target species will eat enough of the bait to consume a lethal dose of the toxicant. Therefore, each toxic bait product must have its own series of laboratory and field trials to establish its efficacy for a target species (and potential hazard to nontargets), since the active ingredient is not the sole determinant of efficacy (Palmateer and McCann 1976; Keith et al. 1985, 1987; Pitt and Sugihara 2008; Pitt et al. 2015). Field efficacy trials for any method targeting mongooses must be designed on a scale large enough to have a statistically significant sample of mongooses, be replicated in several locations, use multiple independent methods to assess the effects on mongoose abundance within the treatment areas, and be compared with a site where no control methods are used.

In a USDA/Wildlife Services study (Keith et al. 1987), both acute and chronic toxicants, including zinc phosphide, warfarin, and diphacinone, were effective against mongooses in laboratory bioassays. Diphacinone was highly effective in low doses (0.18 mg/kg), which would minimize hazards to nontarget species. Subsequent field trials at Hawaii Volcanoes National Park and at James Campbell National Wildlife Refuge using diphacinone mixed in raw hamburger at a concentration of 0.00025% diphacinone, placed in bait stations 125–250 m apart, killed a high percentage of radio-collared mongooses (Keith et al. 1990). In 1991, a Special Local Need registration was approved for 0.1% diphacinone concentrate to be mixed into raw hamburger to make a 0.00025% diphacinone bait to be applied in specially designed bait stations. The design specified a 4-in.-diameter polyvinyl chloride (PVC) pipe in the shape of a T, with entrances in the arms of the T, and bait placed in the supporting arm. This technique proved effective but expensive due to a variety of factors, including bait cost, labor costs associated with bait preparation, bait station construction, logistics involved with ensuring fresh bait over a 12–15-day feeding period required to effect control, and installation and maintenance in remote areas. Given these restrictions, it was impractical to apply to large conserva-tion areas (Stone et al. 1995), and the registration was allowed to expire.

Recent research on mongoose control has focused on identifying nontoxic baits and lures with a large call distance (distance of effective attraction) to stand out in prey-rich environments (Pitt and Sugihara 2008; Pitt et al. 2015). The instructions for mongoose on the current Ramik® Mini Bars SLN describing the spacing and area over which bait stations should be distributed are based upon the results of Pitt et al. (2015), which determined home range sizes and measured the distances mon-gooses traveled to investigate novel food baits. No field trials have been conducted to determine the effectiveness of the current bait station instructions at reducing mongoose abundances.

Mongooses were eliminated from within a portion of the Ka'ena Point Natural Area Reserve on Oahu that is separated from the rest of the island by a predator-proof fence. Ramik® Mini Bars containing 0.005% diphacinone were placed in bait stations in a 25-m grid during construction of the fence, and mongooses were not detected in the area prior to final completion of the fence (Young et al. 2013). Since animals were not radio-collared, and no carcasses were recovered, the exact methods respon-sible for the apparent eradication are unknown. Of two mongooses that subsequently entered the fenced area, one was caught in a leg-hold trap, and mongoose scat was detected on top of a bait station from which bait take was noted (Young et al. 2013).

A variety of live and kill traps are used for mongoose control in Hawaii. The effectiveness of these methods for controlling mongooses primarily depends on the type of bait used, the spacing between traps, and the area over which they are placed (Keith et al. 1987; Pitt et al. 2015). The skill and experience of the individual trapper in trap placement and setting also affect trapping success. However, while mortal-ity numbers have been recorded in preliminary field trials evaluating efficacy of a variety of trapping methods (Peters et al. 2011), no independent monitoring has determined whether overall mongoose abundance is reduced under current mon-goose trapping practices. A variety of multikill, "self-resetting" devices have been developed for use on stoats (*Mustela erminea*) and rats in New Zealand. Limited data are available on the efficacy of self-resetting traps on mongoose control, but VanderWerf (2015) reported poor success for mongoose control relative to other removal methods. No trials using standardized quantitative methods have been con-ducted in Hawaii to evaluate the humaneness of self-resetting traps or their effective-ness at reducing local mongoose populations to levels low enough to protect native species. Furthermore, successful use of such devices would require the use of mon-goose-specific lures and/or toxicants, neither of which are available. The lures cur-rently used in self-resetting devices come in a variety of food-based scents, but scent lures have been shown to be ineffective for mongooses in Hawaii field trials (Pitt and Sugihara 2008). No extensive evaluation comparing the efficacy of different trap types or trapping strategies on mongoose population densities has been performed.

FUTURE CONTROL METHODS

Several commercial rodenticide products are registered for use in rodent control in Hawaii, including chlorophacinone (0.005% a.i.) (Rozol® Pellets; EPA SLN No.

HI-080002), which is approved for field use against rodents in Hawaiian fruit and nut orchards, sugarcane fields, and selected field seed (corn, soybean) crops (Pitt et al. 2011), as well as a pelleted diphacinone bait (Diphacinone-50; EPA Reg. No. 56228-35) for broadcast application in conservation areas. Only one product is registered for use in mongooses: a diphacinone (0.005% a.i.) rodenticide bait block (Ramik® Minibars; EPA Reg. No. 61282-26) for use in bait stations. However, the attractiveness and palatability of these baits to mongooses has not been thoroughly evaluated in controlled feeding trials. In addition, poor bait acceptance and problems with regurgitation (emesis) have been reported from the field. Other registered rodenticide baits are routinely used in and around industrial, commercial, and agricultural food commodity storage, as well as sorting, processing, and packing structures in Hawaii to control rodent infestation, reduce product and structural damage, and address health and sanitation concerns. Mongooses have been known to visit the rodenticide bait stations, but the operational efficacy of these baits for free-ranging mongooses has been questionable or unknown. Preliminary results from recently completed trials conducted at the USDA National Wildlife Research Center (NWRC) Hawaii Field Station suggest that most commercial wax block or pelleted bait formulations were not palatable to mongooses in free-feeding laboratory trials (USDA/NWRC Hawaii Field Station, unpublished data). One exception was bromethalin, which mongooses readily consumed and to which they succumbed after one day's feeding. Incorporating effective registered rodenticide toxicants into a more palatable matrix may result in more effective bait for mongooses, but would require additional research and evaluation. In addition, mongooses have been shown to be highly susceptible to diphacinone, and formulations with fresh ground beef have proved to be efficacious in free-ranging mongooses (Keith et al. 1990).

A relatively new vertebrate pesticide containing the active ingredient para-aminopropiophenone (PAPP) has shown to be effective in Australia (foxes, feral dogs/cats) and New Zealand (stoats, ferrets, and feral cats) (Fisher et al. 2005; Fisher and O'Connor 2007; Eason et al. 2010). In the United States, PAPP is being investigated as a tool for coyote control (Savarie et al. 1983; Young 2014). PAPP, a methemoglobin forming chemical, reduces the oxygen-carrying capacity of the blood, and, at toxic doses, induces a reported humane death (sleepy, lethargic, unconsciousness within a few hours) (Savarie et al. 1983). The NWRC Hawaii Field Station has informally collaborated with researchers in Japan toward the development of control tools, including microencapsulated PAPP for mongooses. Its lower toxicity to rodents has attracted Japanese interest in testing PAPP against mongooses owing to the presence of two species of native rats in Okinawa and Amami Oshima Islands in the Japanese archipelago. Preliminary results with formulated microencapsulated PAPP delivered in fresh minced chicken look promising for mongoose control.

Sodium nitrite, a commonly used food preservative and ingredient in commercial fertilizers, is another methemoglobemia-inducing chemical that has been investigated as a potential vertebrate pesticide. Australian researchers have conducted studies on the use of microencapsulated sodium nitrite to control invasive wild pigs (Cowled et al. 2008; Lapidge et al. 2012). Hog-Gone® pelleted bait containing sodium nitrite was developed in Australia for invasive wild pig control and is being tested in the United States (Campbell et al. 2012). Witmer (2013) evaluated sodium nitrite as a rodenticide, and preliminary results suggest the potential for its use as a toxicant against selected rodent species in the United States. However, preliminary research with mongooses shows some taste aversion at higher concentrations (5%) when formulated in fresh chicken meat (USDA/NWRC, unpublished data). Additional trials at lower sodium nitrite concentrations are planned.

CONCLUSION

The introduction of the small Indian mongoose as a biological control agent has resulted in failure and significant ecological damage throughout much, if not all, of its introduced range. Not only have mongooses failed to provide long-term rodent control, but mongooses have caused irreparable damage to native fauna and become reservoirs for diseases such as rabies, leptospirosis, and in some regions possibly *Salmonella* and *Campylobacter*. Mongoose suppression of viper populations has met with some success in some Croatia (Barun et al. 2010), but snakes, including vipers, did not make up a significant proportion of the mongoose diet in Japan (Abe et al. 1999).

Effective and efficient population control methods for mongooses have yet to be developed. Trap and lethal removal programs such as those practiced on Okinawa may be successful in significantly reducing populations but are labor- and time-intensive, taking decades of consistent effort, with no guarantee of ultimate success. A variety of toxicants have been evaluated that are effective in laboratory trials, but a bait matrix/delivery system, palatable to mongooses and suitable for field application, has remained elusive. Furthermore, numerous strategies for mechanical removal (i.e., live trap/remove, kill traps) are in use with no standardized measure to evaluate efficacy.

The risk of unintentional mongoose introduction to the North American continent (including currently mongoose-free islands of Hawaii) is highlighted by the early reports of mongooses found in Florida and more recent documentation of this invasive pest species on the island of Kauai. With increased interisland travel among Caribbean Islands and between islands and mainland countries, vigilance must be maintained in inspecting cargo and vessels to prevent further mongoose introductions. Additional intensive research into local and large-scale mongoose management strategies is necessary. Future discussions of introducing biological control agents should carefully consider the long-term consequences before taking any action.

REFERENCES

Abe, S. 2013. Eradication project of invasive alien mongooses on Amami-oshima Island, Japan. Page 195 in Ministry of the Environment, Government of Japan and International Union for Conservation of Nature (editors). The 1st Asia Parks Congress, November 13–17, 2013. Sendai, Japan.

Abe, S., Y. Hanada, Y. Abe, Y. Takatsuki and H. Nigi. 1999. Food habits of feral mongoose (*Herpestes* sp.) on Amamioshima, Japan. In G. H. Rodda, Y. Sawai, D. Chiszar and H. Tanaka (eds). *Problem Snake Management: The Habut and the Brown Treesnake*. Cornell University Press, New York, USA, pp. 372–383.

Alicata, J. E. 1958. A study of leptospirosis in Hawaii. *Plantation Health* 8:505–512.

Alicata, J. E. and V. Breaks. 1943. A survey of leptospirosis in Honolulu. *Hawaii Medical Journal* 2:137–142.

Allen, G. M. 1911. Mammals of the West Indies. Bulletin of the Museum of Comparative. *Zoology* 54:175–263.

Anonymous. 1900. Destruction of birds by the pestiferous mongoose. *The Honolulu Republican* July 31, 1900, p. 7, Image 7.

Anonymous. 1921. Cash prizes offered to most successful hunters of mongoose. *The Maui News* September 9, 1921, p. 9, Image 6.

Asdell, S. A. 1964. *Patterns of Mammalian Reproduction*, 2nd ed., Cornell University Press, NY, pp. 477–478, 670 pp.

Baldwin, P. H., C. W. Schwartz and E. R. Schwartz. 1952. Life history and economic status of the mongoose in Hawaii. *Journal of Mammalogy* 33:335–356.

Banko, P. C. 1982. Productivity of wild and captive nene populations. *Proceedings of the Hawaii Volcano National Park National Scientific Conference* 4:12–32.

Banko, P. C. 1992. Constraints on productivity of wild Nene or Hawaiian Geese *Branta sandvicensis*. *Wildfowl* 43:99–106.

Barun, A., I. Budinski and D. Simberloff. 2008. A ticking time-bomb? The small Indian mongoose in Europe. *Aliens* 26:14–16.

Barun, A., C. C. Hanson, K. J. Campbell and D. Simberloff. 2011. A review of small Indian mongoose management and eradications on islands. In C. R. Veitch, M. N. Clout and D. R. Towns (eds). *Island Invasives: Eradication and Management*. IUCN, Gland, Switzerland, pp. 17–25.

Barun, A., D. Simberloff and I. Budinski. 2010. Impact of the small Indian mongoose on native amphibians and reptiles of the Adriatic Islands, Croatia. *Animal Conservation* 13:549–555.

Berentsen, A. R., S. R. Johnson, A. T. Gilbert and K. C. VerCauteren. 2015. Exposure to rabies in small Indian mongooses (*Herpestes auropunctatus*) from two regions in Puerto Rico. *Journal of Wildlife Diseases* 51:896–900.

Berentsen, A. R., S. R. Johnson and K. C. VerCauteren. 2014. Bait matrix flavor preference by mongoose (*Herpestes auropunctatus*) in Puerto Rico: Implications for oral rabies vaccination. *Caribbean Journal of Science* 48:52–58.

Borroto-Páez, R. 2009. Invasive mammals in Cuba: An overview. *Biological Invasions* 11:2279–2290.

Campbell, T. A., M. J. Bodenchuk, J. D. Eisemann, S. J. Lapidge, L. Staples and P. Morrow. 2012. Preliminary assessment of the HogHopper™ for exlucing nontarget wildlife. *Proceedings of the Vertebrate Pest Conference* 12:333–336.

Coblentz, B. E. and B. A. Coblentz. 1985a. Reproduction and the annual fat cycle of the mongoose on St. John, US Virgin Islands. *Journal of Mammalogy* 66:560–563.

Coblentz, B. E. and B. A. Coblentz. 1985b. Control of the Indian mongoose (*Herpestes auropunctatus*) on St. John, US Virgin Islands. *Biological Conservation* 33:281–288.

Collar, N. J., L. P. Gonzaga, N. Krabbe, A. Madrono Nieto, T. A. Parker III and D. C. Wege. 1992. *Threatened Birds of the Americas*. ICBP/ICUN Red Data Book, 3rd Edition, part 2. International Council for Bird Preservation, Cambridge.

Corbet, G. B. and J. E. Hill. 1992. *Mammals of the Indomalayan Region: A Systematic Review*. Oxford University Press, Oxford.

Corn, J. L. and M. J. Conroy. 1998. Estimation of density of mongooses with capture-recapture and distance sampling. *Journal of Mammalogy* 79:1009–1015.

Cowan, D. P., M. T. Craig, G. C. Spencer, D. G. Ainley and D. Zanjac. 2014. Managment Actions to Prevent the Disappearance of the Hawaiian Petrels (*Pterodroma sandwichensis*) and Newell's Sheawaters (*Puffinus newelli*) From West Maui, Hawaii. *Hawaii Conservation Conference*, Hawaii Convention Center, Honolulu Hawaii, July 15–17, 2014. Pp 121–122. Abstract only.

Cowled, B. D., P. Eslworth and S. J. Lapidge. 2008. Additional toxins for feral pig (*Sus scrofa*) control: Identifying and testing Achilles' heels. *Wildlife Research* 35:651–662.

Creekmore, T. E., S. B. Linhart, J. L. Corn, M. D. Whitney, B. D. Snyder and V. F. Nettles. 1994. Field evaluation of baits and baiting strategies for delivering oral vaccine to mongooses in Antigua, West Indies. *Journal of Wildlife Diseases* 30:497–505.

Des Vœux, G. W. 1903. *My Colonial Service in British Guiana, St Lucia, Trinidad, Fiji, Australia, Newfoundland and Hong Kong: With Interludes* vol. 1. J. Murray, London.

Doty, R. E. 1945. Rat control on Hawaiian sugar cane plantations. *The Hawaiian Planters' Record* 49:71–239.

Duffy, D. C. 2007. Forest bird and non-native mammal inventories at Ka'apahu, Haleakala National Park (HALE), Maui, Hawaii. Technical Report 145, Pacific Cooperative Studies Unit, University of Hawaii at Manoa. 31pp.

Dyer, J. L., P. Yager, L. Orciari, L. Greenberg, R. Wallace, C. A. Hanlon and J. D. Blanton. 2014. Rabies surveillance in the United States during 2013. *Journal of the American Veterinary Medical Association* 245:1111–1123.

Eason, C. T., E. C. Murphy, S. Hix and D. B. MacMorran. 2010. Development of a new humane toxin for predator control in New Zealand. *Integrative Zoology* 5:31–36.

Edwards, R. 2006. Habitat selection of the introduced small Indian mongoose (Herpestes javanicus), using radio telemetry and mark-recapture techniques on St. Lucia. *MS Thesis*, University of Exeter, United Kindgom. 62pp.

Engeman, R., D. Whisson, J. Quinn, F. Cano, P. Quiñones and T. H. White, Jr. 2006. Monitoring invasive mammalian predator populations sharing habitat with critically endangered Puerto Rican parrot *Amazonia vittata*. *Oryx* 40:95–102.

Espeut, W. B. 1882. On the Acclimatization of the Indian mungoos in Jamaica. *Proceedings of the Zoological Society of London* 50:712–714.

Everard C. O. R. and J. D. Everard. 1992. Mongoose rabies in the Caribbean. *Annals of the New York Academy of Sciences* 653:356–366.

Everard, C. O. R., A. E. Green and J. W. Glosser. 1976. Leptospirosis in Trinidad and Grenada, with special reference to the mongoose. *Transactions of the Royal Society of Tropical Medicine and Hygiene* 70:57–61.

Fisher, P. and C. O'Connor. 2007. Oral toxicity of *p*-aminopropiophenone to ferrets. *Wildlife Research* 34:19–24.

Fisher, P. M., C. E. O'Connor and E. C. Murphy. 2005. Acute oral toxicity of p-aminopropiophenone to stoats. *New Zealand Journal of Zoology* 32:163–169.

Gorman, M. L. 1975. The diet of feral *Herpestes auropunctatus* (Carnivora: Viverridae) in the Fijian Islands. *Journal of Zoology* 175:273–278.

Gorman, M. L. 1976. Seasonal changes in the reproductive pattern of feral *Herpestes auropunctatus* (Carnivora: Viverridae), in the Fijian Islands. *Journal of the Zoological Society of London* 178:237–246.

Guzmán-Colón, D. and G. J. Roloff. 2014. Small Indian mongoose (*Herpestes auropunctatus*) population abundance and effects of habitat features on trapping success in protected areas of eastern Puerto Rico. *Caribbean Naturalist* 19:1–12.

Hays, W. S. T. 1999. Annual dispersal cycle of the small Indian mongoose (*Herpestes auropunctatus*) (Carnivora: Herpestidae in Hwai'i. *Pacific Science* 53:252–256.

Hays, S. T. and S. Conant. 2003. Male social activity in the small Indian mongoose *Herpestes javanicus*. *Acta Theriologica* 48:485–494.

Hays, S. T. and S. Conant. 2007. Biology and impacts of Pacific Island invasive species. 1. A worldwide review of effects of the small Indian mongoose, *Herpestes javanicus* (Carnivora: Herpestidae). *Pacific Science* 61:3–16.

Henderson, R. W. 1992. Consequences of predator introductions and habitat destruction on amphibians and reptiles in the post-Columbus West Indies. *Caribbean Journal of Science* 28:1–10.

Higa, H. H. and I. T. Fujinaka. 1976. Prevalence of rodent and mongoose leptospirosis on the Island of Oahu. *Public Health Reports* 91:171.

Hinton, H. E. and A. M. S. Dunn. 1967. *Mongooses: Their Natural History and Behaviour.* Oliver and Boyd, Ltd, London. 144pp.

Hoagland D. B., G. R. Horst and C. W. Kilpatrick. 1989. Biogeography and population biology of the mongoose in the West Indies. In C. A. Woods (editor). *Biogeography of the West Indies, Past, Present and Future.* Sandhill Crane Press, Gainesville, Florida, pp. 611–634.

Hodges, C. S. N. and R. J. Nagata. 2001. Effects of predator control on the survival and breeding success of the endangered Hawaiian Dark-rumped Petrel. *Studies in Avian Biology* 22:308–318.

Horst, G. R., D. B. Hoagland and C. W. Kilpatrick. 2001. The mongoose in the West Indies: The biogeography and population biology of an introduced species. in C. A. Woods and F. E. Sergile (eds). *Biogeography of the West Indies: Patterns and Perspectives.* CRC Press, Boca Raton, Florida, USA, pp. 409–424.

Hudson, M. A. 2010. *Optimising Monitoring and Control Strategies for Invasive Alien Mammals: The Small Indian Mongoose in St. Lucia.* University of Bath, United Kingdom. 51pp.

Johnson, S. R., A. R. Berentsen, C. Ellis, A. Davis and K. C. VerCauteren. 2016. Estimates of small Indian mongoose densities: implications for rabies management. *Journal of Wildlife Management* 80:37–47.

Kami, H. T. 1964. Foods of the mongoose in the Hamakua District, Hawaii. *Zoonoses Research* 3:165–170.

Keith, J. O., D. N. Hirata and D. L. Espy. 1985. Control of mongoose predation on endangered Hawaiian birds. Unpublished progress report. USFWS Denver Wildlife Research Center, 37 pp.

Keith, J. O., D. N. Hirata and D. L. Espy. 1987. Control of mongoose predation on endangered Hawaiian birds. Unpublished progress report, 1986. USDA APHIS Denver Wildlife Research Center. 24pp.

Keith, J. A., D. N. Hirata D. L. Espy, S. Greiner and D. Griffin. 1990. Field evaluation of 0.00025% diphacinone bait for mongoose control in Hawaii. Unpublished final report USDA/APHIS/Denver Wildlife Research Center, 49pp.

Kridler, E. 1965. Experimental poisoning of the Hawaiian mongoose with thallium sulfate. U.S. Fish and Wildlife Service, Honolulu, Hawaii. Unpublished report. 12 pp.

Lapidge, S., J. Wishart, L. Staples, K. Fagerstone, T. Cambell and J. Eisemann. 2012. Development of a feral swine toxic bait (Hog-Gone®) and bait hopper (Hog-Hopper™) in Australia and the USA. *Proceedings of the Wildlife Damage Management Conference* 14:19–24.

Lekagul, B. and J.A. McNeely. 1977. *Mammals of Thailand.* Association for the Conservation of Wildlife, Bangkok, Thailand.

León, A. and R. L. Joglar. 2005. La jicotea Puertorriqueña. In: R. L. Joglar (editor). Editorial Instituto de Cultura Puertorriqueña. *Biodiversidad de Puerto Rico: Vertebrados Terrestres y Ecosistemas.* Serie de Historia Natural, San Juan, Puerto Rico, 181–190.

Lewis, C. B. 1953. Rats and the mongoose in Jamaica. *Oryx* 3:170–172.

Lewis, D. S., R. van Veen and B. S. Wilson. 2011. Conservation implications of small Indian mongoose (*Herpestes auropunctatus*) predation in a hotspot within a hotspot: The Hellshire Hills, Jamaica. *Biological Invasions* 13:25–33.

Linhart, S. B., T. E. Creekmore, J. L. Corn, M. D. Whitney, B. D. Snyder and V. F. Nettles. 1993. Evaluation of baits for oral rabies vaccination of mongooses: Pilot field trials in Antigua, West Indies. *Journal of Wildlife Diseases* 29:290–294.

Loope, L. L., O. Hamann and C. P. Stone. 1988. Comparative conservation biology of oceanic archipelagoes: Hawaii and the Galápagos. *BioScience* 38:272–282.

Lowe S., M. Browne, S. Boudjelas and M. De Poorter. 2000. *100 of the World's Worst Invasive Alien Species A Selection from the Global Invasive Species Database.* Published by The Invasive Species Specialist Group (ISSG) a specialist group of the Species Survival Commission (SSC) of the World Conservation Union (IUCN), Invasive Species Specialist Group, Auckland, New Zealand, 12pp.

Lunan, J. 1814. *Hortus Jamaicensis, Or a Botanical Description (according to the Linnean System) and an Account of the Virtues, &c. of Its Indigenous Plants Hitherto Known, as Also of the Most Useful Exotics: Complied from the Best Authorities, and Alphabetically Arranged*, in Two Volumes vol. 2. St. Jago de la Vega Gazette, Jamaica.

Matthias, M. A. and P. N. Levett. 2002. Leptospiral carriage by mice and mongooses on the island of Barbados. *The West Indian Medical Journal* 51:10–13.

Menard, T., K. Gundersen and J. Chapman. 2013. Kaua'i mongoose: Can we eradicate them? *Hawaii Conservation Conference*, July 16–18, 2013, Hawaii Convention Center, p. 30, Abstract only.

Minette, Henri P. 1964. Leptospirosis in rodents and mongooses on the island of Hawaii. *The American Journal of Tropical Medicine and Hygiene* 13:826–832.

Morton, M. N. 2005. Mongoose trapping at Louvet Estate iguana nesting site 2005. Unpublished report. St. Lucia Forestry Department and Durrell Wildlife Conservation Trust. 5pp.

Myers, J. G. 1931. The present position of the mongoose in the West Indies. *Tropical Agriculture* 8:94–95.

Nadin-Davis S. A., G. Torres, M. de Los Angeles Ribas, M. Guzman, R. Cruz de La Paz, M. Morales and A. I. Wandeler. 2006. A molecular epidemiological study of rabies in Cuba. *Epidemiology and Infection* 134:1313–1324.

Nadin-Davis S. A., J. Velez, C. Malaga and A. I. Wandeler. 2008. A molecular epidemiological study of rabies in Puerto Rico. *Virus Research* 131:8–15.

Nellis, D. W. 1982. Mongoose influence in the ecology of islands. *Transactions of the International Congress of Game Biologists* 14:311–314.

Nellis, D. W. 1989. *Herpestes auropunctatus. Mammalian Species* 342:1–6.

Nellis, D. W., N. F. Eichholz, T. W. Regan and C. Feinstein. 1978. Mongoose in Florida. *Wildlife Society Bulletin* 6:249–250.

Nellis, D. W. and C. O. R. Everard. 1983. The biology of the mongoose in the Caribbean. In P. W. Hummenlinck and L. J. Van der Steed (eds). *Studies on the Fauna of Curacao and Other Caribbean Islands*, Foundation for the Scientific Research in Surinam and the Netherlands Antilles, Utrecht, the Netherlands, pp. 1–162.

Nellis, D. W. and V. Small. 1983. Mongoose predation on sea turtle eggs and nests. *Biotropica* 15:159–160.

Palmeeter, S. D. and J. A. McCann. 1976. Relationship of acceptance and mortality of anticoagulant baits to rats. *Bulletin of Environmental Contamination and Toxicology* 15:750–755.

Peacock, D. and K. Abbott. 2010. The mongoose in Australia: Failed introduction of a biological control agent. *Australian Journal of Zoology* 58:205–227.

Pearson, O. P. and P. H. Baldwin. 1953. Reproduction and age structure of a mongoose population in Hawaii. *Journal of Mammalogy* 34:436–447.

Peters, D., L. Wilson, S. Mosher, J. Rohrer, J. Hanley, A. Nadig, M. Silbernagle, et al. 2011. Small Indian mongoose–management and eradication using DOC 250 kill traps, first lessons from Hawaii. In C. R. Veitch, M. N. Clout and D. R. Towns (eds). *Island Invasives: Eradication and Management*. IUCN, Gland, Switzerland, pp. 225–227.

Pimentel, D. 1955a. Biology of the Indian mongoose in Puerto Rico. *Journal of Mammalogy* 36:62–68.

Pimentel, D. 1955b. The control of the mongoose in Puerto Rico. *Americal Journal of Tropical Medicine and Hygiende* 4:147–151.

Pimentel, D., R. Zuniga and D. Morrison. 2005. Update on the environmental and econimic costs associated with alein-invasive species in the United States. *Ecological Economics* 52:273–288.

Pitt, W. C., L. C. Criscoll and R. T. Sugihara. 2011. Efficacy of rodenticide baits for the control of three invasive rodent species in Hawaii. *Archives of Environmental Contamination and Toxicology* 60:533–542.

Pitt, W. C. and R. T. Sugihara. 2008. Evaluation of selected natural and artificial attractants, lures, and bait substrates for attracting small Indian mongooses (*Herpestes auropunctatus*) to traps and activity stations in Hawaii.

Pitt, W. C., R. T. Sugihara and A. R. Berentsen. 2015. Effect of travel distance, home range, and bait on the management of small Indian mongooses, *Herpestes auropunctatus*. *Biological Invasions* 17:1743–1759.

Public Health Reports (1896–1970), vol 30, No. 38, Sept 17, 1915, p. 2834.

Quinn, J. H. and D. A. Whisson. 2005. The effects of anthropogenic food on the spatial behaviour of small Indian mongooses (*Herpestes javanicus*) in a subtropical rainforest. *Journal Zoology* 267:339–350.

Quinn, J. H., D. A. Whisson and F. Cano. 2006. Managing the small Indian mongoose (Herpestes javanicus) in the midst of human recreation: What is the optimal approach? *Proceedings of the Vertebrate Pest Conference* 22:393–398.

Rhynd, K. J. R., P. A. Leighton, D. A. Elcock, P. J. Whitehall, A. Rycroft and S. K. Macgregor. 2014. Prevalence of *Salmonella* spp. and thermophilic *Campylobacter* spp. in the small Asian mongoose (*Herpestes javanicus*) in Barbados, West Indies. *Journal of Zoo and Wildlife Medicine* 45:911–914.

Savarie, P. J., H. Ping Pan, D. J. Hayes, J. D. Roberts, G. L. Dasch, R. Felton and E. W. Schafer, Jr. 1983. Comparative acute oral toxicity of para-aminopropiophenone. *Bulletin of Environmental Contamination and Toxicology* 30:122–126.

Schwartz, A. and R. W. Henderson. 1991. *Amphibians and Reptiles of the West Indies: Descriptions, Distributions, and Natural History*. Univ. Florida Press, Gainesville, 720p.

Seaman, G. A. and J. E. Randall. 1962. The mongoose as a predator in the Virgin Islands. *Journal of Mammalogy* 43:544–546.

Shekhar, K. S. 2003. The status of mongooses in central India. *Small Carnivore Conservation* 29:22–23.

Simberloff, D., T. Dayan, C. Jones and G. Ogura. 2000. Character displacement and release in the small Indian mongoose, *Herpestes javanicus*. *Ecology* 81:2086–2099.

Smith, D. G., J. T. Polhemus and E. A. VanderWerf. 2000. Efficacy of fish-flavored diphacinone bait blocks for controlling small Indian mongoose (*Herpestes auropunctatus*) populations in Hawai'i. *'Elepaio* 60:47–51.

Smith, H. R. and C. L. Remington. 1996. Food specificity in interspecies competition. *BioScience* 46:436–447.

Soy, J. and C. A. Mancina. 2008. Solenodon cubanus. The IUCN Red List of Threatened Species. Version 2015.2. http://dx.doi.org/10.2305/IUCN.UK.2008.RLTS.T20320A9185957.en. Accessed July 22, 2015.

Spencer, H. J. 1950. *Mongoose control research project, Virgin Islands, USA, October–November 1950. United States Fish and Wildlife Service*, unpublished report. 7pp.

Stone, C. P., M. Dusek and M. Aeder. 1995. Use of an anticoagulant to control mongooses in Nene breeding habitat. *'Elepaio* 54:73–78.

Stone, C. P., L. W. Pratt and J. M. Yoshioka. 1994. Hawai'i's plants and animals: biolgocial sketches of Hawaii Volcanoes National Park. University of Hawaii Press, Honolulu, Hawaii, USA.

Thomas, C. 2009. Saving a venomous ghost. *Science* 325:531.

Tierkel, E. S., G. Arbona, A. Rivera and A. de Juan. 1952. Mongoose rabies in Puerto Rico. *Public Health Reports* 67:274–278.

Tomich, P. Q. 1986. *Mammals in Hawaii*, 2nd ed. Bishop Museum Special Publication 76. Bishop Museum Press, Honolulu, Hawaii, 375pp.

Townsend, J. 2006. Predation of a golden swallow (*Tachycineta euchrysea*) nest by the Indian mongoose (*Herpestes javanicus*) in the Sierra de Bahoruco, Dominican Republic. *Journal of Caribbean Ornithology* 19:108–109.

Turvey, S. T., H. M. R. Meredith and R. P. Scofield. 2008. Continued survival of Hispaniolan solenodon *Solenodon paradoxus* in Haiti. *Oryx* 42:611–614.

Urich, R. W. 1914. The mongoose in Trinidad and methods of destroying it. *Board of Agriculture Trinidad and Tobago Circular* 12:5–12.

Urich, R. W. 1931. The mongoose in Trinidad. *Tropical Agriculture* 8:95–97.

VanderWerf, E. 2015. Performance of Goodnature Automated Predator Traps to Protect Forest Bird and Sea Bird Nests on O'ahu. *Hawaii Conservation Conference*, August 3–6, 2015, University of Hawaii at Hilo, p. 97. Abstract only.

Veron, G., M-L. Patou, G. Pothet, D. Simberloff and A. P. Jennings. 2007. Systematic status and biogeography of the Javan and small Indian mongooses (Herpestidae, Carnivora). *Zoologica Scripta* 36:1–10.

Vilella, F. J. 1998. Biology of the mongoose (*Herpestes javanicus*) in a rain forest of Puerto Rico. *Biotropica* 30:120–125.

Vilella, F. J. and P. J. Zwank. 1993. Ecology of the small Indian mongoose in a coastal dry forest of Puerto Rico where sympatric with the Puerto Rican nightjar. *Caribbean Journal of Science* 29:24–29.

Walker, R. L. 1974. 1972 report of Nene restoration program. State Department of Land and Natural Resources. *Elepaio* 34:123–142.

Watari, Y., S. Nishijima, M. Fukasawa, F. Yamada, S. Abe and T. Miyashita. 2008. Evaluating the "recovery level" of endangered species without priori information before alien invasion. *Ecology and Evolution* 3:4711–4721.

Westermann, J. H. 1953. Nature preservation in the Caribbean. A review of literature on the destruction and preservation of flora and fauna in the Caribbean area. Foundation for Scientific Research in Surinam and the Netherlands Antilles, Utrecht, the Netherlands. 107pp.

Wetmore, A. 1927. The birds of Porto Rico and Virgin Islands. *New York Academy of Sciences, Scientific Survey of Porto Rico and the Virgin Islands* 9(3 and 4):254–571.

Williams, C. B. 1918. The food of the mongoose in Trinidad. *Bulletin of the Department of Agriculture of Trinidad and Tobago* 17:167–186.

Witmer, G. 2013. A preliminary evaluation of sodium nitrite as a rodenticide. Unpublished final report, QA-1752. USDA/APHIS/WS/NWRC, 14pp.

Wolcott, G. N. 1953. Food of the mongoose *Heprestes javanicus auropunctatus* in St. Croix and Puerto Rico. *Journal of Agriculture of the University of Puerto Rico* 37:241–247.

Wong, M., A. R. Katz, D. Li and B. A. Wilcox. 2012. Leptospira infection prevalence in small mammal host populations on three Hawaiian islands. *The American Journal of Tropical Medicine and Hygiene* 87:337–341.

Woodworth, J. R. and D. H. Woodside. 1953. Mongoose poison experiment. Territory of Hawaii Division of Fish and Game, unpublished report. 18pp.

Yamada, F. 2002. Impacts and control of introduced small Indian mongoose on Amami Island, Japan. In C. R. Veitch and M. N. Clout (eds). *Turning the Tide: The Eradication of Invasive Species*. IUCN SSC Invasive Species Specialist Group, IUCN, Gland, Switzerland and Cambridge, UK, pp. 389–392.

Young, J. 2014. Efficacy of para-aminopropiophenone formulation with the M-44 for coyote control. QA-2049. 15pp.

Young, L. C., E. A. VanderWerf, M. T. Lohr, C. J. Miller, A. J. Titmus, D. Peters and L. Wilson. 2013. Multi-species predator eradication within a predator-proof fence at Ka'ena Point, Hawai'i. *Biological Invasions* 15:2627–2638.

Zieger, U., D. A. Marson, R. Sharma, A. Chikweto, K. Tiwari, M. Sayyid, B. Lousin et al. 2014. The phylogeography of rabies in Grenada, West Indies, and implications for control. *PLOS Neglected Tropical Diseases* 8(10):e3251.

13 Feral Cats

Christopher A. Lepczyk and David C. Duffy

CONTENTS

INTRODUCTION

People have a long history with domestic cats (*Felis catus*) that spans thousands of years. In fact, findings from the early agricultural village of Quanhucun in Shaanxi, China, indicate that cats have been associated with humans for between 5560 and 5280 years (Hu et al. 2014). These findings from China predate previous estimates by more than a millennium, suggesting an even longer-term coexistence with cats than previously thought. In the millennia since early mutualistic coexistence and domestication, cats have accompanied human society as it has moved around the world. In nearly all such situations, some cats have become feral as people leave locations and the cats stay, are abandoned, or simply leave human settlements. As a result, every continent, save Antarctica, has feral cats present on it today, as well as many islands throughout the world. However, because of their near cosmopolitan distribution, feral cats have also wreaked ecological havoc on many ecosystems in which they are present (e.g., Simberloff 2013).

Feral cats are defined variously in the veterinary medical, animal welfare, and wildlife literature—often depending upon the degree of human contact and owner-ship status (Patronek 1998; Baker et al. 2010). However, it can be challenging to differentiate feral cats, which are under no ownership and are free-ranging on the landscape, from semiferal cats, which have some human contact, such as those living in colonies (Baker et al. 2010) and pet cats allowed outdoors. While ecologically it makes little difference whether or not there is a line in the sand in definitions, in this chapter we focus our attention on feral cats, which we consider those that are unowned and free-ranging in the environment, including those living in cat colonies and cats in urban ecosystems.

As obligate carnivores, cats fill the role of mesopredator in ecosystems in which they occur, or when higher-order predators are absent, they become apex predators, such as on many islands (Hess 2011). In particular, cats depredate a wide range of animal species in order to fulfill their daily dietary requirement of protein (Spotte

2014). Moreover, cats also act as disease reservoirs and vectors, which can negatively affect many other species. Finally, feral cats are often viewed negatively when they occur in urban areas or locations where people regularly work (Lohr and Lepczyk 2014). Hence, feral cats pose both ecological and societal concerns that require management attention, which is the focus of the chapter.

HISTORY OF CAT INTRODUCTIONS IN THE UNITED STATES

Feral cats have been observed in North America for well over a hundred years as evidenced by documents of feral cat predation in wildlife-related literature (e.g., Forbush 1916). Though an exact time point at which domestic cats became feral in North America is unknown, it likely was not long after European settlers brought them to the continent. For instance, in Hawaii, European ships brought domestic cats to the islands in the late 1700s (King 1984), and feral cats were noted in the 1840s, only decades after Europeans arrived in the island archipelago (Brackenridge 1841). Cats most certainly did arrive in continental North America before they arrived to many Pacific islands, simply based upon when Europeans arrived along the eastern coast of the United States and Canada. Because cats were often used to reduce rodent populations within agricultural settings and on ships, separating out timelines when feral cat populations may have been established in North America is likely impossible. Furthermore, because cats were not widely recognized to negatively impact the ecosystem, especially in continental settings, pinpointing their numbers and spread would be challenging, if not impossible. Regardless of the exact time course, feral cats have been present in many locations for multiple decades, if not centuries, in some locations (e.g., warmer climates). Furthermore, as knowledge and societal views have changed, today we have a greater recognition that feral cats may be an ecological problem. For instance, during the 1980s and 1990s, several seminal publications drew renewed attention to the issue of feral and outdoor cat depredation on native bird species (Churcher and Lawton 1987; Coleman and Temple 1993), which prompted many municipalities to enact laws and organizations to develop outreach material aimed at keeping cats indoors. Finally, owned cat numbers have been increasing for decades in the United States (Lepczyk et al. 2010; Figure 13.1), and it is believed that the feral cat population is following a similar trajectory.

From a life history perspective, cats have a number of attributes that have allowed them to be successful invaders. Specifically, cats have been extremely successful in colonizing new ecosystems due to a combination of their generalist diet, the assemblage of other species present, and their reproductive abilities. This combination of factors is particularly advantageous in warmer climates or areas where they are subsidized by people (Figure 13.2).

In terms of diet, cats are opportunistic predators and obligate carnivores (Coman and Brunner 1972; Fitzgerald and Karl 1979) that can survive on preformed and metabolic water in food for months (Prentiss et al. 1959). Cats have also evolved to survive off only animal tissue and have a set of several nutritional adaptations as carnivores. Specifically, cats have a limited ability to regulate enzymes of amino acid metabolism, an inability to synthesize niacin from tryptophan, and an inability to use plant material for conversion to amino acids and vitamins (MacDonald et al.

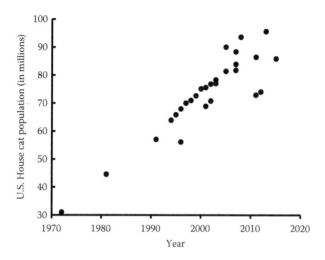

FIGURE 13.1 Estimates of the owned cat population in the United States based upon animal welfare and pet food manufacturing companies.

1984). Hence, while cats do consume plant material (Errington 1936; Jackson 1951), they are dependent on meeting their energetic demands through consumption of a high-protein diet. Dietary analyses have been carried out in cats around the world for over a 100 years (e.g., Forbush 1916), with many studies finding that either birds or small mammals are the dominant prey items, often depending upon the ecosystems

FIGURE 13.2   A cat colony being fed by a caretaker. (Photo courtesy of Grant Sizemore.)

in which the studies were conducted (Coman and Brunner 1972; Pearre and Maass 1998; Fitzgerald and Turner 2000). Such dietary differences across studies are likely a reflection of differences in prey availability among habitats (Dickman 2009). In our own compilation of approximately 260 cat diet studies, we have found over 1000 different species consumed, with the house mouse (*Mus domesticus*), European rabbit (*Oryctolagus cuniculus*), black rat (*Rattus rattus*), house sparrow (*Passer domesticus*), and common blackbird (*Turdus merula*) among the most common (Lepczyk et al., unpublished manuscript). Although many of these species are associated with humans, this is more a reflection of study location than diet preference, as the overall view is that cats will take any available prey that they are physically capable of killing (Duffy and Capece 2012). However, recent findings suggest cats continue to hunt specific species of prey, even when these prey become scarce, and that they can exhibit individual variation in hunting behavior (Dickman and Newsome 2015). Furthermore, cats scavenge carcasses of animals (e.g., Ragg et al. 2000; Olson et al. 2016; Abernethy et al. 2016), including animals larger than they can kill, and consume spoiled and wasted food left by people (Errington 1936; Jones and Coman 1981; Lepczyk et al., unpublished manuscript).

As predators, cats fall into higher trophic levels, depending somewhat on the composition of other species present in the ecosystem. In fact, feral cats occur in both ecosystems that have mesopredators and apex predators, as well as systems devoid of these species. In places where top predators exist, feral cats can be regularly depredated (Crooks and Soulé 1999). However, depredation levels in urban areas based on scat analysis suggested that cats were not a large part of urban coyote diets (Gehrt and Riley 2010). These low rates of predation may be due to how cats and coyotes partition themselves on the landscape to avoid interspecific competition, as was found in Chicago's urban metroparks (Gehrt et al. 2013). However, in places where predators are reduced or absent, feral cat numbers can increase markedly due to either mesopredator release (Crooks and Soulé 1999) or simply by being the only predator, as is the case in a number of island ecosystems. This is not to say that cat numbers will increase exponentially forever, but rather that several key factors that regulate many populations are absent or reduced where cats have become successfully established.

A final factor allowing feral cats to be successful is their high fecundity. Under good nutritional status, female cats can become reproductively active and produce offspring by six months of age (Nutter et al. 2004). However, most cats begin breeding at physiological maturity, which ranges from six to nine months in females and nine to twelve months in males (Say et al. 1999; Nutter et al. 2004). Females have 1.3–1.9 litters per year (though some cats have litters up to three times per year; Warner 1985; Nutter et al. 2004) that typically range from one to nine kittens per litter (Forbush 1916), with an average litter size of between 4 and 4.6 (Warner 1985; Wallace and Levey 2006). As with many terrestrial vertebrates, these fecundity estimates are affected by nutritional status and health of the female, as well as environmental aspects, such as photoperiod. Thus, while reports indicate that ownership status is related to litter size and number of litters per year across feral to fully owned cats (Schmidt et al. 2007), this is almost certainly due to differences in health and nutritional status of the females. As a result, well-fed feral cats very likely have a greater lifetime reproductive output than poorly fed feral cats.

High fecundity does not itself guarantee a growing or rapidly growing population, as survival is a key modulating factor in population growth. Survival differs from birth to maturity notably, depending upon climate, food availability, density of cats, and the composition of other predators in the environment (see Spotte 2014 for a review). Though variable, many reported survival rates of feral cats are over 25% for the first year, meaning that if conservative numbers of litters are considered along with conservative litter sizes, the population is easily at replacement level. Hence, overall population growth rates can vary markedly based on location due to various environmental factors.

Diet breadth, lack of competition or predation, and high reproductive potential, acting alone and in tandem with one another, allow cats to be successful in many ecosystems, particularly in warmer climates where temperatures do not limit the species. In higher latitudes or colder temperatures, feral cats may have lower survival probabilities due to their physiology and may not survive overwinter. Notably, however, comparative estimates of feral cat survival across latitudes are absent in the peer-reviewed literature. That said, feral cats can be found in colder climates if they can find food and shelter, which are often located near human habitation. Thus, urban areas and farms can provide safe havens for feral cats from the winter elements, allowing them to overwinter successfully in cold environments.

SCOPE OF THE PROBLEM

Because feral cats are found not only in rural and relatively wild locations, but also in urban areas, the scope of the problem across the United States is quite large. Numerically, there is no agreed upon estimate of the number of feral cats across the continent, because as with any free-ranging animal, estimating their true population size over such a large space is essentially impossible. That said, a number of different estimates have suggested that the numbers range from 60 million to over a 100 million (Jessup 2004; Winter 2004). More importantly, however, is that most estimates are now well over a decade old, and given that the owned cat population (Lepczyk et al. 2010) has been increasing during this time (Figure 13.1), it is likely that the number of feral cats has been increasing as well. Because of these large numbers and the near cosmopolitan distribution of cats, the problems they pose ecologically, economically, and sociologically are significant.

Ecologically, cats pose a number of concerns. Historically, foremost among these concerns has been the effects of predation on native wildlife (Forbush 1916; Churcher and Lawton 1987; Coleman and Temple 1993; Lepczyk et al. 2004; Loss et al. 2013). Cats depredate essentially any animal that they can handle, and scavenge many more (Lepczyk et al., unpublished manuscript; Olson et al. 2016; Abernethy et al. 2016), due to their generalist nature and high protein requirements. As a result, cats have caused local extirpations of species, exterminated species endemic to oceanic islands, and are responsible for the mortality of billions of birds and mammals annually (Medway 2004; Bonnaud et al. 2011; Medina et al. 2011; Frank et al. 2014; Figure 13.3). In terms of population dynamics, cat depredation can exceed local recruitment, affecting the persistence of native wildlife populations (van Heezik et al. 2010). For instance, nest success and juvenile postfledging survival of gray catbirds (*Dumatella*

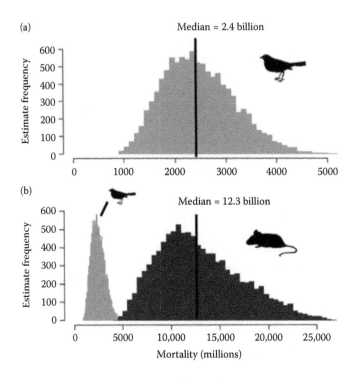

FIGURE 13.3 Estimated annual free-ranging cat depredation on (a) birds and (b) mammals in mainland areas of the contiguous United States. (From Loss, S.R., et al. 2013. *Nature Communications,* 4, 1396. Reprinted with permission from Nature Publishing Group.)

carolinensis) in urban areas was most impacted by predation, which represented 79% of mortality, with cats accounting for the greatest source of predation (Balogh et al. 2011). Although cats certainly depredate a large number of animals—and evidence suggests they have population-level affects in both island and continental areas—it has yet to be determined if cats are ultimately affecting mortality in an additive or compensatory nature in continental locations (e.g., Balogh et al. 2011). Given that cats are unevenly distributed across the landscape and the degree of human subsidies varies, coupled with the fact that prey species have different life history attributes, it is important to note that both additive and compensatory mortality scenarios likely occur, depending upon the location and the type of prey species.

Aside from predation, cats carry a wide range of diseases that pose risks to themselves, other wildlife species, and humans (Moodie 1995; Lepczyk et al. 2015). For instance, felids, including domestic cats, are the definitive hosts of *Toxoplasma gondii*, a parasite that is responsible for a wide range of ailments in a number of species, including humans (Smith and Frenkel 1995; Dubey 2002, 2010). Recent work has shown toxoplasmosis to be associated with a range of neurological and psychiatric disorders in humans such as schizophrenia, epilepsy, altered personality traits, and "aggregate neuroticism" (McAllister 2005; Lafferty 2006; Palmer 2007; Torrey et al. 2007, 2015). Moreover, toxoplasmosis has led to deaths in a number of animal species, including sea otters (*Enhydralutris nereis*; Kreuder et al. 2003), Hawaiian crows (*Corvus hawaiiensis*; Work et al. 2000), Hawaiian geese (*Branta sandvicensis*; Work et al. 2002), Hawaiian monk seals (*Monachus schauinslandi*; Honnold et al. 2005), and red-footed boobies (*Sula sula*; Work et al. 2002). Toxoplasmosis also affects livestock (Dubey and Jones 2008) and oysters (Lindsay et al. 2004; Esmerini et al. 2010). Likewise, cats are also carriers of rabies (Eng and Fishbein 1990) and plague (Craven et al. 1993), two diseases that, while thought of as rare in present-day developed countries, nonetheless continue to occur regularly.

All feral cats pose a risk of disease transmission due to their interactions with one another, with wildlife, and exposure to the environment. Cats can expose humans and other wildlife to disease in a variety of ways: mechanically through bites and scratches, through aerosol passage of pathogens, enterically through contact with feces in soil or water, or by movement of disease-bearing ectoparasites. Many diseases are also of concern where densities of cats are great, as higher densities of cats allow for a greater opportunity for transmission.

Evaluating the damage caused by feral cats in terms of economics is important for both determining the cost of the problem as well as for weighing management options. Although economic evaluations exist for many invasive species (e.g., brown tree snake [*Boiga irregularis*]; Shwiff et al. 2010), there is a paucity of studies with regards to cats. One of the few studies to consider the economic damage posed by feral and outdoor cats in the United States was conducted around the turn of the most recent century and estimated the cost at $17 billion/year (Pimentel et al. 2000). This dollar value is based on the estimated annual mortality to birds in the United States, which at that time they took conservatively to be eight birds per cat per year (from McKay 1996), multiplied by an estimated feral and free-ranging cat population size of 71 million, multiplied by $30 per bird killed (see Pimentel et al. 2000 for details). The most difficult and often challenged of these three values is the cost of a bird. The reality is, however, that birds can be economically evaluated in terms of the amount spent on bird rearing, watching, hunting, and fines for illegal take (Lohr et al. 2013). Though all of these three values are estimates themselves, what is apparent is that even altering the values by an order of magnitude lower would result in a multimillion dollar problem.

Aside from the estimate by Pimentel and colleagues, the peer-reviewed literature is depauperate of economic analyses. However, if we update Pimentel et al.'s estimate with revised estimates of bird mortality alone from unowned cats (i.e., feral cats), which Loss et al. (2013) estimated to range between 952 million and 3.065 billion for the continental United States per year, a more recent cost would be in the range of $28.56–$91.95 billion/year. Importantly, this estimate is only for the continental United States and only for birds. Given that annual mammal mortality estimated by Loss et al. (2013) was five times greater than birds, not to mention mortality on herpetofauna and other species, the economic valuation of just mortality alone is most certainly much higher. Notably, none of the estimates presented or published consider any other forms of cost, such as those borne by animal control and animal welfare organizations, property owners, or human health care costs. In fact, human health care costs are likely far in excess of many other costs based on treatment for diseases like rabies and toxoplasmosis.

Feral cats also pose a challenge to society at large. For instance, cats can be a nuisance to people as they defecate in children's sandboxes and on lawns, pose a threat to people with asthma, leave a urine stench, make noise, and provoke barking by dogs (Proulx 1988; Jarvis 1990; Natoli 1994; Baker 2001). For these reasons, it is perhaps unsurprising that recent work in Hawaii has shown that the vast majority of stakeholders (87%), including the public and animal welfare organizations, want to see feral cats removed from the landscape (Lohr and Lepczyk 2014). Furthermore, there has been growing concern over human health due to cat-borne diseases (see Lepczyk et al. 2015 for a review).

Aside from the nuisance and health concerns, a large sociological aspect of feral cats relates to their management. Specifically, while human dimensions research has indicated a desire to reduce feral cat populations (Lohr and Lepczyk 2014), there has been a great deal of conflict among some members of society in terms of the facts surrounding cat management and what management technique to use (Lepczyk et al. 2011; Peterson et al. 2012). Much of this conflict ultimately relates to what constitutes humane treatment of animals and the no-kill movement.

CONTEMPORARY APPROACHES FOR MANAGING FERAL CATS

Feral cats are not a new problem across North America as illustrated by Forbush's (1916) work a century ago on cat predation. Thus, the issue of feral cat management is an old one to wildlife ecologists and one in which a number of professional societies (e.g., American Bird Conservancy, The Wildlife Society) have taken a stance on. However, because of the growth in feral cat numbers, the rise in scientifically unsupported management approaches, a broader recognition of the problems feral cats pose, and the conflicts among stakeholder groups, cat management has risen to a top issue for vertebrate pest management.

Nearly all cat management options involve either deterring their presence or managing populations through direct or indirect methods (Table 13.1). Deterring cats involves approaches designed to either limit ingress of new individuals or reduce the suitability of a location to support cats. Hence, deterrence does not affect the overall population size necessarily, but simply results in cats occurring in lower densities in certain parts of the landscape. On the other hand, many forms of management are geared at managing cat populations generally through alteration of birth and death rates.

One common form of managing terrestrial vertebrates, whether native or nonnative, is through habitat modification (i.e., altering the environment). In the case of feral cats, the goal of habitat modification is to reduce their presence at a given location on the landscape. Simple approaches to altering the environment that make it less hospitable to cats include keeping garbage and other waste in secure containers and removing large brush piles near homes. Cats do consume refuse and get into garbage (Spotte 2014) and are often found foraging and living in and around waste transfer sites or dumps (Dickman 1996). Thus, removing waste and storing it in secured containers can be effective at reducing animal access to waste (e.g., Barrett et al. 2014). Likewise, keeping sandboxes covered and removing large brush piles can both reduce the suitability of a given location for feral cat habitats.

TABLE 13.1
Linking Management Action to Population Control in Feral Cats

Management Action	Vital Rate Affected	Cats Remain on the Landscape
Habitat modification	None	Yes, but in different locations
Deterrents	None	Yes, but in different locations
Fencing	None	Yes, but cannot access certain locations
Licensing	None	Yes
Adoption	Emigration (Birth)	Maybe
Euthanasia[a]	Death	No
Sanctuaries	Emigration (Birth)	No
Contraception	Birth	Yes
TNR	Birth	Yes
Nothing	None	Yes

[a] Includes techniques such as hunting, trapping followed by euthanasia, etc.

Aside from altering habitat, another method aimed at discouraging cat presence is the use of sonic deterrents. However, the use of sonic deterrents has shown mixed results, with some finding little evidence of deterrence (Mills et al. 2000), whereas others (Nelson et al. 2006) have found some evidence they reduce the probability of a cat intrusion into a site and stronger evidence that such intrusions are of shorter duration when devices are present. Motion-activated sprinklers are a similar type of deterrent that shoot a pulse of water at any animal that moves. Finally, there are chemical repellents that discourage cats, but notably must be applied regularly to the locations where cats are not desired.

A more direct method to keeping feral cats away from a given area is simply to exclude them through the use of fences. A variety of fences have been developed that are resistant to the cat's ability to tolerate electric shock, dig, climb vertical surfaces, and jump at least 1.8 m (Long and Robley 2004). In one recent case, a predator-excluding fence was built after removal of feral cats and has successfully excluded them for years, resulting in an increase in the number of nesting seabirds (Young et al. 2013).

The second main set of management approaches for feral cats are those aimed at controlling a vital rate. In contrast to previously mentioned approaches that might aim at affecting presence or migration, this second set of approaches focuses on managing births and deaths through a variety of techniques (Table 13.1). Most management techniques that address vital rates tend to focus on forms that reduce cat numbers, either through increasing mortality or physically removing cats from unwanted locations.

One of the most commonly used management approaches successfully implemented in a number of locations is eradication (see Campbell et al. 2011 for a review). However, most successful eradication efforts to date have occurred on islands or small isolated populations, although large-scale eradication efforts are currently underway in Australia. In reviewing island eradications, Campbell et al.

(2011) identified 87 successful campaigns across 83 islands, ranging in size from 5 to 29,000 ha, with each successful campaign requiring an average of 2.7 eradication methods. A number of different specific methods exist for eradicating feral cats that entail either lethal methods (see Fisher et al. 2015 for a review) or removal, which includes sanctuaries and shelters. One eradication method is the use of toxic baits, such as sodium monofluoroacetate (1080). Sodium monofluoroacetate is inexpensive and nonpersistent in food chains (Rodríguez et al. 2006), but at present it is not permitted in the United States, and though cats are highly susceptible to it, the degree to which it kills humanely has been questioned (Sherley 2007). A second toxicant that has shown great promise is para-aminopropriophenone (PAPP), which is fast-acting and appears to be "humane and effective" for carnivores (Eason et al. 2010b). In fact, grooming traps specifically designed for cats were recently used to apply a gel containing PAPP to the fur of feral cats which was consumed via grooming and resulted in mortality (Read et al. 2014). Notably, because some birds may be vulnerable to PAPP, the bait must be designed to limit such exposure (Eason et al. 2010a). Furthermore, PAPP is not currently approved for use in North America. When considering use of toxicants, it is important to bear in mind that toxicant efficiency depends on bait size, shape, taste, color, strength, and positioning, both to attract cats and to deter incidental take by other species (Morgan et al. 1996; Wickstrom et al. 1999; Marks et al. 2006; Algar and Brazell 2008). Interestingly, a recent suggestion that diets lacking arginine be tested as an alternative to poison, as it would be host-specific since cats require arginine and its absence from a single meal in a hungry cat can induce hyperammonemia in less than one hour, with death being inevitable (Spotte 2014).

A second form of eradication is via hunting. In many cases, hunting is a preferred technique for removing the last remaining cats at a site (Nogales et al. 2004), though it may be ineffective when cats are at low densities (Bester et al. 2002). However, hunting is often used in combination with other approaches to ensure complete eradication (e.g., Young et al. 2013). Similarly, trapping can be an effective means of removing cats but is often used in combination with other methods (Nogales et al. 2004). Because live traps have to be checked at least daily or require a mechanism to signal if a cat is caught, it is a labor-intensive approach (Benevides et al. 2008; Will et al. 2010). Given the effort required for live trapping, there have also been a variety of kill traps developed that are used in New Zealand, Australia, and the Pacific Islands (e.g., Glen et al. 2014).

A final form of eradication that has been applied in the control of feral cats is the use of disease. Specifically, diseases have been used as a biocontrol method of eradication and have been effective when used in combination with other techniques. For instance, feline panleukopenia virus (FPV) has been used to reduce cat populations on several oceanic islands (Rauzon 1985; Van Rensburg et al. 1987). Feline leukemia virus (FeLV) and feline immunodeficiency virus (FIV), both of which remain effective at low host densities, might serve as effective agents for cat control on islands and have the potential to greatly increase the efficiency of an eradication program when used with other techniques (Courchamp and Sugihara 1999). Notably, though, biological control has the potential to spread to owned cats or closely related species and not be effective in locations where the diseases and consequent natural immunity are already present.

An approach that is akin to eradication is removing feral cats from the environment and housing them elsewhere, such as in sanctuaries. Cat sanctuaries house feral cats in enclosed areas where they are provided with food, shelter, and veterinary services, helping to eliminate disease risk while removing their impact on wildlife (Winter 2003). Sanctuaries may be appropriate for very small populations, but are expensive and have a finite capacity, making them an impractical management tool in many contexts.

In contrast to management approaches that seek to reduce feral cat populations via increasing mortality (or removing cats from the landscape) are approaches that seek to limit recruitment via a reduction in birth rate. One approach that has been increasingly touted by animal welfare organizations in the United States is trap-neuter-return (TNR). Although there are several variations of this technique that go by subtly different acronyms (e.g., trap-vasectomy-hysterectomy-release [TVHR]; McCarthy et al. 2013), they are effectively similar to TNR in that they leave cats on the landscape (Lepczyk et al. 2013), and thus we do not differentiate them here. TNR is an in situ method in which feral cats are captured, sterilized, and returned to the environment. From a theoretical mathematical point of view, TNR is very appealing, as a population that is appropriately sterilized will fall to zero over time as births are eliminated. However, TNR assumes a closed population (i.e., no immigration), which simply does not exist in almost any free-ranging terrestrial vertebrate, and that managers can sterilize a sufficient portion of the population to reduce growth rates below replacement levels. Given the theoretical underpinning and lack of euthanasia, TNR is sold as a valid and humane form of management by organizations such as the Humane Society of the United States, Best Friends, and Maddie's Fund.

Unfortunately, TNR does not eliminate free-ranging cats (Castillo and Clarke 2003; Winter 2004; Foley et al. 2005), and thorough critiques (Dauphiné and Cooper 2009; Longcore et al. 2009; Duffy and Capece 2012; Spotte 2014) have indicated that its goal is to "keep cats alive as long as possible through provisioning, veterinary care, and favorable legislation" (Spotte 2014). In addition to the negative effects of TNR on wildlife (e.g., Longcore et al. 2009; Lepczyk et al. 2010), TNR cat colonies are an ineffective management option for disease risk reduction as they may actually increase risk as they "can increase immigration and kitten recruitment, which would lead to naive populations of cats that would be a source for zoonotic diseases including rabies and toxoplasmosis" (Gerhold and Jessup 2012). Because colonies that do not have adequate and continuing veterinary care could infect free-ranging domestic cats (Murphy et al. 1999), it could "impose a financial and emotional burden on adjacent cat owners that are either not usually considered in discussions of cat colonies, or are dismissed" (Duffy and Capece 2012). Furthermore, having a large number of cats in any colony simply poses a greater potential for disease transmission due to density dependence. Given that TNR is less cost effective (Lohr et al. 2013) and less preferred than removal approaches (Lohr and Lepczyk 2014; Lohr et al. 2014), coupled with the fact that no peer-reviewed studies have demonstrated it to work, the use of TNR for managing feral cats is a poor choice.

THE FUTURE OF MANAGING FERAL CATS

Managing feral cats needs to be conceived of over the long term, as no management options will yield immediate outcomes. Much like changing societal behaviors can take decades, pursuing management actions that addresses feral cats from all fronts is a multiyear or decade prospect in many instances. While small islands or individual properties can be managed effectively using current techniques, broadscale management will likely require using multiple techniques, changing policies, and changing human behavior.

Over the long term, managing feral cats requires managing both owned and unowned cats. In many places throughout the United States, owned cats are unlicensed, lack microchips, are unsterilized, and allowed to roam freely outdoors. As such, owned cats have opportunities to become feral, interact with feral cats, and contribute more individuals to feral populations when unwanted cats are abandoned. Thus, as part of a multipronged approach, one major component of reducing feral cat numbers over the long run is to increase responsible pet ownership. Namely, house cats need to be valued and cared for in a way that benefits both the cat and the environment.

The primary steps needed in owned cat management include increasing sterilization rates and reducing unwanted births. Policies and/or economic incentives can be used to encourage sterilization. Likewise, requiring licensing to breed cats has been considered as an approach to reduce supply and increase value in cats. At present, animal welfare organizations have a supply problem, in that they have more cats than they have space and homes for, which makes cats have low economic value. When people place greater economic value on living or nonliving items, they tend towards more responsible care and ownership of that item.

Aside from requiring that all owned cats be licensed and microchipped, policies are needed to require owned cats be kept indoors or confined to one's property. Although such policies do exist in a number of cities in the United States, enforcement is not always implemented. Concurrently with policies aimed at keeping cats indoors or on one's property, policies aimed at TNR-based approaches need to be challenged as they hold no scientific merit. Without working on owned cat issues in conjunction with feral cat issues, management will always be in a state of triage as new cats continue to flow into the environment.

In terms of managing feral cats in the future, it is critical that scientists and managers continue to challenge policies aimed at keeping feral cats on the landscape (e.g., Lepczyk and Rubinoff 2016). At the same time, there are both developing technologies and policies designed to add to the management toolbox. First, kill traps that use PAPP are currently undergoing efficacy testing in Australia (Read et al. 2014) and may provide an option for field implementation in the future. However, efficacy testing and governmental approval in the United States would be required prior to any field implementation. Second, while nonsurgical long-term or permanent contraception remains a holy grail of vertebrate pest management, recent work in mice has demonstrated the possibility of long-term infertility (Li et al. 2015) and may offer a new direction for managing cats (Williams 2015). In fact, the Michelson Prize has sought to incentivize "a low-cost, permanent, nonsurgical sterilant for male and female cats and dogs" by awarding $25 million to the first entity to accomplish this task. Third, Australia and New Zealand have been leaders in managing many invasive vertebrates, including feral cats, through the scale of their operations and techniques (e.g., traps and poisons) and, if successful, may provide evidence for similar management in the United States. Specifically, after reviews that linked cats to the extinction or decline of many native mammals (Woinarski et al. 2014, 2015), the Australian government swiftly implemented several initiatives focused on recovering threatened species via cat control. In 2014, a Threatened Species Commissioner was appointed to champion practical conservation actions to recover threatened biota, with a key focus on feral cats. A broad threatened species strategy set targets for the culling of two million feral cats, application of control measures over 10 million hectares, and the establishment of cat-free havens both on the continental mainland and on offshore islands (Commonwealth of Australia 2015; see Department of the Environment 2015 for a more detailed plan describing options for cat management). In addition to these initiatives, Aust$29.98 million was allocated to a Threatened Species Recovery Hub in 2015, with the first theme of this facility focusing on reducing the impacts of introduced predators on threatened mammals (http://www.nespthreatenedspecies.edu.au/). As a government-led effort to reduce the impacts of cats on wildlife, these initiatives demonstrate that there is political will and understanding of the problem in such a way as to facilitate change. In particular, Australia can serve as a model for the United States and other nations to follow in terms of implementing feral cat management and funding it.

CONCLUSION

Ultimately, if feral cat management in the United States is to be successful, a variety of approaches are needed in conjunction with enforcement of existing policies and the creation of new policies and management plans as well as funding. Given the continued growth of the owned cat population and laws promoting TNR, the feral cat problem is unlikely to decrease without a prolonged effort. As a result, it is most important to focus attention first on locations of greatest need, such as near protected areas, wildlife refuges, islands, and other locations of biological importance. However, large-scale efforts in both on-the-ground management and policy are needed for broader-scale success. As such, reductions in feral cat numbers need to be considered on more of a human generation time akin to other long-term societal changes.

REFERENCES

Abernethy, E.F., Turner, K.L., Beasley, J.C., DeVault, T.L., Pitt, W.C., and Rhodes, O.E., Jr. 2016. Carcasses of invasive species are predominantly utilized by invasive scavengers in an island ecosystem. *Ecosphere* 7:e01496. doi: 10.1002/ecs2.1496

Algar, D., and Brazell, R.I. 2008. A bait-suspension device for the control of feral cats. *Wildlife Research* 35:471–476.

Frank, A.S.K., Johnson, C.N., Potts, J.M., Fisher, A., Lawes, M.J., Woinarski, J.C.Z. et al. 2014. Experimental evidence that feral cats cause local extirpation of small mammals in Australia's tropical savannas. *Journal of Applied Ecology* 51:1486–1493.

Baker, D. 2001. Cat curfew-Casey City Council. In: *Proceedings of the 10th Urban Animal Management Conference*, 2001, Melbourne.

Baker, P.J., Soulsbury, C.D., Iossa, G., and Harris, S. 2010. Domestic cat (*Felis catus*) and domestic dog (*Canis familiaris*). In: *Urban Carnivores: Ecology, Conflict, and Conservation*. Gehrt, S.D., Riley, S.P.D., and Cypher, B.L. (eds.). Johns Hopkins University Press, Baltimore, MD, pp. 157–172.

Balogh, A.L., Ryder, T.B., and Marra, P.P. 2011. Population demography of Gray Catbirds in the suburban matrix: Sources, sinks and domestic cats. *Journal of Ornithology* 152:717–726.

Barrett, M.A., Telescol, D.J., Barrett, S.E., Widness, K.M., and Leone, E.H. 2014. Testing bear-resistant trashcans in residential areas of Florida. *Southeastern Naturalist* 13:26–39.

Benevides, F.L., Jr., Hansen, H., and Hess, S.C. 2008. Design and evaluation of a simple signaling device for live traps. *Journal of Wildlife Management* 72:1434–1436.

Bester, M.N., Bloomer, J.P., van Aarde, R.J., Erasmus, B.H., van Rensburg, P.J., Skinner, D., Howell, P.G., et al. 2002. A review of the successful eradication of feral cats from sub-Antarctic Marion Island, southern Indian Ocean. *South African Journal of Wildlife Research* 32:65–73.

Bonnaud, E., Medina, F.M., Vidal, E., Nogales, M., Tershy, B., Zavaleta, E., Donlan, S.V. et al. 2011. The diet of feral cats on islands: A review and a call for more studies. *Biological Invasions* 13:581–603.

Brackenridge, W.D. 1841. Journal kept while on the U.S. exploring expedition, 1838–1841. Unpublished manuscript at the Maryland Historical Society.

Campbell, K.J., Harper, G., Algar, D., Hanson, C.C., Keitt, B.S., and Robinson, S. 2011. Review of feral cat eradications on islands. In: *Island Invasives: Eradication and Management*. Veitch, C.R., Clout, M.N., and Towns, D.R. (eds.). IUCN, Gland, Switzerland, pp. 37–46.

Castillo, D., and Clarke, A. 2003. Trap/neuter/release methods ineffective in controlling domestic cat 'colonies' on public lands. *Natural Areas Journal* 23:247–253.

Churcher, P.B., and Lawton, J.H. 1987. Predation by domestic cats in an English village. *Journal of Zoology* 212:439–455.

Coleman, J.S., and Temple, S.A. 1993. Rural residents' free-ranging domestic cats: A survey. *Wildlife Society Bulletin* 21:381–390.

Coman, B.J., and Brunner, H. 1972. Food habits of the feral house cat in Victoria. *Journal of Wildlife Management* 36:848–853.

Commonwealth of Australia. 2015. *Threatened Species Strategy*. Commonwealth of Australia, Canberra.

Courchamp, F., and Sugihara, G. 1999. Modeling the biological control of an alien predator to protect island species from extinction. *Ecological Applications* 9:112–123.

Craven, R.B., Maupin, G.O., Beard, M.L., Quan, T.J., and Barnes, A.M. 1993. Reported cases of human plague infections in the United States, 1970–1991. *Journal of Medical Entomology* 30:758–761.

Crooks, K.R., and Soulé, M.E. 1999. Mesopredator release and avifaunal extinctions in a fragmented system. *Nature* 400:563–566.

Dauphiné, N., and Cooper, R.J. 2009. Impacts of domestic cats (*Felis catus*) on birds in the United States: A review of recent research with conservation and management recommendations. In: *Proceedings of the 4th International Partners in Flight Conference: Tundra to Tropics*, pp. 205–219.

Department of the Environment. 2015. *Threat Abatement Plan for Predation by Feral Cats*. Commonwealth of Australia, Canberra.

Dickman, C.R. 1996. *Overview of the Impacts of Feral Cats on Australian Native Fauna*. Australian Nature Conservation Agency, Canberra.

Dickman, C.R. 2009. House cats as predators in the Australian environment: Impacts and management. *Human–Wildlife Conflicts* 3:41–48.

Dickman, C.R., and Newsome, T.M. 2015. Individual hunting behaviour and prey specialization in the house cat *Felis catus*: Implications for conservation and management. *Applied Animal Behaviour Science* 173:76–87.

Dubey, J.P. 2002. A review of toxoplasmosis in wild birds. *Veterinary Parasitology* 106:121–153.

Dubey, J.P. 2010. *Toxoplasmosis of Animals and Humans*, 2nd ed. CRC Press, Boca Raton, Florida.

Dubey, J.P., and Jones, J.L. 2008. *Toxoplasma gondii* infection in humans and animals in the United States. *International Journal of Parasitology* 38:1257–1278.

Duffy, D.C., and Capece, P. 2012. Biology and Impacts of Pacific Island Invasive Species.7. The Domestic Cat (*Felis catus*). *Pacific Science* 66:173–212.

Eason, C.T., Murphy, E.C., Hix, S., Henderson, R.J., and MacMorran, D. 2010a. *Susceptibility of Four Bird Species to Paraaminopropiophenone (PAPP)*. DOC Res. Dev. Ser. 320. Department of Conservation, Wellington, New Zealand.

Eason, C.T., Murphy, E.C., Hix, S., Henderson, R.J., and MacMorran, D. 2010b. Development of a new humane toxin for predator control in New Zealand. *Integrated Zoology* 5:31–36.

Eng, T.T., and Fishbein, D.B. 1990. Epidemiologic factors, clinical findings, and vaccination status of rabies in cats and dogs in the United States in 1988. *Journal of the American Veterinary Medical Association* 197:201–209.

Errington, P.L. 1936. Notes on food habits of southern Wisconsin house cats. *Journal of Mammalogy* 17:64–65.

Esmerini, P.O., Gennari, S.M., and Pena, H.F. 2010. Analysis of marine bivalve shellfish from the fish market in Santos city, São Paulo state, Brazil, for *Toxoplasma gondii*. *Veterinar Parasitology* 170:8–13.

Fisher, P., Algar, D., Murphy, E., Johnston, M., and Eason, C. 2015. How does cat behavior influence the development and implementation of monitoring techniques and lethal control methods for feral cats? *Applied Animal Behaviour Science* 173:88–96.

Fitzgerald, B.M., and Karl, B.J. 1979. Foods of feral house cats (*Felis catus* L.) in forest of the Orongorongo Valley, Wellington, New Zealand. *New Zealand Journal of Zoology* 6:107–126.

Fitzgerald, B.M., and Turner, D.C. 2000. Hunting behaviour of domestic cats and their impact on prey populations. In: *The Domestic Cat: The Biology of its Behavior*. 2nd ed. Turner, D.C., and Bateson, P. (eds.). Cambridge University Press, Cambridge, pp. 151–175.

Foley, P., Foley, J.E., Levy, J.K., and Paik, T. 2005. Analysis of the impact of trap-neuter-return programs on populations of feral cats. *Journal of American Veterinary Medical Association* 227:1775–1781.

Forbush, E.H. 1916. The domestic cat: Bird killer, mouser, and destroyer of wildlife, means of utilizing and controlling it. *The Commonwealth of Massachusetts State Board of Agriculture Economic Biology Bulletin* 2.

Gehrt, S.D., and Riley, S.P.D. 2010. Coyotes (*Canis latrans*). In: *Urban Carnivores: Ecology, Conflict, and Conservation*. Gehrt, S.D., Riley, S.P.D., and Cypher, B.L. (eds.). Johns Hopkins University Press, Baltimore, MD, pp. 79–96.

Gehrt, S.D., Wilson, E.C., Brown, J.L., and Anchor, C. 2013. Population ecology of free roaming cats and interference competition by coyotes in urban parks. *PLoS ONE* 8(9):e75718. doi:10.1371/journal.pone.0075718.

Gerhold, R.W., and Jessup, D.A. 2012. Zoonotic diseases associated with free-roaming cats. *Zoonoses and Public Health* 60:189–195.

Glen, A.S., Warburton, B., Cruz, J., and Coleman, M. 2014. Comparison of camera traps and kill traps for detecting mammalian predators: A field trial. *New Zealand Journal of Zoology* 41:155–160.

Hess, S.C. 2011. By land and by sea—The widespread threat of feral cats on Hawaiian wildlife. *The Wildlife Professional*, Spring 2011:66–67.

Honnold, S.P., Braun, R., Scott, D.P., Sreekumar, C., and Dubey, J.P. 2005. Toxoplasmosis in a Hawaiian monk seal (*Monachus schauinslandi*). *Journal of Parasitology* 91:695–697.

Hu, Y., S. Hu, W. Wang, X. Wu, F.B. Marshall, X. Chen, L. Hou, et al. 2014. Earliest evidence for commensal processes of cat domestication. *Proceedings of the National Academy of Sciences* 111:116–120.

Jackson, W.B. 1951. Food habits of Baltimore, Maryland, cats in relation to rat populations. *Journal of Mammalogy* 32:458–461.

Jarvis, P.J. 1990. Urban cats as pests and pets. *Environmental Conservation* 7:169–171.

Jessup, D.A. 2004. The welfare of feral cats and wildlife. *Journal of the American Veterinary Medical Association* 225, 1377–1383.

Jones, E., and Coman, B.J. 1981. Ecology of the feral cat, *Felis catus* (L.), in southeastern Australia. I. Diet. *Australian Wildlife Research* 8:537–547.

King, C. (Ed.). 1984. *Immigrant Killers: Introduced Predators and the Conservation of Birds in New Zealand*. Oxford University Press, Auckland.

Kreuder, C., Miller, M.A., Jessup, D.A., Lowenstine, L.J., Harris, M.D., Ames, J.A., Carpenter, T.E., et al. 2003. Patterns of mortality in southern sea otters (*Enhydra lutris nereis*) from 1998–2001. *Journal of Wildlife Diseases* 39:495–509.

Lafferty, K.D. 2006. Can the common brain parasite, *Toxoplasmosis gondii*, influence human culture? *Proceedings of the Royal Society B* 273:2749–2755.

Lepczyk, C.A., Bird, D.M., Calver, M., Conant, S., Duffy, D., Duval, F., Hutchins, M. et al. 2013. Feral cat management: Letter to editor regarding McCarthy et al. *Journal of American Veterinary Medical Association* 243:1301–1392.

Lepczyk, C.A., Dauphiné, N., Bird, D.M., Conant, S., Cooper, R.J., Duffy, D.C., Hatley, P.J., et al. 2010. What conservation biologists can do regarding trap-neuter-return: Response to Longcore et al. *Conservation Biology* 24:627–629.

Lepczyk, C.A., Lohr, C.A., and Duffy, D.C. 2015. A review of cat behavior in relation to disease risk and management options. *Applied Animal Behaviour Science* 173:29–39.

Lepczyk, C.A., Mertig, A.G., and Liu, J. 2004. Landowners and cat predation across rural-to-urban landscapes. *Biological Conservation* 115:191–201.

Lepczyk, C.A., and Rubinoff, D. 2016. Ban on feeding feral cats would be good for the cats, residents, Hawaii. Island Voices editorial. *Honolulu Star-Advertiser*. February 21.

Lepczyk, C.A., van Heezik, Y., and Cooper, R.J. 2011. An issue with all-too-human dimensions. *The Wildlife Professional* Spring 2011, 68–70.

Li, J., Olvera, A.I., Akbari, O.S., Moradian, A., Sweredoski, M.J., Hess, and S., Ha, B.A. 2015. Vectored antibody gene delivery mediates long-term contraception. *Current Biology* 25:R820–R822.

Lindsay, D.S., Collins, M.V., Mitchell, S.M., Wetch, C.N., Rosypal, A.C., Flick, G.J., Zajac, A.M., et al. 2004. Survival of *Toxoplasma gondii* oocysts in eastern oysters (*Crassostrea virginica*). *Journal of Parasitology* 90:1054–1057.

Lohr, C.A., Cox, L.J., and Lepczyk, C.A. 2013. The costs and benefits of trap-neuter-release and euthanasia removal in urban cat programs: The case of O'ahu Hawai'i. *Conservation Biology* 27:64–73.

Lohr, C.A., and C.A. Lepczyk. 2014. Desires and management preferences of stakeholders regarding feral cats in the Hawaiian islands. *Conservation Biology* 28:392–403.

Lohr, C.A., Lepczyk, C.A., and Cox, L.J. 2014. Identifying people's most preferred management technique for feral cats in Hawaii. *Human–Wildlife Interactions* 8:56–66.

Long, K., and Robley, A. 2004. *Cost Effective Feral Animal Exclusion Fencing for Areas of High Conservation Value in Australia*. Department of the Environment and Heritage, Canberra, Australia. http://nla.gov.au/nla.arc-51699.

Longcore, T., Rich, C., and Sullivan, L.M., 2009. Critical assessment of claims regarding management of feral cats by trap-neuter-return. *Conservation Biology* 23:887–894.

Loss, S.R., Will, T., and Marra, P.P. 2013. The impact of free-ranging domestic cats on wildlife of the United States. *Nature Communications* 4:1396.

MacDonald, M.L., Rogers, Q.R., and Morris, J.G. 1984. Nutrition of the domestic cat, a mammalian carnivore. *Annual Review of Nutrition* 4:521–562.

Marks, C.A., Johnston, M.J., Fisher, P.M., Pontin, K., and Shaw, M.J. 2006. Differential particle size ingestion: Promoting target specific baiting of feral cats. *Journal of Wildlife Management* 70:1119–1124.

McAllister, M.M. 2005. A decade of discoveries in veterinary protozoology changes our concept of subclinical toxoplasmosis. *Veterinary Parasitology* 132:241–247.

McCarthy, R.J., Levine, S.H., and Reed, J.M. 2013. Estimation of effectiveness of three methods of feral cat population control by use of a simulation model. *Journal of the American Veterinary Medical Association* 243:502–511.

McKay, G.M. 1996. Feral cats in Australia: Origins and impacts. In: *Unwanted Aliens? Australia's Introduced Animals*. McKay, G.M. (ed.). The Rocks (Australia): Nature Conservation Council of New South Wales, pp. 9–15.

Medina, F.M., Bonnaud, E., Vidal, E., Tershy B.R., Zavaleta, E.S., Donlan, C.J., Keitt, B.S., et al. 2011. A global review of the impacts of invasive cats on island endangered vertebrates. *Global Change Biology* 17:3503–3510.

Medway, D.G. 2004. The land bird fauna of Stephens Island, New Zealand in the early 1890s, and the cause of its demise. *Notornis* 51:201–211.

Mills, D.S., Bailey, S.L., and Thurstans, R.E. 2000. Evaluation of the welfare implications and efficacy of an ultrasonic 'deterrent' for cats. *Veterinary Record* 147:678–680.

Moodie, E. 1995. *The Potential for Biological Control of Feral Cats in Australia*. Australian Nature Conservation Agency, Canberra.

Morgan, D.R., Innes, J., Ryan, C., and Meikle, L. 1996. Baits and baiting strategies for multispecies pest control and feral cats. In: *Science for Conservation*, vol. 40. Department of Conservation, Wellington, New Zealand.

Murphy, F.A., Gibbs, P.J. Studdert, M.J., and Horzinek, M.C. 1999. *Veterinary Virology*. 3rd ed. Academic Press, New York.

Natoli, E. 1994. Urban feral cats (*Felis catus* L.): Perspectives for a demographic control respecting the psycho-biological welfare of the species. Ann. 1st. Super. *Sanita* 30:223–227.

Nelson, S.H., Evans, A.D., and Bradbury, R.B. 2006. The efficacy of an ultrasonic cat deterrent. *Applied Animal Behavior Science* 96:83–91.

Nogales, M., Martín, A., Tershy, B.R., Donlan, C.J., Veitch, D., Puerta, N., Wood, B., et al. 2004. A review of feral cat eradication on islands. *Conservation Biology* 18:310–319.

Nutter, F.B., Levine, J.F., and Stoskopf, M.K. 2004. Reproductive capacity of free-roaming domestic cats and kitten survival rate. *Journal of the American Veterinary Medical Association* 225:1399–1402.

Olson, Z.H., Beasley, J.C., and Rhodes, O.E. Jr. 2016. Carcass type affects local scavenger guilds more than habitat connectivity. *PLoS ONE* 11(2):e0147798. doi:10.1371/journal.pone.0147798.

Palmer B.S. 2007. Meta-analysis of three case controlled studies and an ecological study into the link between cryptogenic epilepsy and chronic toxoplasmosis infection. *Seizure* 16:657–663.

Patronek, G.J. 1998. Free-roaming and feral cats—their impact on wildlife and human beings. *Journal of the American Veterinary Medical Association* 212:218–226.

Pearre Jr., S., and Maass, R., 1998. Trends in the prey size-based trophic niches of feral and House Cats *Felis catus* L. *Mammal Review* 28:125–139.

Peterson, M.N., Hartis, B., Rodriguez, S., Green, M., and Lepczyk, C.A. 2012. Opinions from the front lines of cat colony management conflict. *PLoS ONE* 7(9): e44616. DOI:10.1371/journal.pone.0044616.

Pimentel, D., Lach, L., Zuniga, R., and Morrison, D. 2000. Environmental and economic costs of nonindigenous species in the United States. *Bioscience* 50:53–65.

Prentiss, P.G., Wolf, A.V., and Eddy, H.A. 1959. Hydropenia in cat and dog. Ability of the cat to meet its water requirements solely from a diet of fish or meat. *American Journal of Physiology* 196:625–632.

Proulx, G. 1988. Control of urban wildlife predation by cats through public education. *Environmental Conservation* 15:358–359.

Ragg, J.R., Mackintosh, C.G., and Moller, H. 2000. The scavenging behaviour of ferrets (*Mustela furo*), feral cats (*Felis domesticus*), possums (*Trichosurus vulpecula*), hedgehogs (*Erinaceus europaeus*) and harrier hawks (*Circus approximans*) on pastoral farmland in New Zealand: Implications for bovine tuberculosis transmission. *New Zealand Veterinary Journal* 48:166–175.

Rauzon, M.J. 1985. Feral cats on Jarvis Island: Their effects and their eradication. *Atoll Research Bulletin* 282:1–32.

Read, J., Gigliotti, F., Darby, S., and Lapidge, S. 2014. Dying to be clean: Pen trials of novel cat and fox control devices. *International Journal of Pest Management* 60:166–172.

Rodríguez, C., Torres, R., and Drummond, H., 2006. Eradicating introduced mammals from a forested tropical island. *Biological Conservation* 130:98–105.

Say, L., Pontier, D., and Natoli, E. 1999. High variation in multiple paternity of domestic cat (*Felis catus* L.) in relation to environmental conditions. *Proceedings of the Royal Society of London B: Biological Sciences* 266:2071–2074.

Schmidt, P.M., Lopez, R.R., and Collier, B.A. 2007. Survival, fecundity, and movements of free-roaming cats. *Journal of Wildlife Management* 71:915–919.

Sherley, M. 2007. Is sodium fluoroacetate (1080) a humane poison? *Animal Welfare* 16:449–458.

Shwiff, S.A., Gebhardt, K., Kirkpatrick, K.N., and Shwiff, S.S. 2010. Potential economic damage from introduction of brown tree snakes, *Boiga irregularis* (Reptilia: Colubridae), to the Islands of Hawai'i. *Pacific Science* 64:1–10.

Simberloff, D. 2013. *Invasive Species: What Everyone Needs to Know.* Oxford University Press, New York, NY.

Smith, D.D., and Frenkel, J.K. 1995. Prevalence of antibodies to *Toxoplasma gondii* in wild mammals of Missouri and east central Kansas: Biologic and ecologic considerations of transmission. *Journal of Wildlife Diseases* 31:15–21.

Spotte, S. 2014. *Free-Ranging Cats: Behavor, Ecology, Management.* John Wiley, UK.

Torrey, E.F, Bartko, J.J., Lun, Z.R., and Yolken, R.H. 2007. Antibodies to *Toxoplasma gondii* in patients with schizophrenia: A meta-analysis. *Schizophrenia Bulletin* 33:729–736.

Torrey, E.F., Simmons, W., and Yolken, R.H. 2015. Is childhood cat ownership a risk factor for schizophrenia later in life? *Schizophrenia Research* 165:1–2.

van Heezik, Y., Smyth, A., Adams, A., and Gordon, J., 2010. Do domestic cats impose an unsustainable harvest on urban bird populations? *Biological Conservation* 143:121–130.

Van Rensburg, P.J., Skinner, J.D., and Van Aarde, R.J. 1987. Effects of feline panleucopaenia on the population characteristics of feral cats on Marion Island. *Journal of Applied Ecology* 24:63–73.

Wallace, J.L., and Levy, J.K. 2006. Population characteristics of feral cats admitted to seven trap-neuter-return programs in the United States. *Journal of Feline Medicine and Surgery* 8:279–284.

Warner, R.E. 1985. Demography and movements of free-ranging domestic cats in rural Illinois. *The Journal of Wildlife Management* 49:340–346.

Wickstrom, M., Thomas, M., Henderson, R., and Eason, C.T. 1999. Development and evaluation of baits for feral cat control. *Science for Conservation* 127:67–74.

Will, D., Hanson, C.C., Campbell, K.J., Garcelon, D.K., and Keitt, B.S. 2010. A trap monitoring system to enhance efficiency of feral cat eradication and minimize adverse effects on nontarget species on San Nicolas Island. In: *Proceedings, 24th Vertebrate Pest Conference*. Sacramento, California, pp. 79–85.

Williams, S.C.P. 2015. DNA "vaccine" sterilizes mice, could lead to one-shot birth control. *Science* DOI: 10.1126/science.aad4672.

Winter, L. 2003. Popoki and Hawai'i's native birds. *'Elepaio* 63:43–46.

Winter, L. 2004. Trap-neuter-release programs: The reality and impacts. *Journal of the American Veterinary Medical Association* 225:1369–1376.

Woinarski, J.C.Z., Burbidge, A.A., Harrison, P.L. 2014. *The Action Plan for Australian Mammals 2012*. CSIRO Publishing, Melbourne.

Woinarski, J.C.Z., Burbidge, A.A., Harrison, P.L. 2015. Ongoing unraveling of a continental fauna: Decline and extinction of Australian mammals since European settlement. *Proceedings of the National Academy of Sciences* 112:4531–4540.

Work, T.M., Massey, J.G., Lindsay, D.S., and Dubey, J.P. 2002. Toxoplasmosis in three species of native and introduced Hawaiian birds. *Journal of Parasitology* 88:1040–1042.

Work, T.M., Massey, J.G., Rideout, B.A., Gardiner, C.H., Ledig, D.B., Kwok, O.C.H., and Dubey, J.P. 2000. Fatal toxoplasmosis in free-ranging endangered alala from Hawaii. *Journal of Wildlife Disease* 36:205–212.

Young, L.C., VanderWerf, E.A., Lohr, M.T., Miller, C.J., Titmus, A.J., Peters, D., and Wilson, L., 2013. Multispecies predator eradication within a predator-proof fence at Ka'ena Point, Hawai'i. *Biological Invasions* 15:2627–2638.

14 Feral Goats and Sheep

Steven C. Hess, Dirk H. Van Vuren,
and Gary W. Witmer

CONTENTS

ORIGIN, ANCESTRY, AND DOMESTICATION

Sheep and goats are among the earliest animals domesticated by mankind (Zeder 2009). Both goats and sheep may have made better candidates for domestication than other animals like deer because they follow a single dominant leader, the herdsman (Geist 1971). They now have a nearly ubiquitous worldwide distribution, and they are among the most abundant of all commensal animals. However, they have also become some of the most widespread invasive feral mammals, particularly on the 100 or more islands throughout the world where they have been introduced, causing severe damage to island ecosystems, in some cases for hundreds of years (Rudge 1984; Chynoweth 2013). Problems caused by feral goats and sheep are a subset of the larger problem of domestic livestock and natural systems. Feral goats are perhaps more widespread than feral sheep because goats have not been as highly modified by the process of domestication (Francis 2015).

The Bezoar ibex (*Capra aegagrus*) is the most likely ancestor of domestic goats (*C. hircus*) from both genetic and paleontological evidence (Pidancier et al. 2006). The domestication process started at least 10,000 years ago in highlands of western Iran, beginning with the selective harvesting of subadult males and the transition from hunting to herding of the species (Zeder and Hesse 2000). Multiple independent domestication events may have occurred or domestication may have incorporated multiple ancestral lineages (Pidancier et al. 2006). Traits selected during domestication include behavior, dairy, meat, skins, pelage color, mohair, cashmere, horns, pathogen resistance, and even intestines for catgut. Selection for reduced body size may have been related to the ability to better survive in hot and arid environments (Zeder 2009). A profound reduction in horn size occurred after humans began to control breeding, particularly in males, possibly associated with the absence of selective pressures for large horns used in mate competition (Zeder 2009).

The West Asiatic mouflon (*Ovis orientalis*) is generally recognized as the wild ancestor of domestic sheep (*O. aries*; Zohary et al. 1998). Herd demographics may have been manipulated to maximize harvests in northeastern Iraq and southeastern Turkey as early as 12,000 years ago (Zeder 2009). Further domestication involved breeding for high yields of wool, meat, and milk, and for docility (Ryder 1983). Moreover, the nature of the fleece changed subsequent to domestication; mouflon fleece was an annually molted brown coat consisting of wooly underfur covered by guard hairs, whereas domestic sheep fleece consists only of white wooly underfur that is not molted and grows continuously (Ryder 1987). Further, domestication may have conferred resistance to common parasites and diseases of wild sheep through selective breeding. Lungworm (*Muellerius* spp.) infections typically do not produce clinical signs in domestic sheep (Pugh 2002) but may be more pathogenic in non-adapted hosts such as bighorn sheep (*O. canadensis*; Demartini and Davies 1977; Pybus and Shave 1984) and possibly mouflon (Panayotova-Pencheva and Alexandrov 2010; Powers et al. 2014).

FORAGE AND WATER NEEDS

Domestic sheep are generally considered grazers, but feral sheep are forage generalists, consuming a wide variety of grasses, forbs, and shrubs according to availability (Van Vuren and Coblentz 1987; Wood et al. 1987). Goats are often regarded as browsers; however, preference for grazing or browsing is determined primarily by environmental conditions, such as seasonal and geographic variation of forage, and thus goats may be considered mixed feeding opportunists (Lu 1988). Goats primarily derive water directly from plant foods in many environments (Robbins 1994), but have also been observed drinking salt water (Gould Burke 1988). While domestic goats have a minimum water requirement of 1.0%–1.5% body weight per day, selective pressures may enable feral goats to survive in environments with even less available water (Dunson 1974). This ability to survive in the absence of permanent water sources enables feral goats to persist in remarkably arid environments (Figure 14.1).

LIBERATION AND BECOMING FERAL

Feral sheep have become established at only a few locations in the United States, nearly all of them islands. The discovery of the Hawaiian Islands by Europeans, like many other islands of the Pacific, marked the beginning of introductions of many animals for labor, milk, and meat. Notably among these were domestic goats and sheep to establish strategic resupply outposts for ships on Cook's voyage in 1778–1779 and on Vancouver's voyages in 1793 and 1794 (Tomich 1986). These livestock proliferated without any predators or competitors and quickly became feral. Sheep were reported at the summit of Mauna Kea, the highest peak in the Pacific, only 32 years after their introduction to Hawaii (Ellis 1917). Later introductions to the Hawaiian Islands included European mouflon (*Ovis musimon*) from the Mediterranean Islands, which are small wild sheep closely related to the early

FIGURE 14.1   Feral goats were able to persist in Hawaii Volcanoes National Park for more than a century and a half despite the complete absence of surface water, primarily deriving needed water from plants. (Photo courtesy of Jack Jeffrey Photography.)

ancestors of domestic sheep. Mouflon were first introduced to the island of Lānaʻi in 1954 for sport hunting opportunities. They were then intentionally hybridized with feral sheep and released between 1962 and 1966 on Mauna Kea, Hawaii Island (Giffin 1982; Tomich 1986; Scowcroft and Conrad 1992). A third mouflon population was founded between 1968 and 1974 at the Kahuku Ranch on Mauna Loa, which began to merge with the Mauna Kea population by the end of the twentieth century (Hess et al. 2006; Ikagawa 2014).

Domestic sheep also were introduced to Shackleford Banks, a barrier island off the coast of North Carolina, as early as the late 1700s (Wood et al. 1987), presumably as part of a ranching operation. The sheep were considered semiferal by the 1940s (Engels 1952) and feral by the late 1970s (Wood et al. 1987). Similarly, domestic sheep were introduced to Santa Cruz Island, California, during the 1850s for ranching; the breed was probably Spanish merino (Van Vuren and Bakker 2009). Sheep on Santa Cruz Island were actively ranched during the late 1800s, including the importation of Leicester and Rambouillet rams, presumably to improve breed quality. However, ranching activities declined by the early 1900s, and the sheep became increasingly feral; by the 1930s, they were completely feral. Domesticated livestock that become feral are released from artificial selection imposed by animal husbandry and might be exposed to environmental pressures to which their domesticated traits are ill-suited. For sheep, ever-growing wool that is no longer shorn can cause thermal stress in warm environments; further, unmolted tail wool fouled with feces can increase the risk of fly strike disease. Feral sheep also often face increased intraspecific competition, which might favor a smaller body size than that preferred by sheep breeders. Hence, sheep that have become feral often exhibit traits such as a reduced body size; spontaneous shedding of their wool, especially on the tail; and a reversion to a pigmented fleece (Van Vuren and Bakker 2009). Some of these traits might have commercial, scientific, or aesthetic value (Van Vuren and Hedrick 1989); for instance, Santa Cruz Island sheep are considered a unique heritage breed by the Livestock Conservancy. Although the environmental tolerance of feral sheep has not been determined, they have persisted for long periods at altitudes ranging from sea level to 2900 m, and at latitudes ranging from the equatorial tropics to the subarctic on the Island of St. Kilda (57°N).

Early introductions and long periods of isolation on islands have caused some feral goat populations to experience substantial genetic drift, producing unique recombinations of breeds. Feral goats present on San Clemente Island and Santa Catalina Island, California, are thought to have descended from those brought by the Spanish in the seventeenth century or the English in the eighteenth century (Coblentz 1978); goats had been brought to San Clemente from Santa Catalina Island in 1875. Original breeds included La Blanca Celtiboras, La Castellana Extremenas, and common dairy and meat goats of Spain, the Malagueñas and Murcianas (Dohner 2001). San Clemente Island domestic goats derived from feral populations have been recognized as a unique heritage breed by the Livestock Conservancy. Goats also were introduced in 1592 to the small, dry, Caribbean island of Mona, between Hispaniola and Puerto Rico, by Spanish explorers and were reportedly abundant enough to sustain hunting by 1632. They were hunted continuously during part of the nineteenth century to feed guano miners, but hunting was curtailed in the 1970s to allow populations to rebound. Goats also have been introduced to other parts of the Puerto Rican archipelago and the U.S. Virgin Islands, but were removed from Buck Island by the late 1940s to allow for more natural ecological conditions. Goats have been introduced widely throughout the Hawaiian Islands, have repeatedly escaped captivity and become feral, and are able to persist in some of the most arid environments with minimal vegetation (Chynoweth et al. 2013). Goats were introduced to Guam, now a territory of the United States, during the Spanish colonization and had become feral by the early 1700s, but numbers were decimated by 1801 because of overhunting. A small population persists today on cliffs on the northern portion of the island (Conry 1988). Feral goats altered native forests for more than a century on Sarigan and Anatahan in the Commonwealth of the Northern Mariana Islands, U.S. possessions in the western Pacific, but were eradicated from Sarigan in 1998 and from Anatahan in 2005 (Kessler 2011). Goats had also been introduced to the remote, small Pacific island of Jarvis at an unknown time but were extirpated by 1935 (Hess and Jacobi 2011).

ECOLOGICAL EFFECTS

Impacts of feral goats and sheep on islands are exacerbated by the fact that islands seldom support native large mammals, especially herbivores and carnivores. Hence, plants that evolved on islands may lack, or may have lost, defenses against herbivory, rendering them especially vulnerable to introduced herbivores (Coblentz 1978; Bowen and Van Vuren 1997). Further, these herbivores typically enjoyed a predator-free environment and little competition from other large vertebrate herbivores, potentially explaining why sheep and feral goat populations on islands have reached extraordinarily high densities (Bowen and Van Vuren 1999; Burness et al. 2001).

The introduction of sheep and goats to islands has generally resulted in the degradation of native vegetation, particularly endemic species. Feral sheep on Santa Cruz Island ate most vegetation within reach, and the result in some areas was a grassland of greatly reduced cover and height of herbaceous species and a fourfold increase in the extent of bare ground (Van Vuren and Coblentz 1987). In woody communities, consumption of shrubs and trees by feral sheep altered the growth form of larger plants and prevented shrub regeneration by removal of all seedlings (Hobbs 1980; Van Vuren and Coblentz 1987). On the Island of Hawaii, browsing by feral sheep prevented regeneration of endemic vegetation, including māmane seedlings (*Sophora chrysophylla*; Scowcroft and Giffin 1983), while bark-stripping by sheep caused direct mortality of mature māmane trees (Figure 14.2; Scowcroft and Sakai 1983). On the Puerto Rican island of Mona, 12 of 86 plant species consumed by goats were under protection status and four of the most commonly consumed species needed special conservation attention (Meléndez-Ackerman et al. 2008). Goats were also believed to compete with an endemic iguana species for herbaceous forage

FIGURE 14.2   Feral sheep in Hawaii have been responsible for the degradation of forest environments and the endangerment of native bird species. (Photo courtesy of Jack Jeffrey Photography.)

on Mona, and to be detrimental to plant populations on other Puerto Rican islands (Wiedwandt 1977).

While consumption or trampling results in direct damage and loss of plants, feral goats and sheep also cause numerous indirect effects, including soil erosion (Coblentz 1978). After protective ground layers of vegetation have been removed, bare substrates become exposed to physical disturbance from hoof action, precipitation, and wind, all of which may further exacerbate erosion. Hoof action has been implicated in the large-scale ecosystem process of soil compaction, which makes soils hydrophobic (i.e., reduces soil water penetration), causing greater soil runoff. For example, hoof action by feral sheep on Santa Cruz Island resulted in extensive disturbance to soil surface particles (Van Vuren 1982), a significant increase in soil compaction (Brumbaugh 1980), and the denudation of over 7% of the soil surface due to trail formation (Van Vuren and Coblentz 1987). The combination of vegetation removal and hoof action by feral sheep has resulted in a dramatic increase in erosion, including gully formation and landslides (Brumbaugh 1980; Pinter and Vestal 2005). Further, stream flows were altered; water infiltration into the soil decreased and surface runoff increased, resulting in higher stream flows early in the season (Van Vuren et al. 2001). On Kahoʻolawe Island of the Hawaiian chain, as much as 2.4 m of soil has been lost in some areas and an additional 1.9 million tons were lost annually as a result of goat and sheep impacts (Kramer 1971; Kahoʻolawe Island Conveyance Commission 1993; Loague et al. 1996), and Yocom (1967) estimated that approximately 1.9 m of soil was lost as a result of goat activity in some areas of Haleakalā Crater on Maui. Damage to nearshore marine environments from silt has been attributed to erosion caused by goats and sheep as well (Kahoʻolawe Island Conveyance Commission 1993; Stender et al. 2014).

Defoliation and trampling by feral sheep can alter vegetation structure as well, including depletion of the herbaceous layer and removal of all shrub leaves within reach (Van Vuren and Coblentz 1987). Further, because sheep can completely defoliate low-stature shrubs and prevent regeneration in tall-stature shrubs and trees, wholesale conversion of plant communities has occurred in areas where feral sheep are present. On the Island of Hawaii, feral sheep caused the destruction of the māmane forest and reduced native cover (Warner 1960; Scowcroft 1983). On Santa Cruz Island, sheep caused a major reduction in the extent and density of woody vegetation (Cohen et al. 2009), including a drastic reduction in the coastal sage scrub community (Brumbaugh 1980), which consists of low-growing shrubs palatable to sheep. Moreover, feral sheep grazing promotes invasive grasses over native grasses (Van Vuren and Bowen 2012).

Community-level interactions among plants and herbivores can aggravate changes in ecosystem processes, leading to positive feedback cycles that reinforce community change. Herbivory can promote productivity of grasses and accelerate ecosystem processes by enhancing the light regime of grasses and by the deposition of feces (Frank et al. 1998, 2002). The proliferation of alien grasses further contributes to changes in nutrient cycling, hydrology, fire regimes, and gradual conversion from forest to savanna and grassland environments (D'Antonio and Vitousek 1992). Grasses throughout the tropics are instrumental in determining fire frequency and severity, often resulting in a positive feedback loop known as the "grass/fire cycle" whereby forest and woodland become gradually replaced by grassland (D'Antonio and Vitousek 1992). While feral goats and sheep may reduce fine fuel loads that lead to fires, they also commit ecosystems to directional change from forests to savannas and grasslands.

Alteration of the composition and structure of plant communities by feral goats and sheep has implications for associated species that depend on those communities for habitat, especially birds. For example, degradation of the māmane forest on the Island of Hawaii caused the rarity of the palila (*Loxioides bailleui*), an endangered bird that relies almost entirely on māmane trees for food (Banko et al. 2002; Hess et al. 2014). Even a moderate alteration of shrub community structure by feral sheep on Santa Cruz Island, involving removal of all lower leaves of shrubs and depletion of the herbaceous layer, resulted in a sharp reduction in numbers, species richness, and diversity of birds, especially ground-nesting birds and insular endemic species (Van Vuren and Coblentz 1987).

MANAGEMENT AND ERADICATION

The severe ecological degradation caused by feral goats and sheep was slow to be realized and addressed. Some early control programs emphasized population suppression to reduce damage, but this approach was problematic. Population suppression must be continued indefinitely, at considerable cost, and even a brief cessation of management activities could lead to population recovery. Further, once an island ecosystem has been degraded, even low densities of goats or sheep might be sufficient to continue further degradation (Van Vuren 1992). The concept of eradicating entire populations of destructive nonnative mammals came about as a solution to primarily agricultural or economic problems, but had not been applied to ecological problems until the mid-twentieth century. By the late twentieth century, many biologists came to a consensus on the negative impacts of feral goats and sheep on islands (Coblentz 1978) and began developing techniques to remove entire island populations (Daly 1989). There are now many examples of successful management efforts resulting in the dramatic recovery of native biota.

The first eradication of goats from an island occurred on the Hawaiian island of Ni'ihau. Goats had been established in the early 1900s, and eradication by contract hunting became warranted by 1911 (Kramer 1971). Lāna'i was also affected by excessive browsing, and by 1900, large areas were deforested by sheep and goats introduced in the mid-1800s (Hobdy 1993). Charles Gay began to eradicate goats and sheep from his Lāna'i ranch in 1902 and fenced the summit cloud forest to protect the watershed. The ornithologist George C. Munro came to run Gay's ranch in 1911 and spent much of his first decade there shooting sheep and goats. Feral goats were eventually eradicated from Lāna'i by 1981, and feral sheep were eradicated later in the 1980s, although European mouflon remain numerous on Lāna'i (Hess and Jacobi 2011).

Goats had been periodically culled from Hawaii Volcanoes National Park (HAVO) on Hawaii Island since 1927 but with no lasting effect due to reinvasion of animals from surrounding areas (Figure 14.3; Baker and Reeser 1972). Managers of national parks in Hawaii took further actions on the recommendation of the Leopold et al. (1963) report on wildlife management in national parks, which stated: "A visitor

FIGURE 14.3   Feral goats remained abundant in HAVO for more than a century and a half despite periodic attempts since 1927 to reduce their populations. (NPS photo courtesy of Don Reeser and Bryan Harry.)

who climbs a mountain in Hawaii ought to see mamane trees and silverswords, not goats." Eradication of goats from 554 km² of HAVO took place from 1968 to 1984 (Tomich 1986), demonstrating the technical feasibility of eradicating ungulates from large areas of multitenure islands and providing specific techniques necessary to accomplish the task. For example, the Judas goat method was devised in HAVO to find remaining animals that eluded eradication efforts by exploiting the gregarious behavior of female ungulates (Taylor and Katahira 1988). Radio-collared females that found and associated with surviving remnant groups could easily be rediscovered; those groups could then be captured or shot, but the Judas female would be spared to repeatedly seek out other animals. The method has been repeatedly applied to many other eradication programs (e.g., Keegan et al. 1994; Campbell and Donlan 2005; Cruz et al. 2009). Following eradication, reinvasion was mitigated by dividing areas into fenced units of manageable size, a difficult logistical process at the time for large areas and for dense tropical forests on volcanic substrates (Figure 14.4). After a century and a half of degradation, a previously undescribed endemic plant species, ʻāwikiwiki or *Canavalia kauensis* (now *C. hawaiiensis*), was found growing on the dry lowlands of Kukalauʻula where goats had been excluded from small areas prior to large-scale eradication (Figure 14.5; St. John 1972).

Haleakalā National Park on Maui also had an intense but sporadic goat control program since early in the twentieth century, with more than 10,000 person-days of active hunting over a four-decade period (Kjargaard 1984). Goats were eliminated

FIGURE 14.4 Fences were constructed to divide HAVO into units of manageable size and prevent reinvasion of feral goats from surrounding areas during eradication efforts. (NPS photo courtesy of Don Reeser and Bryan Harry.)

from the 45-km² Kīpahulu District by the late 1980s after 51 km of fence was constructed between 1983 and 1987 (Stone and Holt 1990). Eradication of goats from the entire 137-km² park was completed in 1989 using Judas goat techniques developed in HAVO (L. Loope, pers. comm.). Goats and sheep were also eradicated from nearby Kahoʻolawe Island in 1990 by ground shooting, helicopter hunting, and the use of Judas animals (Kahoʻolawe Island Conveyance Commission 1993).

Santa Cruz Island supported at least 21,000 feral sheep in 1980 (Van Vuren and Coblentz 1989). Sheep removal was complicated by the large size of the island (249 km²), rugged terrain with few roads, and differing land ownership; the western 90% was owned by The Nature Conservancy (TNC) and the eastern 10% was privately owned. TNC chose a combination of fencing and shooting from the ground to remove sheep from their portion of the island. Trapping and shipping of live animals to the U.S. mainland were considered but not adopted because of logistical limitations posed by island ruggedness and lack of access, and because a market for live feral sheep could not be found (Schuyler 1993). TNC first built over 160 km of fencing that partitioned the island into 23 segments, each between 137 and 4517 ha in size, including a fence that separated TNC lands from the eastern end of the island (Schuyler 1993). Shooting from the ground required an extended time, but fencing the island allowed segments to be cleared sequentially without the risk of recolonization from neighboring areas, and also limited density-dependent recovery by sheep to the segment being hunted. Hunting began in late 1981 (Schuyler 1993); by December 1986 more than 31,000 sheep had been shot, and by January 1988 only 40 sheep remained. By June 1988, a total of 37,000 sheep had been shot

FIGURE 14.5   Feral goats had been excluded from small areas prior to large-scale eradication, which resulted in the emergence of a previously undescribed endemic plant species, 'āwikiwiki or *Canavalia kauensis* (now *C. hawaiiensis*) after a century and a half of degradation. (NPS photo courtesy of Don Reeser and Bryan Harry.)

and five sheep remained, which were subsequently found and shot (Schuyler 1993). Feral sheep remained on the eastern 10% of the island, which was acquired by the National Park Service (NPS) in 1997. Because this portion of the island had gentler topography, and given concerns over the opposition to lethal removal by animal rights organizations, NPS employed live-capture and removal (Faulkner and Kessler 2011). Beginning May 1997, sheep were captured by baiting or herding into corrals; as numbers declined, remnant sheep were pursued individually and captured. In December 1999, the last sheep had been captured and removed, totaling 9200 individuals (Faulkner and Kessler 2011).

Feral goats became established on Santa Catalina Island by the mid-1800s and reached numbers as high as 30,000 by the 1930s (Schuyler et al. 2002). Despite ongoing control efforts by sport hunters and island managers, 8000 goats remained by 1990. Control intensified from 1990 to 1994, when about 95% of remaining goats were shot from the ground and from the air, but funding ran out and control was suspended. Shooting resumed in 1996, with an additional 600 goats removed. Concerns for animal welfare caused a shift to removal by live capture in 1999 (121 goats captured), followed by a return to hunting a year later (66 goats shot). Renewed concerns for animal welfare resulted in an attempt to live capture the last 25–30 goats

(Schuyler et al. 2002). With the aid of the Judas goat method, the last goats were removed by live capture in 2002 (Campbell and Donlan 2005).

After the 468-km^2 Kahuku Ranch was acquired by HAVO, a directed volunteer program was initiated in 2004 to reduce the abundance of European mouflon and prevent further degradation to native biota (Stephens et al. 2008). Staff hunting, shooting from helicopters, and the use of forward-looking infrared radar (FLIR) were successful, and eradication became the goal. More than 6600 mouflon were removed, and the number observed during aerial surveys dropped from 1785 in 2004 to 378 in 2014. No mouflon were observed in two intensively managed subunits during the last survey, although small numbers have been periodically detected with game cameras and dispatched since that time (Judge et al. 2016). The Kahuku Unit may soon become the largest area from which mouflon will have been eradicated (Figure 14.6).

One of the few continental areas in the United States where free-ranging sheep have become problematic is Amistad National Recreation Area, Texas (Slade 2005). A single breeding pair of European mouflon entered the park from a neighboring ranch during the mid-1970s and proliferated over the next 20 years, reaching over 400 individuals by the mid-1990s. Population growth then further accelerated, reaching more than 2000 individuals by 2005, competing for forage with native white-tail deer (*Odocoileus virginianus*), and damaging fragile soils. More than 1300 sheep were removed using net guns from helicopters, which reversed ecological degradation

FIGURE 14.6 European mouflon sheep were introduced to the Kahuku Ranch beginning in 1968; Kahuku Ranch was later acquired by HAVO and eradication of mouflon commenced in 2004. (USGS photo.)

(Slade 2005). Other locations in the United States where European mouflon have been introduced and continue to roam freely include: North Haven Island, Maine; Speiden Island, Washington; California; Texas; and New Mexico.

MANAGEMENT FAILURES AND PROTRACTED ERADICATION ATTEMPTS

Many efforts to eradicate feral goats and sheep have not proceeded as anticipated; although intentions have been well justified, challenges to these operations have been sometimes surprising. A well-documented example is on Mauna Kea, where feral sheep have repeatedly reached excessive numbers, devastating the watershed and semiarid subalpine woodland environment. Foresters for the Territory of Hawaii conducted sheep drives starting in 1934 that eliminated tens of thousands of individuals. The Mauna Kea Forest Reserve (MKFR) was fenced in between 1935 and 1937 (Bryan 1937a), and nearly 47,000 sheep were removed in the following 10 years by foresters and Civilian Conservation Corps workers using drives on foot and horseback (Bryan 1937b, 1947). Populations rebounded when sport hunting became a management goal of wildlife biologists after World War II, and by 1960, the dire condition of the Mauna Kea forest was decried but not widely known outside of Hawaii (Warner 1960). Despite this knowledge, hybrid European mouflon x feral sheep were released from 1962 until 1966 to further improve hunting opportunities (Figure 14.7; Giffin 1982). Exclosures, aerial photography, and altered tree size classes all demonstrated the effects of browsing and bark stripping by sheep, goats, and other ungulates on the subalpine vegetation and associated wildlife (Scowcroft 1983; Scowcroft and Giffin 1983; Scowcroft and Sakai 1983). U.S. Federal District court orders of 1979 and 1986 mandated the removal of goats and sheep to protect the endangered palila that feed and raise their nestlings on māmane seed pods. More than 87,000 sheep have been removed from the MKFR over a 75-year period, but sheep have not yet been eradicated (Banko et al. 2014). Patchy recovery of māmane occurred after sheep removals began (Hess et al. 1999); however, habitat loss has been compounded by drought, which has further contributed to the ongoing long-term decline of palila (Banko et al. 2009, 2013). The fence surrounding Mauna Kea has recently been reconstructed to contemporary standards, and sheep removals have accelerated (Hess and Banko 2011). Although the subalpine woodland of Mauna Kea has demonstrated the ability to regenerate after feral sheep and goat populations have been reduced (Scowcroft and Giffin 1983, Reddy et al. 2012), the cumulative degradation to this ecosystem may preclude long-term suitability for dependent native wildlife such as palila.

Another example is on San Clemente Island, where concerted efforts to eradicate the over 20,000 feral goats on the island were initiated in 1972 by the U.S. Navy, the owner of the island, because of the threat that goats posed to several federally listed plants and animals endemic to the island (Keegan et al. 1994). Trapping and shooting removed about 16,000 goats, but many remained (Van Vuren 1992). In the late 1970s, the U.S. Navy decided to shoot the remaining goats from helicopters, but because of a lawsuit by Give Our Animals Time (GOAT), the U.S. Navy was directed to use trapping as a nonlethal means to complete the eradication. Another

FIGURE 14.7 Feral domestic sheep were intentionally hybridized with mouflon and released between 1962 and 1966 on Mauna Kea. Although federal court orders of 1979 and 1986 mandated the removal of all sheep goats from Mauna Kea to protect an endangered bird species, hybrid sheep remain abundant throughout high-elevation areas of the Hawaiian island. (Photo courtesy of Jack Jeffrey Photography.)

3000–4000 goats were removed by trapping, but some goats remained, so shooting was resumed in 1983. However, a series of lawsuits, as well as directives from the Department of Defense, repeatedly interrupted these efforts, and a small remnant population eluded all attempts at removal (Van Vuren 1992). The population was eliminated when the last 263 goats were shot between 1989 and 1991 using the Judas goat technique, totaling 29,000 goats trapped or shot since efforts began in 1972 (Keegan et al. 1994).

ECOSYSTEM RECOVERY AND UNEXPECTED EFFECTS

Removal of feral goats or sheep usually results in a remarkable recovery of native vegetation. On Mauna Kea, exclusion of feral sheep resulted in a rapid increase in regeneration and growth of māmane and other native species (Scowcroft and Giffin 1983); within about 40 years, māmane recovery within exclosures had progressed to the point of potential suitability as habitat for the palila (Reddy et al. 2012). On Santa Cruz Island, feral sheep removal resulted in a nearly two order magnitude increase in native grass biomass (Van Vuren and Bowen 2012); a two- to fourfold increase in density of native shrubs, including the reappearance of low-growing shrubs that had been locally extirpated by sheep grazing (Wehtje 1994; Van Vuren 2012); and the transition of a degraded grassland into a coastal sage scrub community (Beltran et al. 2014). Recovery of vegetation had cascading effects that included major changes in bird density and species richness (Van Vuren 2013) and a drastic reduction in the occurrence of landslides (Pinter and Vestal 2005).

Managers often expect the restoration of the original community after eradicating a feral herbivore, but undesired or unexpected consequences can result from overlooked ecological linkages (Zavaleta et al. 2001; Morrison 2011). The removal of feral goats and sheep can sometimes result in the proliferation of invasive plants that had been suppressed by these generalist herbivores (Zavaleta et al. 2001). In a meta-analysis of vegetation response after goat eradication from islands worldwide, species richness as well as the percentage cover, often including exotic plant species, increased despite the presence of rodents and other mammalian herbivores on some islands; percentage cover increased more on tropical islands than in other locations largely due to exotic plants (Schweizer et al. 2016). On Santa Cruz Island, fennel (*Foeniculum vulgare*), an invasive weed that had been suppressed by feral sheep grazing, increased explosively after sheep eradication (Beatty and Licari 1992; Klinger 2007). Fennel crowded out native vegetation in most of the areas in which it grew, and it has proved problematic to control (Dash and Gliessman 1994). Such undesired or unexpected changes can extend beyond simple two-species interactions (Morrison 2011). Sheep removal and associated vegetation regrowth on Santa Cruz Island led to an increase in some bird species, but not all; surprisingly, some birds declined, especially those that prefer open, less-vegetated habitats, including insular endemic taxa of conservation concern (Van Vuren 2013). The solution is to proceed with eradication using a holistic approach to restoration and comprehensive, strategic planning which may also require the control of other invasive species that proliferate after the removal of sheep and goats (Zavaleta et al. 2001; Morrison 2011).

In some cases, an initial rapid spread of introduced species occurred following the removal of feral goats, but stabilized over longer periods of time, ultimately benefiting native biota (Kessler 2002, 2011). In highly modified ecosystems, such as heavily invaded tropical dry forests, removal of goats also facilitated the short-term proliferation of an invasive plant (Kellner et al. 2011). Long-term studies on the effects of ungulate exclusion indicate that animal removal can release invasive pyrogenic grasses from top-down control and adversely affect native plants (Cabin et al. 2000). However, when invasive grasses have been controlled after ungulate removal, an increase in natural recruitment of native woody seedlings into larger size classes has been observed (Thaxton et al. 2010).

OPPORTUNITIES AND CHALLENGES

Technical impediments to eradication from a biological perspective have now largely been overcome, allowing the eradication of goats from islands as large as Santiago Island (585 km^2), Galápagos (Cruz et al. 2009). Judas techniques and variations have proven to be highly effective (Taylor and Katahira 1988); the Mata Hari technique induces sterilized females into permanent estrus to attract males, thereby simultaneously reducing the number of pregnant females (Campbell 2007; Campbell et al.

2007). Other techniques that include combinations of hunting, trapping, snaring, toxicants, and the strategic use of fences have also been highly effective to complete the eradication of feral goats and sheep from islands or from large areas of islands. The greatest remaining impediments are perhaps the attitudes and conflicting values of residents and stakeholders toward the lethal removal of animals, particularly for large, multitenure islands (Cambell and Donlan 2005).

Animal rights groups often play a major role in eradications, a role that planning should anticipate. For example, the U.S. Navy was sued by an animal rights group when trying to eradicate feral goats on San Clemente Island, which resulted in long delays and density-dependent recovery of goat numbers (Van Vuren 1992). Protests by animal rights groups also altered eradication plans for feral goats on Santa Catalina Island, although eradication was eventually achieved (Schuyler et al. 2002).

Hunting interests can play an important role as well. Sheep eradication efforts on Santa Cruz Island were halted by a lawsuit from the California Wildlife Federation, which sought to maintain the hunting opportunities for feral sheep (Schuyler 1993). TNC was prepared for such a lawsuit and submitted an effective response; the lawsuit was quickly dropped (Schuyler 1993). Because the island had been segmented by fences, the delay caused by the lawsuit had little effect on eradication progress. Hunting advocates had also expressed longstanding opposition to the eradication of sheep from Mauna Kea (Juvik and Juvik 1984). A major objection was the waste of food resources, which was addressed by carcass salvaging operations using helicopters; however, salvage operations diverted resources that would have otherwise been applied to removing additional sheep (Tummons 2012). Management agencies have also attempted to recruit public hunters in eradication efforts. Sport hunters contributed to sheep eradication on TNC lands on Santa Cruz Island, killing 5300 sheep (Schuyler 1993). On Mauna Kea, all hunting limits for sheep were dropped and hunters were guided to areas with abundant sheep to expedite removal; however, the sheep population apparently increased despite these incentives (Banko et al. 2014). At the Kahuku Unit of HAVO, guided volunteer hunters removed a disproportionate number of rams, which promoted a strong female bias and high population growth rates in the remaining population (Stephens et al. 2008).

Differential land ownership also may represent an impediment to successful eradication efforts. On Santa Cruz Island, for example, the eastern 10% of the island was privately owned in the 1980s, and the owner had no interest in eradication, so eradication had to proceed in two stages, the western 90% followed by the eastern 10%. Also, TNC was more willing to face sociocultural challenges and chose shooting as their primary tool, whereas NPS (the eventual owner of the eastern 10%) addressed animal welfare concerns by employing live capture and transport to the mainland.

Eradication of entire populations requires complete commitment of resources to this goal. Adequate resources need to be in place prior to commencing operations; goat eradication on Santa Catalina Island was delayed two years because of inadequate funding (Schuyler et al. 2002). While the majority of a population can be removed quickly, eradication efforts become progressively more difficult as densities become reduced. Managers of Hawaii's national parks found that reducing populations by half cost the same amount regardless of initial abundance. Density reduction without eradication can trigger density-dependent responses that rapidly restore numbers; based on demographic information, Rudge and Smit (1970) estimated that a feral goat population reduced by 80% can recover 90% of original numbers in four years. Further, detecting the last animals can be exceptionally difficult because remaining animals become wary of removal efforts; the result can be a premature declaration of success, termination of the program, and in consequence the "Lazarus effect," the reappearance of animals thought eradicated (Morrison et al. 2007). Careful planning is needed for extended monitoring after the last animal has been detected to verify that eradication has in fact occurred (Morrison et al. 2007).

The future for management of feral goats and sheep will most likely see the eradication of entire populations from increasingly larger, more complex islands and areas with thicker vegetation using more sophisticated techniques and tools (Campbell and Donlan 2005; Hess and Jacobi 2011; Chynoweth et al. 2013). In addition to Judas techniques, next-generation tools including remote sensing imagery, real-time telemetry, and thermal imaging devices such as FLIR may greatly improve the efficiency of locating and removing the last remaining animals that require the greatest effort, as well as posteradication monitoring. The outcomes of future eradication programs will likely result in overwhelmingly positive ecosystem responses, as they have already; however, community and multispecies interactions, other newly introduced invasive species, and climate change may all present new obstacles to recovery.

REFERENCES

Baker, J. K., and D. W. Reeser. 1972. Goat management problems in Hawaii Volcanoes National Park: A history, analysis, and management plan. U.S. Department of the Interior National Park Service Natural Resources Report Number 2. Office of the Chief Scientist, Washington, DC.

Banko, P. C., K. W. Brinck, C. Farmer, and S. C. Hess. 2009. Palila. Pages 513–529 in *Conservation Biology of Hawaiian Forest Birds: Implications for Island Avifauna.* T. K. Pratt, C. T. Atkinson, P. C. Banko, J. D. Jacobi, B. L. Woodworth (eds). Yale University Press, New Haven, CT.

Banko, P. C., R. J. Camp, C. Farmer, K. W. Brinck, D. L. Leonard, and R. M. Stephens. 2013. Response of palila and other subalpine Hawaiian forest bird species to prolonged drought and habitat degradation by feral ungulates. *Biological Conservation* 157: 70–77.

Banko, P. C., S. C. Hess, P. G. Scowcroft, C. Farmer, J. D. Jacobi, R. M. Stephens, and R. J. Camp et al. 2014. Evaluating the long-term management of introduced ungulates to protect the Palila, an endangered bird, and its critical habitat in subalpine forest of Mauna Kea, Hawaii. *Arctic, Antarctic, and Alpine Research* 46: 871–889.

Banko, P. C., L. Johnson, G. D. Lindsey, S. G. Fancy, T. K. Pratt, J. D. Jacobi, and W. E. Banko. 2002a. *Palila (Loxioides bailleui).* The Birds of North America. Number 679.

Beatty, S. W., and D. L. Licari. 1992. Invasion of fennel (*Foeniculum vulgare*) into shrub communities on Santa Cruz Island, California. *Madroño* 39: 54–66.

Beltran, R. S., N. Kreidler, D. H. Van Vuren, S. A. Morrison, E. S. Zavaleta, K. Newton, B. R. Tershy, et al. 2014. Passive recovery of vegetation after herbivore eradication on Santa Cruz Island, California. *Restoration Ecology* 22: 790–797.

Bowen, L., and D. H. Van Vuren. 1997. Insular endemic plants lack defenses against herbivores. *Conservation Biology* 11: 1249–1254.

Bowen, L., and D. H. Van Vuren. 1999. Density, reproduction, and group size in feral sheep on Santa Cruz Island, California, USA. *Mammalia* 63: 21–28.

Brumbaugh, R. W. 1980. Recent geomorphic and vegetal dynamics on Santa Cruz Island, California. Pages 139–158 in *The California Islands: Proceedings of a Multidisciplinary Symposium*. D. M. Power (ed). Santa Barbara Museum of Natural History, Santa Barbara, California.

Bryan, L. W. 1937a. The big fence on the Big Island. *Paradise of the Pacific* 49: (15), 30.

Bryan, L. W. 1937b. Wild sheep in Hawaii. *Paradise of the Pacific* 49: (19), 31.

Bryan, L. W. 1947. Twenty-five years of forestry work on the Island of Hawaii. *Hawaii Planters' Record* 51: 1–80.

Burness, G. P., J. Diamond, and T. Flannery. 2001. Dinosaurs, dragons, and dwarfs: The evolution of maximal body size. *Proceedings of the National Academy of Sciences* 98: 14518–14523.

Cabin, R. J., S. G. Weller, D. H. Lorence, T. W. Flynn, A. K. Sakai, D. Sandquist, and L. J. Hadway. 2000. Effects of a long-term ungulate exclusion and recent alien species control on the preservation and restoration of a Hawaiian tropical dry forest. *Conservation Biology* 14: 439–453.

Campbell, K. J. 2007. Manipulation of the reproductive system of feral goats (*Capra hircus*) to increase the efficacy of Judas goats: Field methods utilising tubal sterilisation, abortion, hormone implants and epididymectomy. *PhD dissertation*, University of Queensland, Gatton, Australia.

Campbell, K. J., G. S. Baxter, P. J. Murray, B. E. Coblentz, and C. J. Donlan. 2007. Development of a prolonged estrus effect for use in Judas goats. *Applied Animal Behaviour Science* 102: 12–23.

Campbell, K., and C. J. Donlan. 2005. Feral goat eradications on islands. *Conservation Biology* 19: 1362–1374.

Chynoweth, M. W., C. M. Litton, C. A. Lepczyk, S. C. Hess, and S. Cordell. 2013. Biology and impacts of Pacific Island invasive species 9: *Capra hircus*, the feral goat. (Mammalia: Bovidae). *Pacific Science* 67: 141–156.

Coblentz, B. E. 1978. The effects of feral goats (*Capra hircus*) on island ecosystems. *Biological Conservation* 13: 279–286.

Cohen, B., C. Cory, J. Menke, and A. Hepburn. 2009. A spatial database of Santa Cruz Island vegetation. Pages 229–244 in *Proceedings of the Seventh California Islands Symposium*. C C. Damiani and D. K. Garcelon (eds). Institute for Wildlife Studies, Arcata, California.

Conry, P. J. 1988. Management of feral and exotic game species on Guam. *Transactions of the Western Section of the Wildlife Society* 24: 26–30.

Cruz, F., V. Carrion, K. J. Campbell, C. Lavoie, and C. J. Donlan. 2009. Bio-economics of large-scale eradication of feral goats from Santiago Island, Galápagos. *Journal of Wildlife Management* 73: 191–200.

Daly, K. 1989. Eradication of feral goats from small islands. *Oryx* 23: 71–75.

D'Antonio, C. M., and P. M. Vitousek. 1992. Biological invasions by exotic grasses, the grass/fire cycle and global change. *Annual Review of Ecology and Systematics* 23: 63–87.

Dash, B. A., and S. R. Gliessman. 1994. Nonnative species eradication and native species establishment: Fennel on Santa Cruz Island. Pages 505–515 in *The Fourth California Islands Symposium: Update on the Status of Resources*. W. L. Halvorson and G. L. Maender (eds). Santa Barbara Museum of Natural History, Santa Barbara, California.

Demartini, J. C., and R. B. Davies. 1977. An epizootic of pneumonia in captive bighorn sheep infected with *Muellerius* sp. *Journal of Wildlife Disease* 13: 117–124.

Dohner, J. V. 2001. *The Encyclopedia of Historic and Endangered Livestock and Poultry Breeds*. Yale agrarian studies, Yale University Press, New Haven, CT.

Dunson, W. A. 1974. Some aspects of salt and water balance of feral goats from arid islands. *American Journal of Physiology* 226: 662–669.

Ellis, W. 1917. *A Narrative of a Tour through Hawaii in 1823*. Hawaiian Gazette Co., Ltd. Honolulu. 367 pp.

Engels, W. L. 1952. Vertebrate fauna of North Carolina coastal islands, II. *Shackleford Banks*. *American Midland Naturalist* 47: 702–742.

Faulkner, K. R., and C. C. Kessler. 2011. Live capture and removal of feral sheep from eastern Santa Cruz Island, California. Pages 295–299 in *Island Invasives: Eradication and Management*. C. R. Veitch, M. N. Clout, and D. R. Towns (eds). Occasional Paper 42, IUCN Species Survival Commission, IUCN, Gland, Switzerland.

Francis, R. C. 2015. *Domesticated: Evolution in a Man-Made World*. W. W. Norton & Company, New York.

Frank, D. A., M. M. Kuns, and D. R. Guido. 2002. Consumer control of grassland plant production. *Ecology* 83: 602–606.

Frank, D. A., S. J. Mc Naughton, and B. F. Tracy. 1998. The ecology of the Earth's grazing ecosystems. *Bioscience* 48: 513–521.

Geist, V. 1971. *Mountain Sheep: A Study in Behavior and Evolution*. University of Chicago Press, Chicago. 383 pp.

Giffin, J. G. 1982. *Ecology of the Mouflon Sheep on Mauna Kea*. State of Hawaii, Department of Land and Natural Resources, Division of Forestry and Wildlife, Honolulu, Hawaii.

Gould Burke, M. 1988. The feral goats of Aldabra: Ecology and population dynamics. *National Geographic Research* 4: 272–279.

Hess, S. C., and P. C. Banko. 2011. Sheep vs. Palila on Mauna Kea: After 200 years of damage, can these native birds recover? *The Wildlife Professional* 5(3): 60–63.

Hess, S. G., P. C. Banko, G. J. Brenner, and J. D. Jacobi. 1999. Factors related to the recovery of subalpine woodland on Mauna Kea, Hawaii. *Biotropica* 31: 212–219.

Hess, S. C., P. C. Banko, L. J. Miller, and L. P. Laniawe. 2014. Habitat and food preferences of the endangered Palila (*Loxioides bailleui*) on Mauna Kea, Hawaii. *Wilson Journal of Ornithology* 126: 728–738.

Hess, S. C., and J. D. Jacobi. 2011. The history of mammal eradications in Hawaii and the United States associated islands of the Central Pacific. Pages 67–73 in *Island Invasives: Eradication and Management*. Veitch, C. R., M. N. Clout, and D. R. Towns (eds). IUCN, Gland, Switzerland.

Hess, S., B. Kawakami Jr., D. Okita, and K. Medeiros. 2006. A preliminary assessment of mouflon abundance at the Kahuku Unit of Hawaii Volcanoes National Park. *US Geological Survey Open File Report OF 2006–1193*.

Hobbs, E. 1980. Effects of grazing on the northern population of *Pinus muricata* on Santa Cruz Island, California. Pages 159–165 in *The California Islands: Proceedings of a Multidisciplinary Symposium*. D. M. Powber (ed). Santa Barbara Museum of Natural History, Santa Barbara, California.

Hobdy, R. 1993. Lāna'i—A case study: The loss of biodiversity on a small Hawaiian island. *Pacific Science* 47: 201–210.

Ikagawa, M. 2014. The distribution and management of a high-threat invasive ungulate, the mouflon sheep, on Hawaii island. Unpublished MS thesis, University of Hawaii at Mānoa.

Judge, S. W., S. C. Hess, J. K. Faford, D. Pacheco, C. R. Leopold, C. Cole, and V. DeGuzman. 2016. Evaluating Detection and Monitoring Tools for Incipient and Relictual Non-native Ungulate Populations. *Hawaii Cooperative Studies Unit Technical Report HCSU-069*.

Juvik, J. O., and S. P. Juvik. 1984. Mauna Kea and the myth of multiple use: Endangered species and mountain management in Hawaii. *Mountain Research and Development* 4: 191–202.

Kaho'olawe Island Conveyance Commission. 1993. Kaho'olawe Island: Restoring a cultural treasure. Kaho'olawe Island Conveyance Commission, Wailuku, HI, USA.

Keegan, D. R., B. E. Coblentz, and C. S. Winchell. 1994. Feral goat eradication on San Clemente Island, California. *Wildlife Society Bulletin* 22: 56–61.

Kellner, J. R., G. P. Asner, K. M. Kinney, S. R. Loarie, D. E. Knapp, T. Kennedy-Bowdoin, E. J. Questad, et al. 2011. Remote analysis of biological invasion and the impact of enemy release. *Ecological Applications* 21: 2094–2104.

Kessler, C. C. 2002. Eradication of feral goats and pigs and consequences for other biota on Sarigan Island, Commonwealth of the Northern Mariana Islands. Pages 132–140 in *Turning the Tide: The Eradication of Invasive Species, Proceedings of the International Conference on Eradication of Island Invasives.* IUCN, Gland, Switzerland.

Kessler, C. C. 2011. Invasive species removal and ecosystem recovery in the Mariana Islands; challenges and outcomes on Sarigan and Anatahan. Pages 320–324 in *Island Invasives: Eradication and Management.* Veitch, C. R., M. N. Clout, and D. R. Towns (eds). IUCN, Gland, Switzerland.

Kjargaard, J. 1984. *Some Aspects of Feral Goat Distribution in Haleakala National Park.* Cooperative National Parks Resource Studies Unit, University of Hawaii at Mānoa.

Klinger, R. C. 2007. Ecosystem engineers and the complex dynamics of non-native species management in California's Channel Islands. Pages 343–365 in *Ecosystem Engineers: Plants to Protists,* K. Cuddington, J. Byers, W. Wilson, and A. Hastings (eds). Academic Press-Elsevier, San Diego, California.

Kramer, R. J. 1971. *Hawaiian Land Mammals.* Charles E. Tuttle Co., Rutland, Vermont, and Tokyo, Japan.

Leopold, A. S., S. A. Cain, C. M. Cottam, I. M. Gabrielson, and T. L. Kimball. 1963. Wildlife management in the national parks. *Transactions of the North American Wildlife Conference* 24: 28–45.

Loague, K., D. Lloyd, T. W. Giambelluca, S. Nguyen, and B. Sakata. 1996. Land misuse and hydrologic response: Kaho'olawe, Hawaii. *Pacific Science* 50: 1–35.

Lu, C. 1988. Grazing behavior and diet selection of goats. *Small Ruminant Research* 1: 205–216.

Meléndez-Ackerman, E., C. Cortés, J. Sustache, M. Morales-Vargas, S. Aragón, M. García-Bermudez, and D. S. Fernandez. 2008. Diet of feral goats in Mona Island Reserve, Puerto Rico. *Caribbean Journal of Science* 44: 199–205.

Morrison, S. A. 2011. Trophic considerations in eradicating multiple pests. Pages 208–212 in *Island Invasives: Eradication and Management.* Veitch, C. R., M. N. Clout, and D. R. Towns (eds). IUCN, Gland, Switzerland.

Morrison, S. A., N. MacDonald, K. Walker, L. Lozier, and M. R. Shaw. 2007. Facing the dilemma at eradication's end: Uncertainty of absence and the Lazarus effect. *Frontiers in Ecology and the Environment* 5: 271–276.

Panayotova-Pencheva, M. S., and M. T. Alexandrov. 2010. Some pathological features of lungs from domestic and wild ruminants with single and mixed protostrongylid infections. *Veterinary Medicine International* 2010: Article ID 741062. http://www.hindawi.com/journals/vmi/2010/741062/

Pidancier, N., S. Jordan, G. Luikart, and P. Taberlet. 2006. Evolutionary history of the genus *Capra* (Mammalia, Artiodactyla): Discordance between mitochondrial DNA and Y-chromosome phylogenies. *Molecular Phylogenetics and Evolution* 40: 739–749.

Pinter, N., and W. D. Vestal. 2005. El Niño-driven landsliding and postgrazing vegetative recovery, Santa Cruz Island, California. *Journal of Geophysical Research* 110: 1–17.

Powers, J. G., C. G. Duncan, T. R. Spraker, B. A. Schuler, S. C. Hess, J. K. J. Faford, and H. Sin. 2014. Environmental conditions associated with lesions of introduced free-ranging sheep in Hawaii. *Pacific Science* 68: 65–74.

Pugh, D. 2002. *Sheep and Goat Medicine.* 1st ed. W. B. Saunders Company, Philadelphia, Pennsylvania.

Pybus, M. J., and H. Shave. 1984. *Muellerius capillaris* (Mueller, 1889) (Nematoda: Protostrongylidae): An unusual finding in Rocky Mountain bighorn sheep (*Ovis Canadensis canadensis* Shaw) in South Dakota. *Journal of Wildlife Diseases* 20: 284–288.

Reddy, E., D. H. Van Vuren, P. G. Scowcroft, J. B. Kauffman, and L. Perry. 2012. Long-term response of the mamane forest to feral sheep management on Mauna Kea, Hawaii. *Pacific Conservation Biology* 18: 123–132.

Robbins, C. T. 1994. *Wildlife Feeding and Nutrition*. Elsevier Science. 2nd ed. Orlando, Florida.

Rudge, M. R. 1984. The occurrence and status of populations of feral goats and sheep throughout the world. Pages 55–84 in *Feral Mammals—Problems and Potential*. P. N. Munton, J. Clutton-Brock, and M. R. Rudge (eds). International Union for Conservation of Nature and Natural Resources, Gland, Switzerland.

Rudge, M. R., and T. J. Smit. 1970. Expected rate of increase of hunted populations of feral goats (*Capra hircus* L.) in New Zealand. *New Zealand Journal of Science* 13: 256–259.

Ryder, M. L. 1983. *Sheep and Man*. Gerald Duckworth and Company, London.

Ryder, M. L. 1987. The evolution of the fleece. *Scientific American* 256: 112–119.

Schuyler, P. T. 1993. Control of feral sheep (*Ovis aries*) on Santa Cruz Island, California. Pages 443–452 in *Third California Islands Symposium: Recent advances in Research on the California Islands*, F. G. Hochberg (ed). Santa Barbara Museum of Natural History, Santa Barbara, California.

Schuyler, P. T., D. Garcelon, and S. Escover. 2002. Control of feral goats (Capra hircus) on Santa Catalina Island, California, USA. Pages 412–413 in *Turning the Tide: The Eradication of Invasive Species*. C. R. Veitch and M. N. Clout (eds). Occasional Paper 27, IUCN Species Survival Commission, IUCN, Gland, Switzerland.

Schweizer, D., H.P. Jones, and N.D. Holmes. 2016. Literature review and meta-analysis of vegetation responses to goat and European rabbit eradications on islands. *Pacific Science* 70: 55–71.

Scowcroft, P. G. 1983. Tree cover changes in māmane (*Sophora chrysophylla*) forests grazed by sheep and cattle. *Pacific Science* 37: 109–119.

Scowcroft, P. G., and C. E. Conrad. 1992. Alien and native plant response to release from feral sheep browsing on Mauna Kea. Pages 625–665 in *Alien Plant Invasions in Native Ecosystems of Hawaii*. C. P. Stone, C. W. Smith, and J. T. Tunison (eds). Cooperative National Park Resources Studies Unit, University of Hawaii, Honolulu.

Scowcroft, P. G., and J. G. Giffin. 1983. Feral herbivores suppress māmane and other browse species on Mauna Kea, Hawaii. *Journal of Range Management* 36: 638–645.

Scowcroft, P. G., and H. F. Sakai. 1983. Impacts of feral herbivores on mamane forests of Mauna Kea, Hawaii: Bark stripping and diameter class structure. *Journal of Range Management* 36: 495–498.

Slade, R. 2005. Removal of mouflon sheep from Amistad National Recreation Area. *The High Bridge Tribune* 3: 7.

Stender, Y., P. L. Jokiel, and K. S. Rodgers. 2014. Thirty years of coral reef change in relation to coastal construction and increased sedimentation at Pelekane Bay, Hawaii. *PeerJ* 2: e300.

St. John, H. 1972. *Canavalia kauensis* (Leguminosae) a new species from the island of Hawaii. Hawaiian Plant Studies 39. *Pacific Science* 26: 409–414.

Stephens, R. M., S. C. Hess, and B. Kawakami Jr. 2008. Controlling mouflon sheep at the Kahuku Unit of Hawaii Volcanoes National Park. *Proceedings of the Vertebrate Pest Conference* 23: 304–309.

Stone, C. P., and R. A. Holt. 1990. Managing the invasions of alien ungulates and plants in Hawaii's natural areas. *Monographs in Systematic Botany from the Missouri Botanical Garden* 32: 211–221.

Taylor, D., and L. Katahira. 1988. Radio telemetry as an aid in eradicating remnant feral goats. *Wildlife Society Bulletin* 16: 297–299.

Thaxton, J. M., T. C. Cole, S. Cordell, R. J. Cabin, D. R. Sandquist and C. M. Litton. 2010. Native species regeneration following ungulate exclusion and nonnative grass removal in a remnant Hawaiian dry forest. *Pacific Science* 64: 533–544.

Tomich, P. Q. 1986. *Mammals in Hawaii*, 2nd Ed. Bishop Museum Press, Honolulu, HI.

Tummons, P. 2012. State pays dearly for salvage of sheep from Mauna Kea hunts. *Environment Hawaii* 22: 12.

Van Vuren, D. H. 1982. Effects of feral sheep on the spatial distribution of artifacts on Santa Cruz Island. *Bulletin of the Southern California Academy of Science* 81: 148–151.

Van Vuren, D. H. 1992. Eradication of feral goats and sheep from island ecosystems. *Proceedings of the Vertebrate Pest Conference* 15: 377–381.

Van Vuren, D. H. 2012. Shrub regeneration after removal of feral sheep from Santa Cruz Island, California. *California Fish and Game* 100: 396–403.

Van Vuren, D. H. 2013. Avian response to removal of feral sheep on Santa Cruz Island, California. *Wilson Journal of Ornithology* 125: 134–139.

Van Vuren, D. H., and V. J. Bakker. 2009. Rapid morphological change in an insular population of feral sheep. *Journal of Zoology* 277: 221–231.

Van Vuren, D. H., and L. Bowen. 2012. Response of grassland vegetation on Santa Cruz Island to removal of feral sheep. *Madroño* 59: 190–195.

Van Vuren, D. H., and B. E. Coblentz. 1987. Some ecological effects of feral sheep on Santa Cruz Island, California, USA. *Biological Conservation* 41: 253–268.

Van Vuren, D. H., and B. E. Coblentz. 1989. Population characteristics of feral sheep on Santa Cruz Island. *Journal of Wildlife Management* 53: 306–313.

Van Vuren, D. H., and P. W. Hedrick. 1989. Genetic conservation in feral populations of livestock. *Conservation Biology* 3: 312–317.

Van Vuren, D. H., M. L. Johnson, and L. Bowen. 2001. Impacts of feral livestock on island watersheds. *Pacific Science* 55: 285–289.

Warner, R. E. 1960. A forest dies on Mauna Kea. *Pacific Discovery* 13(2): 6–14.

Wehtje, W. 1994. Response of a Bishop pine (*Pinus muricata*) population to removal of feral sheep on Santa Cruz Island, California. Pages 331–340 in *The Fourth California Islands Symposium: Update on the Status of Resources*. W. L. Halvorson and G. L. Maender (eds). Santa Barbara Museum of Natural History, Santa Barbara, California.

Wiedwandt, T. A. 1977. Ecology, behavior and management of the Mona Island ground iguana, *Cyclura stejnegeri*. *PhD dissertation*, Cornell University, Ithaca, New York.

Wood, G. M., M. T. Mengak, and M. Murphy. 1987. Ecological importance of feral ungulates at Shackleford Banks, North Carolina. *American Midland Naturalist* 118: 236–244.

Yocom, C. F. 1967. Ecology of feral goats in Haleakala National Park, Maui, Hawaii. *American Midland Naturalist* 77: 418–451.

Zavaleta, E. S., R. J. Hobbs, and H. A. Mooney. 2001. Viewing invasive species removal in a whole-ecosystem context. *Trends in Ecology and Evolution* 16: 454–459.

Zeder, M. A. 2009. The neolithic macro-(r)evolution: Macroevolutionary theory and the study of culture change. *Journal of Archaeological Research* 17: 1–63.

Zeder, M. A., and B. Hesse. 2000. The initial domestication of goats (*Capra hircus*) in the Zagros Mountains 10,000 years ago. *Science* 287: 2254–2257.

Zohary, D., E. Tchernov, and L. K. Horwitz. 1998. The role of unconscious selection in the domestication of sheep and goats. *Journal of Zoology* 245: 129–135.

15 European Starlings

George Linz, Ron Johnson, and James Thiele

CONTENTS

INTRODUCTION

European starlings (*Sturnus vulgaris,* Sturnidae) are native to Europe, southwest Asia, and North Africa and have successfully established populations on every continent but Antarctica (Rollins et al. 2009). In 1890 and 1891, a member of the American Acclimatization Society, Eugene Scheiffelin, released 100 starlings into New York City's Central Park, with the objective of introducing all the birds mentioned in the plays of William Shakespeare to North America (Cabe 1993). He was successful, as 16 pairs survived and reproduced prolifically. Starlings reached the Mississippi River in 1928 and were observed on the West Coast in 1942. In a little over a century, the United States (U.S.) starling population grew to approximately 200 million (Feare 1984; Cabe 1993; Johnson and Glahn 1994), but has now declined to about 140 million (Jernelov 2017). They now inhabit all of North America. Their range extends southward to the Bahamas, Central America, the Yucatan Peninsula, Puerto Rico, Jamaica, and Cuba. There are no subspecies in North America. Genetic analysis indicates that all starlings in North America descended from the New York City colony (Cabe 1993). Outside their native range, starlings are considered to be one of the most destructive invasive bird species worldwide, nominated by the Invasive Species Specialist Group, a science and policy network under the Species Survival Commission of the International Union for Conservation of Nature, to the "100 World's Worst" invaders (Lowe et al. 2004; Rollins et al. 2009).

We propose seven factors contributing to the success of the European starling as an invasive species. Starlings (1) nest in cavities that are protected from weather and predators; (2) compete successfully with native cavity-nesting birds, often taking nest sites from other birds; (3) use a wide range of nest locations, including natural cavities in trees, nest boxes, and holes in buildings, large signs or billboards, and a variety of other structures; (4) show an inclination for juvenile birds to disperse widely after fledging; (5) have bills, eye placement, and flock-foraging behaviors that are well adapted for foraging on grubs and other larvae just under the soil surface; (6) have an omnivorous diet with ability to forage in a wide array of places, including pastures, lawns, urban food and grain-handling areas, waste and landfill sites, and livestock facilities; and (7) are adapted to thrive in cold climates, typically migrating southward in winter only from north of 40° latitude, thus reducing energetic costs and risks associated with predation and foraging in unfamiliar areas.

European starlings are not protected by the Migratory Bird Treaty Act and no state laws in the United States directly protect them. State or local laws, however, may stipulate humane treatment of animals and regulate or prohibit certain management techniques such as harassing, trapping, shooting, or use of toxicants to kill starlings. Regardless, local law enforcement and government agencies should be contacted before attempting to harass starlings.

PHYSICAL DESCRIPTION

During spring and summer, starlings are glossy, dark-colored birds, but during winter, the body feathers are tipped with white speckles (Figure 15.1). In late winter, iridescent hues of green and purple become prominent in males on feathers of the head and neck. Overall, females are duller and less glossy than males. Juveniles are tannish colored until their first prebasic molt in early fall, after which they resemble adults. Starlings measure about 20 cm in length with a rounded body and short tail with females (69–93 g) smaller than males (73–96 g). Their wings (31–40 cm long) have a green or purple sheen.

Both mandible length and coloration are consistent within sexes. From late December through June, both sexes have bright-yellow mandibles measuring about 1.9 cm in length. The lower mandible of females typically has a pale-pink base, whereas males have a pale-blue base. Mandibles in both sexes become dark after June. Adult females often have a light-colored ring that surrounds the iris. In comparison, adult males typically have uniformly brown-colored eyes.

FOODS

Starlings are omnivorous, with a substantial diet of invertebrates (e.g., coleopteran [beetle] and lepidopteron [moths and butterflies]) larvae obtained from fields and

FIGURE 15.1 Female European starling (*Sturnus vulgaris*) in winter plumage. (Photo by Dr. Hays Cummins, Miami University.)

lawns, especially during nesting season and in areas where the ground remains unfrozen and moist. From midsummer to early fall, starlings forage on wild and cultivated fruits and ripening corn, especially sweet corn (Tupper et al. 2014). In the winter, particularly when the ground is snow-covered or frozen, they frequent feedlots, dairies, and urban landfills, where food is abundant and energy laden (Morrison and Caccamise 1990; Caccamise 1991). Starlings require 14–42 g per day of fatty foods in winter, but up to 400 g of berries and grapes, showing that they can efficiently digest fats, but are less efficient at processing carbohydrates (Martinez del Rio et al. 1995).

MIGRATORY PATTERNS

In southern and mid-latitudinal regions of the United States, starling flocks begin to disperse by late January or mid-February, as resident starlings start establishing reproductive territories. At the same time, migrant populations are affected by migratory restlessness, marked by changes in activity areas and longer daily movements (Kessel 1953; Dolbeer 1982). A northeastward spring migration occurs from mid-February to late March, and southwestward fall migration occurs from September to early December. Starlings are short-distance migrators that often travel only 400–500 km to reach reproductive areas; however, some travel 1000–1500 km, especially to escape heavy winter snows that cover food sources. Starlings living in the Midwest and Great Lakes region of North America regularly migrate (Kessel 1957; Dolbeer 1982), whereas starlings nesting south of 40°N rarely migrate (Kessel 1957; Dolbeer 1982; Cabe 1993). In North America, starlings sometimes associate with flocks of blackbirds (Icteridae) in winter, but are not closely related to blackbirds, which are native to North America.

LOCAL MOVEMENTS

Understanding local starling movement patterns is important in developing management options, especially when starling populations conflict with human activities and result in complaints by citizens to local wildlife officials. A variety of options are available to manage such conflicts, with the more suitable approaches depending on the numbers of starlings, the conflict location and situation, and other factors. Knowledge of starling movements, behaviors, and preferred roost sites is critical for developing effective management strategies. To that end, Homan et al. (2008, 2010, 2012, 2013) and Gaukler et al. (2012) used radio-tagging technology in a series of studies to develop a better understanding of starling movements and related activities in rural and urban areas across several states known for large wintering starling populations.

Three starling behavior studies related to daily movement patterns were conducted in the last decade around concentrated animal feeding operations (CAFOs). In one study, Homan et al. (2013) captured and radio-tagged starlings in early fall at dairies in northeastern Ohio. They found that birds visited the dairies where they were initially captured (home sites) on 85% of the days and spent 58% of each day at the dairies. Interchange of radio-tagged birds (n = 40) among dairies located 4.1–11.0 km apart was seven times less than two dairies located 1.3 km apart. In addition to using home-site roosts, these birds also used a distant roost (22 km).

In a second study, Homan et al. (2010) radio-tagged and tracked starlings using three CAFOs in the northern Texas Panhandle. They discovered that fidelity to sites of capture (home feedlots) was different among the three radio-tagged cohorts. Cohorts from Sites A and C were recorded at home feedlots on 48% and 59% of tracking days, respectively, whereas the cohort from Site B was at its home feedlot 95% of days. Use of roost sites appeared to depend on habitat composition surrounding the study feedlots. Site B was agricultural in nature with open fields and pastures prevalent near the feedlot, whereas cohorts from Sites A and C used urban areas and a small CAFO. Homan et al. (2010) concluded that higher habitat heterogeneity reduced rates of daily use of home feedlots for starlings using Sites A and C. This tends to complicate management strategies because starlings may be erratic in their daily use of CAFOs, and thus urban areas, when present, may be used as refuges by birds affected by management toxicants, leading to adverse public exposure to dead and dying birds.

Finally, Gaukler et al. (2012) monitored site use and movements of radio-tagged starlings during the winter months at two CAFOs in central Kansas. Their data showed starling site fidelity was 68% and 55% for Sites A and B, respectively. Minimal exchange (9%) occurred between Feedlots A and B, showing that starlings rarely abandoned the feedlot where they were captured, but they did observe a bird 68 km from their capture site. Gaukler et al. (2012) suggested that reducing bird numbers within the feedlot might lower the risk of spreading pathogens among feedlots.

Three studies on starling movement were also conducted in three urban areas: Omaha, Nebraska; Indianapolis, Indiana; and central New Jersey. Closely monitoring movements and activities of starlings can provide baseline knowledge needed when developing wildlife management options in urban settings. In downtown

Omaha, Homan et al. (2008) captured and radio-tagged starlings that were considered to be a public nuisance. They were able to find roost sites that were previously unknown, including a major roost that was contributing birds to the downtown roost. Some radio-tagged birds were local and never left Omaha. In comparison, other radio-tagged birds made daily trips of 5–16 m to towns and rural areas mostly in western Iowa, returning to roost in Omaha. The birds used food processing plants, grain depositories, industrial parks, feedlots, water treatment facilities, and power plants. Lawns and alleyways of residential areas also received heavy use.

In a second study in central New Jersey, Homan et al. (2012) radio-tagged starlings at three sites to help determine their movements, behaviors, and roosting sites. Biologists needed this information to help find starling carcasses after implementing a successful toxic baiting program. Starlings using a rural study site showed strong site fidelity with birds roosting on site and moving an average of only 2 km during the day. In comparison, starlings in the urban–suburban mosaic showed less fidelity, wandering 4–6 km from the banding site and seldom roosting on site. They found no interaction among roost sites by radio-tagged birds. Homan et al. (2012) predicted that most baited starlings would be found within 6 km of the bait site.

In a third study, Homan (unpublished data) captured and radio-tagged starlings in downtown Indianapolis (n = 11) and 5 km southwest of the downtown area (n = 38). They combined the data from all tagged birds and found that between December 2006 and March 2007, the farthest relocation of a starling was 19 km from the site of capture. Most of the locations were confined to areas relatively close (≤10 km) to the downtown area. Major areas of daytime use were commercial–industrial properties surrounding an airport and a sanitary landfill, and the landfill itself. These sites of activity were about 6–11 km south-southwest of downtown Indianapolis. Besides roosting at several sites in the downtown area (e.g., buildings, monuments, and industrial sites), the radio-tagged birds also roosted on airport grounds. Several smaller satellite roosts were also found in the vicinity of the major roosts.

Data from these six case studies showed that starlings are adaptable and may fly considerable distances from their roost to find a rich food supply. Further, movement patterns likely are dependent on the degree of habitat heterogeneity in the area surrounding the roost site, with more heterogeneity associated with increased movement. Given these results, it seems prudent to use radiotelemetry technology to accurately assess the movements and activity areas associated with a particular roost prior to implementing management actions to reduce human–wildlife conflict. This is particularly important if use of toxic bait is considered for reducing the population.

NESTING

Starlings typically nest in holes in trees, buildings, and nest boxes that are also preferred by native bird species, including eastern bluebirds (*Sialia sialis*), purple martins (*Progne subis*), wood ducks (*Aix sponsa*), and several species of woodpeckers (Cabe 1993). Proper nest box construction reduces starling occupation. For example, most starlings cannot enter a bluebird nest box with a properly sized 3.97-cm-diameter entry hole. Starlings will evict wood ducks, screech owls (*Megascops* spp.), and other cavity nesters from nesting boxes that by necessity must have large-diameter openings. In this case, regular monitoring and nest cleaning are required. Breeding starlings have a high degree of nest site fidelity (Kessel 1957); however, young-of-the-year disperse widely to find new breeding sites. Pairings are socially monogamous, with sexual maturity occurring at one year, but first-year birds may fail in their attempts to establish reproductive territories when there are limited nest sites available and abundant experienced birds. Depending on latitude, the reproductive period lasts from late March through early July. A pair of starlings annually hatches one or two clutches, consisting of four to six pale blue eggs. Incubation is 12 days and primarily by the female. Both parents feed the nestlings a variety of invertebrates for 21 days and continue up to 10 days after the fledglings leave the nest (Tinbergen 1981; Ricklefs and Smeraski 1983; Drent et al. 1985; Craig and Feare 1999).

POPULATION DYNAMICS

The North American starling breeding population is estimated at 57 million (Partners in Flight Science Committee 2013). Starlings are prolific and have a 48%–79% rate of nest success, with 60% of adults surviving annually, but 80% of nestlings failing to survive to reproduce (Kessel 1957; Royall 1966). Mortality rates are greater in fall and winter because of migration, scarcity of natural foods, and inclement weather. Causes of mortality include disease, predation, and starvation; none of these are believed to regulate the population. Each year, 80–100 million starlings die of natural causes and 1–3 million are killed at CAFOs during winter. Parasites and extreme weather events that limit availability of invertebrates can cause mortality of adults and nestlings (Boyd 1951; Gromadzki 1980; Tinbergen 1981). The major population limiting factor, however, could be availability of nesting sites that are shared with 27 native cavity nesters (Koenig 2003). Further, Koenig (2003) suggested that the interaction of native cavity nesters and starlings is complex and warrants additional research at multiple spatial scales.

ROOSTING BEHAVIOR

During summer, fall, and winter, starlings gather in roosts that may range in size from a few hundred to over 10 million birds. Roosts sometimes include an abundance of blackbirds and smaller numbers of other birds (e.g., robins, *Turdus migratorius*; northern cardinals, *Cardinalis cardinalis*) (Heisterberg et al. 1990). Starlings typically leave their roosts at sunrise, departing on a direct route for daytime foraging and resting areas (i.e., activity areas) that are often within 24 km of the roost, but can be up to 50 km away (Dolbeer 1982). Activity areas average 7–10 km² and usually are centered on food sources including landfills, granaries, food processing plants, and CAFOs. The majority of starlings arrive at their activity areas within a couple hours of sunrise.

By late afternoon, starlings begin to return to their roost. Returning flights can take up to two hours to complete, with several foraging stops along the way. Starlings may pass over smaller roosts, some lying closer to the main areas of daily activity, to reach larger roosts lying farther away. Flight lines leading toward the roost become obvious about an hour before sunset. Flocks will often stage near the roosting site using wooded areas, power lines, bridges, industrial superstructures, and other sites with plentiful perching substrates. Forays to nearby open grounds occur during the staging period, with birds briefly feeding. Entry into the roost begins about 30 minutes before sunset. Starlings may use a large roost consistently for weeks or months.

Night roosts can be in rural, urban, and suburban venues where shelter from the wind and cold temperatures can be found on tall buildings, bridges, conifer stands, tree groves, and vegetated wetlands (Homan et al. 2008, 2010, 2012). Urban roosts typically have from 10,000 to 30,000 starlings. Morning departures from urban roosts are difficult to track because starlings leave at first light and break into smaller flocks often going in several directions. Urban starlings use surrounding industrial parks, recreation areas, granaries, landfills, and suburban areas throughout the day. Very few starlings remain within the urban area proper. Outlying agricultural habitats within 40 km of an urban roost may be used. Upon returning to an urban roost, starlings stage in secluded industrial areas and commercial areas within a few kilometers of the roost site.

Urban roost sites in city centers may be spread across several urban features, including multistoried buildings, landscaping (especially evergreens), monuments, signage bracing, superstructures, and overpasses (Homan et al. 2008, 2012). Urban roosts are often satellite roosts, lying within a few kilometers of a bigger, main urban roost that serves as the primary roost source. Main roosts are usually located in secluded urban areas where public access is limited, but may be less than 8 km from a roost in the city center. Birds switch occasionally between satellite roosts and the main roost. Main roosts are found in industrial parks, landscaped commercial complexes, abandoned buildings, recreation areas, railroad yards, woodlots, wetlands, bridges, and wooded buffer zones. Main roosts in urban areas can harbor more than 100,000 birds and be difficult to find. For example, a 100-yard section of a four-lane railroad overpass in Omaha, Nebraska, held approximately 70,000 starlings roosting on the concrete support structure and cable pipes underneath the overpass during January (Jeff Homan, pers. comm., U.S. Department of Agriculture [USDA] Animal and Plant Health Inspection Service [APHIS]). Urban roost sites are devoid of birds throughout the day; however, excretal whitewash on perching sites will indicate that a site could be a major roost.

In suburban areas, starlings roost in conifer and deciduous tree stands in residential and business areas, tree groves in parks and abandoned lots, and in vegetated lowlands (Homan et al. 2012). Suburban roosts are much smaller than urban roosts, consisting of just a few hundred birds. Although suburban roosts are usually smaller than urban roosts, many of them can be scattered throughout the suburban landscape.

In agricultural landscapes, starlings may roost at wildlife refuges, game management areas, private wetlands, and abandoned or accessible buildings such as aircraft hangars and agricultural equipment storage buildings. Wetlands with dense stands of emergents can be a preferred habitat for mixed-species flocks of roosting birds, including blackbirds, robins, and starlings. Thick stands of evergreens also are used. Roost sizes in agricultural landscapes can exceed one million birds during winter and attract flocks from over 50 km. Starlings may also use CAFOs as roosting sites. A CAFO can host a few hundred to a few thousand roosting starlings, depending on its size. Although starlings that roost at CAFOs may also feed there, they may, alternatively, leave the CAFO shortly after sunrise and return in the afternoon.

AGRICULTURAL IMPACTS

Starlings damage apples, blueberries, cherries, figs, grapes, peaches, and strawberries by partially or wholly eating the fruit (Nelms et al. 1990; Tobin et al. 1991; Tracey et al. 2007; Lindell et al. 2012; Anderson et al. 2013) (Figure 15.2). Starlings can begin to damage fruit in May, with early damage done by aggregated family groups that can number 1000 birds. Tobin et al. (1991) assessed bird damage to cherries in New York, and found that early ripening cultivars in the study area suffered the most bird damage and, therefore, might warrant the grower's maximum attention. Two decades later, Lindell (2015) conducted quantitative damage surveys in Michigan and found that bird damage to cherries was highest during low-yield years and in early-ripening varieties (Figure 15.3). Further, damage was highest under power lines, at field edges, near night roosts, and in areas with little human activity. Large fruit is more likely to be partially damaged by pecking and slashing (Tracey et al. 2007). Pecked fruit reduces the quality of the fruit and increases vulnerability to diseases and crop pests (Pritts 2001). Pimentel et al. (2000) estimated that yearly starling damage to fruit and grain crops was US$800 million, based on estimated losses of US$5/ha.

In 2012, Anderson et al. (2013) conducted a survey of Honeycrisp apple, blueberry, cherry, and wine grape growers in California, Michigan, New York, Oregon,

FIGURE 15.2 Bird damage to sweet cherries in late May 2012 in Michigan. (Photo by S. Wieferich.)

FIGURE 15.3 Quantitative damage assessments are needed for economic analyses. (Photo by George M. Linz.)

and Washington to estimate costs of bird damage. Growers ranked starlings either first or second among bird species believed responsible for damaging the five crop types in the survey and estimated that birds annually damaged US$70 million of grapes. Grape producers ranked starlings first among three major grape depredating bird species, which included American robins (*Turdus migratorius*) and wild turkeys (*Meleagris gallopavo*). Other results from the 2012 survey of producers indicated damage from birds of US$51 million to sweet cherries and US$33 million to blueberries. Overall, Anderson et al. (2013) estimated that bird damage costs per hectare ranged from US$104 in Oregon tart cherries to US$7267 in Washington Honeycrisp apples. Aggregate bird damage in the five crops and states was estimated at US$189 million.

In the United States, starlings are not considered serious pests in cereal crops or oilseed crops. Even so, producers of sweet corn in several midwestern states annually complain about starling damage during the ripening period, but the amount of damage to sweet corn caused by starlings has not been documented. Starlings also will pull sprouts of some grain crops, but damage appears to be minor and intermittent.

FEEDLOT AND DAIRY IMPACTS

Large flocks of starlings that sometimes number over 100,000 exploit the resources found at CAFOs during late fall and winter (Figure 15.4). Starling use of CAFOs varies greatly, and ease of dispersing flocks appears dependent on weather, especially

FIGURE 15.4   Starlings at a cattle feedlot. (Photo by J. Thiele.)

temperature and snow cover (Linz et al. 2007). Damage to livestock yards and dairies is greatest during winter months because insects and other natural foods are typically unavailable (Linz et al. 2007). Open feeder systems are ideal for starlings because they provide access to livestock rations and easy escape from human and predator threats. Damage estimates at CAFOs showed that over a period of 60 days in the winter, a flock of 1000 starlings can eat about 1.5 tons of cattle feed, representing a loss up to US$0.92 per feedlot animal (Depenbusch et al. 2011). About 250,000 starlings that were using a feedlot in Kansas increased the cost of feeding a ration of steam-flaked corn by $43 per heifer over a 47-day period between mid-January and March. Costs in lost production (i.e., livestock weight gained per unit feed consumed) over this period was $1.00 per animal. In 1999, three feedlot operators in Kansas estimated a loss of $600,000 from bird damage alone (U.S. Department of Agriculture 2000). Data reported in 1968 from Colorado feedlots indicated the cost of cattle rations consumed during winter by starlings was $84 per 1000 starlings (Besser et al. 1968). With the current cost of feed, the associated losses would certainly be much higher.

In addition to causing economic losses due to eating cattle food, wild birds harbor microorganisms, including *Campylobacter*, *Listeria*, *Salmonella*, Shiga toxin-producing *Escherichia coli* (STEC), *Yersinia*, and *Cryptosporidium* (Feare 1984; Gautsch et al. 2000; Clark and McLean 2003; LeJeune et al. 2008). Starlings have been implicated as sources of pathogens causing disease and economic losses to livestock producers (LeJeune et al. 2008; Carlson et al. 2010, 2011, 2012; Gaukler et al. 2012). Avian salmonellosis (primarily, *Salmonella enterica*) has been documented in starlings and is transmissible to humans, poultry, and livestock (Feare 1984). Starlings also carry *Mycobacterium avium paratuberculosis*, which causes Johne's disease in cattle (also known as paratuberculosis) (Matthews and

McDiarmid 1979; Corn et al. 2005). The bacteria are excreted in feces and milk. Johne's disease costs the United States dairy industry $200–$250 million annually (Ott et al. 1999; Beard et al. 2001). STEC is another disease that might be transmitted by starlings to cattle. In the cattle industry, annual costs of illnesses related to *E. coli* STEC exceeded $267 million (NCBA 2004). Humans may get this disease from consuming tainted food products, especially ground beef. Further research is needed to better clarify the role of starlings and other factors in the transmission or prevalence of disease.

Carlson et al. (2011) refined our understanding of the impacts starlings have on disease transmission to cattle. They conducted a study where starling numbers were reduced using the toxicant compound DRC-1339 (3-chloro-4-methylaniline hydrochloride, also 3-chloro-p-toluidine hydrochloride, 3-chloro-4-methylaniline) to evaluate relationships between starlings and *Salmonella enterica* within CAFOs. Within the starling-reduced CAFO, compared to a control CAFO, epidemiological evidence of *S. enterica* disappeared from feed bunks and substantially declined within water troughs following reduction operations. Further, they found that *Salmonella enterica* contamination of both cattle feed troughs and water troughs was significantly related to numbers of starlings. Carlson et al. (2012) also showed that the interaction between European starlings and ambient air temperature explained the occurrence of *S. enterica* in cattle feed. Specifically, the risk of *S. enterica* contamination of cattle feed by starlings was greatest when winter temperatures were highest ($\geq 10°C$). Thus, they concluded that the risk of *S. enterica* contamination of cattle feed by starlings will be worst on the few winter days when daytime high temperatures are above freezing and large numbers of birds are present. Because these conditions will be most common in the late winter and early spring, Carlson et al. (2012) recommend that starling control operations (population control, habitat management, exclusionary devices, and bird repellents) on feedlots and dairies be conducted as early in the winter as weather conditions allow. We caution, however, that multiple biological, environmental, and facility management factors could influence frequency and duration of *S. enterica* in cattle feces, including herd size and age, manure management and disposal, feed storage, access to bacterially contaminated waters, season, and influx of new cattle.

STEC and *Salmonella* spp. are two important foodborne pathogens in the United States that cause more than one million clinical illnesses each year. Direct medical costs resulting from infections of *E. coli* and *Salmonella* spp. are about US$400 million per year. Other starling-borne diseases that can infect humans and domestic fowl include *Chlamydophila psittaci,* a lethal bacteria that may cause avian chlamydiosis and respiratory psittacosis in humans (including chlamydiosis, psittacosis, ornithosis, parrot fever), usually from inhaling spores that live in dried feces (Grimes 1978; Grimes et al. 1979; Andersen et al. 1997; Conover and Vail 2015).

Economic impacts of avian-borne disease on herd health are substantial and likely annually exceed US$1 billion (Pimentel et al. 2000, 2005). Annual costs in the United States from gastrointestinal diseases in livestock caused by *E. coli* spp. (e.g., scours) and *M. avium* (Johne's disease) are estimated at US$600 million. The average cost of an outbreak of *Salmonella* among dairy cattle is US$4000 per farm per incident. Salmonellosis is a far more common affliction in livestock than either

E. coli or *M. avium*, and annual costs of salmonellosis probably exceed US$600 million. A survey of dairy producers in Pennsylvania in 2009, suggested that veterinary costs at dairies with flocks of starlings numbering between 1000 and 10,000 birds were 38% higher than at dairies without starlings (US$91 vs. US$66) (Shwiff et al. 2012). Although not statistically different, this result is suggestive and warrants further attention.

Starling fecal matter can pass transmissible gastroenteritis (TGE) to swine. For example, during the winter of 1978–1979 in Nebraska, starlings served as apparent vectors for an outbreak of TGE that caused the loss of 10,000 swine in one month (Pilchard 1965; Bohl 1975; Gough et al. 1979; Johnson and Glahn 1994). At current market value, this loss might be over US$1.0 million.

Histoplasmosis, a noncommunicable fungal disease of the lungs caused by *Histoplasma capsulatum*, can be contracted by humans, especially if they breathe spores in disturbed dust at starling roosts (DiSalvo and Johnson 1979; Storch et al. 1980; Chu et al. 2006). Most histoplasmosis cases occur in the mid-Atlantic, central, and southeastern states with 37 cases requiring hospitalization per 1 million persons (Chu et al. 2006). Most infections are asymptomatic and subclinical; between 50% and 80% of people who live in areas where *H. capsulatum* is common show antibody evidence of exposure, yet only 5% develop symptoms severe enough to be categorized as clinically sick. Symptoms include fever, cough, weakness, headaches, and muscle aches. Excreta need to accumulate for more than three years before fungal spore densities reach levels high enough to affect human health. Bird droppings must dry out and then be rewetted before spores can form. Although *H. capsulatum* is associated primarily with soils, it can be found growing inside of and around buildings. In 1990s, histoplasmosis was reported at a manufacturing facility in Nebraska used by starlings (J. Hobbs, Nebraska USDA APHIS Wildlife Services, pers. comm.). People at highest risk of exposure, however, are those working in agriculture, particularly poultry, or those who come in contact with bird or bat roosts that might have been abandoned a decade or more prior to disturbance (DiSalvo and Johnson 1979).

Finally, West Nile virus (WNV) was confirmed in North America in 1999 and since then has spread across the Unietd States. This is a serious and life-threatening disease to humans and wildlife. Sullivan et al. (2006) found that red-winged blackbirds are WNV hosts and can disperse diseases along their migratory routes. The role of starlings in dispersing WNV is unknown, but starlings can act as hosts for the virus (Bernard et al. 2001), and thus may be involved in spreading the disease among vertebrates including, humans, horses, and birds.

URBAN IMPACTS

Starlings sometimes roost in urban environments causing residents to voice concerns about the noise, smell, and unsightliness of starling roosts. Their excreta dirty plate glass, sidewalks, city monuments, landscaping, facades, and entryways, and create unsanitary conditions. Pedestrian shoes easily transfer feces, which potentially could harbor diseases, into buildings. Large deposits of excreta corrode metals, including support structures of buildings and bridges.

Maintenance costs can be substantial when large numbers of starlings roost in towns and cities. In one situation in Omaha, Nebraska, contracts for a single cleaning of windows of a large skyscraper were about US$50,000 (J. Thiel, Nebraska USDA APHIS Wildlife Services, pers. comm.). Finally, urban and suburban starlings commonly use exhaust vents of buildings as nesting sites. Nests can clog vents and create unsafe venting conditions for buildings.

Starling roosts near airports pose an aircraft safety hazard because of the potential for birds to be ingested into jet engines (Barras et al. 2003). Starlings have caused some of the most disastrous bird–aircraft strikes because of their body density and flocking behavior. For example, in Boston in 1960, starlings caused a crash that killed 62 people when they were ingested into engines of a Lockheed Electra aircraft on takeoff. Between 1990 and 2014 in the United States, starlings were identified in 3663 strikes on military and civilian aircraft (Dolbeer et al. 2015). Total costs were estimated at US$7 million, but no human fatalities were recorded.

INTEGRATED STARLING MANAGEMENT

FRIGHTENING DEVICES AND REPELLENTS

Humans have used harassment tactics to protect resources from birds for thousands of years. Simple devices such as human effigies (scarecrows) and loud noises from sticks clanging together have now been supplemented with animated scarecrows, recorded distress calls, propane exploders, battery-operated alarms, pyrotechnics (e.g., bangers, shell-crackers, and screamers), lights (for roosting sites at night), hawk kites, flashing metal coated tape, and ultrasonic devices (Conover 2001; Berge et al. 2007). Effectiveness of these devices varies; for example, ultrasonic devices are not effective because, with few exceptions (e.g., South American oilbirds, *Steatornis caripensis*), birds do not hear ultrasonic sounds. Propane exploders and pyrotechnics are the most popular frightening devices because of relatively low purchasing and operating costs. Outfitted with automatic timers that turn on and off each day, exploders are useful for reducing habituation, coordinating timing of the explosions with periods of heavy foraging, and may help in preventing noise complaints from neighbors. Pyrotechnics have an advantage over propane exploders when larger, more inaccessible areas need protection. In all cases, local law enforcement should be contacted for any necessary permits before using these devices.

Distress calls and alarm calls are used often in combination with visual stimuli (e.g., raptor decoy). Achieving adequate broadcasting coverage often requires expensive electronic systems; consider their cost when defending a large area against starlings. For example, Berge et al. (2007) conducted an evaluation of alarm/distress broadcast calls and found that when incorporated into a bird management plan, these devices helped reduce bird damage. These authors calculated that, assuming average yield and price for the Pinot noir vineyards (6.7 tons/ha, $2200/ton), and considering the cost of the broadcast units ($230 per unit), the estimated savings for adding broadcast calls to conventional methods was $700 per ha. These nonlethal tactics generally produce short-term relief unless their intensity and location are varied.

Using an integrated approach that combines auditory, gustatory, and visual senses is an accepted strategy to thwart habituation. Shooting live ammunition is sometimes used as a stand-alone dispersal technique and to reinforce other scare devices rather than as a method for population management (Bomford and O'Brien 1990). Shooting is labor intensive, requiring diligence and consistency. Starlings, especially during winter, focus their daily activities in relatively confined areas where pursuit and harassment with firearms is practicable. If time is limited, employ frightening devices in early morning and late afternoon, when birds are most actively feeding.

The compound 4-aminopyridine (Avitrol®), a restricted-use chemical frightening agent, is sometimes used (Avitrol Corporation 2013). Avitrol® baits are usually placed on grains or pellets and are diluted with untreated grains or pellets. Birds that eat the treated baits behave erratically and give warning cries that frighten other birds from the area. Birds that eat a 4-aminopyridine–treated particle usually die, as might hawks and owls that swallow grain bait while eating an affected or dead bird.

Bird repellents such as methyl anthranilate (MA) are trigeminal irritants that can be used to repel starlings from feedlots and fruit production facilities (Avian Enterprises, LLC. 2015). In birds, MA acts as a chemosensory repellent that causes irritation to pain receptors associated with the trigeminal nerve in the mouth and nostrils of birds. MA is registered for use on numerous fruit and grain crops. At relatively high concentrations (5000 ppm [0.5%] to 10,000 ppm [1%]), MA is a reliable sensory repellent. MA requires multiple applications because it rapidly degrades in the environment, solubilizes in rain, and requires strong concentrations to reach irritation thresholds. Although multiple applications might be needed, a positive cost-benefit ratio might be achieved when protecting high-value crops, such as cherries, blueberries, grapes, and sweet corn. The majority of field studies with MA applications to fruits, however, have shown either no repellency effect or very short-term effects. On the other hand, variability among test sites often encountered during field experiments makes quantitative assessments challenging.

Polybutenes, formulated under various trade names, are sticky materials that might discourage starlings from roosting on ledges and beams. However, labor costs and longevity might preclude using these compounds on large structures.

EXCLUSION

Nylon or plastic netting can be used to exclude starlings from barns, ledges of multi-storied buildings, undersides of roof beams, rafters, and other perch sites. Although highly effective, the initial investment for nets is high and the nets must be monitored for tears and general degradation. Starlings can be excluded from buildings by installing 10-in.-wide door strips made of either heavy plastic or rubber. Install the strips with gaps less than 5 cm to prevent starling entry into the building. Other tactics include placing 45°-angle coverings of wood, metal, or Plexiglas® over ledges to prevent starlings from perching, nesting, or roosting. Metal protectors or porcupine wires are available for preventing roosting on ledges or roof beams.

Protection of ripening fruit is a high-priority agroeconomic issue for growers. Netting might be cost-effective where bird damage is expected to be high, and particularly cost-effective for preventing damage to high-value grapes (Tobin et al. 1991;

FIGURE 15.5 Netting is effective for protecting high-value crops, such as grapes. (Photo by S. Wieferich.)

Berge et al. 2007; Tracey et al. 2007) (Figure 15.5). Assuming a 10-year life span, the cost of labor, netting, and construction of an application-removal system for large-area netting is about US$1000/ha per year. Wine grapes, which can be valued at nearly US$19,760/ha for some varieties, may justify the use of large-area netting.

Lethal

Reducing the numbers of starlings to a tolerable level is among the variety of options available to manage human–starling conflicts, with the more suitable approaches depending on the numbers of starlings, the conflict location and situation, and other factors. Starlicide Complete™ (EPA Reg. No. 67517-8) is a commercially available pesticide registered for controlling starlings in CAFOs. Starlicide's active ingredient is compound DRC-1339, a slow-acting toxicant originally developed for controlling starlings around livestock and poultry operations (DeCino et al. 1966; Royall et al. 1967; West 1968; Carlson et al. 2012). Compound DRC-1339 Concentrate (also known as Starlicide Technical®) is a powder that can be custom-mixed with several bait substrates, including cracked corn, rolled corn, distiller's grain, milo, rolled milo, poultry pellets, raisins, and French fries. Compound DRC-1339 Concentrate is for use only by USDA-WS employees or those under their direct supervision.

Bird species exhibit a range of sensitivity to DRC-1339 (Eisemann et al. 2003). Starlings are highly sensitive to DRC-1339, with a single treated bait causing death in one to three days. Gulls, icterids, and corvids also are very sensitive to this toxicant (Eisemann et al. 2003). DRC-1339 is a slow-acting toxicant, so dead starlings are generally not found at the bait site. Baited birds can behave normally for several hours after ingesting treated bait, thus most birds succumb to the toxic effects in a night roost or at staging areas, often near water. Poisoned starlings are not dangerous to scavengers or predators as the chemical is quickly metabolized and excreted (Eisemann et al. 2003).

When the best solution is to reduce starling numbers with DRC-1339–treated baits, it is prudent to quickly find and properly dispose the carcasses to ease public angst over discovering carcasses. To that end, USDA WS policy requires that employees notify local government agencies (e.g., city, township, state government officials) of potential lethal management actions (U.S. Department of Agriculture 2009).

Sodium lauryl sulfate (SLS) is a surfactant that can be used only by USDA WS personnel or official cooperators for managing roosts of pest bird species, including starlings. SLS destroys the insulating properties of feathers, causing hypothermia. Generally, this technique has limited effectiveness because of the specific weather conditions needed and equipment logistics. SLS is for use only in winter conditions (<41°F) on upland roosts and cannot be sprayed over bodies of water or in areas of direct runoff. Wetted birds die as soon as 30 minutes after spraying with SLS. Before using SLS, the roosts must be observed for nontarget species. Field trials with SLS were conducted in southeastern Missouri between 2005 and 2007, using ground-based spray systems (Byrd et al. 2009). A pump delivered water at 6 gallons per minute per sprinkler head. Up to 12,000 starlings and 3000 blackbirds were killed at a 50,000-bird roost during a single SLS spray using four sprinkler heads. Other sprays were not successful. Poor results were obtained in three of eight roost sprays conducted in southeastern Missouri, attributed to low water quality that decreased the effectiveness of the SLS and pump malfunction.

Trapping starlings is time consuming, and success varies with time of year, population size, and amount of area needing protection. Starlings seem particularly easy to capture in the winter when food resources are often limited. For example, Thiele et al. (2012) captured nearly 5000 starlings in the greater Omaha, Nebraska area during winter with modified Australian crow traps and drop-in decoy traps stocked with decoy birds. Part of their success was attributed to providing good care for the decoy birds, including providing fresh food, water, and sheltered perching sites. They also replaced their decoy stock with newly caught birds periodically. Use of cage traps, however, at feedlots and dairies may not be cost-effective because of the comparatively low economic value of livestock feed and the large numbers of starlings that may be present. On the other hand, trapping starlings might be cost-effective at vineyards, fruit orchards, and berry farms for preventing starling damage early in the crop season because starlings, especially juveniles, are not trap-wary (Conover and Dolbeer 2007). Decoy traps allow wildlife managers and growers to reduce the numbers of depredating target species while greatly reducing the risks of taking nontarget species.

HABITAT MODIFICATION

On-site management practices are important for mitigating starling damage at CAFOs and in urban settings. The primary goal is to limit the availability of food, water, and roosting sites. Where starlings are roosting, one option is to remove or thin perch sites used by starlings. Day roosts and night roosts may include tree stands, dense vegetation (e.g., evergreens) on lee sides of buildings, and emergent vegetation growing in wetlands and low-lying areas. Thinning young stands by 30%–50% may disperse roosts or prevent roosting. Pruning side branches of roost trees discourages roosting. Limit heat leakage from buildings, especially tall buildings with perch sites. Aquatic herbicides are commercially available to thin dense stands of emergents (Linz and Homan 2011) where applicable. In some regions, wetlands and dense thickets of bottomlands are highly preferred winter roosting sites. These sites may be located several kilometers from sites used for daily activities.

Cultural Practices

Both CAFO operators and urbanites can benefit from removing food sources and cleaning up spilled grains and garbage. Timed automatic-release feeders can be a good option to avoid parts of the day when starlings are likely to be foraging. For example, switch to afternoon or nighttime feeding schedules, if possible. Eliminate unnecessary pools of water; also, lower water levels in water containers to prevent starlings from drinking and bathing. Use feed with large forms of more than 0.5-in. diameter that starlings have difficulty swallowing. Feed losses to starlings can be nearly eliminated by using 0.75-in. × 3-in. extruded pellets.

Starlings naturally fear birds of prey. Thus, falconers can sometimes effectively move birds out of crops. It is labor intensive, requiring the falconer to be on site. It is expensive, costing more than US$500 per day. Most falconers prefer to use their birds in fairly open habitats, where chances of injuries to the falcons are low. Blueberries and other types of high-value fruits with shrubby habits are more fitted to falconry than treed fruits. Installation of nesting boxes and artificial perches for birds of prey at orchards, vineyards, and CAFOs has the potential to provide a low-cost alternative to falconry.

Determining Cost-Benefits

A rule of thumb for evaluating the economic feasibility of a management method involves comparing pretreatment costs of bird damage to the amortized costs of deploying a treatment method, and then assessing potential savings gained from applying the treatment. The resulting savings must be greater than the depreciated costs. A cost-benefit ratio of 1:2 or greater should be expected. Assuming all crop inputs were made before damage occurred, a general formula for agricultural and fruit crops would be as follows (using acres as the areal unit):

$$(A \times B + C/D) - ([A \times B] - [A \times E]) > F$$

where
    A = economic production per acre (i.e., price received at sale)
    B = proportion of anticipated bird damage under no treatment
    C = amortized cost of using method (including labor, equipment, and maintenance)

D = acres of crop protected by method

E = proportion of damage after implementing treatment

F = depreciated value of method or other accounting for lost value and function

RESEARCH NEEDS

The reproductive biology of starlings is well documented, but data on regional migratory patterns and local movements in relation to feedlots, diaries, and urban areas are needed. These data are especially important as the climate changes and starlings (and blackbirds) potentially begin using more northerly winter roost locations (Strassburg et al. 2015). We speculate that these changes could compound conflicts with human endeavors, including issues related to airport safety and disease transfer among CAFOs. Scientists for USDA WS, in collaboration with North Dakota State University, The Ohio State University, and others, are beginning to gather these data. This information will be useful for developing risk assessments and economic impact models that will help determine the overall consequences of starlings. Additional efforts are underway to develop and evaluate better bait carriers for the compound DRC-1339. Better information is needed to determine the role of starlings along with other factors in the prevalence and transmission of diseases. Research that would compare population dynamics and limiting factors of starlings in Europe to those in North America would be instructive, as would research on the potential for enhancing bird predators as a starling management approach (Gaston 2010; Kross et al. 2012). Finally, research to better understand and manage the complex interactions between starlings and native cavity nesters at multiple spatial scales is warranted (Koenig 2003).

ACKNOWLEDGMENTS

We thank all the past and current scientists who contributed thousands of hours studying the European starling and their interactions and conflicts with human endeavors. M. Avery, J. Carlson, and M. Tobin reviewed an earlier version of the manuscript. This manuscript was carried out under the U.S. Department of Agriculture, Animal and Plant Health Inspection Service, Wildlife Services, and National Wildlife Research Center Protocol Number QA-2480.

REFERENCES

Andersen, A. A., J. E. Grimes, P. B. Wyrick et al. 1997. Chlamydiosis. In *Diseases of Poultry*. 10th edition, ed. B. W. Calnek. Iowa State University Press, Ames, Iowa, USA, pp. 333–349.

Anderson, A., C. A. Lindell, K. M. Moxcey et al. 2013. Bird damage to select fruit crops: The cost of damage and the benefits of control in five states. *Crop Protection* 52:103–109.

Avian Enterprises, LLC. 2015. *Avian Control Bird Repellent*. Jupiter, Florida, USA. http://aviancontrolinc.com/wp-content/uploads/2015/12/Avian-Control-Bulk-Label-Pages-2015-EPA-Layout.pdf. Accessed April 12, 2016.

Avitrol Corporation. 2013. Avitrol corn chops. http://www.avitrol.com/pdf/avitrol-us-corn-chops-specimen-label.pdf. Accessed June 26, 2015.

Barras, S. C., S. E. Wright, T. W. Seamans. 2003. Blackbird and starling strikes to civil aircraft in the United States, 1990–2001. In *Management of North American Blackbirds: Proceedings of a Special Symposium of the Wildlife Society 9th Annual Conference*, ed. G. M. Linz. USDA/APHIS Wildlife Services, National Wildlife Research Center, Fort Collins, Colorado, USA, pp. 91–96.

Beard, P. M., M. J. Daniels, D. Henderson et al. 2001. Paratuberculosis infection of nonruminant wildlife in Scotland. *Journal of Clinical Microbiology* 39:1517–1521.

Berge, A., M. Delwiche, W. P. Gorenzel, and T. Salmon. 2007. Bird control in vineyards using alarm and distress calls. *American Journal of Enology and Viticulture* 58:135–143.

Bernard, K. A., J. G. Maffel, S. A. Jones, E. B. Kauffman, G. Ebel, and D. A. P. Dupuis. 2001. West Nile virus infection in birds and mosquitoes, New York State 2000. *Emerging Infectious Diseases* 7:679–685.

Besser, J. F., J. W. DeGrazio, and J. L. Guarino. 1968. Costs of wintering starlings and redwinged blackbirds at feedlots. *Journal of Wildlife Management* 32:179–180.

Bohl, E. H. 1975. Transmissible gastroenteritis. In *Diseases of Swine*, 4th edition, eds. H. W. Dunne and A. D. Leman. Iowa State University Press, Ames, Iowa, USA, pp. 168–188.

Bomford, M., and P. H. O'Brien. 1990. Sonic deterrents in animal damage control: A review of device tests and effectiveness. *Wildlife Society Bulletin* 18:411–422.

Boyd, E. M. 1951. A survey of parasitism of the starling *Sturnus vulgaris* L. in North America. *Journal of Parasitology* 37:56–84.

Byrd, R., J. Cummings, S. Tupper, and J. Eisemann. 2009. Evaluation of sodium lauryl sulfate as a blackbird wetting agent. *Proc. Wildlife Damage Management Conference* 13:191–196.

Cabe, P. R. 1993. European starling (*Sturnus vulgaris*). In *The Birds of North America*, eds. A. Poole and F. Gill, No. 48. The Academy of Natural Sciences, Philadelphia, Pennsylvania and The American Ornithologists Union, Washington, DC, USA.

Caccamise, D. F. 1991. European starling fidelity to diurnal activity centers: Role of foraging substrate quality. *Wilson Bulletin* 103:13–24.

Carlson, J. C., J. W. Ellis, S. K. Tupper, A. B. Franklin, and G. M. Linz. 2012. The effect of European starlings and ambient air temperature on *Salmonella enterica* contamination within cattle feed bunks. *Human–Wildlife Interactions* 6:64–71.

Carlson, J. C. R. M. Engeman, D. R. Hyatt et al. 2011. Efficacy of European starling control to reduce Salmonella enterica contamination in a concentrated animal feeding operation in the Texas panhandle. *BMC Veterinary Research* 7:9.

Carlson, J. C., A. B. Franklin, D. R. Hyatt, S. E. Pettit, and G. M. Linz. 2010. The role of starlings in the spread of Salmonella within concentrated animal feeding operations. *Journal of Applied Ecology* 47:1–7.

Chu, J. H., C. Feudtner, K. Heydon, T. J. Walsh, and T. E. Zaoutis. 2006. Hospitalizations for endemic mycoses: A population-based national study. *Clinical Infectious Diseases* 42:822–825.

Clark, L., and R. G. McLean. 2003. A review of pathogens of agricultural and human health interest found in blackbirds. In *Management of North American Blackbirds: Proceedings of a Special Symposium of the Wildlife Society 9th Annual Conference*, ed. G. M. Linz. USDA/APHIS Wildlife Services, National Wildlife Research Center, Fort Collins, Colorado, USA, pp. 103–108.

Conover, M. R. 2001. *Resolving Human–Wildlife Conflicts: The Science of Wildlife Damage Management*. CRC Press/Taylor & Francis, Boca Raton, Florida.

Conover, M. R., and R. A. Dolbeer. 2007. Use of decoy traps to protect blueberries from juvenile European starlings. *Human–Wildlife Conflicts* 1:265–270.

Conover, M. R., and R. M. Vail. 2015. *Human Diseases from Wildlife*. CRC Press/Taylor & Francis, Boca Raton, Florida.

Corn, J. L., E. J. B. Manning, S. Sreevatsan, and J. R. Fischer. 2005. Isolation of *Mycobacterium avium* subsp. *paratuberculosis* from free-ranging birds and mammals on livestock premises. *Applied and Environmental Microbiology* 71:6963–6967.

Craig, A., and C. Feare. 1999. *The Starling*. Princeton University Press, New Jersey.

DeCino, T. J., D. J. Cunningham, and E. W. Schafer, Jr. 1966. Toxicity of DRC-1339 to starlings. *Journal of Wildlife Management* 30:249–253.

Depenbusch, B. E., J. S. Drouillard, and C. D. Lee. 2011. Feed depredation by European starlings in a Kansas feedlot. *Human–Wildlife Interactions* 5:58–65.

DiSalvo, A. F., and W. M. Johnson. 1979. Histoplasmosis in South Carolina: Support for the microfocus concept. *American Journal of Epidemiology* 109:480–492.

Dolbeer, R. A. 1982. Migration patterns for age and sex classes of blackbirds and starlings. *Journal of Field Ornithology* 53:28–46.

Dolbeer, R. A., S. E. Wright, J. R. Weller, A. L. Anderson, and M. J. Begier. 2015. Wildlife strikes to civil aircraft in the United States 1990–1994. Federal Aviation Administration National Wildlife Strike Database Serial Report Number 21. http://www.faa.gov/airports/airport_safety/wildlife/media/Wildlife-Strike-Report-1990-2014.pdf. Accessed April 7, 2016.

Drent, R. H., J. M. Tinbergen, and D. H. Biebach. 1985. Incubation in the starling (*Sturnus vulgaris*): Resolution of the conflict between egg care and foraging. *Netherlands Journal of Zoology* 35:103–123.

Eisemann, J. D., P. A. Pipas, and J. L. Cummings. 2003. Acute and chronic toxicity of compound DRC-1339 (3-chloro-4-methylaniline hydrochloride) to birds. In *Management of North American Blackbirds: Proceedings of a Special Symposium of The Wildlife Society 9th Annual Conference*, ed. G. M. Linz. USDA/APHIS Wildlife Services, National Wildlife Research Center, Fort Collins, Colorado, USA, pp. 24–28.

Feare, C. J. 1984. *The Starling*. Oxford University Press, New York.

Gaston, K. J. 2010. Valuing common species. *Science* 327:154–155.

Gaukler, S. M., H. J. Homan, G. M. Linz, and W. J. Bleier. 2012. Using radio-telemetry to assess the risk European starlings pose in pathogen transmission among feedlots. *Human–Wildlife Interactions* 6:30–37.

Gautsch, S., P. Odermatt, A. P. Burnens, J. Bille, and R. Ewald (Translated abstract). 2000. The role of common starlings (*Sturnus vulgaris*) in the epidemiology of bacterial, potentially human pathogenic, disease agents. *Schwiez Arch Tierheilkd* 142:165–172.

Gough, P. M., J. W. Beyer, and R. D. Jorgenson. 1979. Public health problems: TGE. *Proc. Bird Control Seminar* 8:136–141.

Grimes, J. E. 1978. Transmission of chlamydia from grackles to turkeys. *Avian Diseases* 22:308–312.

Grimes, J. E., K. J. Owens, and J. R. Singer. 1979. Experimental transmission of *Chlamydia psittaci* to turkeys from wild birds. *Avian Diseases* 23:915–926.

Gromadzki, M. 1980. Reproduction of the starling *Sturnus vulgaris* in Zulawy Wislane, North Poland. *Acta Ornithologica* 17:195–223.

Heisterberg, J. F., J. L. Cummings, G. M. Linz, C. E. Knittle, T. W. Seamans, and P. P. Woronecki. 1990. Field trial of a CPT-Avicide aerial spray. *Proc. Vertebrate Pest Conference* 14:350–360.

Homan, H. J., J. T. LeJeune, D. L. Pearl et al. 2013. Use of dairies by post-reproductive flocks of European starlings. *Journal of Dairy Science* 96:4487–4493.

Homan H. J., G. M. Linz, G. W. Unrein, J. R. Thiele, and J. M. Hobbs. 2008. Movements of European starlings captured at a winter roost in Omaha, Nebraska. *Proc. North American Prairie Conference* 20:79–82.

Homan, H. J., A. A. Slowik, L. B. Penry, and G. M. Linz. 2012. Site use of European starlings wintering in central New Jersey. *Proc. Vertebrate Pest Conference* 25:230–234.

Homan, H. J., A. A. Slowik, L. B. Penry, G. M. Linz, M. J. Bodenchuk, and R. L. Gilliland. 2010. Site use of European starlings captured and radio-tagged at Texas feedlots during winter. *Proc. Vertebrate Pest Conference* 24:250–256.

Jernelov, A. 2017. *The Long-Term Fate of Invasive Species.* Springer International Publishing, Switzerland.

Johnson, R. J., and J. F. Glahn. 1994. European starlings. In *Prevention and Control of Wildlife Damage*, ed. S. E. Hygnstrom, R. M. Timm, and G. E. Larson. University of Nebraska–Lincoln, Lincoln, Nebraska, USA, pp. E109–E120.

Kessel, B. 1953. Distribution and migration of the European starling in North America. *Condor* 55:49–67.

Kessel, B. 1957. A study of the breeding biology of the European starling (*Sturnus vulgaris*) in North America. *American Midland Naturalist* 58:257–331.

Koenig, W. D. 2003. European starlings and their effect on native cavity-nesting birds. *Conservation Biology* 17:1134–1140.

Kross, S. M., J. M. Tylianakis, and X. J. Nelson. 2012. Effects of introducing threatened falcons into vineyards on abundance of Passeriformes and bird damage to grapes. *Conservation Biology* 26:142–149.

Lindell, C. A. 2015. When using bird deterrents in sweet cherries makes sense. http://msue. anr.msu.edu/news/when_using_bird_deterrents_in_sweet_cherries_makes_sense. Accessed June 26, 2015.

Lindell, C. A., R. A. Eaton, E. M. Lizotte, and N. L. Rothwell. 2012. Bird consumption of sweet and tart cherries. *Human–Wildlife Interactions* 6:283–290.

Linz, G. M., and H. J. Homan. 2011. Use of glyphosate for managing invasive cattail (*Typha* spp.) to protect crops near blackbird (Icteridae) roosts. *Crop Protection* 30:98–104.

Linz, G. M., H. J. Homan, S. M. Gaukler, L. B. Penry, and W. J. Bleier. 2007. European starlings: A review of an invasive species with far-reaching impacts. In *Managing Vertebrate Invasive Species: Proceedings of an International Symposium*, ed. G. W. Witmer, W. C. Pitt, and K. A. Fagerstone. USDA/APHIS Wildlife Services, National Wildlife Research Center, Fort Collins, Colorado, USA, pp. 378–386.

LeJeune, J. T., H. J. Homan, G. M. Linz, and D. L. Pearl. 2008. Role of European starlings in the transmission of *E. coli* O157 on dairy farms. *Proc. Vertebrate Pest Conference* 23:31–34.

Lowe, S., M. Browne, S. Boudjelas, and M. DePoorter. 2004. 100 of the world's worst invasive alien species: A selection from the global invasive species database. The Invasive Species Specialist Group (ISSG) a specialist group of the Species Survival Commission (SSC) of the World Conservation Union (IUCN). http://www.issg.org/database/species/reference_files/100english.pdf. Accessed April 7, 2016.

Martinez del Rio, C., M. L. Avery, and K. E. Brugger. 1995. Sucrose as a feeding deterrent for fruit-eating birds. In *Repellents in Wildlife Manage Symposium: Proc. Second Denver Wildlife Research Center Special Symposium*, ed. R. Mason. USDA/APHIS Wildlife Services, National Wildlife Research Center, Fort Collins, USA, pp. 353–369.

Matthews, P. R., and A. McDiarmid. 1979. The production in bovine calves of a disease resembling paratuberculosis with a *Mycobacterium* sp. Isolated from a woodpigeon (*Columba palumbus*). *Veterinary Record* 104:286.

Morrison, D., and D. F. Caccamise. 1990. Comparison of roost use by three species of communal roost mates. *Condor* 92:405–412.

NCBA (National Cattleman's Beef Association). 2004. *A Basic Look at E. coli O157.* National Cattlemen's Beef Association, Centennial, CO, USA. http://www.beefresearch.org/CMDocs/BeefResearch/Safety_Fact_Sheets/E.Coli_Final.pdf. Accessed April 27, 2016.

Nelms, C. O., M. L. Avery, and D. G. Decker. 1990. Assessment of bird damage to early-ripening blueberries in Florida. *Proc. Fourteenth Vertebrate Pest Conference* 14:302–306.

Ott, S. L., S. J. Wells, and B. A. Wagner. 1999. Herd-level economic losses associated with Johne's disease on U.S. dairy operations. *Preventive Veterinary Medicine* 40:179–92.

Partners in Flight Science Committee. 2013. *Population Estimates Database, Version 2013.* http://rmbo.org/pifpopestimates. Accessed on April 7, 2016.

Pilchard, E. J. 1965. Experimental transmission of transmissible gastroenteritis virus by starlings. *American Journal of Veterinary Research* 114:1177–1179.

Pimentel, D., L. Lach, R. Zoniga, and D. Morrison. 2000. Environmental and economic costs of nonindigenous species in the United States. *BioScience* 50:53–65.

Pimentel D., R. Zuniga, and D. Morrison. 2005. Update on the environmental and economic costs of alien-invasive species in the United States. *Environmental Economics* 52:273–288.

Pritts, M. P. 2001. Bye bye birdie: Repelling birds from fruit plantings. *New York Quarterly* 9:5–7.

Ricklefs, R. E., and C. A. Smeraski. 1983. Variation in incubation period within a population of the European starling. *Auk* 100:926–931.

Rollins, L. A., A. P. Woolnough, A. N. Wilton, R. Sinclair, and W. B. Sherwin. 2009. Invasive species can't cover their tracks: Using microsatellites to assist management of starling (*Sturnus vulgaris*) populations in Western Australia. *Molecular Ecology* 18:1560–1573.

Royall, W. C., Jr. 1966. Breeding of the starling in central Arizona. *Condor* 68:195–205.

Royall, W. C., Jr., T. C. DeCino, and J. F. Besser. 1967. Reduction of a starling population at a turkey farm. *Poultry Science* 46:1494–1495.

Shwiff, S. A., J. C. Carlson, J. H. Glass et al. 2012. Producer survey of economic and disease impacts associated with bird-livestock interactions in commercial dairies. *Dairy Science* 95:6820–6829.

Storch, G., J. G. Burford, R. B. George, L. Kaufman, and D. L. Ajello. 1980. Acute histoplasmosis. Description of an outbreak in northern Louisiana. *Chest* 77:38–42.

Strassburg, M., S. M. Crimmins, G. M. Linz, P. C. McKann, and W. E. Thogmartin. 2015. Winter habitat associations of blackbirds and starlings wintering in the south-central United States. *Human–Wildlife Interactions* 9:171–179.

Sullivan, H., G. M. Linz, L. Clark, and D. M. Salman. 2006. West Nile virus antibody prevalence in red-winged blackbirds (*Agelaius phoeniceus*) from North Dakota (2003–2004). *Vectorborne and Zoonotic Diseases* 6:305–309.

Thiele, J. R., H. J. Homan, G. M. Linz, and G. W. Unrein. 2012. Developing an effective management plan for starlings roosting in downtown Omaha, Nebraska. *Wildlife Damage Management Conference* 14:87–90.

Tinbergen, J. M. 1981. Foraging decisions in starlings (*Sturnus vulgaris*). *Ardea* 69:1–67.

Tobin, M. E., R. A. Dolbeer, C. M. Webster, and T. W. Seamans. 1991. Cultivar differences in bird damage to cherries. *Wildlife Society Bulletin* 19:190–194.

Tracey, J., M. Bomford, Q. Hart, G. Saunders, and R. Sinclair. 2007. *Managing Bird Damage to Fruit and Other Horticultural Crops.* Bureau of Rural Sciences, Canberra. Also available at: http://www.dpi.nsw.gov.au/__data/assets/pdf_file/0005/193739/managing_bird_damage-full-version.pdf.

Tupper, S. K., S. J. Werner, J. C. Carlson et al. 2014. European starling feeding activity on repellent treated crops and pellets. *Crop Protection* 63:76–82.

U.S. Department of Agriculture. 2000. *Environmental Assessment: Bird Damage Management at Livestock Feeding Facilities in the Kansas Wildlife Services Program.* U.S. Department of Agriculture, Animal Plant Health Inspection Service, Kansas Wildlife Services Program, Manhattan, Kansas, USA.

U.S. Department of Agriculture. 2009. *Directive 2.401: Pesticide Use.* www.aphis.usda.gov/wildlife_damage/directives/2.401_pestiside_use.pdf. Accessed July 21, 2015.

West, R. R. 1968. Reduction of a winter starling population by baiting its preroosting areas. *Journal of Wildlife Management* 32:637–640.

16 Monk and Rose-Ringed Parakeets

Michael L. Avery and Aaron B. Shiels

CONTENTS

INTRODUCTION

Worldwide, there are about 350 species of parrots and parakeets (order: Psittaciformes). According to the analyses of Cassey et al. (2004), 54 of these species have been introduced to areas outside their native ranges, and 38 species have become established in the nonnative range. Humans exhibit ambivalent feelings toward parrots and parakeets. Many of these birds are strikingly beautiful and highly prized as companion animals, while others are banned because of potential agricultural damage or competition with native species. Many parrot species are afforded special protection because they are endangered in their native habitats, but often these same species are considered crop pests and persecuted by farmers (e.g., Tella et al. 2013).

The United States was once home to two species of native parrots, the Carolina parakeet (*Conuropsis carolinensis*) and the thick-billed parrot (*Rhynchopsitta pachyrhyncha*). Carolina parakeets were found throughout southeastern United States, as far north as New York and North Dakota, and as far west as Colorado and Texas (Snyder and Russell 2002). Thick-billed parrots occupied northern Mexico and portions of the bordering states of Arizona and New Mexico (Snyder et al. 1999). These two native parrots were lost during the twentieth century. The demise of the Carolina parakeet was probably due to combined effects of disease, shooting (for sport, crop protection, and millinery), and habitat loss (Snyder and Russell 2002). Extirpation of the thick-billed parrot from the southwestern United States was most likely due to hunting (Snyder et al. 1994). A reintroduction program that began in 1986 for the thick-billed parrots in Arizona did not result in a self-sustaining population, although the species persists in Mexico (Snyder et al. 1994).

While there is an absence of native parrots in the United States today, at least nine species of introduced parrots are currently recognized as being established in the United States by the American Ornithologists' Union (Chesser et al. 2015). Nineteen additional free-flying, introduced parrot species are recognized, but not considered established (Chesser et al. 2015). At least five species are established in the state of Hawaii (Runde et al. 2007; Pyle and Pyle 2009).

The monk parakeet (*Myiopsitta monachus*) and rose-ringed parakeet (*Psittacula krameri*) are undoubtedly the world's most successful introduced parrot species. Each species now enjoys a broad nonnative range where conflicts with human activity include crop damage (Conroy and Senar 2009; Gaudioso et al. 2012), competition with native species (Strubbe and Matthysen 2007), and property damage (Avery et al. 2006). Each species exemplifies invasiveness through its capacity to adapt to new conditions and to exploit opportunities created by human activity. Biologists and resource managers are challenged to develop and implement effective strategies that not only protect resources from these invasive species, but also account for public opinions which often favor the charismatic avian invaders.

MONK PARAKEET (FAMILY: PSITTACIDAE)

The monk parakeet, also known as Quaker parakeet, is a medium-sized parrot (110–130 g body mass, approximately 28 cm total length). On average, adult males are slightly larger than adult females except during the breeding season when body mass of females increases slightly. Adult males and females are identical in plumage. The plumage is green on the back and tail, and grayish on the underside. The wings are dull green with the outer wing feathers blue (Spreyer and Bucher 1998).

Monk parakeets primarily eat a variety of fruit, seeds, buds, and flowers. In its native range within South America, the species is regarded as a major pest to crops such as sorghum, sunflower, and rice. The monk parakeet inhabits open woodlands, savannah, agricultural areas, and disturbed habitats from southern Bolivia, through Paraguay, southern Brazil, Uruguay, to southern Argentina (Spreyer and Bucher 1998).

FIGURE 16.1   Electric utility facilities, including substations as pictured here, are favorite nest sites for monk parakeets in the United States. (USDA photo.)

Monk parakeets are monogamous. One clutch of four to eight eggs is produced annually during the well-defined spring breeding season. The female incubates the eggs and broods the nestlings while the male contributes nest materials and brings food to the female. Monk parakeets are unique among psittacines as they use sticks and twigs to construct bulky nests which house from one to many individual nesting chambers (Spreyer and Bucher 1998). The nest structure is the focus of the para- keets' social system as the birds live in their nests year round, not just during the breeding season (Figure 16.1).

Breeding adults and nonbreeding subadults defend and maintain their nests throughout the year, and instances of nonbreeding parakeets helping to feed nest- lings or recent fledglings have been documented (Bucher et al. 1990; Eberhard 1998). Nesting season and molt cycle of the species in the United States are shifted six months with respect to the native range (Avery et al. 2012).

ORIGIN IN THE UNITED STATES

In 1966, the U.S. Bureau of Sport Fisheries and Wildlife implemented regulations requiring declarations (Form 3–177) of wildlife imported into the United States (Banks 1970). Psittacines, however, were initially exempted from this reporting requirement, so the first full tabulations of monk parakeet importations date from 1968. During the five-year period 1968–1972, over 63,000 monk parakeets were imported into the United States. Most, perhaps all, of these imports were from Paraguay. Importations from Paraguay were suspended in 1973 due to concern over Newcastle's disease. Imports resumed in 1974 when 608 birds were brought in (Greenhall 1977). In 1978, Uruguay and Argentina became the main sources of monk parakeets imported into the United States. Since 1981, importation records have been maintained and compiled by the Convention on International Trade in Endangered Species of Wild Fauna and Flora (CITES). Data on exports and imports by species can be found online from the CITES trade database (http://trade.cites.org/). Importation of monk parakeets essentially ended with the passage of the Wild Bird Conservation Act of 1992, which ensures that exotic bird species are not harmed by international trade and encourages wild bird conservation in countries of origin. Monk parakeets continue to be available in the pet trade because they are readily bred and raised in captivity.

ESTABLISHMENT IN THE UNITED STATES

Monk parakeets in New York/New Jersey. Free-flying parakeets were first reported from the New York/New Jersey metropolitan area in 1967, and nest construction was observed in 1970. These earliest populations in the New York/New Jersey area probably originated from escaped pet birds imported from Paraguay. The Audubon Christmas Bird Count (CBC) is the only long-term, structured source of survey data (http://netapp.audubon.org/CBCObservation/). Monk parakeets first appeared on the CBC for New York in 1970, and they appeared on the New York CBC every year thereafter, except 1983. Data from the CBC support the notion that many early populations in the New York/New Jersey area did not persist. One exception is Brooklyn, where parakeet numbers on the CBC were low, but consistent, throughout the 1970s and 1980s. Parakeet numbers in the New York/New Jersey area have increased slowly since then and are at their highest level ever, according to the 2014 CBC results.

Monk parakeets in Connecticut. In Connecticut, two monk parakeets were recorded in 1971 and again in 1972 (Neidermyer and Hickey 1977). In 1973, the Connecticut Department of Environmental Protection reported 34 monk parakeets in the state at three locations (Olivieri and Pearson 1992). Monk parakeets first appeared on the CBC in Connecticut in 1974 (three birds). As in New York/New Jersey, these first parakeets probably originated from Paraguay. Monk parakeets next appeared on the Connecticut CBC in 1985, and they have been in the count results ever since. The CBC counts peaked in 2005 and have slowly declined since.

Monk parakeets in Florida. The monk parakeet was first recorded breeding in Miami in 1969, and since the early 1970s, the species has been firmly established (Owre 1973). Florida monk parakeets have been on the CBC since 1974. The count numbers in Florida peaked in 2002, followed by a steady annual decline until 2014 when numbers recorded rose again. As judged by the CBC, Florida has the most monk parakeets of any state.

EXPANSION AND POPULATION GROWTH

The U.S. Fish and Wildlife Service initiated a nationwide monk parakeet control and removal program in the early 1970s because the species had a reputation as an agricultural pest in South America and there was concern for the impacts it might have on crops in the United States (Neidermyer and Hickey 1977). Other concerns included possible transmission of psittacosis and interspecific competition with native wildlife. During 1970–1975, participants in this removal program recorded 367 confirmed sightings of monk parakeets from 30 states (Neidermyer and Hickey 1977). As a result, 163 birds were removed from 16 states, mostly New York (88) and California (35). The removal program ended in 1975 and was considered a success in reducing the growth and spread of monk parakeet populations (Neidermyer and Hickey 1977).

The monk parakeet has thrived in the United States since 1975 and has become an urban/suburban species with no obvious factor limiting population growth. Nationwide, monk parakeets have exhibited exponential growth, buoyed principally by the Florida population (Van Bael and Pruett-Jones 1996). Declines in Florida and Connecticut populations since 2003 ended the exponential growth pattern, but monk parakeet populations are growing elsewhere, particularly in Texas (e.g., Reed et al. 2014), which now is second only to Florida in numbers recorded on the CBC (Figure 16.2). In Illinois, researchers hypothesize that actual parakeet numbers might not be decreasing, but instead their geographical distribution might be changing (Pruett-Jones et al. 2012).

Monk parakeets are well established outside of the United States. The first published records of escaped monk parakeets in Spain are from 1975, when the species established in Barcelona (Batllori and Nos 1985). Largely due to importation of thousands of birds from South America for the pet trade, the species quickly expanded to other cities in Spain and established populations in other countries such as Belgium, Italy, and England (Sol et al. 1997; Butler 2002; Strubbe and Matthysen 2009a). Monk parakeets appear to be spreading in Mexico where they were first reported in 1999 (MacGregor-Fors et al. 2011). In recent years (2010–2014), CBC records from Puerto Rico and Bahamas also have included monk parakeets.

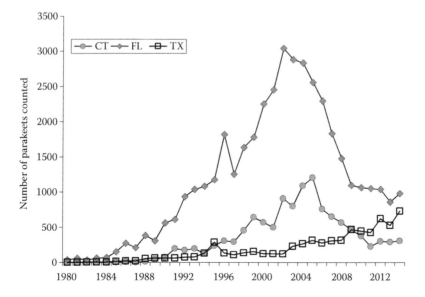

FIGURE 16.2 Monk parakeet population trends in Florida, Connecticut, and Texas, 1980–2014. (Data from Audubon CBC (http://netapp.audubon.org/CBCObservation/.)

BEHAVIORAL SHIFTS BETWEEN NATIVE AND INTRODUCED RANGE

The monk parakeet is considered a serious native crop pest in South America, where it inflicts damage to field crops such as sorghum, sunflower, rice, and corn (Aramburú 1995). The extent and severity of the damage varies across the region, but crop losses can be severe locally (Bucher 1992). The concerns regarding crop damage that were initially expressed when monk parakeets first appeared in the United States have yet to materialize. To date, these birds have demonstrated a propensity for life in urban and suburban areas, not rural or agricultural areas which they inhabit in their native range. Thus, monk parakeets in the United States have seldom been implicated in damage to crops (Tillman et al. 2001). However, if their population increases and habitat requirements are not met in urban environments, the birds could conceivably move into agricultural landscapes and cause crop damage, as has already occured in Spain (Senar et al. 2016).

Monk parakeets in south Florida build nests principally on man-made structures such as stadium light poles, cell towers, and electric utility facilities (Newman et al. 2008). Similar nest-site selection preference was documented in Texas, where 75% of monk parakeet nests were on electric utility structures (Reed et al. 2014). In Argentina, monk parakeets nest on electric utility poles in areas where there are no trees available for nesting (Bucher and Martín 1987).

In south Florida, crop contents of nestlings (n = 26) consisted mostly of sunflower (43%) and millet (32%) seeds (Tillman, unpublished data). These two seeds are the most common constituents in bird seed mixes sold for backyard feeders. The use of feeders exemplifies the behavioral flexibility of the monk parakeet. Such feeding opportunities are probably uncommon in the native range where monk parakeets are generally considered crop pests, rather than welcome additions to the local avifauna.

To a much greater extent than in the native range, resource acquisitions by monk parakeets in the United States have become subsidized by human activities. Boarman (2003) observed that human activities provide wildlife with inadvertent sources of food, water, and other resources. Such resources tend to be more stable and predictable than a natural environment, and animals that take advantage of them are able to prosper and expand their range, often to the detriment of competitors and species they prey upon. His remarks were in the context of common raven (*Corvus corax*) management, but the same concept is applicable to invasive species such as the monk parakeet. That is, these invasives are thriving in the United States and in other countries in large part because they are able to adapt to human activities and exploit feeding and nesting opportunities provided by these activities.

PREDATORS AND DISEASE IN INTRODUCED RANGE

Adverse effects of predation, diseases, and parasites on monk parakeets have not been documented in the United States. Predation on eggs and nestlings is a substantial source of mortality for parakeets in their native range (Navarro et al. 1992), but there are no similar observations in the United States. On two occasions, one of us (MLA) observed fish crows (*Corvus ossifragus*) at parakeet nest structures. One time, a crow stuck its head inside the nest entrance but emerged with nothing. The second time, a crow landed on the nest structure, pulled a stick from the outside of the nest, and flew off with it. Predation on adult parakeets most likely occurs in the United States, but we are not aware of any documented instances.

Millsap et al. (2002) reported the death of a fledgling bald eagle (*Haliaeetus leucocephalus*) in Pinellas County, Florida due to *Chlamydophilia psittaci* infection. Possible sources of this contamination were monk parakeets which had built a nest of their own at the base of the eagle nest. Monk parakeets also build nests at the base of osprey (*Pandion haliaetus*) nests (Pranty 2009; M.L. Avery, unpublished data), so if the parakeets in the area are carrying *C. psittaci*, then potential cross-species contamination exists there as well.

RANGE OF DAMAGE

Agricultural. The only documented agricultural damage that we are aware of in the United States is from south Florida, where monk parakeets and other bird species inflict localized damage to tropical fruit crops such as longans and lychee (Tillman et al. 2001). However, the potential for significant damage by monk parakeets to other fruit crops (citrus, blueberry, grapes, etc.) as well as to field crops such as sunflower, corn, rice, and sorghum remains great. Several states, including Hawaii, California (Dana et al. 1974), and Kansas (Buhler et al. 2001), have banned monk parakeets because of their reputation for causing agricultural damage in their native range (http://mrbeanva.tripod.com/legalq.htm).

Human health and safety. Parakeets and other psittacines are known to carry bacteria that can cause psittacosis in humans (Raso et al. 2014).

Natural resources. We are unaware of any damage to natural resources attributable to monk parakeets. Davis (1974) reported that, in New Jersey, parakeets killed blue jays (*Cyanocitta cristata*) and an American robin (*Turdus migratorius*), but this report could not be verified.

Property damage. Wherever monk parakeets occur in the United States, utility companies must cope with parakeets building nests on electrical utility facilities. Parakeets also build nests on stadium light poles and in trees, but utility structures are preferred in many locations (Newman et al. 2008; Reed et al. 2014). The utility facilities include distribution poles, which are familiar sights along neighborhood streets throughout the country; transmission line towers, which support high-voltage power lines traversing the countryside; and electrical substations, where power lines enter and the high voltages are decreased for distribution to residences and businesses (Avery and Lindsay 2016). The bulky nests of sticks and branches constructed by parakeets create hazards for safe, reliable electrical service on each of these types of facilities. When wet nest materials come in contact with energized components, a short circuit is created and results in a power outage. Such events cause a financial burden to customers who lose power, a health and safety concern, and the utility companies lose revenue and must replace or repair damaged equipment (Avery et al. 2002).

METHODS TESTED TO CONTROL MONK PARAKEETS

In the monk parakeet's native range, lethal control with toxic chemicals is the primary means of population management for crop damage reduction (Linz et al. 2015). Workers spread a paste containing carbofuran or other toxic pesticides around the entrance to the parakeets nest, and the birds succumb from ingesting the toxin as they preen the paste from their feathers.

For electric utility companies in the United States, the most common management approach is to remove nests from power poles or substations when the nests become large enough to constitute a threat to service reliability (Avery et al. 2006). Nest removal is a short-term strategy as the nest occupants almost immediately begin to rebuild at the same site. Further, it might be counterproductive because taking down a nest structure containing nest chambers of several pairs likely will cause the pairs to disperse, and if each pair initiates nest building at a new site, then the original single nest structure eventually becomes several.

Various control methods have been tested to determine their usefulness in keeping monk parakeets from nesting on utility facilities (Avery et al. 2006). Visual scare devices that were tested but were not effective include models of owls, rubber snakes, scare-eye balloons, and a taxidermic parakeet effigy suspended near nests of a substation (Avery et al. 2002). Loud noises were also ineffective and are not compatible with the residential location of most of the parakeet nesting locations. Parakeets were sensitive to a hand-held red laser; birds were repeatedly flushed from their nests at a substation, but the effect was only temporary. Even after a week of laser harassment, the birds still returned to the nest sites at the substation (Avery et al. 2002).

Attempts to remove parakeets from substations using the chemical irritant methyl anthranilate (MA) also proved unsuccessful. The MA was formulated to be dispensed as a fog, and although there were signs (head shaking and bill wiping) that the fog was irritating to the birds, the treatment did not dissuade them to leave the substation (Avery et al. 2006). Conceivably, a system of aerosol or fog dispensers could be devised and installed on a substation to deliver MA in an effective treatment, but currently this methodology remains unproven.

Monk parakeets can be trapped on distribution poles, particularly at night when the birds are in their nest. One successful technique involves use of a long-handled net to cover the nest opening and to catch the birds as they fly out of the nest (Martella et al. 1987). Netters can stand on the ground and use a long pole, or a bucket truck can be used to raise netters to the level of the nest. Once in place, the netters then employ long-handled nets and cover one or two openings to catch birds as they attempt to flee. If attempted during daylight, birds depart before the net can be properly placed. On distribution poles in south Florida, the nest was usually removed immediately or shortly after netting. Findings revealed that higher capture rates at the nest sites greatly retarded the rate at which those sites were reoccupied (Tillman et al. 2004).

Trapping at the nest is generally not appropriate at substations where access to nests is very difficult due to large amounts of high-voltage equipment. Also, substation nests are quite close together, so the disturbance at the first nest where netting is attempted causes the other birds at the site to leave their nests prematurely and avoid capture. Monk parakeets are wary of traps and not easily captured, even with extensive prebaiting and the use of decoy birds (Avery et al. 2002; Tillman et al. 2004). A passive, unattended trap is not sufficient to capture the parakeets occupying a substation. It is possible to lure birds into a very spacious, open trap with end doors that can be slammed shut via remote control. Also, remotely triggered traps mounted on platform feeders within a substation have been used successfully to capture small groups of parakeets attracted to the food placed there (Avery and Lindsay 2016).

METHODS UNTESTED

To our knowledge, there has been minimal evaluation of aural deterrents for dispersing parakeets and keeping them from nesting on utility structures. In particular, sound deterrents that mimic or reproduce parakeet alarm or distress calls should be investigated, singly and in combination with lasers or other visual scare tactics.

FEASIBILITY OF ERADICATION

The nationwide monk parakeet "retrieval" program initiated by the U.S. Fish and Wildlife Service in the early 1970s failed to eradicate the species. Due to public opposition, a similar broadscale eradication effort has not been attempted because in states where parakeets are now established, the goal of management efforts is not to eradicate monk parakeets, but rather to ensure reliable delivery of electric service. To this end, methods development is focused on the problem of minimizing parakeet nesting on utility structures. This approach not only offers the best chance for achieving the management objective, but also helps to assuage concerns raised by parties interested in the welfare of the parakeet population. When incipient populations arise in states where parakeets are not tolerated, the birds can be readily removed by shooting or trapping.

FUTURE TOOLS AND TECHNIQUES BEING DEVELOPED OR NEEDED

Reproductive control. Considerable efforts have been expended in the development of a safe, effective contraceptive approach for monk parakeet population management (Yoder et al. 2007; Avery et al. 2008). The technique using diazacon as the active ingredient has been tested in cage and field studies, and the approach has been shown effective in reducing parakeet productivity and safe for nontarget species (Yoder 2011). Lack of Environmental Protection Agency (EPA) registration currently prevents diazacon fertility control to be implemented as a population management technique. It is not known at this time whether the necessary funding to perform the remaining EPA-mandated studies will be forthcoming.

Structural modifications. For nesting, monk parakeets display definite preferences for types of electric utility structures and for specific locations on the structures (Avery et al. 2006; Newman et al. 2008; Reed et al. 2014). The actual design will vary, but the objective is to eliminate structural features which allow parakeets to obtain a firm base from which to begin their nest construction. One example of the effects of structural design involves two types of transmission line supports. The

FIGURE 16.3   Example of an electric utility pole with many angles and surfaces suitable for monk parakeet nests. (USDA photo.)

older, multicircuit design consists of two vertical supports connected by horizontal and diagonal cross pieces which provide parakeets with numerous nesting opportunities (Figure 16.3). The updated standard design is now a single vertical pole with narrow cross members supporting insulators (Figure 16.4). This structure design eliminates the substrates suitable for the parakeets to initiate nest building.

FIGURE 16.4   Example of an electric utility pole affording no opportunity for monk parakeets to establish and maintain a nest. (USDA photo.)

Another option is to eliminate the acute angles that occur on transmission line support structures and in substations. The angles are formed where a horizontal beam is intersected by a diagonal support piece. Inserting a triangular block in the acute angle would create a right angle that presumably the birds would find less appealing as a nest site. Another approach is to eliminate the flat surfaces upon which the parakeets begin nest-building activity. This could be accomplished through installation of rounded, semicircular covers on the flat surfaces of beams. The more smooth and slippery the surface, the more effective a nesting deterrent it will be. The challenge to implementing structural modifications is that so many angles need to be eliminated and so many surfaces need to be rounded that it is unrealistic to expect they could all be done. But perhaps certain key parts could be retrofitted or replaced to help reduce nesting activity.

ROSE-RINGED PARAKEET (FAMILY: PSITTACULIDAE)

Rose-ringed parakeets (*Psittacula krameri*) are medium to large parakeets (40+ cm, 110–182 g; Butler 2003) that are native to Asia and Africa. These birds have tails approximately equal in length to their bodies, and are of bright green plumage with red bills. The adult males have a dark ring (sometimes reddish) around their neck, which justifies the naming of this species as the rose-ringed, or ring-necked, parakeet (Figure 16.5).

Juvenile males cannot be readily distinguished from adult or juvenile females. These cavity-nesting birds are highly social, foraging, roosting, and loosely nesting in the same areas.

FIGURE 16.5 Male rose-ringed parakeet. (Photo by Dick Daniels.)

In their introduced range, rose-ringed parakeets generally establish night roosts on tree branches or on palm fronds (Gaudioso et al. 2012; Sheehey and Mansfield 2015). They rarely excavate a new cavity for nesting, but instead often widen the opening of an already established cavity, and they can displace other cavity nesters during this process (Strubbe and Matthysen 2007). Large-diameter trees with ample surrounding shrub and tree cover appear to be the preferred nesting conditions in both the United Kingdom (Butler 2003) and Hawaii (Gaudioso et al. 2012). The median clutch size is four eggs, yet two eggs are generally fertile, and two fledglings per nest are common (Butler 2003; Lambert et al. 2009).

Rose-ringed parakeets are opportunistic granivores-frugivores, largely consuming dry and fleshy fruits and seeds, but they are also known to consume nectar and flower buds (Ali and Ripley 1969; Clergeau and Vergnes 2011). In their native range, they are well-known pests of agricultural crops, particularly corn (*Zea mays*) and sorghum (*Sorghum bicolour*) (Ali and Ripley 1969). Crop damage in the United States has not been well documented aside from damage to corn (Gaudioso et al. 2012) and tropical fruit (Bukoski, pers. comm.) on Kauai in Hawaii.

Introduced Range

Due to their large popularity as pets (caged birds), rose-ringed parakeets have established feral populations in at least 35 countries (Butler 2003), which makes this species the most widely introduced parrot in the world. In fact, the Invasive Species Compendium (2012) reports that 76 countries currently have rose-ringed parakeets or had them in the past. Of the four recognized subspecies of *Psittacula krameri* (two are from sub-Saharan Africa, including Senegal, Uganda, Sudan, Somalia, and Ethiopia; and two are from Asia, including India, Sri Lanka, Nepal, and Pakistan), the majority of the invasive rose-ringed parakeets are from northern India and Pakistan (*P. k. borealis*; Jackson et al. 2015). Temperature apparently provides some limitation to where they can become established (Roscoe et al. 1976; Butler 2005), yet they have successfully colonized tropical, subtropical, and temperate environments. In Europe, the main established populations are in the United Kingdom (ca. 10,000 individuals in 2004), Belgium (ca. 7000 in 2005), the Netherlands (5400 in 2004), and Germany (5700 in 2003) (Butler 2005; Strubbe and Matthysen 2007).

Establishment in the United States

In the United States, rose-ringed parakeets have become well established in parts of Florida, Hawaii, and southern California, and additional, more temporary populations have been reported in Virginia, Texas, Louisiana, and Alabama (Invasive Species Compendium 2012; Sheehey and Mansfield 2015). Escapees from the pet trade probably account for most of these established populations in the United States, and intentional releases or large storm events such as hurricanes that damage aviaries are additional pathways that facilitated some rose-ringed parakeet establishment in the United States (Gaudioso et al. 2012; Sheehey and Mansfield 2015). According to the CITES trade database (http://trade.cites.org/), there have been approximately 60,000 live rose-ringed parakeets imported into the United States for the period of

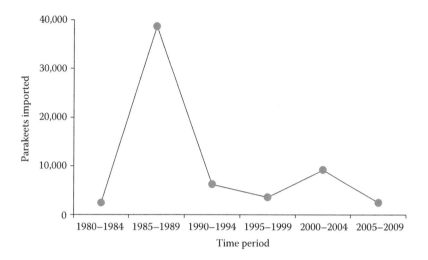

FIGURE 16.6   Rose-ringed parakeets imported to the United States, 1980–2009. (The data are from http://trade.cites.org/.)

1980–2007, and the majority of these (>40,000) were imported during 1985–1990 (Figure 16.6). Apparently, the majority of the rose-ringed parakeets imported into the United States were wild-caught because a very small portion (approximately 3%) of the CITES-listed imports stated they had originated from stocks bred in captivity.

When introduced to New York City, rose-ringed parakeets did not establish, possibly due to insufferably cold winter weather (Roscoe et al. 1976). According to the Florida Fish and Wildlife Conservation Commission, rose-ringed parakeets were introduced as early as the 1930s in Florida, and populations at various times occupied 15 counties presumably through repeated introductions or escapees rather than by breeding in the wild (http://myfwc.com/wildlifehabitats/nonnatives/birds/rose-ringed-parakeet/). The species has declined in the state during recent years and currently is restricted to the Naples area in southwestern Florida where a small population persists (Pranty and Garrett 2011). There are several discrete populations of rose-ringed parakeets in southern California. With introductions as early as 1977, the estimated population in the greater Bakersfield area was approximately 3000 individuals in 2012, and additional smaller populations have been reported in San Diego, Anaheim, Santa Cruz, Malibu, and Pasadena (Sheehey and Mansfield 2015). The state of Hawaii also has rose-ringed parakeets established on at least two islands, with over 2000 individuals on Kauai (Gaudioso et al. 2012) and at least 500–1000 on Oahu (Kalodimos, unpublished data).

Expansion and Population Growth

It took 130 years for rose-ringed parakeets in the United Kingdom to establish a self-sustaining population (Lever 1987; Tayleur 2010). In Kauai, rose-ringed parakeets were first released in the 1960s in the south part of the island. By 1982, the population had grown to 50 birds, and a second pair of rose-ringed parakeets had been released after Hurricane Iwa. By 1994, late 2000s, and 2011, the population was estimated at 150–200 birds, 500–1000 birds, and over 2000 birds, respectively (Gaudioso et al. 2012; U.S. Department of Agriculture [USDA], Wildlife Services [WS], National Wildlife Research Center [NWRC], unpublished data). The size of the current rose-ringed parakeet population on Kauai is unknown, but USDA staff estimated the population in 2015 to be at least threefold greater than the 2011 estimate even after accounting for removal of 100–200 birds each year around agricultural fields.

In southern California, there was a self-sustaining population estimated at 60 individuals in 1997 (Garrett 1997; Butler 2005), and yet by 2012 there were 3000 individuals in Bakersfield alone (Sheehey and Mansfield 2015). Population growth of introduced rose-ringed parakeet in the United Kingdom varied from 15% per year on the Isle of Thanet to approximately 30% per year in the Greater London area (Butler 2003). Rose-ringed parakeet populations are relatively sedentary, and they expand their range slowly as evidenced by just 400 m/year expansion in the United Kingdom (Butler 2003). Although no formal measurements of geographical expansions are available for the United States, the rose-ringed parakeet populations appear to spread slowly and probably at comparable rates as those reported for populations in the United Kingdom. The relatively slow geographic expansion of rose-ringed parakeets is probably partly attributable to their release near human habitations, which offer an abundance of resources and few predators, and that the original birds released were often pets and therefore habituated to humans. Additionally, Strubbe and Matthysen (2007) suggest that dispersal may be limited because of the species' communal roosting behavior.

BEHAVIORAL SHIFTS BETWEEN NATIVE AND INTRODUCED RANGE

The habitats where rose-ringed parakeets reside in their native range are largely woodlands and grasslands. In their introduced range, they thrive in a wider range of habitats, including highly disturbed urban areas to natural or seminatural areas. In Belgium, the greatest densities of rose-ringed parakeets are in forested areas where there is an abundance of cavities for nest sites (Strubbe and Matthysen 2011); however, in Kauai, nearly 1000 birds roost outside of administrative buildings in downtown Lihue (Gaudioso et al. 2012). Behavior of male rose-ringed parakeets may differ between their native and introduced range; in India, males were found tending the young at the nest sites (Hossain et al. 1993), whereas observations of nests in the United Kingdom and Belgium revealed that the breeding males left the nest to return to the communal roost just after sunset (Butler 2003; Strubbe and Matthysen 2011). Further observations of male rose-ringed parakeets in Brussels, Belgium, revealed that males would feed their female mate and young at the nest site during the day, then briefly stop to forage on the way from the nest to the roost just before dark (Strubbe and Matthysen 2011). Another difference between rose-ringed parakeets in their native (India) range and in their introduced range is that nesting in India can occur in rock crevices and buildings (Ali and Ripley 1969; Roberts 1991; Juniper and Parr 1998), whereas in the United Kingdom, they only nest in tree cavities (Pithon and Dytham 1999; Butler 2003).

Individuals travel more than 6 km a day in the United Kingdom, presumably for forage (Butler 2003); foraging distances in native India are several kilometers a day

(Chakravarthy 1998). Rose-ringed parakeets in Kauai have home ranges that vary widely (0.11–6437 ha), and the average home range (1771 ha; n = 16) of rose-ringed parakeets on Kauai was almost 13–24 times greater (average 75–86 ha; n = 5) than rose-ringed parakeets in Brussels (Strubbe and Matthysen 2011). In Brussels, orchards, parks, and gardens were the most preferred habitats of rose-ringed parakeets, probably a result of the abundance of high-quality food in those environments relative to the less favored and visited habitats of deciduous forests, coniferous forests, and (nonorchard) agricultural lands (Strubbe and Matthysen 2011). In fact, Strubbe and Matthysen (2011) concluded that, at least in Brussels, rose-ringed parakeets do not forage in agricultural areas.

Frequent activity and foraging in suburban areas by the rose-ringed parakeets may also be influenced by backyard bird feeders. Clergeau and Vergnes (2011) studied four radio-collared rose-ringed parakeets in Paris, France, and observed that half of the parakeets' feeding time was at backyard bird feeders. Possibly backyard feeders, which can be reliable food sources, assist rose-ringed parakeet survival through the winter (Butler 2003; Clergeau and Vergnes 2011).

PREDATORS AND DISEASE IN INTRODUCED RANGE

Predation of rose-ringed parakeets has not been well documented in their introduced range, though rodents represent the most-likely threats when the parakeets are nesting in cavities, while raptors would be able predators of parakeets when they are flying or roosting. Predation of rose-ringed parakeets in their native range of India is generally attributed to owls, crows, and snakes (Lamba 1966; Shivanarayan et al. 1981; Hossain et al. 1993; Dhanda and Dhindsa 1998). In their introduced range, the only available records of rose-ringed parakeet predation are those by rodents, and such predation events seem relatively uncommon. In Italy, the native red squirrel (*Sciurus vulgaris italicus*) was documented entering a tree hole and emerging 10–15 seconds later with a living featherless rose-ringed parakeet in its mouth. While the squirrel was inside the nest cavity, adult parakeets stayed outside and vocalized loudly until the squirrel had exited the cavity and made off with the chick (Mori et al. 2013a). In the United Kingdom, introduced grey squirrels (*Sciurus carolinensis*) are the only reported predators of rose-ringed parakeets (Schwartz et al. 2009).

Diseases that psittacines can contract, and are known to negatively affect these birds, include beak and feather disease, avian bornavirus, Newcastle disease virus, avian pox virus, avian influenza, avian psittacosis, and pulmonary disease (England 1974; Tozer 1974). It is also possible for some of these diseases to transfer from the parakeets to poultry or wild birds, as well as humans (see below). In Kauai, all 15 rose-ringed parakeet individuals tested were negative for avian influenza virus and avian psittacosis (Gaudioso et al. 2012).

RANGE OF DAMAGE

Agriculture. In their native range, many farmers consider rose-ringed parakeets to be the most serious avian pest because of the heavy damage they cause to agricultural crops (e.g., corn, sorghum, rice, safflower, sunflower, fleshy fruit) and stored grains

(Shivanarayan et al. 1981; Dhindsa and Saina 1994; Mukherjee et al. 2000). In fact, rose-ringed parakeet reduced crop yields of corn and sorghum in India by 74%–81% (Reddy 1998, 1999). Gizzard and crop contents of nestlings and adults in India consisted primarily of sorghum, corn, and sunflower (Shivanarayan et al. 1981; Sani et al. 1994). In Pakistan, rose-ringed parakeets are a serious pest of corn, sunflower, rape seed, as well as fruit crops such as mangos, citrus, and guava (Bashir 1979). Bashir (1979) reported that the estimated annual loss in corn seed from rose-ringed parakeets in Pakistan was 97,000 tons, which was equivalent at the time to US$15 million.

In their introduced range, crop damage is much less common than in their native range, aside from the extensive damage to both corn and tropical fruit crops on the island of Kauai (Gaudioso et al. 2012). Seasonal availability of cereal and fruit crops in north temperate latitudes (northern United Kingdom and United States) does not coincide with the early breeding season of the rose-ringed parakeet (e.g., February; Lambert et al. 2009; Strubbe and Matthysen 2011). In Kauai, rose-ringed parakeets leave their roost after sunrise and begin feeding in corn fields. These birds "sample" the corn as it is ripening by clipping at the tassels, before feeding in a particular corn field; this type of damage is unique to the rose-ringed parakeets on Kauai, as other (native and nonnative) birds may also feed on the corn in Kauai (Gaudioso et al. 2012). When rose-ringed parakeets feed on the corn, they feed on the kernels of the corncobs just prior to the harvest stage (Gaudioso et al. 2012).

Human health and safety. Presence and expansion of rose-ringed parakeet populations around airports in the United Kingdom (Fletcher and Askew 2007) and Hawaii (W. Bukoski, pers. comm.) have caused concern for human safety associated with potential aircraft strikes with these birds. At Heathrow Airport, Fletcher and Askew (2007) reported that one of the 54 bird strikes in 2005, and two of the 44 bird strikes in 2006, involved rose-ringed parakeets. Although no airstrikes with rose-ringed parakeets have yet been reported on Kauai, scare techniques and occasional removal are practiced by USDA WS (W. Bukoski, pers. comm.). As rose-ringed parakeet populations grow around urban and suburban areas, including airports, concerns for air traffic and human safety may become more pressing. Additionally, airstrikes with birds are costly, and have been reported to be UK£20,000 per bird strike (Tayleur 2010).

Rose-ringed parakeets are potential vectors of diseases that can negatively affect both poultry and humans. These diseases include Newcastle disease (Butler 2003), cryptosporidium (Morgan et al. 2000), and psittacosis (Fletcher and Askew 2007; Raso et al. 2014). Exposure to disease from bird droppings deposited below roosting sites, which can be in public places, is a human health concern that does not require direct contact with the birds. Although less severe for humans than contracting a disease from rose-ringed parakeets, residents and tourists in Hawaii have complained about the loud, shrill calls commonly made by these birds (Gaudioso et al. 2012).

Natural resources. The growing number and sizes of introduced rose-ringed parakeet populations has raised concern for their potential involvement in loss of native biodiversity. Rose-ringed parakeets feed on both native and nonnative plants (Cramp 1985; Strubbe and Matthysen 2007; Clergeau and Vergnes 2011), and by doing so they probably destroy the majority of the seeds that they consume. In the United

Kingdom, they are known to feed on berries of native holly (*Ilex* spp.) and elder (*Sambucus* spp.), and seeds of native hornbean (*Carpinus betulus*) and ash (*Fraxinus excelsior*) (Cramp 1985). Additional native plants were consumed by rose-ringed parakeets that had colonized Paris (Clergeau and Vergnes 2011). Despite their consumption of native plants in Europe, no reports yet exist indicating that rose-ringed parakeets are altering the natural or seminatural environment through their feeding ecology (Tayleur 2010). In Australia, however, rose-ringed parakeets are known to strip bark, which has killed some trees and possibly shifted the local tree community composition and structure (Fletcher and Askew 2007). Reports of damage to native species in the United States are lacking, though they have been observed feeding on some native plants in Hawaii, including endemic palms (*Pritchardia* spp.) (W. Bukoski, pers. comm.). Additionally, there is potential for rose-ringed parakeets to spread nonnative or invasive plants if seeds are small enough or otherwise remain viable after passing through the bird. Partially intact seeds of the nonnative invasive plant *Passiflora edulis* were removed from the crop/gizzard of an individual captured in Kauai (Gaudioso et al. 2012). If parakeets continue to grow in numbers and locations throughout the United States, they have the potential to alter natural areas through their feeding behavior.

Because of their cavity-nesting behavior, rose-ringed parakeets may threaten or otherwise competitively exclude native birds and bats that rely upon cavities. This has been a serious concern in the United Kingdom and other parts of Europe, as rose-ringed parakeets are thought to have a competitive advantage over several native breeding birds, including kestrel (*Falco tinnunculus*), stock dove (*Columa oenas*), jackdaw (*Corvus monedula*), Eurasian nuthatch (*Sitta europaea*), and common starling (*Sturnus vulgaris*) (Butler 2003; Fletcher and Askew 2007; Strubbe and Matthysen 2007). Newson et al. (2011) evaluated the potential for population-level impacts of rose-ringed parakeets on all native cavity-nesting species in the parakeet's current range in the United Kingdom and found no evidence for a significant impact through competition. In contrast, and through a series of correlative analyses that included six native cavity-nesting birds (stock dove, jackdaw, Eurasian nuthatch, common starling, and two primary-cavity nesting woodpeckers) in Belgium, Strubbe and Matthysen (2007) found the nuthatch was negatively associated with rose-ringed parakeet abundance. Through further experimental study where rose-ringed parakeets were blocked from nest cavities, Strubbe and Matthysen (2009b) demonstrated that indeed rose-ringed parakeets competitively exclude some Eurasian nuthatches from nesting cavities. Furthermore, rose-ringed parakeets apparently begin to breed earlier in the season than nuthatches, which may enable them to claim the best cavities before the nuthatches (Strubbe and Matthysen 2007). Based on modeling geographic spread and the potential for rose-ringed parakeets to usurp nests, Strubbe et al. (2010) concluded that, at most, one-third of the Eurasian nuthatch population would be at risk to the ill effects of the rose-ringed parakeet.

Rose-ringed parakeets also reportedly attack and kill little owls (*Athene noctua*) and red squirrels (*Sciurus vulgaris*) in Europe (Mori et al. 2013b; Menchetti and Mori 2014). In possible defense of their nest, three rose-ringed parakeets killed an adult red squirrel (Japiot 2005). Additionally, rose-ringed parakeets were reported injuring and killing a small bat (*Nyctalus leisleri*) that was roosting or possibly hibernating in a tree cavity in Italy (Menchetti et al. 2014). There have been no reports in the United States involving the effects of rose-ringed parakeets on native fauna. However, due to the diversity of native cavity nesters that appear to be vulnerable to the negative effects of rose-ringed parakeets, future studies should prioritize investigation of the interactions between local native cavity nesters and rose-ringed parakeets.

Property damage. Unlike other parrots that cause frequent damage to vehicles and dwellings (e.g., native kea, *Nestor notabilis*, in New Zealand; Brejaart 1994) or electrical outages from their nest-building behavior (e.g., introduced monk parakeets in Florida; Avery et al. 2002), there is little property damage associated with rose-ringed parakeets other than crop damage and airplane collisions. A more subtle example of property damage induced by rose-ringed parakeets is the large amounts of droppings produced at high-density roosts. Populations of rose-ringed parakeets in both Kauai and California form such high-density roosts in public areas, such as in palms lining shopping malls and public buildings (Gaudioso et al. 2012; Sheehey and Mansfield 2015). As with many of the negative impacts caused by parrots, we expect those attributable to the rose-ringed parakeet to become more common as more populations become established and the sizes of established populations increase.

Methods Tested to Control Rose-Ringed Parakeets

For rose-ringed parakeets, a suite of scare tactics have been used historically to reduce the negative impacts on crops. Auditory scare tactics have included shooting of guns and propane-gas cannons, and the use of loudspeakers. Additionally, rose-ring parakeets are shot or netted in agricultural settings (Koopman and Pitt 2007; Gaudioso et al. 2012). These methods have been largely unsuccessful because the birds become accustomed to the tactic, and the methods are costly and can require humans patrolling the sites multiple times a day (Bashir 1979; Gaudioso et al. 2012). Incorporating scare and lethal methods in cornfields of Hawaii has been estimated to cost corn companies hundreds of thousands of dollars each year (Koopman and Pitt 2007). Live-trapping rose-ringed parakeets has been tried with variable success. Large box traps baited with fruit or seed, or a live conspecific parakeet as a lure, were used in Pakistan to reduce rose-ringed parakeet damage to sunflower crop (Bashir 1979). When this method was replicated in Kauai, the traps were never visited by any parakeets despite their frequent flights over the trap and roosting above the trap (Gaudioso et al. 2012).

Shooting of rose-ringed parakeets, both for harassment and harvest at roost or in fields when foraging, has been the dominant control method continued in Kauai, where these parakeets cause persistent damage to seed corn crops. Crop damage appears less severe when shooters patrol the cornfields, but the rose-ringed parakeet population has not noticeably declined on Kauai (Gaudioso et al. 2012; W. Bukoski, pers. comm.). Although rose-ringed parakeet control methods have historically been practiced to combat parakeet damage to crops (e.g., in native range, Bashir 1979; in introduced range of Kauai, Gaudioso et al. 2012), licenses have been issued since 2010 in the United Kingdom to harvest rose-ringed parakeets in order to prevent serious damage or disease, preserve air safety, preserve public health and safety, and to conserve native flora and fauna (Tayleur 2010).

Fertility control is a nonlethal technique that is often more appealing to the public for long-term population management. In the United Kingdom, the fertility control agent diazacon has been tested and proven effective against captive rose-ringed parakeets (Lambert et al. 2010). Diazacon reduces blood cholesterol and cholesterol-dependent hormones, and has therefore proven useful for reducing reproductive output in birds (Lambert et al. 2010). A suitable formulation and delivery system is needed before diazacon can be an effective method for rose-ringed parakeet population control. A challenge for rose-ringed parakeet control in Kauai has been a difficulty in successfully drawing the parakeets to bait (Gaudioso et al. 2012); therefore, if diazacon is formulated in palatable bait, it may still be difficult to deliver to the birds in such areas where alternative food sources are abundant or more desirable. Effects on nontarget species and potential secondary toxicity to predators feeding on dosed parakeets also must be evaluated.

METHODS UNTESTED

Effectively reducing rose-ringed parakeet populations would appear easiest at their roosts because they are stationary (sleeping), accessible, and generally in large numbers. However, most of the roosts are located in urban or suburban areas (e.g., rugby club, cemetery, shopping mall, residential area), which means the high visibility of such an effort, even practiced at night, would have to be considered (Butler 2003). In urban and suburban areas of Kauai, rose-ringed parakeet flocks departing their roosts often number 175 individuals (Gaudioso et al. 2012). Therefore, the use of mist nets to capture a large number of birds leaving the roost may be worth investigating. Long poles, or a bucket truck, could be used to raise nets after sundown to the level of the roosting birds, and the nets could easily be lowered to collect entangled individuals immediately after the birds leave the roost (typically sunrise).

Because there are no remaining native parrots in the United States, it would seem a parrot-specific disease may be considered as a method of biological control; however, we know of no such diseases currently available, and even psittacosis can be passed to nonpsittacine birds. One type of control effort that seems to be lacking in rose-ringed parakeet management is the use of toxicants. Although avian toxicants are field tested as a means of controlling agricultural pests in the United States (Linz 2013), their use is generally unwelcome by the public. Additionally, testing appropriate delivery devices would be critical to ensure that nontargets, such as native birds, would not be negatively affected by toxicants targeted for rose-ringed parakeets. Furthermore, in Kauai, where rose-ringed parakeets are damaging agricultural crops, there has been low success in luring individuals into traps (Gaudioso et al. 2012).

FUTURE CHALLENGES TO ERADICATION/CONTROL

A prominent barrier to rose-ringed parakeet eradication or control efforts is human fondness for this and other parrot species. In addition to humans having rose-ringed parakeets as pets, many people enjoy seeing the birds in the wild, including at parks and backyard bird feeders. Lambert et al. (2010) stated that eradication of established rose-ringed parakeets in the United Kingdom is unlikely, largely because a substantial portion of the human population welcomes the colorful addition to the avifauna. The amount of effort required to curtail the population would be considerable, and Butler (2003) suggested that a 30% reduction in rose-ringed parakeet population would be needed to prevent further increase in the Greater London population. Population control and local eradication of rose-ringed parakeets are controversial, but the very destructive nature of these birds to a diversity of cereal and fruit crops, their potential to spread disease and to outcompete native birds, and to adversely affect native plant communities warrant consideration of suppression of existing rose-ringed parakeet populations and prevention of expansions.

SYNTHESIS OF FUTURE CHALLENGES (BOTH SPECIES)

The impacts of invasive monk parakeets and rose-ringed parakeets are undeniable, but often there is little empathy for wildlife managers when they are up against charismatic birds which otherwise seem a pleasant addition to the local avifauna. Many residents may not care or are unaware that a parakeet is nonnative or not. Thus, one challenge is to increase education and public awareness with regard to the potential negative consequences of invasive wildlife. This could be difficult to achieve because, in some cases, the invasive species is part of the local culture. For example, the monk parakeet is the only species of parrot successfully established in the northern United States, where it is a colorful subtropical curiosity and a particularly welcome sight where winters can be long and cold. In some cities, such as Brooklyn and Chicago, monk parakeets have for many years been the subject of study by college students and amateur birders, so any disruption of the birds will be readily noticed and widely publicized in local media. Social controversies surrounding monk parakeet management are often fomented by organized animal rights advocacy groups. Leery of adverse publicity, utility companies adversely affected by the parakeets are reluctant to initiate management programs in the face of such opposition.

Availability of resources, especially funding, for development and implementation of new management tools represents a major challenge. Often the information needed for cost-benefit analysis is not available, so it can be difficult to justify the expense required to develop a new technology, such as a contraceptive for parakeet population management or a repellent for crop protection. Time is on the side of the invader, so while the necessary background information is collected, populations increase, disperse to new areas, and the problems they cause become that much more difficult to address.

Another challenge is to remain vigilant to impacts of the invasive species. Just because impacts are not readily apparent does not mean they are not occurring or will not occur eventually. For example, to date, there is no evidence that monk parakeets compete with native species for essential resources. Nor have monk parakeets lived up to their South American reputation as serious crop depredators. Rose-ringed parakeets have demonstrated no adverse impact on native species in the United

States. We suggest, however, that a complacent attitude regarding management of the monk parakeet or rose-ringed parakeet is not appropriate. Instead, aggressive, science-based management strategies are needed to address specific current problems and to minimize future negative impacts that stem from these birds.

REFERENCES

Ali, S., and S. D. Ripley. 1969. *Handbook of the Birds of India and Pakistan*, vol. 3. Oxford: Oxford University Press.

Aramburú, R. M. 1995. Feeding ecology of the monk parakeet (*Myiopsitta monachus monachus*) in Buenos Aires, Province, Argentina (Aves: Psittacidae). *Physis. Seccion los Continentes y los Organismos Terrestres* 53:29–32.

Avery, M. L., E. C. Greiner, J. R. Lindsay, J. R. Newman, and S. Pruett-Jones. 2002. Monk parakeet management at electric utility facilities in south Florida. *Vertebrate Pest Conference* 20:140–145.

Avery, M. L., and J. R. Lindsay. 2016. *Monk Parakeets. Wildlife Damage Management Technical Series*. USDA, APHIS, WS National Wildlife Research Center, Ft. Collins, Colorado. 11p.

Avery, M. L., J. R. Lindsay, J. R. Newman, S. Pruett-Jones, and E. A. Tillman. 2006. Reducing monk parakeet impacts to electric utility facilities in south Florida. In *Advances in Vertebrate Pest Management*, volume IV, eds. C. J. Feare, and D. P. Cowan, pp. 125–136. Fürth, Federal Republic of Germany: Filander Verlag.

Avery, M. L., E. A. Tillman, K. L. Keacher, J. E. Arnett, and K. J. Lundy. 2012. Biology of invasive monk parakeets in south Florida. *Wilson Journal of Ornithology* 124:581–588.

Avery, M. L., C. A. Yoder, and E. A. Tillman. 2008. Diazacon inhibits reproduction in invasive monk parakeet populations. *Journal of Wildlife Management* 72:1449–1452.

Banks, R. C. 1970. Birds imported into the United States in 1968. *Spec. Sci. Rep.-Wildl.* 136. Washington, DC: U.S. Fish Wildl. Serv.

Bashir, E. A. 1979. A new "parotrap" adapted from the MAC trap for capturing live parakeets in the field. *Bird Control Seminar* 8:167–171.

Batllori, X., and R. Nos. 1985. Presencia de la cotorrita gris (*Myiopsitta monachus*) y de la cotorrita de collar (*Psittacula krameri*) en el area metropolitana de Barcelona. *Miscellania Zoologica* 9:407–411.

Boarman, W. I. 2003. Managing a subsidized predator population: Reducing common raven predation on desert tortoises. *Environmental Management* 32:205–217.

Brejaart, R. 1994. Aspects of the ecology of Kea, Nestor notabilis (Gould), at Arthur's Pass and Craigieburn Valley. *MS Thesis*, Canterbury, NZ: Lincoln University.

Bucher, E. H. 1992. Neotropical parrots as agricultural pests. In *New World Parrots in Crisis: Solutions from Conservation Biology*, eds. S. R. Beissinger, and N. R. F. Snyder, pp. 201–219. Washington, DC: Smithsonian Inst. Press.

Bucher, E. H., and L. F. Martín. 1987. Los nidos de cotorras (*Myiopsitta monachus*) como causa de problemas en lineas de transmision electrica. *Vida Silvestre Neotropical* 1:50–51.

Bucher, E. H., L. F. Martín, M. B. Martella, and J. L. Navarro. 1990. Social behavior and population dynamics of the monk parakeet. *Proc. International Ornithol. Congr.* 20:681–689.

Buhler, R. D., J. D. Lambley, and T. Sim IV. 2001. Monk parakeet *Myiopsitta monachus*. Unpublished report. Topeka, KS: Kansas Department of Agriculture.

Butler, C. 2002. Breeding parrots in Britain. *British Birds* 95:345–348.

Butler, C. J. 2003. Population biology of the introduced rose-ringed parakeet *Psittacula krameri* in the UK. *PhD Diss.*, University of Oxford.

Butler, C. J. 2005. Feral parrots in the continental United States and United Kingdom: Past, present, and future. *Journal of Avian Medicine and Surgery* 19:142–149.

Cassey, P., T. M. Blackburn, K. E. Jones, and J. L. Lockwood. 2004. Mistakes in the analysis of exotic species establishment: Source pool designation and correlates of introduction success among parrots (Psittaciformes) of the world. *Journal of Biogeography* 31:277–284.

Chakravarthy, A. F. 1998. Feeding behavior of parakeets on rice in the hill region of Karnatka. In *Birds in Agricultural Ecosystems*, eds. M. S. Dhindsa, P. S. Rao, and B.M. Parasharya, pp. 71–74. Hyderabad: Society for Applied Ornithology.

Chesser, R. T., R. C. Banks, K. Burns et al. 2015. AOU Checklist of North and Middle American Birds (7th Edition and Supplements). http://www.americanornithology.org/content/aou-checklist-north-and-middle-american-birds-7th-edition-and-supplements#sthash.pegZ4n0e.dpuf (accessed August 20, 2015).

Clergeau, P., and A. Vergnes. 2011. Bird feeders may sustain feral rose-ringed parakeets *Psittacula krameri* in temperate Europe. *Wildlife Biology* 17:248–252.

Conroy, M. J., and J. C. Senar. 2009. Integration of demographic analyses and decision modeling in support of management of invasive monk parakeets, an urban and agricultural pest. In *Modeling Demographic Processes in Marked Populations*, eds. D. L. Thomson, E. G. Cooch, and M. J. Conroy, pp. 491–510. New York: Springer Science+Business Media.

Cramp, S. 1985. *Handbook of theBirds of Europe, the Middle East and North Africa. The Birds of the Western Palearctic*, vol. IV. Oxford: OxfordUniversityPress.

Dana, R., R. Thompson, A. Bischoff, D. Clark, L. Davis, and M. Keffer. 1974. Pest evaluation – monk parakeet *Myiopsitta monachus*. Unpublished report. Sacramento: California Department of Food and Agriculture.

Davis, T. R. 1974. The monk parakeet: A potential threat to agriculture. *Vertebrate Pest Conference* 6:253–256.

Dhanda, S. K., and M. S. Dhindsa. 1998. Nest-site selection and other aspects of breeding ecology of the rose-ringed parakeet *Psittacula krameri*. In *Birds in Agricultural Ecosystems*, eds. M. S. Dhindsa, P. S. Rao, and B. M. Parasharaya, pp. 85–102. Hyderabad: Society of Applied Ornithology.

Dhindsa, M. S., and H. K. Saina. 1994. Agricultural ornithology: An Indian perspective. *Journal of Bioscience* 19:391–402.

Eberhard, J. R. 1998. Breeding biology of the monk parakeet. *Wilson Bulletin* 110:463–473.

England, M. D. 1974. Feral populations of parakeets. *British Birds* 67:393–394.

Fletcher, M., and N. Askew. 2007. Review of the status, ecology, and likely future spread of parakeets in England. Unpublished report. York: Central Science Laboratory.

Garrett, K. L. 1997. Population status and distribution of naturalized parrots in southern California. *Western Birds* 28:181–195.

Gaudioso, J. M., A. B. Shiels, W. C. Pitt, and W. P. Bukowski. 2012. Rose-ringed parakeet impacts on Hawaii's seed crops on the island of Kauai: Population estimate and monitoring of movements using radio telemetry. Unpublished report QA 1874, Hilo, HI: USDA National Wildlife Research Center.

Greenhall, A. M. 1977. *Birds Imported into the United States in 1973 and 1974*. U.S. Fish Wildl. Serv., Wildl. Leaflet 511. Washington, DC: U.S. Department of the Interior.

Hossain, M. T., K. Z. Husain, and M. K. Rahman. 1993. Some aspects of the breeding biology of the rose-ringed parakeet, *Psittacula krameri borealis* (Neumann). *Bangladesh Journal of Zoology* 21:77–85.

Invasive Species Compendium. 2012. *Psittacula krameri (Rose-Ringed Parakeet)*. Wallingford, UK: CABI Publishing. www.cabi.org

Jackson, H., D. Strubbe, S. Tollington, R. Prys-Jones, E. Matthysen, and J. J. Groombridge. 2015. Ancestral origins and invasion pathways in a globally invasive bird correlate with climate and influences from bird trade. *Molecular Ecology* 24:4269–4285. DOI: 10.1111/mec.13307.

Japiot, X. 2005. Psittacides en villes d'Europe. Mairie de Paris, Direction des Parcs, Jardin and Espaces Verts, Service de l'Ecologie Urbaine, Section Etudes et Prospectives Environnementales, Pole Biodiversitè (Psitaccids in European towns. Paris city council, Parks board, Gardens and green spaces, Urban Ecology department, Unit of environmental studies and issues, Biodiversity center).

Juniper, T., and M. Parr. 1998. *Parrots: A Guide to the Parrots of the World.* Robertsbridge: Pica.

Koopman, M. E., and W. C. Pitt. 2007. Crop diversification and leads to diverse bird problems in Hawaiian agriculture. *Human–Wildlife Conflicts* 1:235–243.

Lamba, B. S. 1966. Nidification of some common Indian birds: 10. The rose-ringed parakeet. *Psittacula krameri scopoli. Proceedings of the Zoological Society, Calcutta* 19:77–85.

Lambert, M. S., G. Massei, J. Bell, L. Berry, C. Haigh, and D. P. Cowan. 2009. Reproductive success of rose-ringed parakeets *Psittacula krameri* in a captive UK population. *Pest Management Science* 65:1215–1218.

Lambert, M. S., G. Massei, C. A. Yoder, and D. P. Cowan. 2010. An evaluation of Diazacon as a potential contraceptive in non-native rose-ringed parakeets. *Journal of Wildlife Management* 74:573–581.

Lever, C. 1987. *Naturalised Birds of the World.* Harlow: Longman Publishing.

Linz, G. M. 2013. Blackbird population management to protect sunflower: A history. *Wildlife Damage Management Conference* 15:42–53.

Linz, G. M., E. H. Bucher, S. B. Canavelli, E. N. Rodriguez, and M. L. Avery. 2015. Limitations of population suppression for protecting crops from bird depredation. *Crop Protection* 76:46–52.

MacGregor-Fors, I., R. Calderón-Parra, A. Meléndez-Herrada, S. López-López, and J. E. Schondube. 2011. Pretty, but dangerous! Records of non-native monk parakeets (*Myiopsitta monachus*) in Mexico. *Revista Mexicana de Biodiversidad* 82:1053–1056.

Martella, M., J. Navarro, and E. Bucher. 1987. Método para la captura de cotorras (*Myiopsitta monachus*) en sus nidos. *Vida Silvestre Neotropical* 1:52–53.

Menchetti, M., and E. Mori. 2014. Worldwide impact of alien parrots (Aves: Psittaciformes) on native biodiversity and environment: A review. *Ethology Ecology and Evolution* 26:172–194.

Menchetti, M., R. Scalera, and E. Mori. 2014. First record of a possibly overlooked impact by alien parrots on a bat (*Nyctalus leisleri*). *Hystrix, the Italian Journal of Mammalogy* 25:61–62.

Millsap, B., T. Breen, E. McConnell et al. 2002. Comparative fecundity and survival of bald eagles fledged from suburban and rural natal areas. Final Performance Report. Tallahassee, FL: Florida Fish and Wildlife Conservation Commission.

Morgan, U. M., L. Xiao, J. Limor et al. 2000. *Cryptosporidium meleagridis* in an Indian ring-necked parakeet (*Psittacula krameri*). *Australian Veterinary Journal* 78:182–183.

Mori, E., L. Ancillotto, M. Menchetti, C. Romeo, and N. Ferrari. 2013a. Italian red squirrels and introduced parakeets: Victims or perpetrators? *Hystrix, the Italian Journal of Mammalogy* 24:195–196.

Mori, E., M. Di Febbraro, M. Foresta et al. 2013b. Assessment of the current distribution of free-living parrots and parakeets (Aves: Psittaciformes) in Italy: A synthesis of published data and new records. *Italian Journal of Zoology* 80:158–167.

Mukherjee, A., C. K. Borad, and B. M. Parasharya. 2000. Damage of rose-ringed parakeet, *Psittacula krameri* Bordeat, to safflower, *Carthamus tinctorius* L. *Pavo* 38:15–18.

Navarro, J. L., M. B. Martella, and E. H. Bucher. 1992. Breeding season and productivity of monk parakeets in Cordoba, Argentina. *Wilson Bulletin* 104:413–424.

Neidermyer, W. J., and J. J. Hickey. 1977. The monk parakeet in the United States, 1970–1975. *American Birds* 31:273–278.

Newman, J. R., C. M. Newman, J. R. Lindsay, B. Merchant, M. L. Avery, and S. Pruett-Jones. 2008. Monk parakeets: An expanding problem on power lines and other electrical utility structures. In *Environmental Concerns in Rights-of-Way Management, Eighth International Symposium*, eds. J. W. Goodrich-Mahoney, L. Abrahamson, J. Ballard, and S. Tikalsky, pp. 355–363. Palo Alto: Electric Power Research Institute.

Newson, S. E., A. Johnston, D. Parrott, and D. I. Leech. 2011. Evaluation the population impact of an invasive species, ring-necked parakeet *Psittacula krameri*, on native avifauna. *Ibis* 153:509–516.

Olivieri, A., and L. Pearson. 1992. Monk parakeets in Bridgeport, Connecticut. *Connecticut Warbler* 12:104–111.

Owre, O. T. 1973. A consideration of the exotic avifauna of southeastern Florida. *Wilson Bulletin* 85:491–500.

Pithon, J. A., and C. Dytham. 1999. Census of the British ring-necked parakeet *Psittacula krameri* population by simultaneous count of count of roosts. *Bird Study* 46:112–115.

Pranty, B. 2009. Nesting substrates of monk parakeets (*Myiopsitta monachus*) in Florida. *Florida Field Naturalist* 37:51–57.

Pranty, B., and K. L. Garrett. 2011. Under the radar; "non-countable" exotics in the ABA area. *Birding* 43:46–58.

Pruett-Jones, S., C. W. Appelt, A. Sarfaty, B. van Vossen, M. A. Leibold, and E. S. Minor. 2012. Urban parakeets in northern Illinois: A 40-year perspective. *Urban Ecosystems* 15:709–719.

Pyle, R. L., and P. Pyle. 2009. *The Birds of the Hawaiian Islands: Occurrence, History, Distribution, and Status*. Honolulu, HI, USA: B.P. Bishop Museum. Version 1 (December 31, 2009), http://hbs.bishopmuseum.org/birds/rlp-monograph/.

Raso, T. F., V. L. Ferreira, L. N. Timm, and M. T. Abreu. 2014. Psittacosis domiciliary outbreak associated with monk parakeets (Myiopsitta monachus) in Brazil: Need for surveillance and control. *JMM Case Reports* 2014 1. doi: 10.1099/jmmcr.0.003343.

Reddy, V. R. 1998. Bird damage to maize crop on student's research farm on Rajendernager, Hyderabad, Andhrapradesh. *Pavo* 36:77–78.

Reddy, V. R. 1999. Intensive damage to maize crop by rose-ringed parakeet at a student's farm in a maize cultivated crop in Rajendernager, Hyderabad, Andhrapradesh. *Indian Journal of Forestry* 3:16–18.

Reed, J. E., R. A. McCleery, N. J. Silvy, F. E. Smeins, and D. J. Brightsmith. 2014. Monk parakeet nest-site selection of electric utility structures in Texas. *Landscape and Urban Planning* 129:65–72.

Roberts, T. J. 1991. *The Birds of Pakistan, Volume 1: Regional Studies and Non-Passeriformes*. Oxford: Oxford University Press.

Roscoe, D. E., W. B. Stone, L. Petrie, and J. L. Renkavinsky. 1976. Exotic psittacines in New York State. *New York Fish and Game Journal* 23:99–100.

Runde, D. E., W. C. Pitt, and J. T. Foster. 2007. Population ecology and some potential impacts of emerging populations of exotic parrots. In *Managing Vertebrate Invasive Species: Proceedings of an International Symposium*, eds. G. W. Witmer, W. C. Pitt, and K. A. Fagerstone, pp. 338–360. Fort Collins, CO: USDA National Wildlife Research Center.

Sani, H. K., M. S. Dhindsa, and H. S. Toor. 1994. Food of the rose-ringed parakeet, *Psittacula krameri*, a quantitative study. *Journal of Bombay Natural History Society* 91:96–103.

Schwartz, A., D. Strubbe, C. J. Butler, E. Matthysen, and S. Kark. 2009. The effect of enemy-release and climate conditions on invasive birds: A regional test using the rose-ringed parakeet (*Psittacula krameri*) as a case study. *Diversity and Distributions* 15:310–318.

Senar, J. C., J. Domènech, L. Arroya, I. Torre, and O. Gordo. 2016. An evaluation of monk parakeet damage to crops in the metropolitan area of Barcelona. *Animal Biodiversity and Conservation* 39:141–145.

Sheehey, A., and B. Mansfield. 2015. *Wild Rose-Ringed Parakeets Psittacula krameri*. *Nature Alley*. Weldon, California, USA: Nature Alley, unpaginated. http://www.natureali.org/roserings.htm

Shivanarayan, N., K. S. Babu, and M. H. Ali. 1981. Breeding biology of the rose-ringed parakeet *Psittacula krameri* at Maruteru. *Pavo* 19:92–96.

Snyder, N. F., E. C. Enkerlin-Hoeflich, and M. A. Cruz-Nieto. 1999. Thick-billed parrot (*Rhynchopsitta pachyrhyncha*). In *The Birds of North America Online*, ed. A. Poole. Ithaca: Cornell Laboratory of Ornithology. http://bna.birds.cornell.edu/bna/species/406. doi:10.2173/bna.406.

Snyder, N. F. R., S. E. Koenig, J. Koschmann, and H. A. Snyder. 1994. Thick-billed parrot releases in Arizona. *Condor* 96:845–862.

Snyder, N. F. R., and K. Russell. 2002. Carolina parakeet (*Conuropsis carolinensis*). In *The Birds of North America*, eds. A. Poole, and F. Gill, number 667. Philadelphia: The Birds of North America, Inc.

Sol, D., D. M. Santos, E. Feria, and J. Clavell. 1997. Habitat selection by the monk parakeet during colonization of a new area in Spain. *Condor* 99:39–46.

Spreyer, M., and E. Bucher. 1998. Monk parakeet (*Myiopsitta monachus*). In *The Birds of North America*, eds. A. Poole, and F. Gill, number 322. Philadelphia: The Birds of North America, Inc.

Strubbe, D., and E. Matthysen. 2007. Invasive ring-necked parakeets *Psittacula krameri* in Belgium: Habitat selection and impact on native birds. *Ecography* 30:578–588.

Strubbe, D., and E. Matthysen. 2009a. Establishment success of invasive ring-necked and monk parakeets in Europe. *Journal of Biogeography* 36:2264–2278.

Strubbe, D., and E. Matthysen. 2009b. Experimental evidence for nest-site competition between invasive ring-necked parakeets (*Psittacula krameri*) and native nuthatches (*Sitta europaea*). *Biological Conservation* 142:1588–1594.

Strubbe, D., and E. Matthysen. 2011. A radiotelemetry study of habitat use by the exotic ringed-necked parakeet *Psittacula krameri* in Belgium. *Ibis* 153:180–184.

Strubbe, D., E. Matthysen, and C. H. Graham. 2010. Assessing the potential impacts of invasive rose-necked parakeets *Psittacula krameri* on native nuthatches *Sitta europeae* in Belgium. *Journal of Applied Ecology* 47:549–557.

Tayleur, J. R. 2010. A comparison of the establishment, expansion and potential impacts of two introduced parakeets in the United Kingdom. *Proceedings of the Conference The Impacts of Non-Native species*, British Ornithologists' Union, Peterborough, UK. https://www.bou.org.uk/bouprocnet/impacts-of-non-native-species/.

Tella, J. L., A. Rojas, M. Carrete, and F. Hiraldo. 2013. Simple assessments of age and spatial population structure can aid conservation of poorly known species: The globally endangered red-fronted macaw as a case study. *Biological Conservation* 167:425–434.

Tillman, E. A., A. C. Genchi, J. R. Lindsay, J. R. Newman, and M. L. Avery. 2004. Evaluation of trapping to reduce monk parakeet populations at electric utility facilities. *Vertebrate Pest Conference* 21:126–129.

Tillman, E. A., A. van Doorn, and M. L. Avery. 2001. Bird damage to tropical fruit in south Florida. *Wildlife Damage Management Conference* 9:47–59.

Tozer, R. B. 1974. Feral parakeets and control of introductions. *British Birds* 67:484–485.

Van Bael, S., and S. Pruett-Jones. 1996. Exponential population growth of monk parakeet in the United States. *Wilson Bulletin* 108:584–588.

Yoder, C. A. 2011. Secondary hazards of DiazaCon™ in American crows and ferrets. Unpublished report. QA-1393. Fort Collins, CO: National Wildlife Research Center.

Yoder, C. A., M. L. Avery, K. L. Keacher, and E. A. Tillman. 2007. Use of DiazaCon™ as a reproductive inhibitor for monk parakeets (*Myiopsitta monachus*). *Wildlife Research* 34:8–13.

17 Introduction History, Impacts, and Management of House Sparrows in North America

Julian D. Avery and Julie L. Lockwood

CONTENTS

INTRODUCTION

During the mid-1800s, the house sparrow (*Passer domesticus*) became an unwitting participant in a global phenomenon of species redistribution. House sparrows, native to much of Europe and northern Asia, were deliberately introduced to such far-flung places as North America, Australia, South America, and Hawaii (Long 1981). Today, house sparrows occupy every continent on Earth except Antarctica; and given global warming trends, they may eventually find themselves there, particularly along the Antarctic Peninsula where summer temperatures can rise above freezing. House sparrows were introduced to various locations via the earnest attempts of humans to surround themselves with familiar species, or in an effort to use house sparrows to control agricultural pests. There is some evidence that house sparrows were also introduced to locations as accidental stowaways on oceangoing ships and trains, which is a reasonable assumption given the sparrow's propensity to nest in the nooks and crannies of human habitations, including ship decks and rigging, and to forage near human infrastructure, including grain transport railcars (Summers-Smith 1963; Anderson 2006). Given their ubiquity, and the occasional concern they generated in terms of their agricultural damage, researchers began compiling information on house sparrow introduction dates, geographical range expansions, and local and rangewide physiological, morphological, and behavioral adaptations. As a result of this intense research focus, house sparrows have provided a staggering amount of information on the mechanisms driving contemporary evolution, as well as the conditions that facilitate successful invasions. Here we focus on their impacts to native species, human economies, and health, as well as on the varied efforts to control or mitigate these impacts.

House sparrows are tightly associated with humans, and in many areas are considered obligate human commensals (Saetre et al. 2012). This association has, no doubt, contributed to their success as an invasive species. Such a broad distribution also provides researchers with the ability to investigate how and why some house sparrow populations differ from others in terms of their morphology, genetic diversity, population size, and impact on co-occurring native species. However, this high degree of commensalism also complicates how scientists should interpret their research on house sparrows relative to other exotic species, particularly when considering how information on house sparrows translates to understanding noncommensal taxa and exotic species found in nonurban systems. These qualities, and the impacts of house sparrows in their exotic ranges, intersect in a paradoxical fashion. During the spread of house sparrows outside their native range in the late 1800s and early 1900s, the sparrow came to be viewed as an economic threat (Brodhead 1971). The literature from this era reflects this viewpoint, with more studies focused on chemical control and deterrence methods. Over time, however, less has been written about sparrow control in Europe and North America, and there has been a surge of interest in the house sparrow from an evolutionary and population genetics point of view.

In this review, we will discuss the initial sparrow introductions into North America and how these releases set the house sparrow on course for one of the most impressive range expansions in recorded history. We will discuss aspects of the house sparrow's life history that led to population expansion, including reviewing relevant behavioral and population genetic data that inform how we view species invasions in general. We will follow this with a discussion of the impacts these birds have on ecological systems and human society. There is a rich literature on house sparrow control, and we review this information with an eye toward distinguishing which methods show promise in achieving management goals. Finally, we suggest several avenues of further investigation into the impacts and management of house sparrows.

PROVENANCE AND INTRODUCTION HISTORY

The house sparrow is a widespread species complex that includes 12 recognized subspecies that naturally inhabit all of Europe, much of the Mediterranean, and Asia. These subspecies were thought to belong to two larger geographical groupings: a northern Palearctic group called *domesticus*, and a southerly oriental group called *indicus*. Along with marked morphological differences, the *indicus* group contains populations that exhibit migratory behavior, whereas the *domesticus* group does not, and *indicus* relies more heavily upon rural habitats than did the *domesticus* group. However, recent molecular evidence has called this division into question, and it now appears likely that all native house sparrow subspecies originated from a single population in the Middle East that expanded its geographical range northward about 10,000 years ago, right at about the same time that agriculture spread out from the same region. This single population underwent rapid diversification after range expansion, eventually splitting into the various now-recognized subspecies (Anderson 2006; Saetre et al. 2012). Human-assisted introductions of house sparrows outside of their native range have involved individuals predominately from the *domesticus* group. The one exception is an introduction of individuals from the *indicus* group to South Africa, along with nearby islands in the Indian Ocean.

The introduction of sparrows into North America was originally thought to have taken place in Brooklyn, New York in 1852 or 1853, with a founding size of approximately 100 individuals, all of whom originated from England. An earlier introduction in 1851 into the city, involving just 16 individuals, was thought to have failed, but there is no evidence that these birds actually perished (Moulton et al. 2010). This sequence of events was called into question by Moulton et al. (2010) who found no actual evidence that the later introduction of 100 individuals ever occurred. Moulton et al. (2010) instead showed that the first 16 individuals released into New York City were likely all that was needed to establish house sparrows, and that they did not perish as previously assumed, but instead likely dispersed away from the location of release and survived despite not being resighted. No matter which interpretation of historical records is correct, after 1852 there were several additional introductions into North America from Europe. There were at least four reported initial releases from Europe during the late 1850s in upstate New York, Maine, Rhode Island, and Massachusetts (Barrows 1889). From 1860 to around 1880, people gathered individual house sparrows from these now established eastern populations and introduced them across the continent while sparrows were also still being imported from Europe. During this time, house sparrows were purposefully released in Kentucky, San Francisco (California), and Salt Lake City (Utah), among other locations (Figure 17.1). By 1886, sparrows had established in 32 states plus Hawaii and several Canadian provinces. The combined forces of self-colonization and widespread deliberate introductions resulted in the house sparrow achieving a transcontinental distribution by 1915 (Figure 17.1).

The rapid colonization of sparrows across North America was by all accounts staggering (Robbins 1973; Anderson 2006), and this expansion was no doubt aided by the multiple introduction events taking place across the continent. Estimates of range expansion vary from 96 km/year to 190 km/year during the period 1886–1910 and 1868–1888, respectively, with the greatest rate of expansion occurring in the mid-central portion of North America (Anderson 2006). At a smaller scale, rates of 16 km/year were observed in Kansas once sparrows had arrived in Topeka from the eastern United States (Anderson 2006). It is clear that the house sparrow's ability to capitalize upon human infrastructure is tightly linked to these observed rates of spread. The conditions that led to rapid colonization of North America by house

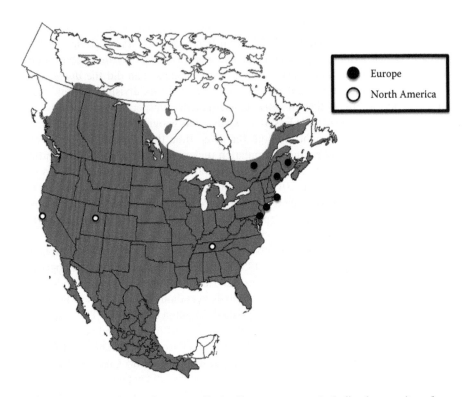

FIGURE 17.1   North American range limit of house sparrows, including known sites of purposeful releases and the origin of the sparrows released (either from Europe or previously established North American populations). (From The Birds of North America, https://birdsna.org, maintained by the Cornell Lab of Ornithology.) (Introduction data from Long, J. L., *Introduced Birds of the World*, London: David and Charles, 1981.)

sparrows include human reliance upon horses for transport and the large quantities of available food from horse feed and agricultural waste (Robbins 1973). The switch from horses to cars, and a reduction in the prevalence of small-scale agricultural systems, may be why some populations of house sparrows have shown very slow range expansion rates after about 1950 (Bergtold 1921; Anderson 2006) and why other populations declined in the 1920s (Robbins 1973).

MECHANISMS OF INVASION SUCCESS IN NORTH AMERICA

LIFE HISTORY

Given the house sparrow's success at establishing exotic populations worldwide, one could be forgiven for assuming that the species possesses extraordinary physiological or reproductive traits. However, there is no single trait, other than its propensity to exist within human settlements, which seems to set it apart from other less-successful exotic birds. For example, house sparrows exhibit a high degree of sedentary behavior across all age classes (Anderson 2006). Adults rarely leave a breeding location, with individuals tending to forage close to their breeding and roosting colonies. Furthermore, natal dispersal distances are generally short (Anderson 2006). Such traits seem to counter the large-scale observations of house sparrows spreading rapidly across the North American continent. How can a sedentary bird colonize that much space so quickly? It is possible that sedentary behavior could increase the likelihood of a newly released population establishing a self-sustaining population. If released individuals fail to disperse away from a single site, or if these individuals disperse only short distances away from this newly formed population, they are more likely to encounter other individuals for breeding. This trait keeps local population levels high enough to avoid Allee effects (e.g., failure to find a mate at low density), or reductions in genetic diversity that can accompany long periods of very small effective population sizes (e.g., inbreeding depression). Overcoming these limitations of small population size certainly helps ensure successful establishment in exotic birds (Blackburn et al. 2009).

Another unremarkable aspect of house sparrow life history is their reproductive behavior. Like the majority of other passerines, house sparrows attempt to breed the first breeding season following birth, and they are socially monogamous, even though extra-pair paternity is regularly observed. Onset of nesting is negatively correlated with latitude and, like other passerines, typically extends from the middle of the spring to late summer, although individuals that nest in heated buildings will initiate egg laying sooner than individuals in unheated locations (Skorka et al. 2009). The ability to capitalize upon human infrastructure may give them an advantage over other passerines, and may contribute to their ability to initiate four broods in a season in some locations (Lowther and Cink 2006). In some cases, they have also been observed initiating a nest up to eight times in a season when experiencing nest failure (Lowther and Cink 2006). It remains to be seen if reproductive output or persistence is comparatively higher than other species, and whether this translates to superior fitness and invasability. For example, eastern bluebirds at lower latitudes will also attempt four broods in a season (Gowaty and Plissner 1998). House sparrows nest colonially and may exhibit nest clumping, possibly as a counter measure to cold climates (McGillivray 1980). Some studies have attempted to determine if kin selection is a factor in the house sparrow's colonization success. However, winter flocks do not seem to contain an unusual number of relatives, implying that birds do not necessarily segregate themselves with respect to kin (Liker et al. 2009).

Egg laying in house sparrows follows expectations based upon optimality theory, with a mid-season maximum clutch size. First-year females initiate laying later than older females and they lay smaller clutches than older individuals. Furthermore, clutch size increases with latitude and pronounced seasonality. In other words, more variable climates result in larger clutches per nesting attempt in house sparrows. Urban environments tend to support slightly smaller clutches than rural habitats. Two or more clutches per breeding season are to be expected with multiple renesting attempts, and four broods are occasionally observed (Lowther and Cink 2006).

With respect to body size, Liker et al. (2008) showed that urban birds tend to be smaller and in poorer condition than rural sparrows. These differences persisted after birds were enclosed in aviaries and given unlimited resources, implying that conditions during the nestling period or adaptations to urban living are the cause of body size differences. It should be noted, however, that there is no support for molecular divergence at putatively neutral microsatellite loci between urban and rural sparrows (they do not segregate into distinct clusters) (Vangestel et al. 2012). Another study supported the finding that urban birds were smaller (Bokony et al. 2012) but used multiple indices to determine that urban habitats were not inherently more stressful than rural habitats. Urban and rural sparrows were in approximately equal body condition during the nonbreeding period, and morphological traits thought to indicate condition did not differ.

ENVIRONMENT

While urbanism in general plays a large role in the success of species introduced to islands (Møller et al. 2015), and human commensalism is often cited as the driver of house sparrow introduction success, it is also clear that house sparrows can thrive under very different conditions than those that occur in their native range (Anderson 2006; Martin et al. 2015). For instance, the nominate subspecies of house sparrows has successfully established exotic populations in environments as disparate as highly seasonal temperate latitudes in North American and tropical latitudes with little climatic fluctuation in South America (Anderson 2006).

Researchers have attempted to explain the broad geographical success of house sparrows by examining the enemy-release hypothesis and individual-level responses to stressors. The enemy-release hypothesis states that exotic populations tend to be successful in part because the coevolved pathogens, predators, and parasites from their native ranges are absent in the new host range, thereby "releasing" the exotic population from this form of regulation (Schrey et al. 2011). There is support for the enemy-release hypothesis from a global analysis of haemosporidian parasites in house sparrow blood. Individual house sparrows from the native range support a greater diversity and higher prevalence of parasites than individuals in the exotic range (Figure 17.2) (Schrey et al. 2011). Lee et al. (2005) showed that individual house sparrows generate a smaller response to immunological challenge than less-successful exotic populations of tree sparrows (*Passer montanus*), which equates to greater energy savings for house sparrows and presumably more energy available for other pursuits, such as egg laying or foraging. While this release may in fact contribute to establishment success, we caution against calling this an adaptation because bottlenecks are known to reduce diversity for immunity genes (Martin et al. 2015).

BEHAVIOR

If there is a single aspect of house sparrow biology that likely determines their invasion success, it is their behavioral flexibility. Individual house sparrows in their exotic range have shown a range of innovative behaviors and behavioral adaptations that may explain their success when introduced. Neophobia (fear of new things) is generally not different between house sparrows and co-occurring native species. However, Liebl and Martin (2014) show that individual house sparrows that live at the edge of their exotic range will readily ingest novel foods that individuals living in the "core" or well-established populations will not (Figure 17.3). These researchers hypothesized that being attracted to new foods and objects helped house sparrows be

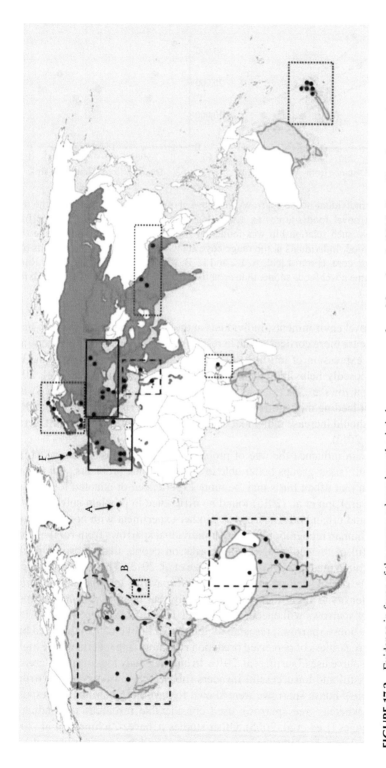

FIGURE 17.2 Evidence in favor of the enemy-release hypothesis in house sparrows from Marzal et al. (2011). Sampling sites are shown as dark dots, and the number of *Plasmodium* species detected in each location is shown by box types (solid, dashed, dotted). Dark green denotes native range and light green, exotic range.

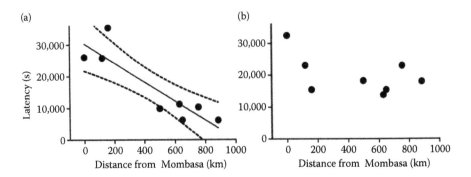

FIGURE 17.3 Individual house sparrows in Kenya show a marked reduction in the time they wait to eat novel food sources as a function of population age. (a) Distance from Mombasa. (b) No such relationship was found for the time individuals wait before they approach novel food. Individuals at the range edge show less neophobio to new foods than those at the range core. (From Liebl, A. L., and L. B. Martin, Living on the edge: Range edge birds consume novel foods sooner than established ones, 2014, by permission of Oxford University Press.)

successful in novel environments. Individuals at the exotic range edge are also more exploratory, release more corticosterone in response to less-predictable stressors, and exhibit greater expression of toll-like receptors linked to the immune system. The latter trait supposedly helps individuals surveil for pathogens (Martin et al. 2014, 2015). House sparrows are, on average, leaner in urban areas (Liker et al. 2008), and exhibit constant feeding motivation regardless of predation risk (Tsurim et al. 2010). This behavior should increase individual energetic efficiency (Martin and Fitzgerald 2005; Sol et al. 2013).

Group size can influence the rate of problem solving in house sparrows (Liker et al. 2009), with larger groups better able to solve complex problems, and there is some indication that urban birds may be more experienced or adapted for problem solving. However, Papp et al. (2015) found no difference in problem-solving ability between rural and urban house sparrows. Another experiment with house sparrows and individual human recognition found that individual sparrows from rural habitats were less fearful of "friendly"-appearing masks on people than "hostile" masks, whereas urban birds made no distinction (Vincze et al. 2015). These authors suggest that there may be diminished rewards to individual sparrows in avoidance behaviors when human density is high, making it less likely that humans will notice house sparrows or that sparrows will encounter the same human more than once.

Particular to house sparrows, researchers found that individuals maintained high foraging effort regardless of perceived predation risk, implying that they have higher efficiency in resource use (Tsurim et al. 2010). In another study looking at differences between successful and unsuccessful invaders (house sparrows and tree sparrows; *Passer montanus*), house sparrows were shown to have no inflammatory response to challenges, whereas tree sparrows used considerable resources responding to immune challenges (Lee et al. 2005). Other studies (Chavez-Zichinelli et al. 2010) have found no relationship between stress responses in urban and suburban habitats, implying that house sparrows have considerable capacity to modulate responses to different conditions. Finally, in a study looking at causal mechanisms underlying population declines in Europe, researchers found that house sparrows do not minimize starvation risk during winter (by increasing fat stores), but instead appear to trade off starvation risk for predation risk (i.e., staying lighter in order to avoid predation) (Macleod et al. 2006). This pattern may help them avoid predation, possibly contributing to increased survival in introduced regions, but it likely also contributes to decreased survival when resources are variable or limited as agricultural practices change (Hole et al. 2002).

GENETICS

Support for historical observations that most house sparrows in North America are derived from native populations in England and Germany comes from Schrey et al. (2011), who show that individuals sampled in North America are genetically very similar to those sampled in England and Germany but not elsewhere in Europe. Schrey et al. (2011) also show that North American house sparrows have lower heterozygosity and higher inbreeding coefficients than individuals from their native source population, but allelic diversity and other measures of variation are similar between exotic and native populations. This result suggests that the multiple release events that occurred in North America (see the "Provenance and Introduction History" section) may have prevented extended population bottlenecks from occurring in the North American exotic population (Roman and Darling 2007; Dlugosch and Parker 2008). For this reason, we should not expect that loss of genetic diversity has played a strong role in limiting the house sparrow's geographical expansion in North America, and all available evidence shows that spread across the continent was quite rapid (see the "Provenance and Introduction History" section).

In support of the idea that house sparrows are highly sedentary, Jensen et al. (2013) found that across European island and mainland populations, genetic differentiation increased with increasing distance between populations. This genetic information reinforces observations that individual house sparrows truly do not disperse far from their natal site or previous nesting site, although mainland populations show stronger isolation by distance patterns than those on islands. Vangestel et al. (2012) also show that urban populations of European house sparrows have a finer degree of differentiation than rural populations, indicating that urban areas attract sparrows rather than act as sources for adjacent populations (i.e., there is unidirectional gene flow into urban areas).

Information on genetic divergence between exotic populations can inform how house sparrows are using available habitats in North America, and thus efforts to control their spread or impact. Kekkonen et al. (2011a) show that recent declines in house sparrow abundance in Europe acted to increase local-scale genetic differentiation between populations, but that, in general, house sparrows are panmictic across the landscape (Kekkonen et al. 2011b). This result suggests that populations separated by inhospitable areas in North America have the potential to show local differentiation, perhaps through adaptation to specific environmental differences (Johnston and Selander 1964). It also suggests that we can use any identified genetic divergence between populations at the local scale as a guide to how widely to implement control efforts. The idea here is that genetically well-mixed population "units" should be considered the target of control. Genetic evidence that this targeted unit is not tightly tied to neighboring units via dispersal suggests that any reduction in abundance will be longer lasting than if control efforts failed to take population connections into account (e.g., Russell et al. 2010).

Finally, researchers are beginning to focus on epigenetic factors that may compensate for reduced genetic or phenotypic variation resulting from population bottlenecks (Martin et al. 2015). Reductions in variability are thought to limit the potential for adaptation in new populations, and the timescale over which invasions occur does not permit time for mutation and recombination to generate new genetic variation. Epigenetic mechanisms present a potential path animals can use to cope with new challenges (Pérez et al. 2006). Epigenetic mechanisms can alter the individual phenotype in ways that typical genetic variation cannot, generating short-term adaptations and permitting genomic–environment interactions. Both Schrey et al. (2012) and Liebl et al. (2013) have found increased expression of epigenetic variation (in the form of DNA methylation) in recently founded populations of house sparrows in Kenya. Schrey et al. (2012) found that this pattern held true when compared to an established exotic population in Florida, and that overall epigenetic diversity was similar despite even more reduced genetic variation in Kenya. Importantly, Liebl et al. (2013) found epigenetic variation to be negatively correlated with increasing genetic diversity in house sparrows, and positively correlated with increases in the inbreeding coefficient. These may be compensatory mechanisms to cope with decreasing genomic variability. A recent study by Wetzel et al. (2012) found that house sparrow heterozygosity was positively related to clutch size, a potentially important relationship in invasion success. Overdominance (essentially heterozygote advantage) has been proposed as an explanation for the relationship (it can maintain genetic diversity), and new selective regimes or events can alter overdominance, allowing rapid responses or changes in traits. House sparrows have continually been subjected to new environmental conditions in North America and worldwide, and this mechanism could possibly contribute to their colonization success, even in situations where the number of founding individuals (propagule size) is low. It is unknown if house sparrows possess greater capacity than other species to use these mechanisms during the invasion process.

IMPACTS IN NORTH AMERICA

Even though house sparrows are not listed as one of the 100 worst invaders by the Invasive Species Specialist Group (2010) (www.issg.org), the species is still notorious for a variety of ecological and economic impacts. From competition for nest sites (Newton 1994) to acting as a reservoir for pathogens (Wheeler et al. 2009), house sparrows have been implicated in many processes that threaten the well-being of humans and animals in rural and urban settings. However, house sparrows are also being used to monitor the presence and persistence of environmental contaminants (Roux and Marra 2007) and are used as a model organism for the study of biological invasions (Martin et al. 2015).

COMPETITION

There is ample evidence that house sparrows are superior competitors to numerous other avian species under a range of environmental conditions (Barrows 1889; Anderson 2006; Invasive Species Specialist Group 2010). House sparrows are secondary cavity nesters, and they typically rely upon holes and crevices in buildings and other structures including nest boxes built for other species. They are competitive for nesting sites as secondary cavities are naturally a limited resource, and house sparrows are known to usurp the nests of other species in North America, including cliff swallows (*Petrochelidon pyrrhonata*), eastern bluebirds (*Sialia sialis*), and purple martins (*Hirundo rusticus*) (Anderson 2006). Forceful interactions between North American native birds and house sparrows over use of secondary cavities can often lead to mortality for an individual of the native species (Crowell 1962; Gowaty 1984; McGrath 1987). House sparrows have also been observed in North American urban areas displacing desert (lesser goldfinch, *Spinus psaltria*), tropical (white-collared seedeater, *Sporophila torqueola*), and temperate native birds (house finch, *Haemorhous mexicanus*) from food resources (Kalinoski 1975; Chavez-Zichinelli et al. 2010; Shochat et al. 2010).

In North America, researchers have noted concurrent declines in house sparrow numbers as house finches have increased in prevalence (Anderson 2006), and it has been postulated that finches are outcompeting sparrows. This makes competitive interactions where house sparrows are dominant to house finches particularly interesting (Evenden 1957; Giesbrech and Ankney 1998). However, there is mixed support for this theory (Cooper et al. 2007; McClure et al. 2011), with correlational evidence for displacement by finches present in the northeastern United States and absent in the southeastern United States. Giesbrech and Ankney (1998) present evidence that house sparrows behaviorally dominate house finches in controlled settings. There are other examples where house sparrows have been noted *not* to compete with native species for resources (Hahn et al. 2011; Baker et al. 2014), but this may stem from the fact that house sparrows typically colonize urban areas where native species are already rare. It should be noted that house sparrows have invaded areas where house finches were already present (western United States), and that house finches have invaded areas where house sparrows were already established (northeastern states and areas such as Long Island and the eastern seaboard). Moulton and Ferris (1991) showed that in Oahu, where both species were introduced around the same time, house sparrows ingested discarded human food, whereas house finches ingested primarily forb seeds. It is possible that there may be dietary shifts in places where they co-occur, leading to lack of competitive interactions.

There is some evidence that competition from house sparrows can influence broader community metrics such as species richness. MacGregor-Fors et al. (2010) found that house sparrow–invaded communities had lower species richness and evenness, potentially a result of competitive exclusion of native bird species by house sparrows. This observation is in contrast to studies on the impact of other exotic birds, which have determined that native and exotic species rarely interact, and that less-competitive species do well when introduced because they are primarily adapted for urban environments, whereas very few native birds exploit urban resources (Sol et al. 2012). Shochat et al. (2010) take the view that exotic species depress biodiversity in urban areas through exclusion, keeping native species out. Shochat et al. (2010) hypothesized that the black-throated sparrow (*Amphispiza bilineata*), a bird occasionally seen in suburban habitats of Phoenix, Arizona, would persist in urban desert habitat in the absence of house sparrows.

In some locations, actions intended to help boost native species persistence may in fact benefit house sparrows. On the island of Bermuda, for example, eastern bluebirds have diverged and become morphologically distinct from mainland populations (Avery et al. 2013, 2014). This island bluebird population is threatened with extinction due in part to competition from house sparrows over the limited secondary cavities available for their nest sites. Local agencies have gone to great lengths to implement nest box programs for bluebirds by either giving away nest boxes or subsidizing the cost of purchasing nest boxes in order to saturate private land with potential bluebird nest sites. Because box owners do not often clean out nest boxes after house sparrows have used them as nesting sites, the influx of boxes may have unintentionally boosted house sparrow numbers on the island. Bluebirds will not nest in a box that has an old house sparrow nest within it. However, house sparrows will readily reuse the same nest and box for another breeding attempt.

DISEASE TRANSMISSION

The propensity for house sparrows to occur in urban and agricultural habitats, sometimes in large densities, places them in close proximity to people and their livestock. This close association means that any diseases that house sparrows carry can be transferred with varying levels of ease to people and livestock. Thus, house sparrows can be a reservoir or vector for a variety of pathogens and bacteria (for a partial summary, see appendix in Anderson 2006).

It is estimated that house sparrows are involved in the spread of approximately 29 human and wildlife diseases (Pimentel et al. 2005), although the exact role played by sparrows is unknown for many of them. These include several arboviruses (e.g., West Nile virus and equine encephalitis), protozoans like *Plasmodium* that cause avian malaria, bacteria such as *Salmonella* and *E. coli* (Fitzwater 1994), and numerous other endo- and ectoparasites. The impacts of many of these diseases on house sparrows themselves are unknown, but some immunocompetence studies have shown that sparrows in poor body condition are more susceptible and less able to mount a defense (Navarro et al. 2003). In a study of bird feeders and mortality during the winter period, Brittingham and Temple (1986) found that house sparrows were by far the most frequently reported dead or dying bird by survey respondents.

In Nebraska, house sparrows have been shown to raise the prevalence of an arbovirus called Buggy Creek virus, which inhabits the swallow bug (*Oeciacus vicarius*) (Brown et al. 2012; O'Brien and Brown 2012). In cliff swallow (*Petrochelidon pyrrhonota*) colonies that have been inhabited by house sparrows, the virus is more virulent than in colonies not invaded by house sparrows. This is a case of a virus switching hosts and becoming prevalent through the invader's presence. Interestingly, cliff swallows show little response to Buggy Creek virus, whereas it induces significant mortality in house sparrow nestlings (O'Brien and Brown 2012).

The threat of West Nile virus transmission to humans has generated considerable interest in determining which species act as reservoirs for the disease. It is well known that house sparrows act as an amplifying host for West Nile virus (Nemeth et al. 2009b). However, little is known about how the virus overwinters and is reintroduced to vectors such as mosquitoes (Anderson 2006). Some authors suspect that house sparrows fill this role, although the overall evidence remains equivocal. Infected house sparrows can develop humoral immunity (Nemeth et al. 2009a), and this immunity can result in zero transmittance of the virus to mosquito vectors. However, this immunity can disappear, leading to relapse of virulence in house sparrows, making them (again) competent West Nile virus hosts (Nemeth et al. 2009b). Surveys of house sparrow tissues did not support the theory of recrudescence, but did support the theory that predators could become infected by ingesting infected house sparrows, thus allowing the virus to persist through the winter in other hosts (Nemeth et al. 2009b). In California, house sparrows were commonly infected with West Nile virus, and mortality of experimentally infected birds was very high (Wheeler et al. 2009). Duggal et al. (2014) has shown that house sparrows may have become resistant to earlier strains of West Nile virus, but that newer strains are more virulent. The role of house sparrows in transmitting and permitting recrudescence of arboviruses and other pathogens means that sparrows can be, and often are, used as a sentinel species for future outbreaks in human populations (Anderson 2006).

Avian malaria is widespread globally, and is of considerable concern for endangered island bird species. In a study of avian malaria expansion in France, researchers determined that climatic variables were related to parasite prevalence in house sparrows, and that under warming climate the parasite would spread considerably (Loiseau et al. 2013). In Hawaii, it is suspected that house sparrows may have played a role in amplifying and facilitating the spread of avian malaria into populations of native Hawaiian honeycreepers, many species of which have very high mortality after malarial infection (Anderson 2006).

POISONING

A relatively common way of controlling house sparrow population has been to apply poison to food resources that kills the individual sparrow that ingests it, increases its chances of dying at the hands of a predator (i.e., stupefies the individual, making it unable to carry out antipredator behaviors), or serves to deter the individual from foraging on that food source again (see below). Some of these poisons are well known to be lethal to several different vertebrate species if they ingest enough of it, which is true, for example, with strychnine, organophosphates, or anticoagulants. The application of these poisons raises two interrelated questions. First, does the application of these poisons to control house sparrows have nontarget effects in the sense that other species may easily ingest the same food and suffer the same consequences (death)? Second, can predators or scavengers that ingest a poisoned house sparrow also die?

Research on either question is scant, with the impacts of poisoned sparrows upon local wildlife particularly understudied in North America. There is considerably more literature on secondary exposure in places like New Zealand, where poisons have been used extensively to remove vertebrate invasive species such as rats (Dowding et al. 1999), and the effects upon nontarget scavenger and predatory species are of high concern since these species are often themselves threatened with extinction. It is generally accepted that predators are at risk from consuming prey, such as house sparrows, that have been directly poisoned (Brakes and Smith 2005) or that have ingested poison that was not intended for them (Dowding et al. 1999). As an example of the latter situation, Elliott et al. (2014) observed house sparrows entering bait chambers intended for the lethal control of rats and ingesting rodenticide. These individuals almost certainly left the chamber and later died, creating an unintended source of nontarget food web effects for rodenticide. There is limited direct evidence that predators in North America, such as American kestrels (*Falco sparverius*), will preferentially capture poisoned house sparrows, often because they behave abnormally or are slower than control prey items (Hunt et al. 1992). It is not clear what effect these poisoned sparrows have on kestrels; however, the chemicals themselves are extremely lethal (Rattner et al. 2011).

In North America, poison is rarely used to control house sparrow populations, thereby minimizing the food web impacts of such programs. However, another type of food web impact can come from urban-dwelling house sparrows. Sparrows living in urban areas have been found to accumulate more heavy metals in their tissues than birds that live in nonurban areas (Kekkonen et al. 2012; Herrera-Duenas et al. 2014). Because house sparrows are one of the most common vertebrates in urban areas, they become a principal food source for a variety of urban predator species. This raises the possibility that house sparrows serve as an intermediate link in bioaccumulation chains of heavy metals in urban ecosystems. There is very little research on this topic in North America (or anywhere else), but it is clear that house sparrows have the potential to transmit contaminants to other wildlife. It is important to note in this context that house sparrows are increasingly being used as indicators of environmental pollution (Anderson 2006; Roux and Marra 2007; Bichet et al. 2013; Herrera-Duenas et al. 2014; Millaku et al. 2015) and chemical application (Martínez-Haro et al. 2007). Sparrows have also been used in tests of agricultural chemical toxicity to determine effects upon farmland birds and humans (reviewed in Anderson 2006), and in tests of taste preference to create pesticide delivery methods that are avoided by wildlife (Gionfriddo and Best 1996; Anderson 2006; Clapperton et al. 2012). Thus, there is a growing body of literature on what metals and chemicals urban house sparrows ingest; we just do not yet know what the food web effects of these accumulations may be.

AGRICULTURAL

After their initial introduction into North America, and subsequent rapid range expansion and increase in abundance, house sparrows were viewed as serious agricultural pests. They consume a wide variety of crops including oats, wheat, rice, barley, and corn; chicken feed; a variety of fruits (mostly when they are young buds); and young vegetables (Invasive Species Specialist Group 2010; Fitzwater 1994). They are also known to transmit pathogens to poultry (Fitzwater 1994), and they cause sanitary problems and contaminate livestock feed through their excrement (Witmer et al. 2007). A few researchers have attempted to quantify the extent of crop yield loss due to house sparrow foraging, and most peg the losses at around 1%–5%, with some losses ranging up to 25% (Anderson 2006). The percentage yield lost depended on the size of the farmed field, with smaller fields suffering heavier crops losses than larger ones due to the propensity of house sparrows to forage near forest edges or other habitat cover (e.g., hedgerows; Anderson 2006).

There have been few studies attempting to quantify the economic costs associated with house sparrows, but those that exist are centered around agricultural damage (Anderson 2006). By 1889, the U.S. Department of Agriculture had recognized the house sparrow as a threat to native birds and agriculture (Barrows 1889; Fitzwater 1994). Between 1990 and 1997, 48 states in the United States requested some form of monetary aid from the federal government for house sparrow control and management (Bergman et al. 2000). To put total damages in perspective, one needs to account for two other species in North America that are considered serious agricultural pests: the rock dove (*Columba livia*) and European starling (*Sturnus vulgaris*). These two species accounted for 93% of a total $28 million price tag for invasive bird damage between 1990 and 1997 (Bergman et al. 2000). House sparrow economic costs comprise an unknown percentage of the remaining $1.96 million that was reported to the U.S. Department of Agriculture's Animal and Plant Health Inspection Service during that same time period (Bergman et al. 2000). Total damages caused by invasive species in the United States is estimated at $120 billion (Pimentel et al. 2005). As a point of comparison, Pimentel et al. (2005) estimated the cost of pigeon damages in the United States per year as $1.1 billion, and this does not include costs associated with the spread of pathogens. Therefore, it could be expected that house sparrows cause considerable economic hardship on an annual basis because of their food preferences and their high densities in urban and rural habitats.

CONTROL OF HOUSE SPARROWS

Largely due to their perceived impacts on agricultural yields, there have been several mechanisms developed for control of house sparrow populations. Given the role of edge habitat in increasing the loss of crops to house sparrow foraging, there have been efforts to manage habitat around farms to reduce sparrow damage, such as removing shelterbelts and removing plants that produce seed resources (MacLeod et al. 2011). But the four main forms of control that are regularly used today are exclusion, repellents, poisons, and traps (Fitzwater 1994; Gorenzel et al. 2000; Anderson 2006). Exclusion efforts concentrate on reducing the number of locations where house sparrows can nest or roost via plugging small holes in and around buildings, and affixing physical barriers to surfaces where house sparrows roost. Repellents involve creating tactile and sound environments that house sparrows actively avoid (e.g., creating sticky roost surfaces, broadcast of loud noises) and spraying food or crops with chemical substances that house sparrows find unpalatable (e.g., spraying fungicide on stored grain) (Anderson 2006; Clapperton et al. 2012).

Poisons have included broad-spectrum agents such as strychnine and organochlorides, which were sprayed onto grain or other bait. Once individual sparrows ingested the poisoned bait, they died within a few days. These poisons are lethal to humans, as well as other vertebrate animals, and thus their use has largely been phased out. Nevertheless, there are a few legal poison control options available today, and these are highly regulated and must be administered by a professional. Live traps have also seen use, and there are a variety of bird traps that are available on the market that can be used to capture house sparrows. Once these individuals are caught, it is up to the person who catches them to decide what to do next. Where house sparrows are exotic, like in the United States, there is no law that prevents one from killing the captured sparrows. The caught sparrows can also be relocated and released alive. Although not often used, other forms of lethal control include shooting sparrows with low-caliber rifles and systematically destroying house sparrow nests.

Repellents and exclusion devices are of some value to homeowners or farmers that have a location- or crop-specific problem with house sparrows, but they are not effective when trying to reduce local population sizes of house sparrows. The use of poisons, traps, and other forms of lethal control are more effective in reducing numbers of house sparrows in a local area if they can be administered to a reasonable fraction of individuals in a local population. Overall, the control of house sparrows is a notoriously difficult prospect because they are so abundant in most habitats and can learn to avoid trapping and hunting methods; additionally, they can identify foods laced with poisons and thus avoid them (Clapperton et al. 2012). Complete eradication of house sparrow populations is often never attainable, even when local reductions in sparrows have occurred, in large part because the site of control is quickly recolonized by individuals coming from neighboring locations (Bednarczuk et al. 2010). Even Barrows' (1889) discussion on house sparrow control with poisons, while outdated, has some valuable insight. He makes the point that poisons do not work to reduce population numbers, and they are dangerous to other species, including humans.

One location where wildlife managers have had success in controlling the geographical range expansion of house sparrows is Australia. House sparrows were introduced to Australia from Great Britain in 1863, and they currently occupy all of eastern Australia and western reaches of the state of South Australia; however, they have been unable to colonize the state of Western Australia. This lack of spread into Western Australia could be due to biogeographic barriers that prohibit their dispersal distances going west, and the typically low density of human infrastructure in Western Australia. The Western Australia government also implements a biosurveillance scheme and stringent lethal biocontrol protocol to keep house sparrows out of their state (Government of Western Australia Department of Agriculture and Food: https://www.agric.wa.gov.au/birds/sparrows). Ships that dock in Perth (and other ports) from Southeast Asia are regularly searched for stowaway house sparrows, and traps are deployed along the state's border with South Australia. When house sparrows are found, they are immediately dispatched. The house sparrow's reliance upon humans may lend itself to more effective prevention measures in this case. Trapping only needs to be focused in urban areas or near shipping ports in Western Australia, and heavy monitoring and removal of house sparrows from these areas alone may be all that is needed to control their spread.

Along these same lines of reasoning, it may be possible to more effectively control the spread of house sparrows in any situation if more is known about their dispersal biology. Species that disperse evenly across a landscape are much harder to detect and encounter because one would have to search every habitat equally, and any one habitat would hold only a small fraction of all dispersing individuals. The house sparrow's high degree of commensalism thus may ultimately provide a mechanism for managers to access a large fraction of the dispersing population, and through lethal control, these dispersal pathways can be largely shut down. For example, D'amico et al. (2013) suggest that inner city roads are where dispersal most often occurs in urban areas, and they suggest that control efforts should target these corridors to prevent spread, although they do not offer suggestions as to how to implement control measures. It has also been suggested that trains were instrumental in driving the expansion of house sparrows in North America (Robbins 1973; Anderson 2006). The amount of rail lines increased dramatically during the same time period when house sparrows were expanding in North America (Figure 17.4) (Stover 1997). Railway men would close train cars, entrapping sparrows enclosed within that were subsequently transported to new areas. It is unclear if trains continue to move large numbers of house sparrows; however, finding these human-made dispersal corridors may be one of the best ways to limit house sparrow range expansion.

There have been some efforts mounted to eradicate sparrows, largely from oceanic islands that are home to native species that may suffer some impact from house sparrow presence. For example, conservation managers on Round Island (Indian Ocean) have explored ways to eradicate (or at least control) house sparrows so as to boost the success of a proposed reintroduction of the endangered Mauritius fody (*Foudia rubra*) (Bednarczuk et al. 2010). Multiple lethal control methods were employed, including glue sticks, shooting, mist netting, funnel trapping, poisoning, and direct removal of nests. Of the methods employed, funnel trapping accounted for the majority of captures (Figure 17.5). Mist-netting was not appropriate due to

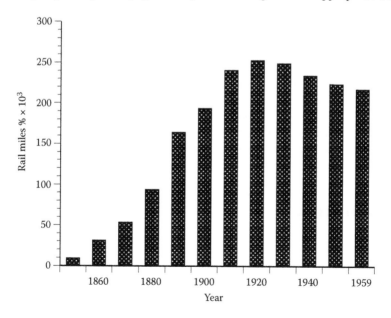

FIGURE 17.4 Data showing 1000s of rail miles by year. (From Stover, J. F., *American Railroads*, 2nd ed., University of Chicago Press, Chicago, IL, 1997. With permission.)

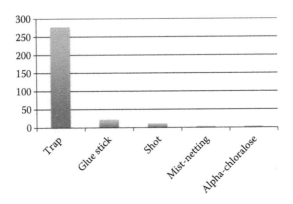

FIGURE 17.5 Eradication of house sparrows on Round Island (Mauritius). Bednarczuk et al.'s (2010) data show that the most efficient way to capture house sparrows was through traps. Despite over 300 individual sparrows being killed in this study, the authors have no evidence that the size of the local population was reduced. (From Bednarczuk, E. et al., *Conservation Evidence*, 7, 75–86, 2010.)

high winds and the effort required per individual caught. Glue sticks required heavy investment of monitoring, making them less efficient in terms of captures and kills than funnel traps, although glue sticks did account for approximately 10% of all captures (Bednarczuk et al. 2010). The authors conclude that closing funnel traps after catch rates decline and deploying them in new prebaited areas could contribute greatly to the success of eradication along with greater use of a trained marksman. While 320 sparrows were killed using all methods, there was no detectable decline in the number of sparrows observed on the island. House sparrows on Round Island were unusual in that they did not rely upon anthropogenic habitats and buildings or human food sources. It is unknown how much of a role the altered behavior of house sparrows on the island influenced eradication efforts, but it is clear that a robust method needs to be implemented for measuring success. Without a standard census method, the authors were unsure if trapping was causing the remaining birds to collect in flocks, creating an illusion of greater or consistent abundance. It should be noted that what worked in an island environment may not translate to other areas or mainland habitats.

RESEARCH NEEDS

IMPACT

During the last 100 years, concerns over the impacts of house sparrows on human populations and native ecosystems have shifted from a focus on agriculture to their role in urban ecosystems. In particular, house sparrows are now regularly considered to be vital links in the spread of human disease. This focus is not surprising given that the majority of the world's human population now lives within urban areas, and house sparrows are one of the most abundant wild vertebrates living in the same urban systems. Nevertheless, the role of house sparrows in the emergence and lethality of human disease is understudied in our estimation. In particular, we see a need to more fully understand how house sparrows utilize urban habitats and how this usage influences their competence as reservoirs of human disease. For example, there is some evidence that the high abundance of house sparrows in residential areas, especially when a wetland of any size is nearby, renders these locations as potential loci for human contraction of West Nile virus (WNV) (Figure 17.6). This effect is due in large part to wetlands sustaining high populations of vector species (mosquitos) that then have access to competent hosts (house sparrows), ultimately all occurring in areas regularly traversed by people (Johnson et al. 2012). There are likely other synergisms between urban habitats, house sparrows, disease vectors, and human disease prevalence that can shed light on ways to mitigate the impact of house sparrows. This information would also be of considerable help in deciding when, and where, to apply lethal house sparrow control. So far, most lethal control has occurred in rural settings, but if house sparrows prove to be a substantive cog in the disease transmission mechanism, their lethal control via poison application in urban areas may be warranted. At present, we have no understanding of how such control measures would influence nontarget native species in urban areas.

Equally compelling (and related) is research on the role of house sparrows as central figures in urban food webs, especially in circumstances where house sparrow impacts are strong enough to warrant lethal control via poison applications, or where

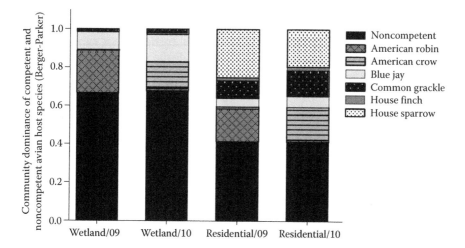

FIGURE 17.6   Urban landscapes. Within this landscape type, house sparrows reach relatively high abundance in residential areas, but not in wetlands embedded in these areas. The higher abundance of house sparrows, which are highly competent WNV hosts, and the proximity of residential areas to wetlands that support WNV vectors (mosquitos), make these locations more likely to be locations where humans contract WNV. (With kind permission from Springer Science+Business Media, *Urban Ecosystems*, The roles of mosquito and bird communities on the prevalence of West Nile virus in urban wetland and residential habitats, 15, 2012, 513–531, Johnson, B. J., K. Munafo, L. Shappell, N. Tsipoura, M. Robson, J. Ehrenfeld, and M. V. K. Sukhdeo.)

heavy metals are likely to be bioavailable. Urban ecosystems provide habitat for a surprising number of native species in North America and worldwide (Chace and Walsh 2006; McKinney 2008; Dearborn and Kark 2010). House sparrows in many of these urban centers form a core piece of the food web, serving as primary consumers in most cases (i.e., consuming primary producers and themselves being prey to secondary consumers). Elucidating the role of house sparrows in these urban food webs can provide critical feedback on their likelihood of serving as bioaccumulators, transmitting heavy metals and pesticides up urban food chains.

MANAGEMENT

If there is a clear message from the above review, it is that the control of house sparrow populations is no simple task. Individually, house sparrows show diverse and adaptive behaviors that make most conventional control measures temporary fixes, at best. House sparrow life history is unremarkable in the context of passerine birds, but it does allow them to attain very high densities in some habitats, and high enough population growth rates such that acceptable forms of lethal control are ineffective.

Although eradication of house sparrows is unlikely in most (if not all) situations, there may be better success in local control of sparrow populations through better delineation of population boundaries and commonly used dispersal corridors. We alluded to the usefulness of research on each of these factors above. Here we want to emphasize that the basic research question addressed by each is the same: how interconnected are sparrow populations across a landscape? Are house sparrow populations in inner cities connected to populations in the suburban fringe of these cities via dispersal, and if so, where are the dispersal corridors? Are rural farmland populations connected to each other, and if so, what do the spatial patterns of dispersal between farms look like? The answers to these questions inform what the correct spatial extent of control methods should be so as to realize substantive drops in local house sparrow populations while also maximizing the length of time before the site is recolonized from outside sources. It also informs efforts to slow recolonization rates, or limit the spread of house sparrows in locations where they are not yet established, by elucidating the dispersal pathways used by individual house sparrows. These aspects of control are receiving increasing attention among all the efforts to control other invasive vertebrates, and the return in terms of increased efficiency and effectiveness of control is substantial (Mack et al. 2000; Hulme et al. 2008). The same returns should be expected for similar research on house sparrows.

We end by noting that, despite the house sparrow's wide exotic distribution, it is becoming a species of conservation concern across Europe and Asia, as it has exhibited dramatic population declines (upward of 60%) over the past two decades (Hole et al. 2002; Atkinson et al. 2005; Shaw et al. 2008). Thus, research effort expended to inform management and impacts of exotic house sparrow populations will almost certainly have a positive influence on efforts to prevent further reductions in sparrow populations in their native range. For instance, Peach et al. (2008) identify correlates of urban house sparrow decline, and replicate studies in declining and increasing introduced populations could greatly inform rescue efforts in Europe. Such conservation efforts may seem counterintuitive, since clearly house sparrows as a species are not threatened with extinction. However, the subspecific phenotypic and genetic diversity across the house sparrows' native range is vastly greater than seen across all exotic populations, and this diversity is worth recognizing and conserving if possible.

REFERENCES

Anderson, T. R. 2006. *Biology of the Ubiquitous House Sparrow: From Genes to Populations.* Oxford University Press, Oxford, UK.

Atkinson, P. W., A. J. Baker, R. M. Bevan, N. A. Clark, K. B. Cole, P. M. Gonzalez, J. Newton, L. et al. 2005. Unravelling the migration and moult strategies of a long-distance migrant using stable isotopes: Red Knot *Calidris canutus* movements in the Americas. *Ibis* 147:738–749.

Avery, J. D., P. Cassey, and J. L. Lockwood 2014. Contemporary divergence of island bird plumage. *Journal of Avian Biology* 45:291–295.

Avery, J. D., D. M. Fonseca, P. Campagne, and J. L. Lockwood 2013. Cryptic introductions and the interpretation of island biodiversity. *Molecular Ecology* 22:2313–2324.

Baker, J., K. J. Harvey, and K. French 2014. Threats from introduced birds to native birds. *Emu* 114:1–12.

Barrows, W. B. 1889. *The English Sparrow (Passer domesticus) in North America: Especially in its Relations to Agriculture.* US Government Printing Office, Washington, DC.

Bednarczuk, E., C. J. Feare, S. Lovibond, V. Tatayah, and C. G. Jones 2010. Attempted eradication of house sparrows *Passer domesticus* from Round Island (Mauritius), Indian Ocean. *Conservation Evidence* 7:75–86.

Bergman, D. L., M. D. Chandler, and A. Locklear 2000. The economic impact of invasive species to wildlife services" cooperators. In *Human Conflicts with Wildlife: Economic Considerations*, Paper 21.

Bergtold, W. H. 1921. The English sparrow (*Passer domesticus*) and the motor vehicle. *Auk* 38:244–250.

Bichet, C., R. Scheifler, M. Coeurdassier, R. Julliard, G. Sorci, and C. Loiseau 2013. Urbanization, trace metal pollution, and malaria prevalence in the house sparrow. *PLoS One* 8:10.

Blackburn, T. M., P. Cassey, and J. L. Lockwood 2009. The role of species traits in the establishment success of exotic birds. *Global Change Biology* 15:2852–2860.

Bokony, V., A. Kulcsar, Z. Toth, and A. Liker 2012. Personality traits and behavioral syndromes in differently urbanized populations of house sparrows (*Passer domesticus*). *PLoS One* 7:11.

Brakes, C. R., and R. H. Smith 2005. Exposure of nontarget small mammals to rodenticides: Short-term effects, recovery and implications for secondary poisoning. *Journal of Applied Ecology* 42:118–128.

Brittingham, M. C., and S. A. Temple 1986. A survey of avian mortality at winter feeders. *Wildlife Society Bulletin (1973–2006)* 14:445–450.

Brodhead, M. J. 1971. Elliott Coues and the sparrow war. *New England Quarterly* 44:420–432.

Brown, C. R., A. T. Moore, and V. A. O'Brien 2012. Prevalence of Buggy Creek virus (Togaviridae: Alphavirus) in insect vectors increases over time in the presence of an invasive avian host. *Vector-Borne and Zoonotic Diseases* 12:34–41.

Chace, J. F., and J. J. Walsh 2006. Urban effects on native avifauna: A review. *Landscape and Urban Planning* 74:46–69.

Chavez-Zichinelli, C. A., I. MacGregor-Fors, P. T. Rohana, R. Valdez, M. C. Romano, and J. E. Schondube 2010. Stress responses of the house sparrow (*Passer domesticus*) to different urban land uses. *Landscape and Urban Planning* 98:183–189.

Clapperton, B. K., R. E. R. Porter, T. D. Day, J. R. Waas, and L. R. Matthews 2012. Designer repellents: Combining olfactory, visual or taste cues with a secondary repellent to deter free-ranging house sparrows from feeding. *Pest Management Science* 68:870–877.

Cooper, C. B., W. M. Hochachka, and A. A. Dhondt 2007. Contrasting natural experiments confirm competition between house finches and house sparrows. *Ecology* 88:864–870.

Crowell, K. L. 1962. Reduced interspecific competition among the birds of Bermuda. *Ecology* 43:75–88.

D'Amico, M., C. Rouco, J. C. Russell, J. Roman, and E. Revilla 2013. Invaders on the road: Synanthropic bird foraging along highways. *Oecologia Australis* 17:86–95.

Dearborn, D. C., and S. Kark 2010. Motivations for conserving urban biodiversity. *Conservation Biology* 24:432–440.

Dlugosch, K. M., and I. M. Parker 2008. Founding events in species invasions: Genetic variation, adaptive evolution, and the role of multiple introductions. *Molecular Ecology* 17:431–449.

Dowding, J. E., E. C. Murphy, and C. R. Veitch 1999. Brodifacoum residues in target and nontarget species following an aerial poisoning operation on Motuihe Island, Hauraki Gulf, New Zealand. *New Zealand Journal of Ecology* 23:207–214.

Duggal, N. K., A. Bosco-Lauth, R. A. Bowen, S. S. Wheeler, W. K. Reisen, T. A. Felix, and B. R. Mann et al. 2014. Evidence for co-evolution of West Nile virus and house sparrows in North America. *PLoS Negl Trop Dis* 8:e3262.

Elliott, J. E., S. Hindmarch, C. A. Albert, J. Emery, P. Mineau, and F. Maisonneuve 2014. Exposure pathways of anticoagulant rodenticides to nontarget wildlife. *Environmental Monitoring and Assessment* 186:895–906.

Evenden, F. G. 1957. Observations on the nesting behavior of the house finch. *Condor* 59:112–117.

Fitzwater, W. D. 1994. House sparrows. *Prevention and Control of Wildlife Damage*, University of Nebraska, Lincoln.

Giesbrecht, D. S., and C. D. Ankney 1998. Predation risk and foraging behaviour: An experimental study of birds at feeders. *Canadian Field-Naturalist* 112:668–675.

Gionfriddo, J. P., and L. B. Best 1996. Grit color selection by house sparrows and Northern Bobwhites. *The Journal of Wildlife Management* 60:836–842.

Gorenzel, W. P., T. P. Salmon, and A. C. Crabb 2000. A national review of the status of trapping for bird control. In *Nineteenth Vertebrate Pest Conference, Proceedings*. Salmon, T. P., and A. C. Crabb (eds). Vertebrate Pest Conference, Davis Proceedings - Vertebrate Pest Conference, San Diego, CA, pp. 5–21.

Gowaty, P. A. 1984. House sparrows kill eastern bluebirds. *Journal of Field Ornithology* 55:378–380.

Gowaty, P. A., and J. H. Plissner 1998. Eastern Bluebird (*Sialia sialis*). In *The Birds of North America*, No. 381, Poole, A., and F. B. Gill (eds). The Birds of North America, Inc., Philiadelphia, PA.

Hahn, I. J., P. M. Vergara, and U. Römer 2011. Importance of nest attributes in the conservation of endemic birds of the Juan Fernández Archipelago, Chile. *Bird Conservation International* 21:460–476.

Herrera-Duenas, A., J. Pineda, M. T. Antonio, and J. I. Aguirre 2014. Oxidative stress of House Sparrow as bioindicator of urban pollution. *Ecological Indicators* 42:6–9.

Hole, D. G., M. J. Whittingham, R. B. Bradbury, G. Q. A. Anderson, P. L. M. Lee, J. D. Wilson, and J. R. Krebs 2002. Agriculture: Widespread local house-sparrow extinctions. *Nature* 418:931–932.

Hulme, P. E., S. Bacher, M. Kenis, S. Klotz, I. Kühn, D. Minchin, W. Nentwig et al. 2008. Grasping at the routes of biological invasions: A framework for integrating pathways into policy. *Journal of Applied Ecology* 45:403–414.

Hunt, K. A., D. M. Bird, P. Mineau, and L. Shutt 1992. Selective predation of organophosphate-exposed prey by American kestrels. *Animal Behaviour* 43:971–976.

Invasive Species Specialist Group 2010. www.issg.com.

Jensen, H., R. Moe, I. J. Hagen, A. M. Holand, J. Kekkonen, J. Tufto, and B.-E. Sæther 2013. Genetic variation and structure of house sparrow populations: Is there an island effect? *Molecular Ecology* 22:1792–1805.

Johnson, B. J., K. Munafo, L. Shappell, N. Tsipoura, M. Robson, J. Ehrenfeld, and M. V. K. Sukhdeo 2012. The roles of mosquito and bird communities on the prevalence of West Nile virus in urban wetland and residential habitats. *Urban Ecosystems* 15:513–531.

Johnston, R. F., and R. K. Selander 1964. House sparrows: Rapid evolution of races in North America. *Science* 144:548–550.

Kalinoski, R. 1975. Intra- and interspecific aggression in house finches and house sparrows. *The Condor* 77:375–384.

Kekkonen, J., I. K. Hanski, H. Jensen, R. A. Väisänen, and J. E. Brommer 2011a. Increased genetic differentiation in house sparrows after a strong population decline: From panmixia towards structure in a common bird. *Biological Conservation* 144:2931–2940.

Kekkonen, J., I. K. Hanski, R. A. Väisänen, and J. E. Brommer 2012. Levels of heavy metals in house sparrows (*Passer domesticus*) from urban and rural habitats of southern Finland. *Ornis Fennica* 89:91–98.

Kekkonen, J., P. Seppa, I. K. Hanski, H. Jensen, R. A. Vaisanen, and J. E. Brommer 2011b. Low genetic differentiation in a sedentary bird: House sparrow population genetics in a contiguous landscape. *Heredity* 106:183–190.

Lee, K., L. Martin, II, and M. Wikelski 2005. Responding to inflammatory challenges is less costly for a successful avian invader, the house sparrow (*Passer domesticus*), than its less-invasive congener. *Oecologia* 145:243–250.

Liebl, A. L., and L. B. Martin 2014. Living on the edge: Range edge birds consume novel foods sooner than established ones. *Behavioral Ecology* 25:1089–1096.

Liebl, A. L., A. W. Schrey, C. L. Richards, and L. B. Martin 2013. Patterns of DNA methylation throughout a range expansion of an introduced songbird. *Integrative and Comparative Biology* 53:351–358.

Liker, A., V. Bokony, A. Kulcsar, Z. Toth, K. Szabo, B. Kaholek, and Z. Penzes 2009. Genetic relatedness in wintering groups of house sparrows (*Passer domesticus*). *Molecular Ecology* 18:4696–4706.

Liker, A., Z. Papp, V. Bokony, and A. Z. Lendvai 2008. Lean birds in the city: Body size and condition of house sparrows along the urbanization gradient. *Journal of Animal Ecology* 77:789–795.

Loiseau, C., R. J. Harrigan, C. Bichet, R. Julliard, S. Garnier, Á. Z. Lendvai, O. Chastel, and G. Sorci 2013. Predictions of avian *Plasmodium* expansion under climate change. *Scientific Reports* 3:1126.

Long, J. L. 1981. *Introduced Birds of the World*. David and Charles, London.

Lowther, P. E., and C. L. Cink 2006. House sparrow (*Passer domesticus*). In *The Birds of North America Online*. Poole, A., (ed). Cornell Lab of Ornithology; retrieved from the birds of North America online: http://bna.birds.cornell.edu/bna/species/012, Ithaca.

MacGregor-Fors, I., L. Morales-Pérez, J. Quesada, and J. Schondube 2010. Relationship between the presence of House Sparrows (*Passer domesticus*) and Neotropical bird community structure and diversity. *Biological Invasions* 12:87–96.

Mack, R. N., D. Simberloff, W. M. Lonsdale, H. Evans, M. Clout, and F. A. Bazzaz 2000. Biotic invasions: Causes, epidemiology, global consequences, and control. *Ecological Applications* 10:689–710.

MacLeod, C. J., D. M. Tompkins, K. W. Drew, and N. Pyke 2011. Does farm-scale habitat composition predict pest-bird numbers and distribution? *Wildlife Research* 38:464–474.

Macleod, R., P. Barnett, J. Clark, and W. Cresswell 2006. Mass-dependent predation risk as a mechanism for house sparrow declines? *Biology Letters* 2:43–46.

Martin, L. B., C. A. C. Coon, A. L. Liebl, and A. W. Schrey 2014. Surveillance for microbes and range expansion in house sparrows. *Proceedings of the Royal Society B—Biological Sciences* 281:8.

Martin, L. B., and L. Fitzgerald 2005. A taste for novelty in invading house sparrows, *Passer domesticus*. *Behavioral Ecology* 16:702–707.

Martin, L. B., A. L. Liebl, and H. J. Kilvitis 2015. Covariation in stress and immune gene expression in a range expanding bird. *General and Comparative Endocrinology* 211:14–19.

Martínez-Haro, M., J. Viñuela, and R. Mateo 2007. Exposure of birds to cholinesterase-inhibiting pesticides following a forest application for tick control. *Environmental Toxicology and Pharmacology* 23:347–349.

Marzal, A., R. E. Ricklefs, G. Valkiūnas, T. Albayrak, E. Arriero, C. Bonneaud, G. A. Czirják et al. 2011. Diversity, loss, and gain of malaria parasites in a globally invasive bird. *PLoS One* 6:e21905.

McClure, C. J. W., L. K. Estep, and G. E. Hill 2011. A multi-scale analysis of competition between the house finch and house sparrow in the southeastern United States. *Condor* 113:462–468.

McGillivray, W. B. 1980. Communal nesting in the house sparrow. *Journal of Field Ornithology* 51:371–372.

McGrath, J. E. 1987. Some notes on the removal of house sparrow nests in the vicinity of eastern bluebird nests. *The Jack Pine Warbler* 65:40.

McKinney, M. L. 2008. Effects of urbanization on species richness: A review of plants and animals. *Urban Ecosystems* 11:161–176.

Millaku, L., R. Imeri, and A. Trebicka 2015. Bioaccumulation of heavy metals in tissues of house sparrow (*Passer domesticus*). *Research Journal of Environmental Toxicology* 9:107–112.

Møller, A. P., M. Díaz, E. Flensted-Jensen, T. Grim, J. D. Ibáñez-Álamo, J. Jokimäki, R. Mänd, et al. 2015. Urbanized birds have superior establishment success in novel environments. *Oecologia* 178:943–950.

Moulton, M. P., W. P. Cropper, Jr., M. L. Avery, and L. E. Moulton 2010. The earliest House Sparrow introductions to North America. *Biological Invasions* 12:2955–2958.

Moulton, M. P., and D. K. Ferris 1991. Summer diets of some introduced Hawaiian finches. *The Wilson Bulletin* 103:286–292.

Navarro, C., A. Marzal, F. De Lope, and A. P. Møller 2003. Dynamics of an immune response in house sparrows *Passer domesticus* in relation to time of day, body condition and blood parasite infection. *Oikos* 101:291–298.

Nemeth, N. M., P. T. Oesterle, and R. A. Bowen 2009a. Humoral immunity to West Nile virus is long-lasting and protective in the house sparrow (*Passer domesticus*). *The American Journal of Tropical Medicine and Hygiene* 80:864–869.

Nemeth, N., G. Young, C. Ndaluka, H. Bielefeldt-Ohmann, N. Komar, and R. Bowen 2009b. Persistent West Nile virus infection in the house sparrow (*Passer domesticus*). *Archives of Virology* 154:783–789.

Newton, I. 1994. The role of nest sites in limiting the numbers of hole-nesting birds: A review. *Biological Conservation* 70:265–276.

O'Brien, V. A., and C. R. Brown 2012. Arbovirus infection is a major determinant of fitness in house sparrows (*Passer domesticus*) that invade Cliff Swallow (*Petrochelidon pyrrhonota*) colonies. *The Auk* 129:707–715.

Papp, S., E. Vincze, B. Preiszner, A. Liker, and V. Bókony 2015. A comparison of problem-solving success between urban and rural house sparrows. *Behavioral Ecology and Sociobiology* 69:471–480.

Peach, W. J., K. E. Vincent, J. A. Fowler, and P. V. Grice 2008. Reproductive success of house sparrows along an urban gradient. *Animal Conservation* 11:493–503.

Pérez, J., M. Nirchio, C. Alfonsi, and C. Muñoz 2006. The biology of invasions: The genetic adaptation paradox. *Biological Invasions* 8:1115–1121.

Pimentel, D., R. Zuniga, and D. Morrison 2005. Update on the environmental and economic costs associated with alien-invasive species in the United States. *Ecological Economics* 52:273–288.

Rattner, B. A., K. E. Horak, S. E. Warner, D. D. Day, C. U. Meteyer, S. F. Volker, J. D. Eisemann, et al. 2011. Acute toxicity, histopathology, and coagulopathy in American kestrels (Falco sparverius) following administration of the rodenticide diphacinone. *Environmental Toxicology and Chemistry* 30:1213–1222.

Robbins, C. S. 1973. Introduction, spread and present abundance of the house sparrow in North America. *Ornithological Monographs* 14:3–9.

Roman, J., and J. A. Darling 2007. Paradox lost: Genetic diversity and the success of aquatic invasions. *Trends in Ecology & Evolution* 22:454–464.

Roux, K. E., and P. P. Marra 2007. The presence and impact of environmental lead in Passerine birds along an urban to rural land use gradient. *Archives of Environmental Contamination and Toxicology* 53:261–268.

Russell, J. C., S. D. Miller, G. A. Harper, H. E. MacInnes, M. J. Wylie, and R. M. Fewster 2010. Survivors or reinvaders? Using genetic assignment to identify invasive pests following eradication. *Biological Invasions* 12:1747–1757.

Saetre, G. P., S. Riyahi, M. Aliabadian, J. S. Hermansen, S. Hogner, U. Olsson, M. F. Gonzalez Rojas, et al. 2012. Single origin of human commensalism in the house sparrow. *Journal of Evolutionary Biology* 25:788–796.

Schrey, A. W., C. A. C. Coon, M. T. Grispo, M. Awad, T. Imboma, E. D. McCoy, H. R. Mushinsky, et al. 2012. Epigenetic variation may compensate for decreased genetic variation with introductions: A case study using house sparrows (*Passer domesticus*) on two continents. *Genetics Research International* 2012:7.

Schrey, A. W., M. Grispo, M. Awad, M. B. Cook, E. D. McCoy, H. R. Mushinsky, T. Albayrak et al. 2011. Broadscale latitudinal patterns of genetic diversity among native European and introduced house sparrow (*Passer domesticus*) populations. *Molecular Ecology* 20:1133–1143.

Shaw, L. M., D. Chamberlain, and M. Evans 2008. The House Sparrow *Passer domesticus* in urban areas: Reviewing a possible link between post-decline distribution and human socioeconomic status. *Journal of Ornithology* 149:293–299.

Shochat, E., S. B. Lerman, J. M. Anderies, P. S. Warren, S. H. Faeth, and C. H. Nilon 2010. Invasion, competition, and biodiversity loss in urban ecosystems. *Bioscience* 60:199–208.

Skorka, P., M. Lenda, and J. Skorka 2009. Supermarkets—A wintering habitat for house sparrow *Passer domesticus l. Polish Journal of Ecology* 57:597–603.

Sol, D., I. Bartomeus, and A. Griffin 2012. The paradox of invasion in birds: Competitive superiority or ecological opportunism? *Oecologia* 169:553–564.

Sol, D., O. Lapiedra, and C. Gonzalez-Lagos 2013. Behavioural adjustments for a life in the city. *Animal Behaviour* 85:1101–1112.

Stover, J. F. 1997. *American Railroads*, 2nd ed. University of Chicago Press, Chicago, IL.

Summers-Smith, J. D. 1963. *The House Sparrow*. Collins, London.

Tsurim, I., B. P. Kotler, A. Gilad, S. Elazary, and Z. Abramsky 2010. Foraging behavior of an urban bird species: Molt gaps, distance to shelter, and predation risk. *Ecology* 91:233–241.

Vangestel, C., J. Mergeay, D. A. Dawson, T. Callens, V. Vandomme, and L. Lens 2012. Genetic diversity and population structure in contemporary house sparrow populations along an urbanization gradient. *Heredity* 109:163–172.

Vincze, E., S. Papp, B. Preiszner, G. Seress, A. Liker, and V. Bókony 2015. Does urbanization facilitate individual recognition of humans by house sparrows? *Animal Cognition* 18:291–298.

Wetzel, D. P., I. R. K. Stewart, and D. F. Westneat 2012. Heterozygosity predicts clutch and egg size but not plasticity in a house sparrow population with no evidence of inbreeding. *Molecular Ecology* 21:406–420.

Wheeler, S. S., C. M. Barker, Y. Fang, M. V. Armijos, B. D. Carroll, S. Husted, W. O. Johnson, et al. 2009. Differential impact of West Nile virus on California birds. *The Condor* 111:1–20.

Witmer, G. W., P. W. Burke, W. C. Pitt, and M. L. Avery 2007. *Management of Invasive Vertebrates in the United States: An Overview.* USDA APHIS Wildlife Services, Fort Collins, CO.

18 Conclusions, Challenges, and Research Needs

Gary W. Witmer, William C. Pitt, and James C. Beasley

CONTENTS

At least 392 species of introduced vertebrate species (excluding fish) occur in the United States and its territories (Witmer and Fuller 2011). Many of these species can be truly considered "invasive," causing various types of economic, social, and ecological harm. In this book, we have presented examples of many invasive vertebrates in the United States, the impacts they cause, and the methods used to control or eradicate them in various parts of the country.

While progress has been and is being made in the management of invasive vertebrates in the United States, there are still many challenges and issues to resolve (Pimentel et al. 2005; National Invasive Species Council 2008), many of which stem from a lack of sufficient funding for public education, prevention, control, and eradiation of invasive vertebrates. Whereas the majority of attention and funding for invasive species in the United States has focused on plants, insects, and pathogens (Pimentel et al. 2005; Pimentel 2011), relatively little effort and few resources have been directed to vertebrate control, with the notable exceptions of brown tree snakes in Guam and wild pigs across several states. Nonetheless, successful eradications of some invasive species have occurred on a handful of islands; however, for mainland areas, the focus continues to be on long-term control.

Public perception also has hindered efforts to manage or eradicate vertebrate species in the United States, as it has elsewhere in the world (National Invasive Species Council 2008). Knowledge of invasive species and the harm they can cause is relatively limited among the general public unless they are personally impacted (Conover 2002; National Invasive Species Council 2008). Furthermore, many members of the public do not readily distinguish between native and nonnative species: as long as an animal is charismatic and not threatening people or causing undue harm, the public tends to view invasive species equally to native animals (Wittenberg and Cock 2001; Witmer and Fuller 2011). In most cases, the ultimate goal in invasive species control is the eradication of all individuals in a given area; however, much of the public has a strong dislike for the killing of animals (Conover 2002). Certain species such as feral cats, feral dogs, feral horses, and introduced primates are particularly sensitive species to control via lethal means. Furthermore, much of the public has a strong fear and distrust of chemicals, and toxicants in particular. Hence, the management of invasive vertebrates, like all wildlife, is being conducted in an increasingly complex arena (Conover 2002; Fall and Jackson 2002).

Exotic pets are very popular with a sizable portion of the public, and thus the pet industry is a major pathway for the introduction of vertebrates into the United States (Kraus 2003; Jenkins 2007). Very few vertebrate species are prohibited from entry into the United States, with a prevailing attitude of "innocent until proven guilty" or a "gray" list approach to species imports (Witmer and Lewis 2001; Pitt and Witmer 2007). However, the development of a more inclusive prohibited species or "black list" approach is needed (Witmer and Lewis 2001; Fowler et al. 2007; Pitt and Witmer 2007). Only after it is determined that a species will not harm the environment, agriculture, or human resources should it be determined safe or added to a "white" list. Such shifts in policy will require greater cooperation, enforcement, and regulation of the pet industry. However, the pet industry is a well-organized, large, and influential industry in the United States, and thus future shifts in policy of this nature will undoubtedly be challenging (Ginsburg 2004).

Once invasive species become established, access to all relevant land and properties is essential for the implementation of successful management and eradication campaigns. However, invasive species mangers often face the situation where access to a wide array of jurisdictions and ownerships is needed. Getting permission to implement control in all these areas rarely occurs and can prevent the success of even a well-planned, well-funded eradication effort. Furthermore, land management mandates and regulations of federal and state agencies vary considerably. This affects the types of management activities (burning, chemical use), vehicles, and tools (leg-hold traps, firearms, toxicants) that can be used on certain properties. Some laws actually protect invasive vertebrate species, such as the Wild Horse and Burro Act and the Migratory Bird Treaty Act. The latter was amended to exclude some nonnative bird species in the United States, such as the mute swan. Finally, there tends to be inadequate coordination and cooperation among invasive species managers across jurisdictions and agencies of all levels of government in the United States. One of the goals of the National Invasive Species Management Plan (National Invasive Species Council 2008) is to enhance interagency coordination of invasive species management. Once established, eradication of an invasive vertebrate species is a complex challenge that in many cases can be difficult to attain. Careful planning is needed, along with adequate resources, public and agency buy-in, highly trained and motivated personnel, contingency plans, and a sustained effort (Broome 2005). Each situation is unique in one or more ways; hence, a cookbook approach cannot be used, even within management efforts for the same species in different areas (Broome 2005).

With the possible exception of rodents and ungulates, the methods and strategies used for management—and especially eradication—of invasive vertebrates need improvement (Wittenberg and Cock 2001; Witmer and Fuller 2011). Much research needs to be conducted to improve detection and eradication methods. Attractants are needed to draw individuals to traps, bait stations, and detection stations. Method improvements are needed to ensure effective and safe delivery of toxicants, vaccines, and fertility control agents. Trained, rapid response teams and early detection systems are needed for many more invasive species. Accessible databases on potential invasive species are needed to summarize species identification, biology, ecology, and effective detection and management methods to identify future hazards. The databases should also identify expertise and literature that can be consulted. Although a variety of databases and websites exist (Sellers et al. 2005), it would be very useful if these could be centralized and standardized (Sellers et al. 2004). Finally, risk assessments are needed to determine on which species effort and resources should be focused to be most cost effective (Hayes 2003).

One can surmise that invasive vertebrate species will continue to challenge resource managers, ecologists, and biologists for a long time to come. It also is likely the list of invasive vertebrate species in the United States will continue to grow, but hopefully, some species will also be removed from the list through implementation of successful management regimes. In the United States, there have been some successes with invasive species management and eradications, especially on islands, but also within some areas of the mainland. As a result of this, along with collaborations with international colleagues and a growing interest and involvement by the public and agencies, we are becoming more knowledgeable and proactive in responding to invasive vertebrate species. However, challenges associated with invasive species management are extensive, and undoubtedly efforts to combat their establishment and spread will continue for the foreseeable future.

REFERENCES

Broome K. 2005. Rat eradication: How to get it right without a recipe. *Proceedings of the Australasian Vertebrate Pest Conference* 13:152–157.

Conover M. 2002. *Resolving Human—Wildlife Conflicts.* Boca Raton, FL: CRC Press.

Fall, M. and Jackson, W. 2002. The tools and techniques of wildlife damage management. *International Biodeterioration and Biodegradation* 49:87–91.

Fowler A, Lodge D, and Hsia J. 2007. Failure of the Lacey Act to protect US ecosystems against animal invasions. *Frontiers in Ecology and the Environment* 5:353–359.

Ginsburg J. 2004. Dinner, pets, and plagues by the bucketful. *The Scientist* 18:28–29.

Hayes K. 2003. Biosecurity and the role of risk assessment. In: Ruiz G, Carlton J, eds. *Invasive Species: Vectors and Management Strategies.* Washington, DC: Island Press, pp. 382–414.

Jenkins P. 2007. *Broken screens: The Regulation of Live Animal Imports in the United States.* Washington, DC: Defenders of Wildlife.

Kraus F. 2003. Invasion pathways for terrestrial vertebrates. In: Ruiz G, Carlton J, eds. *Invasive Species: Vectors and Management Strategies.* Washington, DC: Island Press, pp. 68–92.

National Invasive Species Council. 2008. *2008–2012 National Invasive Species Management Plan.* Washington, DC: National Invasive Species Council.

Pimentel D. 2011. *Biological invasions: Economic and Environmental Costs of Alien Plant, Animal, and Microbe Species.* 2nd Ed. Boca Raton, FL: CRC Press.

Pimentel D, Zuniga R, and Morrison D. 2005. Update on the environmental and economic costs associated with alien-invasive species in the United States. *Ecological Economics* 52:273–288.

Pitt W. and Witmer G. 2007. Invasive predators: A synthesis of the past, present, and future. In: Elewa A, ed. *Predation in Organisms: A Distinct Phenomenon.* Heidelberg, Germany: Spring-Verlag, pp. 265–293.

Sellers E, Muir R, and Simpson A. 2004. *Summary Report on the Experts Meeting on Implementation of a Global Invasive Species Information Network*. Oak Ridge, Tennessee: National Biological Information Infrastructure, Information International Associates, Inc.

Sellers E, Simpson A, and Curd-Hetrick S. 2005. *List of Invasive Alien Species (IAS) Online Databases and Data Bases Containing IAS Information*. Oak Ridge, TN: National Biological Information Infrastructure, Information International Associates, Inc.

Witmer, GW. and Fuller PL. 2011. Vertebrate species introductions in the United States and its territories. *Current Zoology* 57:559–567.

Witmer G. and Lewis J. 2001. Introduced wildlife in Oregon and Washington. In: Johnson D, O'Neil T, ed. *Wildlife-Habitat Relationships in Oregon and Washington*. Corvallis, OR: Oregon State University Press, pp. 423–443.

Wittenberg R. and Cock M. 2001. *Invasive Alien Species: A Toolkit of Best Prevention and Management Practices*. Wallingford, United Kingdom: CAB International.

Index

Printed and bound by CPI Group (UK) Ltd, Croydon, CR0 4YY

17/10/2024

01775709-0012